Engineering Physics

Engineering Physics

As per the latest syllabus
JNTU, Hyderabad
For first year BE/BTech. Students
(Common to all Branches)

FOURTH EDITION

S O Pillai
Former Professor
Department of Physics
Anna University
Chennai, Tamil Nadu

Sivakami
Former Tutor in Physics
Hindu College, Nagarcoil and
Alagappa Arts & Science College
Karaikudi, Tamil Nadu

NEW AGE
JNTU

An Imprint of

NEW AGE INTERNATIONAL (P) LIMITED, PUBLISHERS
New Delhi • Bangalore • Chennai • Cochin • Guwahati
Hyderabad • Kolkata • Lucknow • Mumbai
Visit us at **www.newagepublishers.com**

Published by New Age International (P) Ltd., Publishers
First Edition: 2006
Fourth Edition: 2014

BRANCHES

- **Bangalore** 37/10, 8th Cross (Near Hanuman Temple), Azad Nagar, Chamarajpet, Bangalore-560 018
Tel.: (080) 26756823, Telefax: 26756820, **E-mail: bangalore@newagepublishers.com**
- **Chennai** 26, Damodaran Street, T. Nagar, Chennai-600 017
Tel.: (044) 24353401, Telefax: 24351463, **E-mail: chennai@newagepublishers.com**
- **Cochin** CC-39/1016, Carrier Station Road, Ernakulam South, Cochin-682 016
Tel.: (0484) 2377004, Telefax: 4051303, **E-mail: cochin@newagepublishers.com**
- **Guwahati** Hemsen Complex, Mohd. Shah Road, Paltan Bazar, Near Starline Hotel, Guwahati-781 008
Tel.: (0361) 2513881, Telefax: 2543669, **E-mail: guwahati@newagepublishers.com**
- **Hyderabad** 105, 1st Floor, Madhiray Kaveri Tower, 3-2-19, Azam Jahi Road, Near Kumar Theater
Nimboliadda, Kachiguda, Hyderabad-500 027, Tel.: (040) 24652456, Telefax: 24652457
E-mail: hyderabad@newagepublishers.com
- **Kolkata** RDB Chambers (Formerly Lotus Cinema) 106A, 1st Floor, S.N. Banerjee Road, Kolkata-700 014
Tel.: (033) 22273773, Telefax: 22275247, **E-mail: kolkata@newagepublishers.com**
- **Lucknow** 16-A, Jopling Road, Lucknow-226 001
Tel.: (0522) 2209578, 4045297, Telefax: 2204098, **E-mail: lucknow@newagepublishers.com**
- **Mumbai** 142C, Victor House, Ground Floor, N.M. Joshi Marg, Lower Parel, Mumbai-400 013
Tel.: (022) 24927869, Telefax: 24915415, **E-mail: mumbai@newagepublishers.com**
- **New Delhi** 22, Golden House, Daryaganj, New Delhi-110 002
Tel.: (011) 23262368, 23262370, Telefax: 43551305, **E-mail: sales@newagepublishers.com**

ISBN: 978-81-224-3651-8

₹ 225.00

C-13-08-7089

Printed in India at Ajit Printers, Delhi.
Typeset at Akriti Graphics, Delhi.

PUBLISHING FOR ONE WORLD
NEW AGE INTERNATIONAL (P) LIMITED, PUBLISHERS
7/30 A, Daryaganj, New Delhi-110002
Visit us at **www.newagepublishers.com**

Preface to the Fourth Edition

In this edition, apart from the minor changes in the figures of some chapters, the misprints and errors detected in the previous edition have been corrected. Interestingly a new chapter on optics has been brought in as per the recently revised syllabus. The readers may aware that all the topics including the newly added one have been treated with simplicity, in lucid language which may in turn help the students who studied the subjects in their mother tongue at the plus two level. The authors will be highly thankful to those who report the shortcomings of this work for improvement in the next edition.

With Seasonal Greetings

Dr. S.O. Pillai
Mrs. Sivakami

Preface to the First Edition

The first author Dr. S.O. Pillai is immensely happy to point out that the feedback of the fifth edition of his main book on Solid State Physics from students, faculty members and academicians revealed that the said volume is widely and unsparingly used by post-graduate students of physical sciences and certain branches of engineering for doing their course work and also for seeking admission in IITs for higher studies through GATE. Candidates aspiring employment in universities and research institutes through written tests do use this book liberally.

However, one common hue and cry from the undergraduate students doing B.Sc. and BE courses in colleges affiliated to universities spread all over this country is that the above said book is too heavy for them and the plea was to bring out an edited version of the main book. The unique representation from the students of JNTU made the authors to write a volume exclusively to cater the needs of the undergraduate engineering students of JNTU. Nevertheless the adoption of this book by students of other universities may help them to adapt for the situations that arise during their course work just like the amphibians. An analysis of the feedback from undergraduate students made Dr. Pillai to feel that the epitomized version of any popular book may not bring out the anticipated benefits; hence he went for an open surgery. A consolidation and rearrangement of lecture notes and teaching materials and their revised versions aided the second author to prepare the handwritten manuscript in the present format and bring out the book in record time to cater the needs of the said category of readers.

In developing a sound pedagogical approach to teaching the material in this text to students over more than three decades, Dr. Pillai has drawn inevitably on the work of others. What is the difference between this and other related texts is not so much the analytic treatment of common subject matter, but the specific choice of material presentation sequence, and assorted explanations. As it is said, "there is nothing new under the sun", so it is not the intent of the first author to claim that the treatment of particular topics in the current text is necessarily unique though he feels half-heartedly unique. The credit in preparing the figures and solutions to a good number of problems goes to the second author. The special features of this task are:

1. Inclusion of short questions and answers to some of them is the other newly introduced concept.
2. Review questions, inclusion of tables, problems with solutions are the other features tried for the first time.
3. *Forget not to read the Bluish writing without pictures of the authors in the third or fourth page of every chapter under the head "Special Information".*
4. Other prominent features of the work if any:

You will find it in the next edition on the basis of your feedback.

Dr. S.O. Pillai
Mrs. Sivakami

Syllabus

JAWAHARLAL NEHRU TECHNOLOGICAL UNIVERSITY, HYDERABAD

Ist YEAR B. TECH.	L	T/P/D	C
	2	1/-/-	4

ENGINEERING PHYSICS

UNIT-I

1. Crystallography: Ionic Bond, Covalent Bond, Metallic Bond, Hydrogen Bond, Vander-Waal's Bond, Calculation of Cohesive Energy of diatomic molecule- Space Lattice, Unit Cell, Lattice Parameters, Crystal Systems, Bravais Lattices, Atomic Radius, Co-ordination Number and Packing Factor of SC, BCC, FCC, Miller Indices, Crystal Planes and Directions, Inter Planar Spacing of Orthogonal Crystal Systems, Structure of Diamond and NaCl.

2. X-ray Diffraction & Defects in Crystals: Bragg's Law, X-Ray diffraction methods: Laue Method, Powder Method: Point Defects: Vacancies, Substitutional, Interstitial, Frenkel and Schottky Defects, line defects (Qualitative) & Burger's Vector.

UNIT-II

3. Principles of Quantum Mechanics: Waves and Particles, de Broglie Hypothesis, Matter Waves, Davisson and Germer's Experiment, Heisenberg's Uncertainty Principle, Schrödinger's Time Independent Wave Equation - Physical Significance of the Wave Function — Infinite square well potential extension to three dimensions.

4. Elements of Statistical Mechanics & Electron Theory of Solids: Phase space, Ensembles, Micro Canonical, Canonical and Grand Canonical Ensembles - Maxwell-Boltzmann, Bose-Einstein and Fermi-Dirac Statistics (Qualitative Treatment), Concept of Electron Gas, Density of States, Fermi Energy-Electron in a periodic Potential, Bloch Theorem, Kronig-Penny Model (Qualitative Treatment), E-K curve, Origin of Energy Band Formation in Solids, Concept of Effective Mass of an Electron, Classification of Materials into Conductors, Semiconductors & Insulators.

UNIT-III

5. Dielectric Properties: Electric Dipole, Dipole Moment, Dielectric Constant, Polarizability, Electric Susceptibility, Displacement Vector, Electronic, Ionic and Orientation Polarizations and Calculation of Polarizabilities: Ionic and Electronic-Internal Fields in Solids, Clausius-Mossotti Equation, Piezo-electricity and Ferro-electricity.

6. Magnetic Properties & Superconducting Properties: Permeability, Field Intensity, Magnetic Field Induction, Magnetization, Magnetic Susceptibility, Origin of Magnetic Moment, Bohr Magneton,

Classification of Dia, Para and Ferro Magnetic Materials on the basis of Magnetic Moment, Domain Theory of Ferro Magnetism on the basis of Hysteresis Curve, Soft and Hard Magnetic Materials, Properties of Anti-Ferro and Ferri Magnetic Materials and their Applications, Superconductivity, Meissner Effect, Effect of Magnetic field, Type-I & Type-II Superconductors, Applications of Superconductors.

UNIT-IV

7. Optics: Interference-Interference in thin films (Reflected light), Newton rings experiment- Fraunhofer diffraction due to single slit, N-slits, Diffraction grating experiment, Double refraction-construction and working, Nicol's Prism.

8. Lasers & Fiber Optics: Characteristics of Lasers, Spontaneous and Stimulated Emission of Radiation, Einstein's Coefficients and Relation between them, Population Inversion, Lasing Action, Ruby Laser, Helium-Neon Laser, Semiconductor Diode Laser, Applications of Lasers- Principle of Optical Fiber, Construction of fiber, Acceptance Angle and Acceptance Cone, Numerical Aperture, Types of Optical Fibers: Step Index and Graded Index Fibers, Attenuation in Optical Fibers, Application of Optical Fiber in communication systems.

UNIT-V:

9. Semiconductor Physics: Fermi Level in Intrinsic and Extrinsic Semiconductors, Calculation of carrier concentration in Intrinsic & Extrinsic Semiconductors, Direct and Indirect Band gap semiconductors, Hall Effect- Formation of PN Junction, Open Circuit PN Junction, Energy Diagram of PN Diode, Diode Equation, I-V Characteristics of PN Junction diode, Solar cell, LED & Photo Diodes. Acoustics of Buildings & Acoustic Quieting, Reverberation and Time of Reverberation, Sabine's Formula for Reverberation Time, Measurement of Absorption Coefficient of a Material, Factors Affecting The Architectural Acoustics and their Remedies.

10. Nanotechnology: Origin of Nanotechnology, Nano Scale, Surface to Volume Ratio, Quantum Confinement, Bottom-up Fabrication: Sol-gel, Top-down Fabrication: Chemical Vapour Deposition, Characterization by TEM.

Contents as per latest JNTU Syllabus

Contents

Chapter 1

UNITS 1 & 2

Bonding in Solids, Crystal Structures, X-ray Diffraction and Defects in Crystals

1.1 BONDING IN SOLIDS

1.1.1 INTRODUCTION

The most convenient basis for solid state theory is a classification scheme based on the character of the *interatomic bonding forces* in various classes of crystalline materials. According to this scheme of classification, all solids fall into one of the five general categories: *molecular*, *ionic*, *covalent*, *metallic* and *hydrogen bonded crystals*. The distinction is not a sharp one, because some may belong to more than one class.

It is a too fundamental fact that inert gases like helium, neon, argon, etc. exist in the atomic form and not easily combine with other atoms to form new compounds because of the magic number 2 in Helium and 8 in other cases. *Interatomic force* among the atoms in a solid is one of the tools to classify the solids and then to study their physical properties.

1.1.2 FORCES BETWEEN ATOMS AND BOND ENERGY

The attractive electrostatic interaction between the electrons and positive charge of the nuclei is totally responsible to group them and then for holding together. Gravitational forces and magnetic forces are negligible compared with the said interaction. When two atoms (or ions) come close to one another, there will be repulsion between negatively charged electrons of both atoms. The *repulsive force* increases very rapidly as the distance of separation decreases. However when the distance of separation is large, there is attraction between positive nucleus and negative electrons. At some optimum distance (say $r = r_0$), the attractive and repulsive forces just balance and hence the resultant force becomes zero. Hence the potential energy at this *equilibrium spacing*, r_o becomes minimum. This magnitude of minimum energy is called *bond energy*. It is usually expressed in kJ/kmol. The value for primary bond ranges from 100×10^3 kJ/kmol to 1000 kJ/kmol and that of secondary bond ranges from 1×10^3 to 60×10^3 kJ/kmol. The centre to centre distance of two bonding atoms is called the *bond length*. The bond length increases on heating. The bond energy and *dissociation energy* will be equal but opposite in sign. A crystal can only be stable if its total energy (K.E + P.E) is lower than the total energy of the atoms or molecules when they are free. The '*cohesive energy*' is defined as the difference of free atom energy and crystal energy.

The variation of interatomic forces and potential energy vs distance of separation is shown in Fig. 1.1. The general equation that represents the force between two atoms or molecules or ions is

$$F = \frac{A}{r^M} - \frac{B}{r^N} \quad \text{...(1.1.1)}$$

A, *B*, *M* and *N* are constants; *r* is the *interatomic distance*.

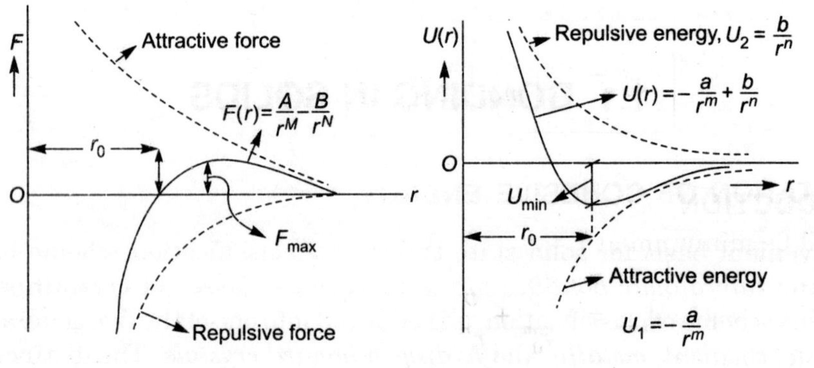

Fig. 1.1 Interatomic forces and potential energy vs distance of separation

From the above equation, one can arrive at the equation for the potential energy of the system:

i.e.,
$$U(r) = \int F\,dr = -\left[\frac{A}{M-1}\right]\left[\frac{1}{r^{M+1}}\right] + \left[\frac{B}{N-1}\right]\left[\frac{1}{r^{N+1}}\right] + C$$

$$= -\frac{a}{r^m} + \frac{b}{r^n} + C$$

When $U = 0$ for $r = \infty$, *C* will be zero.

Thus

$$U(r) = -\frac{a}{r^m} + \frac{b}{r^n} \quad \text{...(1.1.2)}$$

Here a, b, m and n are different constants. The first term of the above equation refers to attraction and the second term repulsion. n is greater than m with n of the order of 9 and $m = 2$ for ionic structures. The forces of attraction result from interaction between outer electrons of two atoms.

The forces of repulsion are from interpenetration of outer electronic shells. Only these forces decide the nature of bonds in solids. The *equilibrium distance* r_0 may be determined as follows:

$$\left[\frac{dU}{dr}\right]_{r=r_0} = 0 = \frac{am}{r_0^{m+1}} - \frac{bn}{r_0^{n+1}}$$

i.e.,

$$r_0^{n-m} = \left\{\frac{nb}{ma}\right\}$$

$$r_0 = \left\{\frac{nb}{ma}\right\}^{\frac{1}{n-m}}$$

Also

$$\boxed{r_0^n = r_0^m\left[\frac{nb}{ma}\right]} \qquad \text{...(1.1.3)}$$

1.1.3 COMPUTATION OF COHESIVE ENERGY

The energy will be minimum at $r = r_0$

i.e.,

$$[U(r)]_{\min} = -\frac{a}{r_0^m} + \frac{b}{r_0^n}$$

Using $r_0^n = r_0^m\left[\frac{nb}{ma}\right]$, from Eqn. 1.1.3, we get

$$U_{\min} = -\frac{a}{r_0^m} + \frac{mab}{nbr_0^m}$$

i.e.,

$$\boxed{U_{\min} = -\frac{a}{r_0^m}\left[1 - \frac{m}{n}\right]} \qquad \text{...(1.1.4)}$$

Conclusion: All stable arrangements of atoms in solids are such that the potential energy is minimum. This is the only way of explaining the cohesion of atoms in solid aggregates.

1.1.4 IONIC BONDING

Ionic crystals are those compounds in which the valence electrons are completely transferred from one atom to the other; the final result being a crystal that is composed of positively and negatively charged ions. Atom whose outermost shell has only a few electrons given up these electrons so that it is left with completely filled outermost shell. This process is termed ionisation. On the other hand if the outermost shell is slightly short of eight electrons, then the atom has a tendency to receive the electrons and become an ion.

An atom of sodium metal has one electron in its outermost shell and it can be easily released with a sodium positive ion left with. This electron can be easily added to the outermost shell of chlorine with seven electrons already. This chlorine atom becomes a negative ion. Because of the mutual attraction between positive and negative ions, a bond is developed between these two ions of opposite charges. The outermost electron of sodium is removed by supplying 495×10^3 kJ/kmol (called first ionisation potential of sodium).

$$Na + E_1 \rightarrow Na^+ + e^{-1}$$

This released electron will now move to occupy the outermost shell of chlorine (with already seven electrons) and produce a negatively charged ion.

$$Cl + e^{-1} \rightarrow Cl^-$$

The electron affinity of chlorine is 370×10^3 kJ/kmol. Thus net increase in potential energy for transferring an electron from sodium to chlorine is 125×10^3 kJ/kmol. The chemical reactions in the formation of NaCl at the *equilibrium spacing* are summarized below:

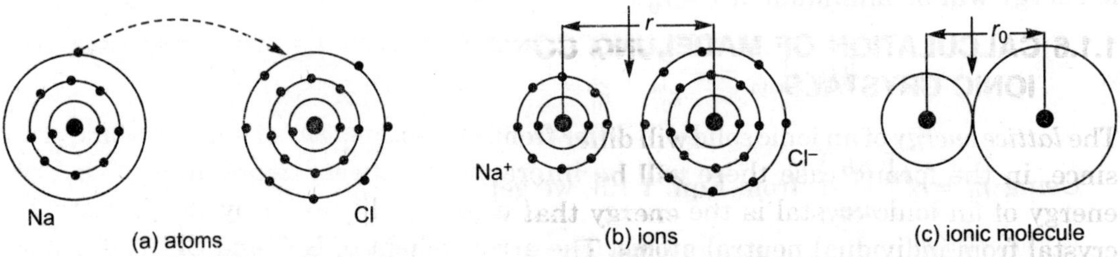

Fig. 1.2 Schematic representation of the formation of ionic molecule of sodium chloride

$$Na + Cl \rightarrow Na^+ + Cl^- \rightarrow NaCl$$

Since chlorine exists as molecules, the chemical reaction must be written as

$$2Na + Cl_2 \rightarrow 2Na^+ + 2Cl^- \rightarrow 2NaCl$$

1.1.5 BOND ENERGY OF NaCl MOLECULE

NaCl is one of the best examples of ionic compound and let the sodium and chlorine atoms be free at infinite distance of separation. The energy required to remove the outer electron from the Na atom (ionisation energy of sodium atom), leaving it a Na^+ ion is about 5 eV; or

$$Na + 5\text{ eV} \rightarrow Na^+ + e^-$$

Similarly when the removed electron from sodium atom is added to chlorine atom, about 3.5 eV of energy (electron affinity of chlorine) is released.

$$Cl + e^- \rightarrow Cl^- + 3.5$$

Thus a net energy of (5.0 – 3.5) = 1.5 eV is spent in creating a positive sodium ion and a negative chlorine ion at infinity.

i.e., $$Na + Cl + 1.5 \text{ eV} \rightarrow Na^+ + Cl^-$$

Now the electrostatic attraction between these ions brings them to the equilibrium spacing r_o = 0.24 nm and the energy released in the formation of NaCl molecule is called the *bond energy* of the molecule and it is obtained as follows:

$$V = -\frac{e^2}{4\pi\epsilon_0 r_0} = -\frac{(1.6\times10^{-19})^2}{4\pi(8.85\times10^{-12})(0.24\times10^{-9})}$$

$$= -2.56\times0.0096\times10^{-16}\text{ J} = -\frac{2.56\times0.0096\times10^{-16}}{1.6\times10^{-19}}\text{eV}$$

$$\boxed{V = -6\text{ eV}} \quad ...(1.1.5)$$

Thus the energy released in the formation of NaCl molecule from the neutral Na and Cl atoms is (5.0 – 3.5 – 6) = – 4.5 eV.

Schematically,

$$\underset{\leftarrow 0.24\text{ nm} \rightarrow}{Na^+ + Cl^- \rightarrow Na + Cl + 4.5\text{ eV}} \rightarrow NaCl$$

This means that to dissociate one NaCl molecule 4.5 eV energy is required.

1.1.6 CALCULATION OF MADELUNG CONSTANT AND LATTICE ENERGY OF IONIC CRYSTALS

The *lattice energy* of an ionic solid will differ from the bond energy of diatomic ionic molecules since, in the former case there will be interactions between more atoms. The cohesive energy of an ionic crystal is the energy that would be liberated by the formation of the crystal from individual neutral atoms. The arrangement of Na^+ and Cl^- ions in a sodium chloride crystal is shown in Fig. 1.3.

Each sodium ion in sodium chloride is subject to the attractive potential due to 6 chlorine ions each at a distance *r*. Thus the attractive potential at the sodium ion by the chlorine ion is

$$U_1 = -6\frac{e^2}{4\pi\epsilon_0 r}$$

The next nearest neighbours are 12 of the same kind of ions at a distance $r(\sqrt{2})$. Thus

$$U_2 = \frac{12 e^2}{4\pi\epsilon_0 r(\sqrt{2})} \text{ and so on.}$$

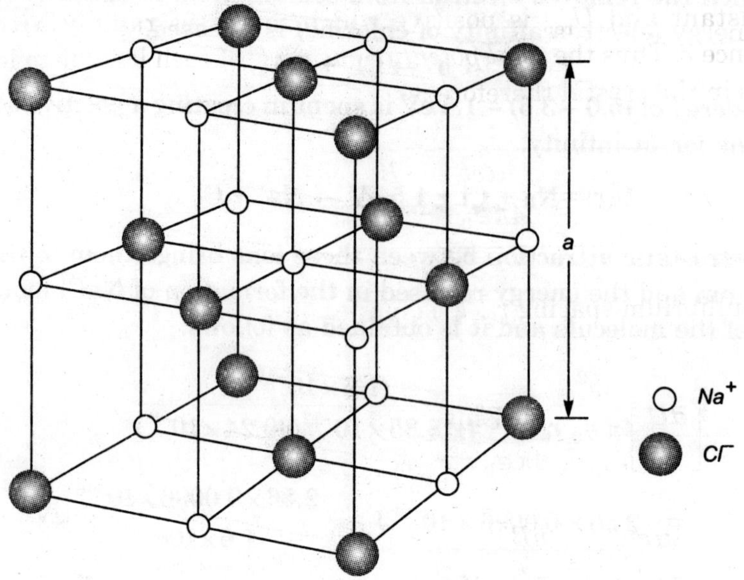

Fig. 1.3 Sodium chloride structure

Thus the total potential energy is

$$U_{coub} = -\frac{6e^2}{4\pi\epsilon_0 r} + \frac{12e^2}{4\pi\epsilon_0 r(\sqrt{2})} - \frac{8e^2}{4\pi\epsilon_0 r(\sqrt{3})} + ...$$

$$= -\frac{e^2}{4\pi\epsilon_0 r}\left\{6 - \frac{12}{\sqrt{2}} + \frac{8}{\sqrt{3}} - \frac{6}{2} + ...\right\}$$

Let $$\alpha = \left\{6 - \frac{12}{\sqrt{2}} + \frac{8}{\sqrt{3}} - \frac{6}{2} + ...\right\}$$

All these relations are applicable only to univalent alkali halides (*i.e.*, $Z_1 = Z_2 = 1$).

According to the scheme of *Evjen*, $\alpha = 1.75$ which is close to the accurate value of 1.747558 obtained for NaCl structure by the Ewold method. This constant is called *Madelung constant.*

Thus

$$U_{\text{coub}} = -\alpha \frac{e^2}{4\pi\epsilon_0 r}$$

The potential energy contribution of the *short-range repulsive forces* can be expressed as

$$U_{\text{rep}} = \frac{B}{r^n}$$

where B is a constant and U_{rep} is positive which increases rapidly with decreasing internuclear distance r. Thus the *total potential energy U* of each ion due to its interaction with all other ions in the crystal therefore is

$$U(r) = -\frac{\alpha e^2}{4\pi \in_0 r} + \frac{B}{r^n} \quad \text{...(1.1.6)}$$

But at the equilibrium spacing r_0, U is minimum and hence $\frac{dU}{dr} = 0$

i.e.,
$$\left[\frac{dU}{dr}\right]_{r=r_0} = \frac{\alpha e^2}{4\pi \in_0 r_0^2} - \frac{nB}{r_0^{n+1}} = 0$$

or
$$\frac{\alpha e^2}{4\pi \in_0 r_0} = \frac{nB}{r_0^n}$$

i.e.,
$$B = \frac{\alpha e^2 r_0^{n-1}}{4\pi \in_0 n} \quad \text{...(1.1.7)}$$

Substituting this value of B in equation (1.1.6) with $r = r_0$, one gets

$$U_{min} = -\frac{\alpha e^2}{4\pi \in_0 r_0} + \frac{\alpha e^2 r_0^{n-1}}{4\pi \in_0 n r_0^n}$$

$$= -\frac{\alpha e^2}{4\pi \in_0 r_0} + \frac{\alpha e^2}{4\pi n r_0 \in_0}$$

or
$$U_{min} = -\frac{\alpha e^2}{4\pi \in_0 r_0}\left[1 - \frac{1}{n}\right] \quad \text{...(1.1.8)}$$

Thus for one kmol of the crystal, *the coulomb energy* is

$$U_0 = -\frac{\alpha e^2 N_A}{4\pi \in_0 r_0}\left[1 - \frac{1}{n}\right] \quad \text{...(1.1.9)}$$

There are two other contributions to the cohesive energy of an ionic crystal that have not been considered, (*a*) the *van der Waals attraction* between adjacent ions, which increases the *cohesive energy* and (*b*) the *zero-point oscillations* of their equilibrium positions which decrease the cohesive energy. Both of these effects being very small in ionic crystals and are usually neglected.

Properties of Ionic Crystals:

(*i*) Most of the ionic crystals have fine crystalline structure. It has been found that the constituents of these crystals are ions and not atoms.

(*ii*) Ionic solids have high melting and boiling points because high external energy is required to overcome the electrostatic forces existing between the ions in ionic solids.

(*iii*) Pure and dry ionic solids are good insulators because no free electrons are available. But at high temperature, the elecrostatic forces between the ions are greatly reduced so that some of the ions may transport charges.

(*iv*) Ionic crystals dissolve easily in solvents like water, the electrostatic force considerably weakened (by 80 times) due to high permittivity of water. The result is the ions become free and wander about in the solution. If now a field is applied, these ions will themselves carry the charge in the solution (electrolysis).

1.1.7 THE COVALENT BOND

The existence of compounds such as H_2, N_2 and O_2 as well as bonds in atomic crystals such as diamond cannot be accounted for by ionic bonds because atoms of only one kind cannot produce ions of opposite kind by transfer of electrons. The bonds that exist between atoms of the same kind are called *covalent bonds.* In a covalent bond the electrons from the outermost shell are shared by both the atoms. This may happen between two atoms of the same type or different types. It may also be noted in this bond, the shared outer electrons belong to both the atoms and not to this one or the other. Therefore, both the atoms are neutral even after the formation of bond. It is due to this reason that solids formed by such bond are usually called *non-polar* substances.

The balance between the attractive and repulsive forces in hydrogen molecule occurs at a separation of 0.074 nm. Hence some energy must be spent to break the covalent bond in a hydrogen molecule into hydrogen atoms. About 4.5 eV energy is required to break one bond between the hydrogen atoms. i.e.,

$$H_2 + 4.5 \text{ eV} \rightarrow H + H$$

The covalent bond between the two hydrogen atoms in a hydrogen molecule is represented as:

$$H : H$$

It is desired to make clear that one electron has come from each atom, it is now indicated as:

$$H \times H, \text{ or } \dot{H} + \dot{H} \rightarrow H{:}H \rightarrow H \times H \rightarrow H\text{—}H; H_2$$

Properties of Covalent Substances

Properties of covalent compounds : (*i*) Covalent compounds may be solids, liquids or gases. Generally those substances which have high molecular exist. Covalent solids are hard as well as brittle.

(*ii*) Covalent solids have crystalline structure *i.e.*, atoms or molecules are arranged in some regular repeatable pattern in three dimensions.

(*iii*) Pure covalent solids are good insulators. The reason is that all the electrons are tightly held in the covalent bonds. However when certain impurities are added to such solids, they become semiconductors.

(*iv*) Since covalent bonds are comparatively weak, therefore, covalent solids have low melting and boiling points.

(*v*) Covalent solids are readily soluble in water. However, they are easily soluble in organic solvents like benzene.

1.1.8 METALLIC BOND AND PHYSICAL PROPERTIES OF METALS

Metallic elements have low ionisation energies and hence, in this bonding, atoms of the same element or different elements give their valence electrons to form an electron cloud or say "electron gas" throughout the space occupied by the atoms. Having given up their valence electrons, the atoms are in reality positive ions. These are held together by forces that are similar to those of ionic bond in that they are primarily electrostatic, but are between the ions and the electrons. Most of the atoms in metals have one or two valence electrons.

These electrons are loosely held by their atoms and therefore can be easily released to the common pool to form an *electron cloud*. The electrostatic interaction between the positive ions and the electron gas holds the metal together. The high electrical and thermal conductivities of metals follow from the ability of the free electrons to migrate through their crystal lattices while all of the electrons in ionic and covalent crystals are bound to particular atoms.

Unlike other crystals, metals may be deformed without fracture, because the electron gas permits atoms to slide fast one another by acting as a *lubricant*. As we have seen carbon can exist in the covalent form and so it is an extremely poor conductor. However, it may also exist in an alternate form as graphite. In this case, bonds are formed in which covalency is not fully achieved and these bonds can break and reform fairly as in metallic

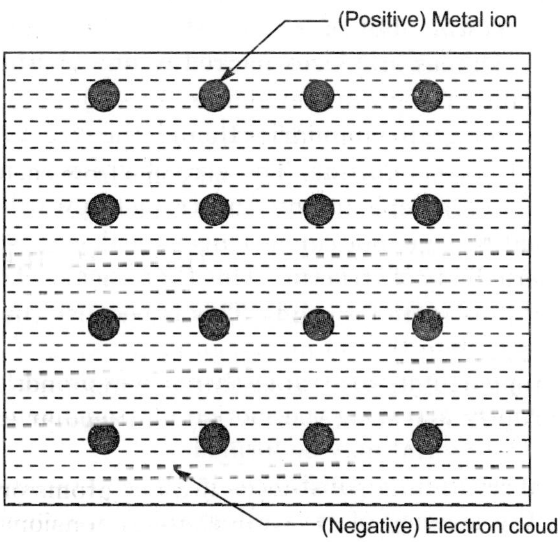

Fig. 1.4 Bonding in metals

bond. For this reason graphite is a conductor. If a p.d is applied between any two points in a metal piece, the electron gas flows from negatively charged part to the positively charged part, constituting electric current. The same free electrons are responsible for transferring thermal energy from places of high temperature to places of low temperature due to their mobility. This accounts for high thermal conductivity of metals. The diffuse nature of electrons in the metallic materials make them opaque. The free electrons absorb light and do not permit it to pass through. For the same reason the atomic planes can slide over each other without causing bonds to rupture thus making the materials ductile.

Metals like sodium, copper and silver have these bonds. The transition metals such as iron, nickel, tungsten, ... have predominantly metallic bond with a significant fraction of covalent bonds.

1.2 CRYSTAL STRUCTURES AND X-RAY DIFFRACTION

1.2.1 INTRODUCTION

On the basis of structure, the solid state of matter can be put into two broad categories: the *crystalline* and the *non-crystalline* (or amorphous). The distinction between the two does not, however, depend on their external appearance or macroscopic form; instead, they are distinguished from one another primarily by the degree of order exhibited by the arrangement of the fundamental particles – atoms, molecules or ions comprising them. In crystalline solids, the atoms or ions (which may be compared with '*building blocks*') are stacked in a regular manner, just like the soldiers do on the parade ground, thus forming a three dimensional pattern which may be obtained by a three-dimensional repetition of a *building block* or '*pattern unit*'. When the regularity of the pattern extends throughout a certain piece of solid, then it can be treated for many purposes a '*single crystal*' and would be so called. However, most of the solids of technical interest such as metals, ceramics, ionic salts, are not single crystals, but often consist of a large number of small crystal sections (grains) of various shapes and sizes packed to one another in a quite irregular way along the interfaces called the grain boundaries. In such materials the regularity or periodicity is interrupted at the grain boundaries though the structure may be more stable. These materials are called *polycrystalline*. The size of grain in which the structure is periodic may vary from submicroscopic (many Angstroms) upwards and when it becomes comparable to the size of pattern unit, one can no longer speak of crystals, since the essential feature of a crystal is its periodicity of structure. One then speaks of *amorphous substances*.

Nature favours crystalline state for solids. This is because the energy of the ordered atomic arrangement is lower than that of an irregular packing of atoms. However when the atoms are not given an opportunity to arrange themselves properly, e.g., the temperature of a liquid is dropped abruptly arresting the motion of atoms before they can rearrange themselves, the amorphous material may be formed.

There are over 20000 crystalline substances in the crystal kingdom. Modern theory of solids is founded on the science of crystallography, which is concerned with the enumeration and classification of the actual structures of various crystalline substances.

Such enumeration and classification provide a natural starting point for mathematical description of crystals. Thus in this chapter we begin our study of the physical properties of crystalline solids by introducing the concept of crystal lattice and its application to the classification of solids.

1.2.2 LANGUAGE OF CRYSTALS

Crystals are identified by their geometrical appearance; external surfaces are usually plane faces. But how do atoms arrange themselves around each other in a crystal? A crystal may be considered as a regular array of units-smallest groupings that repeat themselves in all directions in space. This repeating unit obeys the *translation operation.* The collection of all translation operations constitutes the *translation group*. The translation operation (vector) $\vec{T}$ is defined in terms of three translation vectors $\vec{a}, \vec{b}, \vec{c}$ as follows:

$$\boxed{\vec{T} = n_1\vec{a} + n_2\vec{b} + n_3\vec{c}} \quad ...(1.2.1)$$

where n_1, n_2, n_3 arbitrary integers. By letting n's assume all integral values, we can reach an infinite, regular repeating array of units, that is, a crystal lattice. The *translation vectors* $\vec{a}, \vec{b}, \vec{c}$ are often called crystal axes and their directions are usually along the axes of the *primitive unit cell* – the smallest repeating unit of the crystal containing lattice points only at its corners. When the operation, $\vec{T}$ is applied to any point $\vec{r}$ in the material, the resulting point $r'(r' = \vec{r} + \vec{T} = \vec{r} + n_1\vec{a} + n_2\vec{b} + n_3\vec{c})$ must be identical in all respects with the original point r. If $\vec{r}'$ is not identical with $\vec{r}$ for any arbitrary choice of n_1, n_2, n_3, the vectors $\vec{a}, \vec{b}, \vec{c}$ are not translation vectors. The geometry array of (lattice) points thus generated constitutes the *lattice*. In order to convert this lattice into a crystal structure, we must locate atoms, ions, group of atoms/ions, or molecules on the *lattice points.* The repeating unit assembly–atom, ion, molecules, radical, etc. – that is located at each lattice point is called the *basis.* The crystal structure is thus given by two specifications: *lattice* + *basis.* Lattice points are an array of points generated by the *translation operation.* The crystal is generated by applying to the basis the translation group of operations for the type of lattice under consideration.

Illustrations

Let us now consider the case of *two dimensional array of points* as shown in Fig. 1.5. It is obvious from the figure that the environment about any two points is the same; hence it represents a space lattice. In a mathematical form, the space lattice may be defined as follows:

We choose any arbitrary point O as origin and consider the *position vector* $\vec{r}_1$ and $\vec{r}_2$ of any two lattice points by joining them to O. If the distance $\vec{T}$ of the two vectors $\vec{r}_1$ and $\vec{r}_2$ satisfies the following relation,

$$\vec{T} = n_1\vec{a} + n_2\vec{b}$$

where n_1, n_2 are integers and $\vec{a}$ and $\vec{b}$ are fundamental *translation vectors* characteristics of the array, then the array of points is a *two dimensional lattice.* For three dimensional lattice,

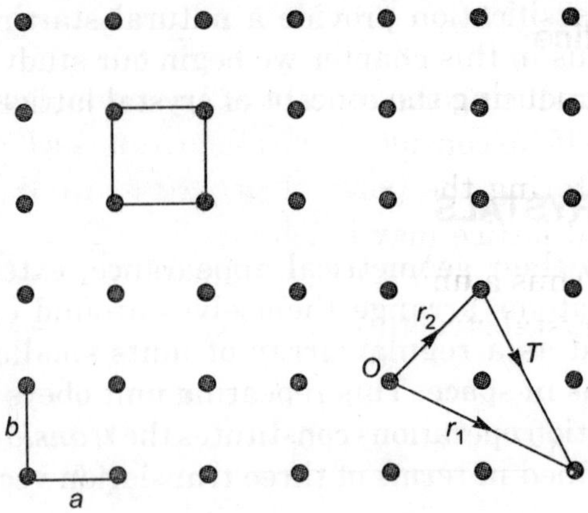

Fig. 1.5 Two dimensional array of points

$$\vec{T} = n_1\vec{a} + n_2\vec{b} + n_3\vec{c}$$

Hence it should be remembered that a crystal lattice refers to the geometry of a set of points in space whereas the structure of the crystal refers to the actual ordering of its constituent ions, atoms and molecules in the space.

The Basis and Crystal Structure

For lattice to represent a crystal structure, we associate every lattice point with one or more atoms (*i.e.*, a unit assembly of atoms or molecules identical in composition) called the *basis* or *pattern*. When the basis is repeated with correct periodicity in all directions, gives the actual crystal. The crystal structure is real, while the lattice is imaginary.

Thus lattice + basis = crystal structure

Figure 1.6 shows the *basis* or *pattern* representing each *lattice point*. It is observed from the figure that the basis consists of three different atoms. It can also be observed that basis is identical in composition, arrangement and orientation. In crystals like aluminium and sodium the basis is a single atom; in NaCl, KCl, etc. the basis is diatomic whereas in crystals like CaF_2 the basis is triatomic.

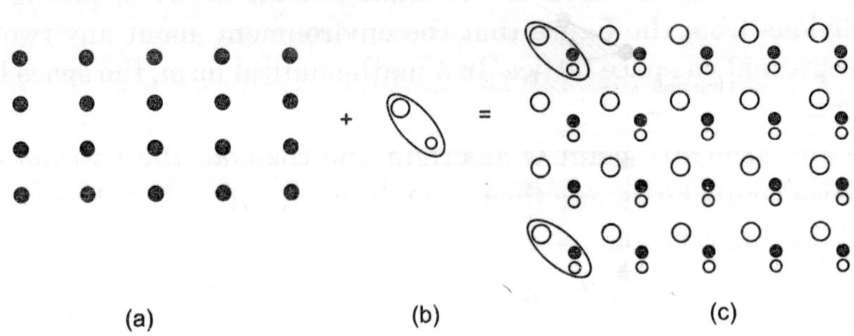

Fig. 1.6 Space lattice, basis and crystal structure

Unit Cell and Crystal Lattice

For discussing the unit cell, let us consider a two dimensional crystal in which the atoms are arranged as in Fig. 1.7. If we consider a parallelogram such as *ABCD* with sides *AB* = *a* and *AD* = *b*, then by rotating the parallelogram by any integral multiple of vector $\vec{a}$ and $\vec{b}$, the whole crystal lattice may be obtained. In this way the fundamental unit *ABCD* is called a unit cell. Thus a unit cell is the smallest geometric figure, the repetition of which gives the actual crystal structure.

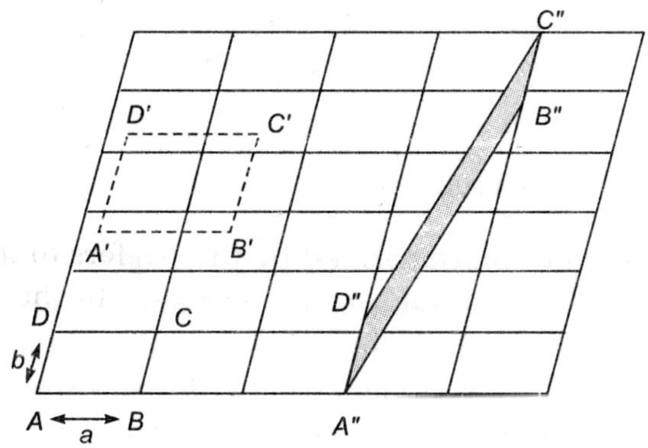

Fig. 1.7 Two dimensional arrangement of atoms

The unit cell may also be defined as the *fundamental elementary pattern* of minimum number of atoms, molecules or group of molecules – which represent fully all the characteristics of the crystal. It should be noted that the choice of a unit cell is not unique but it can be constructed into a number of ways as *A'B'C'D'* or *A"B"C"D"* shown in Fig. 1.7. The unit cell should be chosen in such a way that is conveys the symmetry of crystal lattice and makes the mathematical calculations easier.

For a three dimensional case, the same procedure may be adopted. A three dimensional unit cell is shown in Fig. 1.8(a).

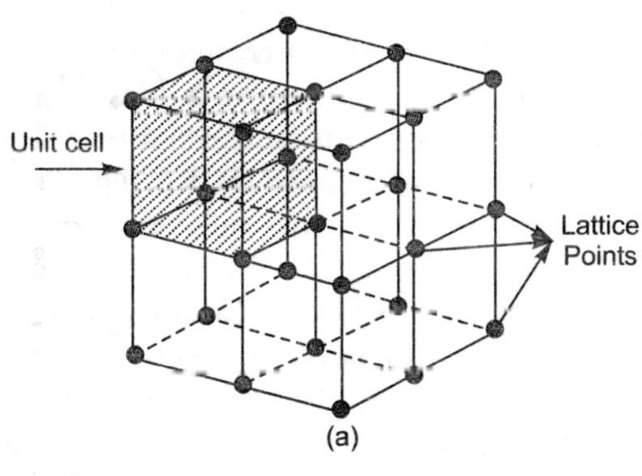

Fig. 1.8(a) Three dimensional arrangement of atoms

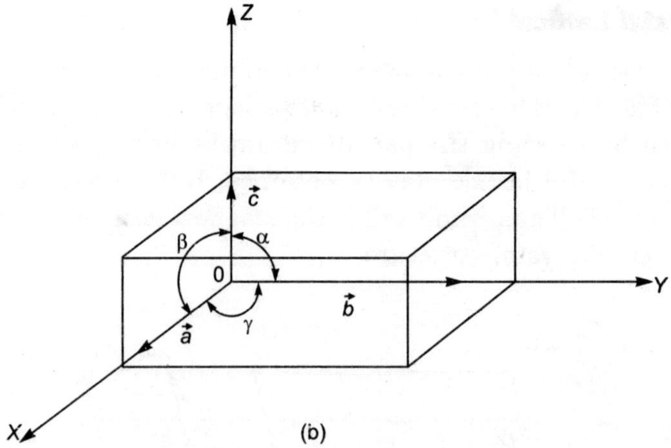

Fig. 1.8(b) Three dimensional unit cell

The unit cell is a parallelepiped formed by the basis vectors $\vec{a}, \vec{b}, \vec{c}$ as concurrent edges and including angles α, β, γ between $\vec{b}$ and $\vec{c}$, $\vec{c}$ and $\vec{a}$ and $\vec{a}$ and $\vec{b}$ respectively as shown in Fig. 1.8(b). Thus in general, a unit cell may be defined as that volume of a solid from which the entire crystal may be constructed by *translation repetition* in three dimensions. The entire lattice structure of a crystal is found to consist of *identical blocks* or *unit cells.*

1.2.3 CRYSTAL SYSTEMS AND BRAVAIS LATTICES

The crystalline structure is often referred to as *latticed structure*. A space lattice is an infinite, three dimensional array of points in which every point has surroundings with that of every other point as discussed earlier. The points with identical surroundings are called *lattice points.* Since the regular pattern of atoms in a space lattice is regular, the atomic arrangements must be described completely by specifying atom positions in some repeating unit of the space lattice. This repeating unit of the space lattice is called a *unit cell*. The unit cell may be regarded as the fundamental building block of a crystal, when several of such unit cells are placed with common surfaces of contact result into a crystal. The edges of unit cell must be vectors connecting any two lattice points.

The *Bravais lattices* describe the different arrangement in which lattice point can be arranged in unit cells. Mathematicians have shown a long time ago that there are only fourteen independent ways of arranging points in three dimensional space such that each arrangement conforms to the definition of a space lattice. Thus there are fourteen Bravais lattices. Each space lattice can be described with reference to its unit cell. Each unit cell can be represented by three intersecting edges a, b, c and angles between a, b and c such that the angle between b and c is α, that between c and a is β and that between a and b is γ. This configuration is shown in Fig. 1.9(a), seven different systems of crystals are obtained by assigning different values to the edges a, b, c and angles α, β and γ. For example the simplest of all these, cubic crystal system is obtained when $a = b = c$ and $\alpha = \beta = \gamma = 90^\circ$, such a system is shown in Fig. 1.9(b). Though not shown in the figure it may be remembered that an atom will be sitting at each corner of the cube in Fig. 1.9(b).

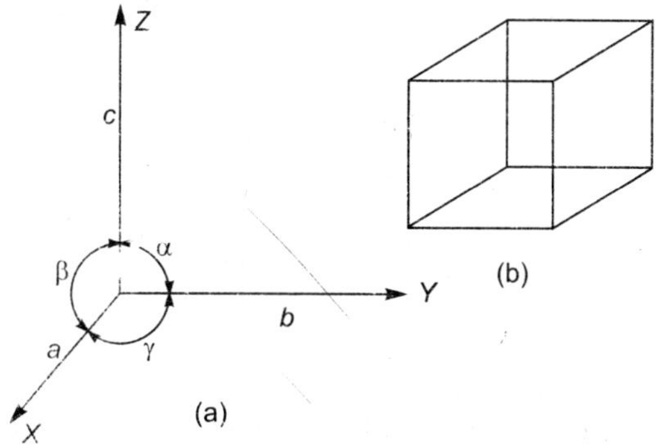

Fig. 1.9 The axes and angles which describe a unit cell (b) A cubic unit cell

When, only corners are occupied by atoms the system is known as *primitive* denoted by *P*, some variations on this primitive system are also possible. For example, additional atom may sit at the centre of all the faces, in which case the system is known as *face-centred*, denoted by *F*. If, however, the centres of only two opposite faces are occupied additionally the system is known as *base centred* denoted by *C*. Yet another variation of the crystal system is when an additional atom occupies the centre of the body of the unit cell denoted by *I*. The seven crystal systems and fourteen Bravais lattices are described in Table 1.1 and shown in Fig. 1.10.

Table 1.1 Crystal systems and Bravais lattices

Sl. No	*System*	*Axes*	*Angles*	*Bravais lattices*	*Lattice symbol*	*Number of lattices in the system*
1.	Cube	$a = b = c$	$\alpha = \beta = \gamma = 90°$	Simple, body centred, face centred	P I F	3
2.	Tetragonal	$a = b \neq c$	$\alpha = \beta = \gamma = 90°$	Simple, body centred	P I	2
3.	Rhombohedral (Trigonal)	$a = b = c$	$\alpha = \beta = \gamma = 90°$	Simple	P	1
4.	Hexagonal	$a = b \neq c$	$\alpha = \beta = 90°$ and $\gamma = 120°$	Simple	P	1
5.	Orthorhombic (Orthogonal)	$a \neq b \neq c$	$\alpha = \beta = \gamma = 90°$	Simple, base centred, body centred, face centred	P C I F	4
6.	Monoclinic	$a \neq b \neq c$	$\alpha = \beta = 90° \neq \gamma$	Simple, base centred	P C	2
7.	Triclinic	$a \neq b \neq c$	$\alpha \neq \beta \neq \gamma \neq 90°$	Simple	P	1
	7					14

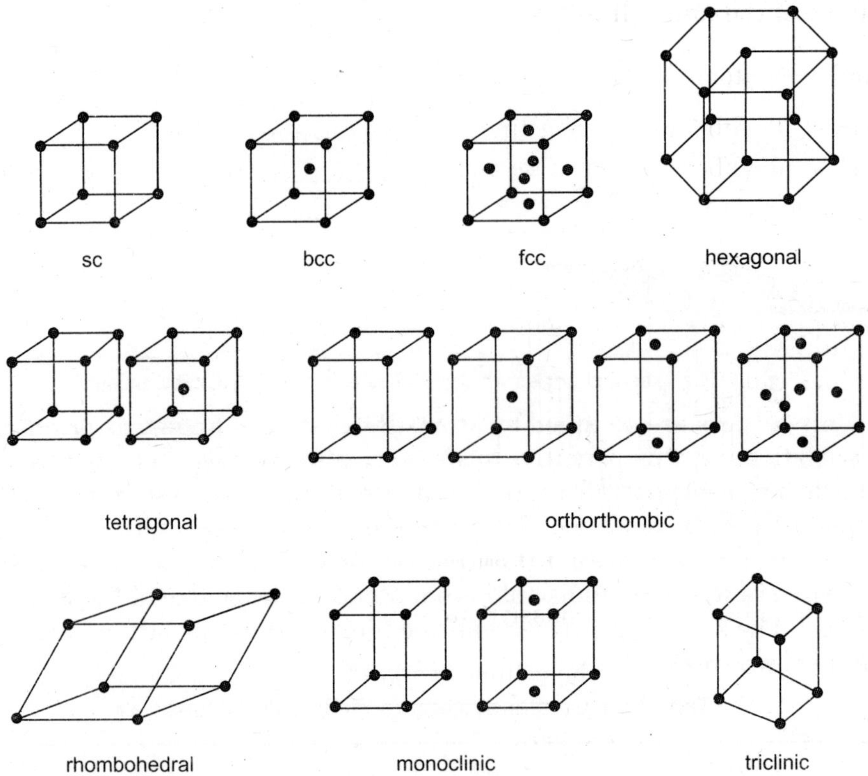

Fig. 1.10 The fourteen Bravais lattices falling in the seven systems of crystals

1.2.4 STRUCTURE OF METALLIC CRYSTALS

Many important properties of solids depend on their crystal structures. Fortunately, relatively few of these need concern us: about three-quarters of the elements (mostly metals and a few semiconductors) used right from transmission lines to micro-instruments in spaceship are metals or semiconductors. These elements crystallize in one of the following structures: *face centred cubic* (fcc), *body centred cubic* (bcc), *hexagonal close-packed* (hcp), or *diamond cubic crystal structure* (dc). Though only one well known element polonium crystallizes in the simple cubic structure, a minor study of this structure theoretically will aid us to extend the frame work to calculate the packing fraction of other structures. The simple study of this simple structure is now discussed.

1. *Simple cubic structure (Primitive)*

The structure shown in the above figure teaches that eight atoms are arranged in a regular cube, one atom at each corner. It is a cubic unit cell of parameter a. The arrangement of atoms is repeated in all X, Y and Z directions. It can also be observed that each corner atom is shared by 8 lattices. *i.e.*, one corner of lattice atom shares effectively 1/8 atom to the unit cell under consideration. From the geometry of the cell it is clear that side of the cube is equal to twice the radius of the atom. *i.e.*, $a = 2r$. Thus

(1) volume of the unit cell is $V = a^3$

(2) number of atoms in a cell is, $n = \frac{1}{8} \times 8 = 1$

(3) nearest neighbours (or co-ordination number) = 6

(4) nearest neighbour distance $2r = a$

(5) the lattice parameter $a = 2$ r

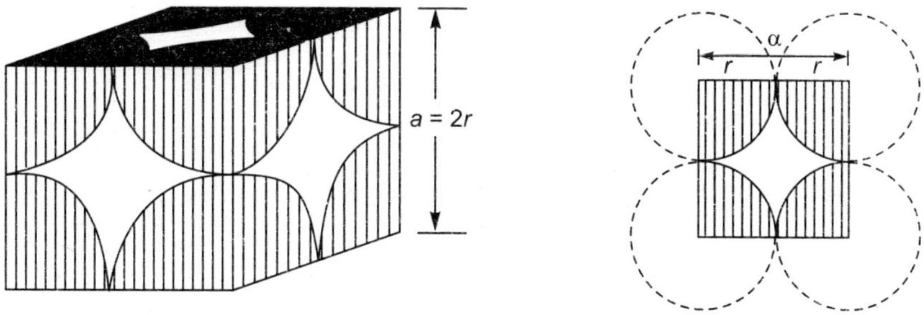

Fig. 1.11 Simple cubic structure

Thus the volume occupied by massive atoms in unit volume is called *packing fraction*, and that is equal to $\frac{v}{V}$. [Here v is the volume of all the atoms in a unit cell and it is $1 \times \frac{4}{3}\pi r^3$

i.e.,
$$PF = \frac{v}{V} = \frac{1 \times \frac{4}{3}\pi r^3}{a^3} = \frac{4}{3}\pi \frac{r^3}{(2r)^3}$$

$$= \frac{4\pi}{3 \times 8} = \frac{\pi}{6} = 0.52$$

$$\boxed{PF = 0.52 \text{ or } 52\%} \quad ...(1.2.2)$$

2. *Body centred cubic lattice (bcc)*

Iron is the most common metal used to fabricate iron curtain and to manufacture different kinds of steel used in many non-basic areas such as engineering and medicine crystallizes in bcc structure. Hence it is absolutely essential to have a geometrical sketch of the unit cell and an environmental study of the corner atoms and the atom sited at the body centre in order to calculate its packing fraction.

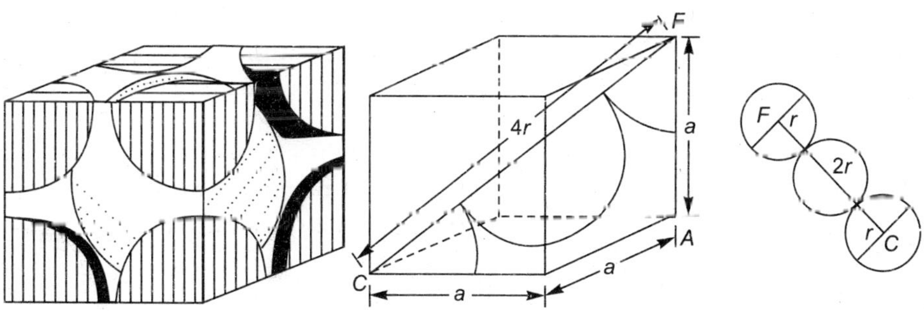

Fig. 1.12 Body centred cubic

Each corner atom in this structure is shared by eight cells an additional atom sits at the body centre well within the unit cell touching the eight corner atoms along the body diagonal and not along the edges as discussed in sc structure. The packing fraction of this structure is computed using the following data and by having a look of Fig. 1.12.

$$(CF)^2 = (CA)^2 + (AF)^2$$

$$(4r)^2 = 2a^2 + a^2 = 3a^2$$

$$16\,r^2 = 3a^2;\; a = \frac{4r}{\sqrt{3}}$$

Thus

(1) nearest neighbour distance, $2r = \dfrac{a(\sqrt{3})}{2}$

(2) the lattice parameter, $a = \dfrac{4r}{\sqrt{3}}$

(3) number of atoms in the unit cell, $2\left(=1+\dfrac{1}{8}\times 8\right)$

(4) volume of all the atoms in the unit cell, $v = 2\times\dfrac{4}{3}\pi r^3$

(5) volume of unit cell, $V = a^3 = \dfrac{64r^3}{3\sqrt{3}}$

(6) coordination number, $CN = 8$

Now

Density of packing or packing fraction,

$$P.F = \frac{v}{V} = \frac{\frac{8}{3}\pi r^3}{64r^3/3\sqrt{3}} = \frac{8\times 3\sqrt{3}\pi r^3}{64\times 3\times r^3}$$

$$P.F = \frac{\pi\sqrt{3}}{8} = 0.68$$

$$\boxed{P.F = \frac{\pi\sqrt{3}}{8} = 68\%} \qquad \text{...(1.2.3)}$$

Chromium and barium are the other examples. Compared with sc structure, the packing fraction of this structure is 16% more. Hence this bcc structure is called a *closely packed structure* while the other one is loosely packed structure.

3. *Face centred cubic lattice (fcc)*

There are two forms of closest packing: face centred cubic (fcc) space lattice and hexagonal close packed structure. Figure 1.13 deals with fcc structure and helps us to find a way out to compute the packing fraction following the general definitions made in the previous cases.

Here the lattice points (or atoms) are situated at all the eight corners as in the previous two cases, but also at the centres of the six faces. The face centre atom shares with two unit cells. Hence the number of atoms in a unit cell is $\left(\frac{1}{8}\times 8+3\right) = 4$. This lattice is also known as a *cubic F lattice*. The atoms touch along the face diagonals. The packing fraction is now calculated by the usual method.

(1) the lattice parameter, $a = \frac{4r}{\sqrt{2}}$

(2) nearest neighbour distance, $2r = \frac{a(\sqrt{2})}{2}$

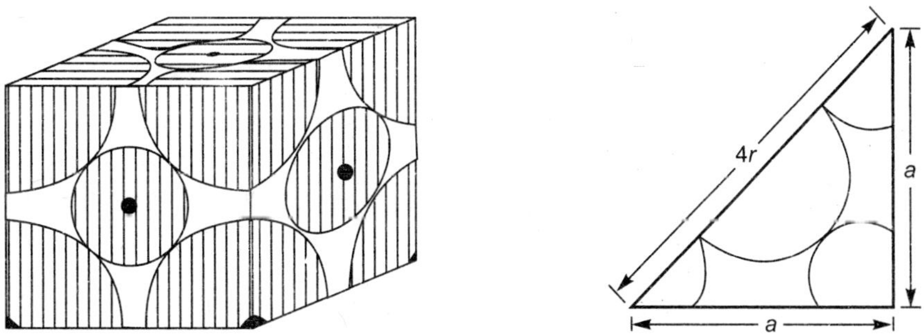

Fig. 1.13 Face centred cubic structure

(3) number of atoms in the unit cell = 4

(4) volume of all the atoms in the unit cell, $= 4\times\frac{4}{3}\pi r^3 = v$

(5) volume of unit cell, $V = a^3$

(6) coordination number, $CN = 12$

Thus density of packing fraction,

$$P.F = \frac{v}{V} = \left[\frac{16\pi r^3}{3}\right] = \left[\frac{2\sqrt{2}}{(4r)^3}\right]$$

$$\boxed{P.F = \frac{\sqrt{2}\pi}{6} = 0.74 \text{ or } 74\%} \qquad ...(1.2.4)$$

copper, aluminium and lead are some examples of this structure.

Table 1.2 Comparison of cell properties of some crystal lattices

Sl. No.	*Properties*	*Diamond cube (dc)*	*Simple cube (sc)*	*Body centred cube (bcc)*	*Face centred cube (fcc)*	*Hexagonal close packed structure (hcp)*
1.	Volume of unit cell	a^3	a^3	a^3	a^3	$\frac{3}{2}\sqrt{3}(a^2c)$
2.	Number of atoms per cell	8	1	2	4	6
3.	Number of atoms per unit volume	$\frac{8}{a^3}$	$\frac{1}{a^3}$	$\frac{2}{a^3}$	$\frac{4}{a^3}$	$\frac{4}{\sqrt{3}(a^2c)}$
4.	Number of nearest neighbours	4	6	8	12	12
5.	Nearest neighbour distance (2r)	$\frac{a\sqrt{3}}{4}$	a	$\frac{a\sqrt{3}}{2}$	$\frac{a\sqrt{2}}{2}$	a
6.	Atomic radius	$\frac{a\sqrt{3}}{8}$	$\frac{a}{2}$	$\frac{a\sqrt{3}}{4}$	$\frac{a\sqrt{2}}{4}$	$\frac{a}{2}$
7.	Atomic packing factor	$\frac{\pi\sqrt{3}}{16} = 0.34$	$\frac{\pi\sqrt{3}}{6} = 0.52$	$\frac{\pi\sqrt{3}}{8} = 0.68$	$\frac{\pi\sqrt{2}}{6} = 0.74$	$\frac{\pi\sqrt{2}}{6} = 0.74$
8.	Examples	Germanium Silicon and Diamond	Polonium	Sodium, Iron, Chromium	Aluminium, Copper and Lead	Magnesium, Zinc and Cadmium

1.2.5 MILLER INDICES

Various properties of crystals, mainly mechanical properties are connected with the structure of the crystal through the help of crystal directions and planes. In a crystal there exists directions and planes which contain a large concentration of atoms. X-rays are easily diffracted through planes containing large concentration of atoms. Miller evolved a method to designate a set of parallel planes in a crystal by three numbers $(h\ k\ l)$ known as Miller indices.

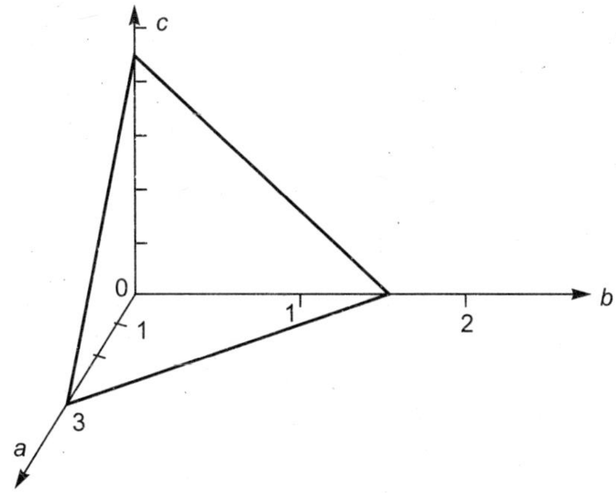

Fig. 1.14 The (3 6 2) plane

Consider the plane shown in Fig. 1.14. The plane makes intercepts on all the axes-at 3a units on the a-axis, $1\frac{1}{2}$ b units on the b-axis and $4\frac{1}{2}$ c on the c-axis. The Miller indices of the plane are found by taking the reciprocals of the axial intercepts (ignoring the lattice repeat units a, b, c): $\frac{1}{3}, \frac{1}{1\frac{1}{2}}, \frac{1}{4\frac{1}{2}}$, or $\frac{1}{3}, \frac{2}{3}, \frac{2}{9}$. Multiplying these three terms by 9, we get the integers (3 6 2), which are called *Miller indices* of the plane. Negative intercepts are written with a bar over them; thus $(\bar{1}\ \bar{2}\ \bar{3})$. If a plane is parallel to a given axis, then its intercept is at infinity and its reciprocal is therefore zero; so a zero in the Miller indices of a plane means that it must be parallel to that axis. For example, the (0 0 1) plane is parallel to a-and b-axes and intercepts the c-axis.

In short *Miller indices* are defined as the reciprocals of the intercepts made by the plane, on the crystallographic axes when reduced to smallest numbers.

Crystal directions

The crystal directions are usually given in a square bracket like [ ]. It is easier to specify crystal direction on a unit cell. In Fig. 1.15, three different directions are shown in the orthorhombic lattice. The direction [1 1 1] is the line passing through the origin and point P. The point P is at a unit cell distance from each axis. The direction [1 0 0] is the line passing through origin and point Q. The point Q is at a distance 1 0 0 from X, Y, and Z axis. Direction [1 0 1] is the line passing through origin and R. The point R is at distance of 1 0 1 form X, Y, and Z axes.

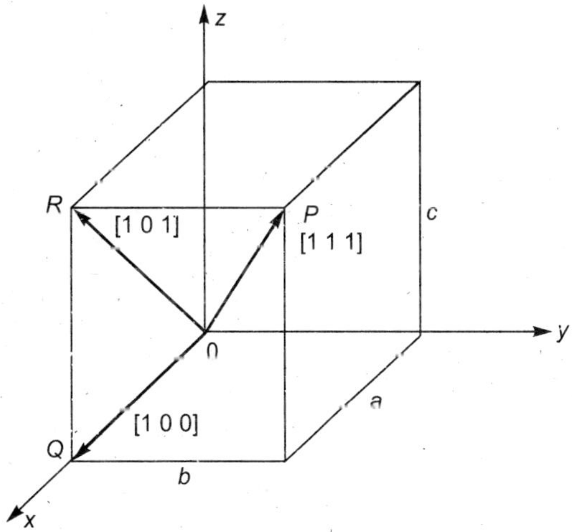

Fig. 1.15 Crystal directions

Conclusions

(*i*) All the parallel equidistant planes have the same Miller indices. Thus the Miller indices define a set of parallel planes.

(*ii*) A plane parallel to one of the co-ordinate axes has intercept of infinity.

(*iii*) If the Miller indices of two planes have the same ratio; *i.e.*, (8 4 4) or (4 2 2) or (2 1 1), then the planes are parallel to each other.

(*iv*) If ($h\ k\ l$) are the Miller indices of a plane, then the plane cuts the axes into h, k and l equal segments.

(*v*) In specifying crystal direction, crystal axes are taken as base directions [3 3 3] or [2 2 2] direction are identical to [1 1 1] direction. In such cases lowest combination of integers *i.e.*, [1 1 1] is used to specify direction.

Interplanar spacings

In a cubic system, the distance between two successive planes can be calculated by using the relation

$$d_{hkl} = \frac{a}{\sqrt{h^2 + k^2 + l^2}}$$

where h, k and l are the indices of the plane and a is the lattice constant. This important equation may be obtained as follows:

($h\ k\ l$) represents the Miller indices of the plane ABC. This plane belongs to a family of planes whose Miller indices are ($h\ k\ l$) because Miller indices represent a set of planes. See Fig. 1.16.

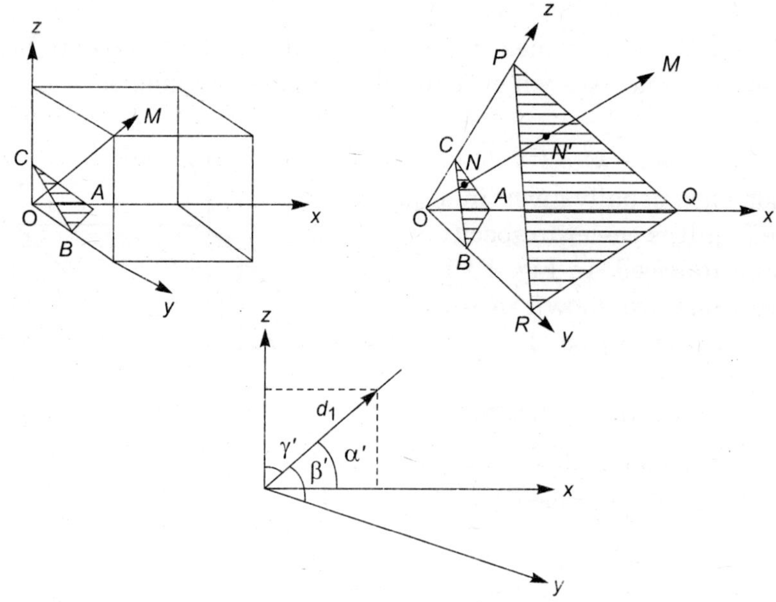

Fig. 1.16 Interplanar spacing in cubic crystal

Let α', β' and γ' (different from interfacial angles α, β, and γ) be the angles between coordinates axes *XYZ* respectively on *ON* and *ON'*.

Thus $$OA = \frac{a}{h}, OB = \frac{a}{k} \text{ and } OC = \frac{a}{l}$$

Also $$\boxed{\cos\alpha' = \frac{d_1}{OA}, \cos\beta' = \frac{d_1}{OB} \text{ and } \cos\gamma' = \frac{d_1}{OC}} \quad ...(1.2.5)$$

Again $$(ON)^2 = x^2 + y^2 + z^2 \text{ with } ON = d_1$$

and $$\cos\alpha' = \frac{x}{d_1}, \cos\beta' = \frac{y}{d_1} \text{ and } \cos\gamma' = \frac{z}{d_1}$$

Thus $$d_1^2 = d_1^2(\cos^2\alpha' + \cos^2\beta' + \cos^2\gamma')$$

$$d_1 = d_1[\cos^2\alpha' + \cos^2\beta' + \cos^2\gamma']^{1/2}$$

In the *Cartesian system of co-ordinates* the sum of the squares of the direction cosines is equal to 1.

Substituting this in eqn. (1.2.5), we get

$$\left[\frac{d_1}{OA}\right]^2 + \left[\frac{d_1}{OB}\right]^2 + \left[\frac{d_1}{OC}\right]^2 = 1$$

$$\left[\frac{d_1 h}{a}\right]^2 + \left[\frac{d_1 k}{b}\right]^2 + \left[\frac{d_1 l}{c}\right]^2 = 1$$

For cubic system,

$$\frac{d_1^2}{a^2}(h^2 + k^2 + l^2) = 1$$

$$d_1^2 = \frac{a^2}{(h^2 + k^2 + l^2)}$$

$$d_1 = \frac{a}{\sqrt{(h^2 + k^2 + l^2)}}$$

Let $ON' = d_2$ be the perpendicular of the next plane *PQR* parallel to *ABC*. The intercepts of the plane on these crystallographic axes are:

$$OA' = \frac{2a}{h}; OB' = \frac{2b}{h} \text{ and } OC' = \frac{2c}{l}$$

(with $OA' = OQ$, $OB' = OR$ and $OC' = OP$)

and $$\boxed{\cos\alpha' = \frac{d_2}{OA'}, \cos\beta' = \frac{d_2}{OB'} \text{ and } \cos\gamma' = \frac{d_2}{OC'}} \quad ...(1.2.6)$$

$$(ON')^2 = d_2^2 \cos^2\alpha' + d_2^2 \cos^2\beta' + d_2^2 \cos^2\gamma' \text{ with } ON' = d_2$$

But $$\cos^2\alpha' + \cos^2\beta' + \cos^2\gamma' = 1$$

Substituting this in Eqn. 1.2.6 and simplifying one gets

$$\left[\frac{d_2h}{2a}\right]^2 + \left[\frac{d_2k}{2b}\right]^2 + \left[\frac{d_2l}{2c}\right]^2 = 1; \text{ for cubic system}$$

$$\frac{d_2^2}{(2a)^2}(h^2 + k^2 + l^2) = 1$$

$$d_2 = \frac{2a}{\sqrt{(h^2 + k^2 + l^2)}}$$

Thus the *interplanar distance* between two adjacent planes of Miller indices $(h\ k\ l)$ in a cubic lattice is given by

$$\boxed{d = (d_2 - d_1) = \frac{a}{\sqrt{h^2 + k^2 + l^2}}} \quad ...(1.2.7)$$

1.2.6 X-RAY DIFFRACTION

A fine beam of X-rays fall on the single crystal. The crystal is rotated in such way that, at some position of the crystal, Bragg's condition is satisfied. By measuring the value of angle θ, the interplanar spacing can be calculated from Bragg's equation.

Bragg's Law

The concept of reciprocal lattice may now be applied to understand the actual situation under which X-rays are diffracted by crystal planes and thereby the Bragg's law. Bragg's law is given by equation

$$\sin\theta_{h\,k\,l} = \frac{1/d_{h\,k\,l}}{2/\lambda} \quad ...(1.2.8)$$

Now let us consider a circle of radius $1/\lambda$ and construct a right angled triangle with its diameter as hypotenuse and one of the sides of length equal to $1/d_{h\,k\,l}$. This is shown in Fig. 1.17. The angle opposite to the side of length equal to $1/d_{h\,k\,l}$ will be equal to θ as evidenced by equation (1.2.8). Now Fig. 1.17 explains Bragg's law graphically.

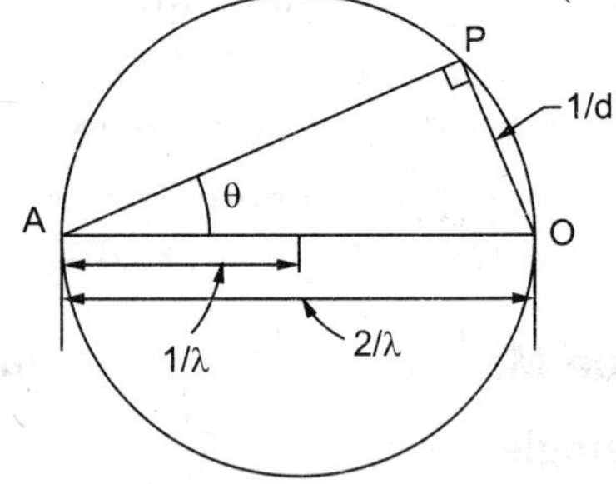

Fig. 1.17 X-ray diffraction

We will try to understand the Bragg's law from Fig. 1.17 by making the latter resemble to a real practical situation. Let us suppose the direction of incident X-ray beam to coincide with the diameter *AO* in the figure. Since *PAO* is equal to $\theta_{h\,k\,l}$, the Bragg angle, the side *AP* of the triangle *APO* should have the same slope as those of the set of parallel planes which diffract X-rays. Let us try to go more closer to the real situation by modifying Fig. 1.17 to Fig. 1.18 where one may imagine one of the set of parallel planes of Miller indices (*h k l*), diffracting X-rays being put at the centre making an angle θ with *AO* and *AO* is the direction of the incident X-ray beam. In Fig. 1.18, *OP* is normal both to the crystal plane (when extended) and *AP*. Hence $\angle OCP$ is equal to $2\theta_{h\,k\,l}$ making *CP* the direction of the diffracted X-ray beam. Hence Fig. 1.18 is a graphical representation of Bragg's law in terms of the reciprocal lattice vector (since *OP* is equal to $1/d_{h\,k\,l}$ it represents the reciprocal lattice vector). Thus the following conditions may be emphasised to explain Bragg's law with the aid of Fig. 1.18.

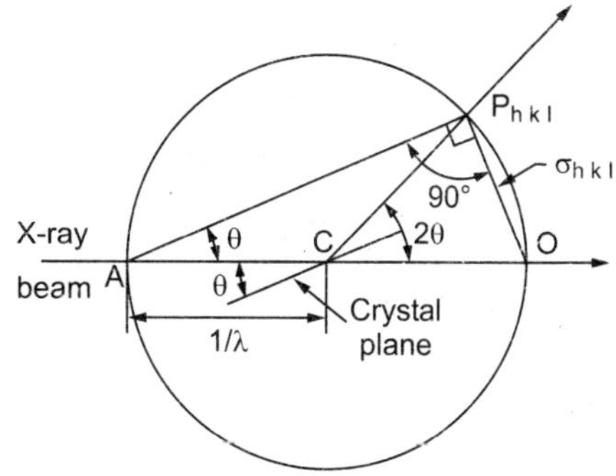

Fig. 1.18 X-ray diffraction

1. The crystal has to be thought of being placed at the centre of a circle of radius $1/\lambda$. In real cases it should be imagined to be at the centre of a sphere of radius $\frac{1}{\lambda}$.
2. Origin of the reciprocal lattice vector is the point *O*, where the X-ray beam leaves the circle (sphere in three dimensions).
3. The point *P* has to be the position of the reciprocal lattice point (at the end of the reciprocal lattice vector $\sigma_{h\,k\,l}$ or **G** of length $1/d_{h\,k\,l}$) or in other words it is sufficient to understand that whenever a reciprocal lattice point lies on the circle (sphere in three dimensions) Bragg's law will be satisfied and the diffracted X-ray beam passes along *CP*. In three dimensions this sphere of radius $1/\lambda$ is also called Ewald's sphere of reflection or simply sphere of reflection or simply Ewald's sphere. Thus reciprocal lattice may be used in conjunction with the sphere of reflection to explain X-ray diffraction experiment. As it is easy to imagine the reciprocal lattice point centred at the point of emergence of the X-ray beam from the sphere of reflection, a construction like Fig. 1.18 can be used to analyse the different ways in which a crystal can be made to satisfy Bragg's condition for its various planes. It is important to note that the X-ray diffraction takes place only when a reciprocal lattice point intersects or lies on the sphere of reflection. Hence a photographic record of the diffracted X-ray beams is nothing but a photographic record of the reciprocal lattice. It is precisely because one normally records the reciprocal lattice of crystals in X ray diffraction experiments that the concept of reciprocal lattice is so essential to their straight forward interpretation.

Laue Method to Study Crystal Structure

A single crystal is placed in the path of X-rays beam and the glancing angle θ is kept constant. Each set of plane will pick up various values of wavelengths; which can satisfy

Bragg's law. Thus each set of plane will produce a spot which fulfils Bragg's law. This method is called Laue technique. Laue method is one of the principal methods used for the analysis of crystal structure. A polychromatic X-ray beam strikes at 90° on the plane of a crystal under study. While passing through the crystal, the rays meet Bragg's planes having different interplaner spacings d. These different sets of planes make different angles θ with the direction of X-rays. Certain combinations θ, and d satisfy the conditions of Bragg's law, with the resulting increase in intensity of diffracted X-rays.

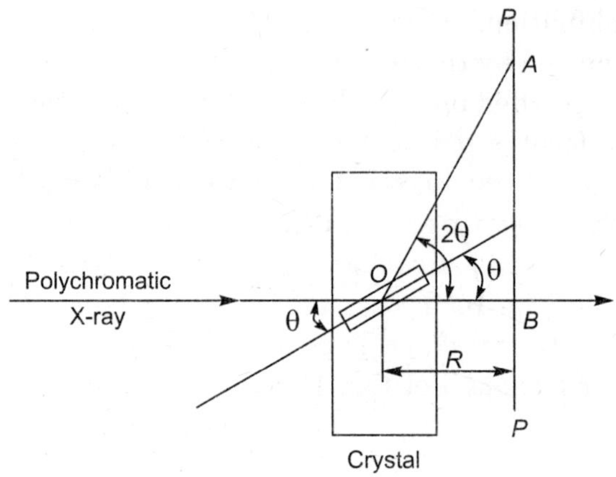

Fig. 1.19 X-ray diffraction (Laue method)

When primary rays pass through the crystal, they produce a black spot at point B on the photographic plate PP. As different wave lengths are included in the primary X-ray, they will produce spots, which are less pronounced, around the central spot as shown in Fig. 1.19. One of the planes considered as the plane for diffraction makes θ with the incident ray. Let A be the spot obtained on the photographic plate.

Thus $$\tan 2\theta = \frac{AB}{R}; \; AB = R \tan 2\theta$$

By measuring AB and R, we can find the value of angle θ for corresponding plane. This method is usually used for single large size crystal.

Powder Crystal Method (Debye–Sherrer Method)

This technique is used when single crystals of large size are not available. The crystalline material is available either in the form of very small crystals or in powder form. In the powder technic, a sample of material is ground into fine powder. The powder is then placed in a capillary tube or pasted on thin wire or pressed and cemented into a thin spindle. This powder specimen is placed at the centre around which a strip of circular photographic film is positioned as shown in Fig. 1.20(a). A narrow beam of monochromatic X-rays is allowed to strike the specimen (in capillary tube or wire or spindle) through a small hole. The specimen (spindle, etc) is slowly revolved inside the specially constructed circular camera called the *powder camera*. The X-ray beam enters through hole a, passes through the specimen and the unused part of beam leaves through hole b.

Let the incident rays make an angle θ with a set of parallel crystal planes. There will be reflection, if Bragg condition is satisfied. As there are a large number of randomly oriented crystals in the powder, thus, there are many possible orientations of this set of planes for same angle θ. Consequently the reflected rays will not be in the form of parallel beam but they (reflected rays) will lie on the surface of a cone (like generators of a cone). The apex of the cone is the point of contact of x-rays with the specimen. For a fixed value of n (say $n = 1$), there are so many combinations of d and θ, that would satisfy Bragg condition. For each combination, one cone of reflection will be there and hence so many cones of reflection are emitted by the specimen. As the film is of narrow width only a part of each reflected cone is recorded on the film which is positioned around the camera. The recorded lines are in the form of pair of arcs as shown in Fig. 1.20(*b*).

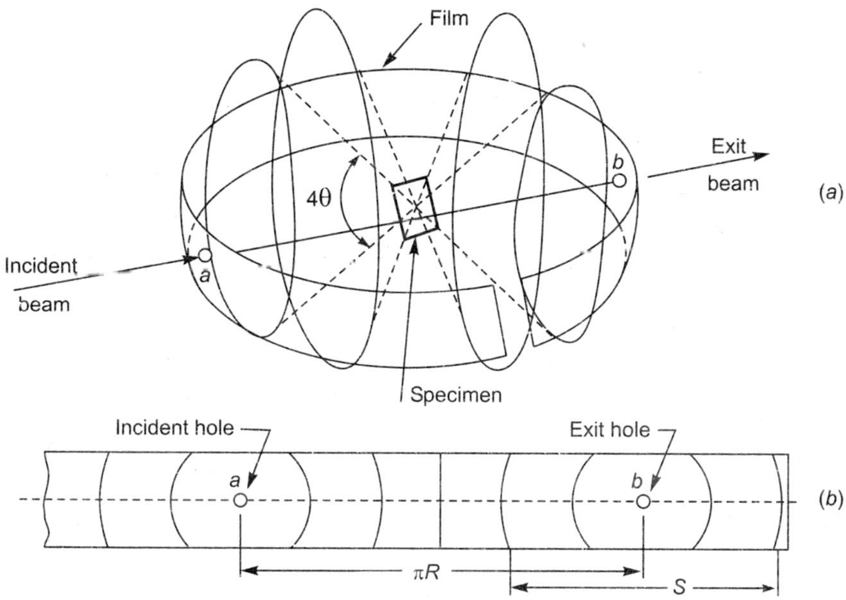

Fig. 1.20 X-ray diffraction (a) (Powder technique) (b) photographic film (record)

The angle between the surface of cone i.e., reflected ray and exist of beam is 2θ or we can say that apex angle is 4θ. In case the Bragg angle is 45°, the cone will take the form of circle. This circle will interest the film strip at midpoints between the incident hole a and exist hole b. When the Bragg angle is greater than 45°, back reflection takes place, the reflected cones are formed toward incidence hole a.

The first pair of arc on each side of exit hole is due to smaller value of θ. The other pairs of arcs beyond this pair are due to larger value of θ. The greater the value of θ, the farther will the arcs from the exist hole, the smaller will be the interplanar spacings according to $d = \dfrac{\lambda}{2\sin\theta}$. If the distance between a pair of arc is S, the value of S is related with θ as below:

$$S \propto 4\theta$$

$$2\pi R \propto 360^\circ$$

$$\frac{S}{2\pi R} = \frac{4\theta}{360^o} = \frac{\theta}{90^o}$$

$$S = \frac{2\pi R\theta}{90^o}$$

The standard diameter of camera is 57.3 mm, then $2\pi R$ is equal to $57.3 \times 3.14 = 180^o$

$$S = 2\theta \text{ degree}$$

The distance S is measured in mm. The half of this distance gives the magnitude of angle θ in degree. The value of θ, along with other known data (n and λ) is substituted in Bragg equation and the interplanar spacing can be calculated. This X-ray diffraction technique is suitable for (*i*) identification of elements and compounds, because the powder patterns are unique for each substance, (*ii*) determination of orientation of a single crystal (*iii*) determination of structures of crystals of high symmetry (*iv*) estimation of grain – size of polycrystalline materials.

The experimental arrangement of powder crystal method is shown in Fig. 1.21. X-rays from the tube are allowed to pass through filter F which absorbs all except one wavelength of incident X-ray beam of monochromatic one. This monochromatic beam is now passed through two fine slits S_1 and S_2 cut in lead plates and then fall on powder crystals C. A photo film is arranged around the inner surface of the camera as shown in the figure.

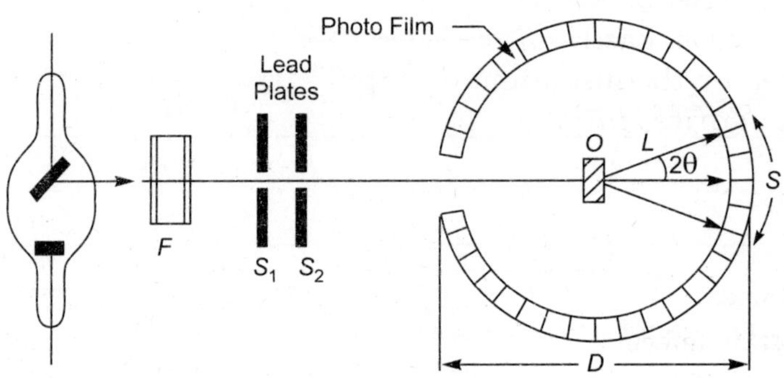

Fig. 1.21 Experimental arrangement (Powder crystal method)

By measuring the intensities of arcs with the help of diffraction meter, the following information can be obtained.

(*a*) By taking the photograph before and after heat treatment or cold working of a crystal, the changes in the lattices can be studied.

(*b*) The dimensions of a unit cell and type of lattice can be determined for simple cases.

(*c*) We can distinguish between crystalline and amorphous substances, as crystalline substances produce lines (arcs) on photo film while amorphous substances do not.

1.3 DEFECTS IN CRYSTALS

1.3.1 IMPERFECTIONS IN CRYSTALS

The mathematically perfect crystal is an exceedingly useful concept. In actual crystals, however, imperfections or defects are always present and their nature and effects are often very important in understanding the properties of crystals. This is because some properties of materials such as stiffness, density and electrical conductivity which are termed *structure-insensitive*, are not affected by the presence of defects in crystals while many of the properties of greatest technical importance such as mechanical strength, ductility, crystal growth, magnetic hysteresis, dielectric strength, conduction in semiconductors, which are termed *structure-sensitive* are greatly affected by the relatively minor changes in crystal structure caused by defects. Ordinary materials, however, contains imperfections of various kinds that produce both useful properties and also such undesirable effects as causing the strength to decrease below that of a perfect crystal. Some of the common defects found in crystals are discussed below:

The structural imperfections can be classified on the basis of their geometry as *point, line, surface,* and *volume imperfections*. The common types of point defects in solid materials can be divided into three classes:

(1) *lattice-site defects*, such as *vacancies* and *interstitialcies*;

(2) *compositional defects*, that is, impurity atoms; and

(3) *electronic defects*. *Line defects* may be classified as (1) *edge dislocation* and (2) *screw dislocation*. Lastly *surface defects* may be the defects on (1) *external surface* represented by the discontinuity at the surface and (2) *internal surface, e.g., grain boundaries, tilt boundaries, twin boundaries, stacking fault*, etc. To summarise, crystallographic defects are:

1. Thermal vibrations
2. Point defects:
   - (*a*) Vacancies
     Schottky defect
   - (*b*) Interstitialcies
     Frenkel defect
   - (*c*) Compositional defects
     (*i*) Substitutional impurity (*ii*) Interstitial impurity
   - (*d*) Electronic defects
3. Line defects:
   - (*a*) Edge dislocation
   - (*b*) Screw dislocation
4. Surface defects:
   - (*a*) Grain boundaries
   - (*b*) Twin boundaries
   - (*c*) Tilt boundaries
   - (*d*) Stacking fault
5. Volume defects.

Thermal Vibrations

Atoms on a mathematically perfect lattice should occupy exactly the sites to which they are attached. It has been observed that no crystal is perfectly rigid, since it can be deformed by finite forces. Hence, it is possible to displace atoms from their ideal sites with a finite expenditure of energy due to thermal vibrations. The frequency of vibration is almost independent of temperature, but the amplitude increases with increasing temperature. When the vibration becomes strong enough the atoms may break the bonds between them and this corresponds to the solid actually melting and becoming liquid, the corresponding temperature is melting temperature. Since the atoms interact with one another, they tend to vibrate in synchronism; that is, groups of atoms tend to move in the same direction, somewhat as waves on the ocean. Because the energy of a particular type of wave can be measured in units known as phonons, the thermal energy of the entire solid specimen is represented quite simply by a certain collection of phonons, each one of which is pictured as a special kind of 'particle'. The thermal vibration of atoms of a solid, as it is, does not seriously disturb the perfection of the crystal. Every atom on the average is, in its proper position. Each atom, therefore, has requisite number of nearest neighbours at proper spacing.

Point Defects

Point imperfections are lattice errors at isolated lattice points. As the name implies, they are imperfect point-like regions in the crystal and, therefore, they may be referred to as zero-dimensional imperfections. A point defect extends its influence only a few atom diameters beyond its lattice position. A point imperfection comes about, as a rule, because of the absence of a matrix atom (an atom that would be present in a perfect crystal), the presence of an impurity atom or a matrix atom in the wrong place. The most common point defects in a crystal of a pure element are described below.

Vacancies

A vacancy is the simplest point defect in a crystal. This refers to a missing atom or vacant atomic site. Such defects may arise either from imperfect packing during the original crystallisation or from thermal vibrations of the atoms at high temperatures. In the latter case, when the thermal energy due to vibration is increased there is an increased probability that individual atoms will jump out of their positions of lowest energy. For most crystals, this thermal energy is of the order of 1 eV per vacancy. Vacancies may be single or two or more of them may condense into a di-vacancy or tri-vacancy [Fig. 1.22 (*a*)].

(*a*) (*b*) (*c*)

Fig. 1.22 Point defects in crystals

Interstitialcies

In a close-packed arrangement of atoms if the atomic packing factor is low, an extra atom may be lodged within the crystal structure. This is known as *interstitials*. An atom can enter the interstitial void or space between the regularly positioned atoms only when it is substantially smaller than the parent atoms [Fig. 1.22 (*a*)], otherwise it will produce atomic distortion. Interstitialcies may also be mono-interstitial, di-interstitials, and tri-interstitials. The vacancy and interstitialcy are, therefore, inverse phenomena.

Schottky Defects and Frenkel Defects

In non-metallic crystals, the formation of a vacancy involves a local readjustment of charge in the surrounding crystal such that charge neutrality is maintained in the crystal as a whole. Thus, if in an ionic crystal there is a vacancy in a positive-ion site, charge neutrality may be achieved by creating vacancy in a neighbouring negative-ion site. Such a pair of vacant sites is called *Schottky* defect [Fig. 1.22 (*b*)]. This is closely related to vacancies.

On the other hand, if the charge neutrality is maintained by having a positive ion in an interstitial position, the pair constitutes a *Frenkel* defect [Fig. 1.22 (*c*)] closely related to interstitialcies.

Close packed structures have fewer interstitialcies and *Frenkel defects* than vacancies and *Schottky defects*, as additional energy is required to force the atoms in their new positions.

In spite of the care taken in the preparation of crystals, vacancies are always present in all crystals. In fact, as a result of *thermal fluctuations*, vacancies are produced and destroyed constantly in the crystal. Formally such a defect might be produced by plucking an interior atom out of its regular lattice site and placing it on the surface. This act requires energy. Moreover, the disorder increases resulting in an increase in the entropy. In thermal equilibrium, a certain number of lattice vacancies are always present which can be estimated as follows:

If E_v is the energy required to take an atom from a lattice site inside the crystal to a lattice site on the surface (called the *enthalpy of formation of the point imperfection*), nE_v is the increase in energy associated with the generation of n isolated vacant sites. The total number of ways in which we can pick up n-atoms from the crystal consisting of N atoms is given by:

$$P = \frac{N!}{(N-n)!\,n!}$$

Since disorder increases due to the creation of n vacancies the corresponding increase in entropy is given by:

$$S = k_B \log P$$

$$S = k_B \log \frac{N!}{(N-n)!\,n!}$$

This inturn produces a change in free energy F,

$$F = U - TS$$

$$= nE_v - k_B T \log \frac{N!}{(N-n)!\,n!}$$

The second term on the right hand side can be simplified by using *Stirling* approximation:

$$\log x! \cong x \log x - x$$

Consequently,

$$F = nE_v - k_B T[N \log N - (N-n)\log(N-n) - n \log n]$$

Free energy in thermal equilibrium at constant volume must be minimum with respect to changes in n;

i.e., $$\left[\frac{\delta F}{\delta n}\right]_T = 0 = E_v - k_B T \log\left[\frac{(N-n)}{n}\right]$$

Thus, $$\frac{E_v}{k_B T} = \log \frac{(N-n)}{n}$$

$$n \cong (N-n)\exp(-E_v/k_B T)$$

If $n << N$, we can neglect n in $(N-n)$; so that,

$$n \approx N \exp(-E_v/k_B T) \quad \text{...(1.3.1)}$$

If $E_v = 1$ eV and $T = 1000\ K$; then,

$$\frac{n}{N} = e^{-11.6}$$

$$\frac{n}{N} = 9.1 \times 10^{-6} \cong 10^{-5}$$

The equilibrium concentration of vacancies decreases as the temperature decreases.

Table 1.3 Enthalpy of formation of vacancies in some crystals

Crystal	E_v *[kJ/kmol] in* 10^3	E_v *in eV/vacancy*
Kr	7.7	0.08
Cd	38.0	0.39
Pb	48.0	0.50
Zn	49.0	0.51
Mg	56.0	0.58
Al	68.0	0.70
Ag	106.0	1.10
Cu	120.0	1.24
Ni	168.0	1.74

In ionic crystals, the formation of paired vacancies is most favoured *i.e.*, an equal number of positive and negative ion vacancies are produced. The formation of pairs makes it possible to keep the surface of the crystal electrostatically neutral. The number of pairs can be related to the total number of atoms present in the crystal on following the same procedure as adopted in pure atomic crystals.

The different ways in which n separated pairs can be formed are:

$$P = \left[\frac{N!}{(N-n)!\,n!}\right]^2$$

Increase in entropy is given by;

$$S = k_B \log P$$

$$= k_B \log \left[\frac{N!}{(N-n)!\,n!}\right]^2$$

with corresponding change in free energy:

$$\boxed{\begin{aligned} F &= U - TS \\ &= nE_p - k_B T \log \left[\frac{N!}{(N-n)!\,n!}\right]^2 \end{aligned}}$$

where, E_p is the energy of formation of a pair.

As before we now apply *Stirling's* approximation to simplify the factorial term.

i.e.,
$$\log \left[\frac{N!}{(N-n)!\,n!}\right]^2 = 2[\log N! - \log(N-n)! - \log n!]$$

$$= 2[N \log N - N - (N-n)\log(N-n) + (N-n) - n \log n + n]$$

$$= 2[N \log N - (N-n)\log(N-n) - n \log n]$$

Thus, free energy,

$$F = nE_p - 2k_B T[N \log N - (N-n)\log(N-n) - n \log n]$$

Differentiating the above equation with respect to n, we get:

$$\left[\frac{\delta F}{\delta n}\right]_T = E_p - 2k_B T\,[0 + \log(N-n) + 1 - \log n - 1]$$

$$= E_p - 2k_B T \log \left[\frac{(N-n)}{n}\right]$$

At equilibrium, the free energy is constant, so that:

$$E_p - 2k_B T \log \left[\frac{(N-n)}{n}\right] = 0$$

$$\frac{E_p}{2k_B T} = \log \left[\frac{(N-n)}{n}\right]$$

$$\frac{N-n}{n} = \exp(E_p/2k_BT)$$

$$\boxed{n = N\exp(-E_p/2k_BT)} \qquad \text{...(1.3.2)}$$

Provided $n << N$. In NaCl crystal, $E_p = 2.02$ eV and at random temperature:

$$\boxed{\frac{n}{N} = \exp(-E_p/2k_BT)}$$

$$= \exp(-2.02/2 \times 0.025)$$
$$= e^{-40.4}$$

i.e.,
$$\frac{n}{N} = 2.8 \times 10^{-18}$$

Proceeding in the same way as in the case of *Schottky* defect, we can calculate the number of *Frenkel* defects in equilibrium at a temperature T. Let in a perfect crystal,

E_i be the energy required to displace an atom from a regular lattice site to an interstitial position.

N_i be the number of interstitial atoms.

N be the number of atoms.

Now the total number of ways in which n *Frenkel* defects can be formed will be given by:

$$P = \left[\frac{N!}{(N-n)!\,n!}\right] \times \left[\frac{N_i!}{(N_i-n)!\,n!}\right]$$

The corresponding increase in entropy due to the creation of *Frenkel defect* is given by:

$$S = k_B \log\left[\frac{N!}{(N-n)!\,n!} \cdot \frac{N_i!}{(N_i-n)!\,n!}\right]$$

which in turn produces a change in free energy:

$$F = U - TS$$

$$= nE_i - k_BT\log\left[\frac{N!}{(N-n)!\,n!} \cdot \frac{N_i!}{(N_i-n)!\,n!}\right]$$

Using *Stirling's* approximation for logarithmic term, we get:

$$\log\left[\frac{N!}{(N-n)!\,n!} \cdot \frac{N_i!}{(N_i-n)!\,n!}\right] = \log\frac{N!}{(N-n)!\,n!} + \log\frac{N_i!}{(N_i-n)!\,n!}$$

$$\cong N\log N + N_i\log N_i - (N-n)\log(N-n)$$
$$- (N_i - n)\log(N_i - n) - 2n\log n$$

Substituting this value of a logarithmic term in the expression for free energy and then differentiating with respect to n, we get:

$$\left[\frac{\delta F}{\delta n}\right]_T = E_i - k_B T \log \frac{(N-n)(N_i - n)}{n^2}$$

At equilibrium, the free energy is constant, so that:

$$\left[\frac{\delta F}{\delta n}\right]_T = 0$$

or
$$E_i = k_B T \log \frac{(N-n)(N_i - n)}{n^2}$$

$$\cong k_B T \log \frac{NN_i}{n^2},$$

taking $N >> n$ and $N_i >> n$.

Thus,

$$E_i = k_B T [\log (NN_i) - 2 \log n]$$

or
$$\log n = \frac{1}{2} \log (NN_i) - \frac{E_i}{2k_B T}$$

$$\boxed{n = (NN_i)^{1/2} \exp(- E_i/2k_B T)}$$

showing that n should be proportional to $[NN_i]^{1/2}$

Compositional Defects

Compositional defects arise from impurity atoms during original crystallisation. Impurity atoms considered as defects in a perfect lattice are responsible for the functioning of most semiconductor devices. They occur on a lattice point as a *substitutional impurity* or as an *interstitial impurity*.

A substitutional impurity is created when a foreign atom substitutes for or replaces a parent atom in the lattice [Fig. 1.23 (*a*)]. In brass, zinc is a substitutional atom in the copper lattice.

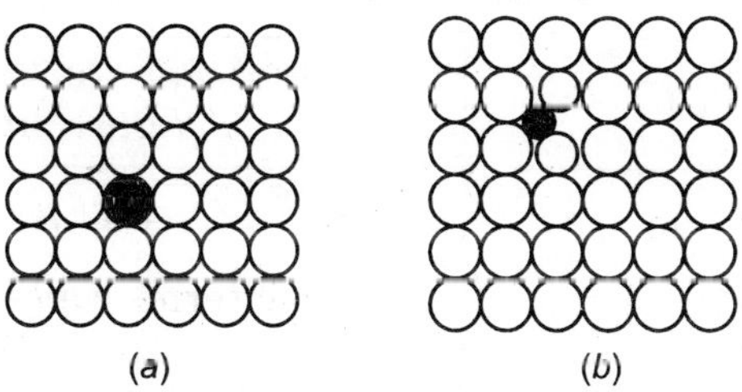

Fig. 1.23 Compositional defects: (a) Substitutional impurity, (b) Interstitial impurity

An interstitial impurity is a small-sized foreign atom occupying an interstice or space between the regularly positioned atoms [Fig. 1.23 (*b*)].

Electronic Defects

Errors in charge distribution in solids are termed electronic defects. This does not imply that deviations from smooth uniformity of charge within the volume of a unit cell are defects. Far from it, the charge distribution is actually non-uniform. However, departures from the normal regularity of charge distribution or energy are electronic defects. Deviations of spatial distribution or charge commonly accompany the geometrical or structural crystal defects just discussed. For example, an impurity atom may have a charge quite different from that of the host atoms and hence may produce a local electronic disturbance. Similarly, a vacancy or interstitial is sure to produce electrical-charge deviations. In addition to these, the electron in a geometrically perfect lattice may themselves move in such a way as to produce local fluctuations in charge. Furthermore, the electrons may absorb varying amounts of thermal energy so that their motion through the lattices is altered. These so-called *electronic imperfections* are primarily necessary to explain electrical conductivity and related phenomenon in solids. Perhaps the most prominent example of this is the creation of positive and negative charge carriers. This effect is responsible for the operation of *p-n* junction and transistors.

Production of Point Defects

In addition to point defects formed by thermal fluctuations, point defects may be created by other means also. Table 1.4 illustrates how the fraction of vacant lattice sites in a metal increases rapidly with temperature. One method of producing an excess number of point defects at a given temperature is by quenching (quick cooling) from a higher temperature. Another method of creating excess defects is by severe deformation of the crystal lattice, for example, by hammering or rolling. While the lattice still retains its general crystalline nature, numerous defects are introduced.

A third method of creating excess point defects is by external bombardment by atoms or high-energy particles, for example, from the beam of the cyclotron or the neutrons in a nuclear reactor. The first particle collides with the lattice atoms and displaces them, thereby forming a point defect. The number produced in this manner is not dependent on temperature but depends only on the nature of the crystal and on the bombarding particles.

Table 1.4 Equilibrium vacancies in a metal

Temperature (°C)	*Approximate fraction of vacant lattice sites*
500	1×10^{-10}
1000	1×10^{-5}
1500	1×10^{-4}
2000	1×10^{-3}

Line Imperfections

If a plane of atoms lies only partway through a crystal, the edge of such a plane is a defect in the form of a line, and is known as *dislocations*. There are in general, two types of dislocations: (*a*) Edge of dislocation and (*b*) Screw dislocation. Both these types are formed in the process of their solidification or crystallization of the metal mainly in the process of their deformation. These dislocations are the most striking defects and are responsible for the useful property of ductility in metals, ceramics and crystalline polymers.

Discussing the edge dislocation first, Fig. 1.24 (*a*) shows a perfect crystal, the top sketch depicting a three-dimensional view and the bottom one showing the atoms on the front face. The perfect crystal is considered to be made up of vertical planes parallel to one another and to the side faces. If one of these vertical planes does not extend from the top to the bottom of the crystal but ends partway within the crystal, as the Fig. 1.24 (*b*), a dislocation is present. In the perfect crystal, the atoms are in equilibrium positions and all the bond lengths are of the equilibrium value. In the imperfect crystal on the right, just above the edge of the incomplete plane, the atoms are squeezed and are in a state of compression. The bond lengths have been compressed to smaller than the equilibrium value. Just below the edge, the atoms are pulled apart and are in a state of tension. The bond lengths have been stretched to above the normal values. The distorted configuration extends all along the edge into the crystal. Thus as the region of maximum distortion is centred around the edge of the incomplete plane, this distortion represents a line imperfection and is called an *edge dislocation*.

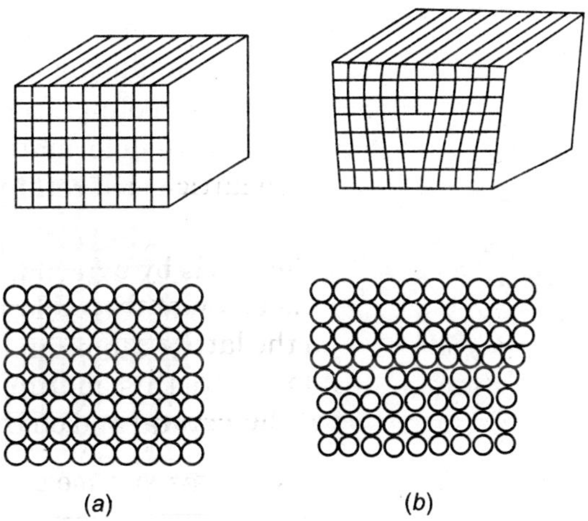

(*a*) (*b*)

Fig. 1.24 An incomplete plane in a crystal results in an edge dislocation

It should be noted that the dislocation of Fig. 1.25 (*a*) represents a defect that runs completely through the crystal from front to back. This extra plane indicated in Fig. 1.25 (*a*) may be either above or below the slip plane shown by the dashed line *AC* as in the 1.25 (*b*). In the former case, the dislocation (*PQ*) is said to be positive and is denoted by the symbol ⊥; in the latter case, it (*P′Q′*) is said to be negative and is denoted by the symbol , where the horizontal line in the symbol represents the slip plane and the vertical line the incomplete plane. The magnitude and direction of the displacement are determined by a vector called the *Burgers vector* which is perpendicular to the dislocation line.

Screw Dislocation

The second basic type of dislocation is the screw or *Burgers dislocation*. In this, the atoms are displaced in two separate perpendicular to each other. Figure 1.26 is a simple example of screw dislocation. The plane *ABCD* is the slipped area. The upper portion of the crystal has been sheared by an atomic distance (as shown by the shaded area) to the right relative to the lower portion. No slip has taken place to the right of *AD* and *AD* is a dislocation line. Here, the dislocation is parallel to its Burgers vector or shear vector. The designation 'screw' for this lattice defect is derived form the fact that the lattice planes of the crystal spiral the dislocation line *AD*.

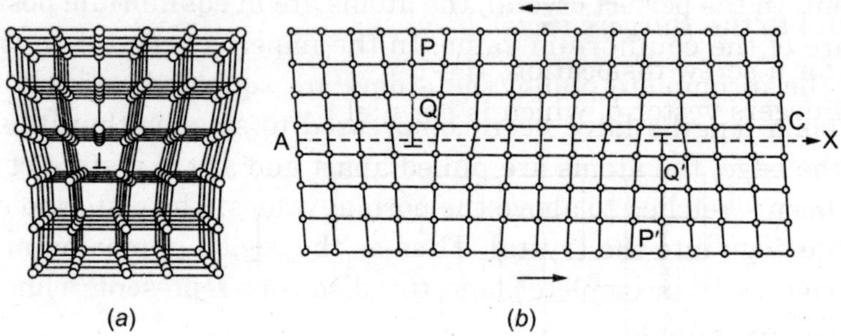

Fig. 1.25 Positive and negative edge dislocations

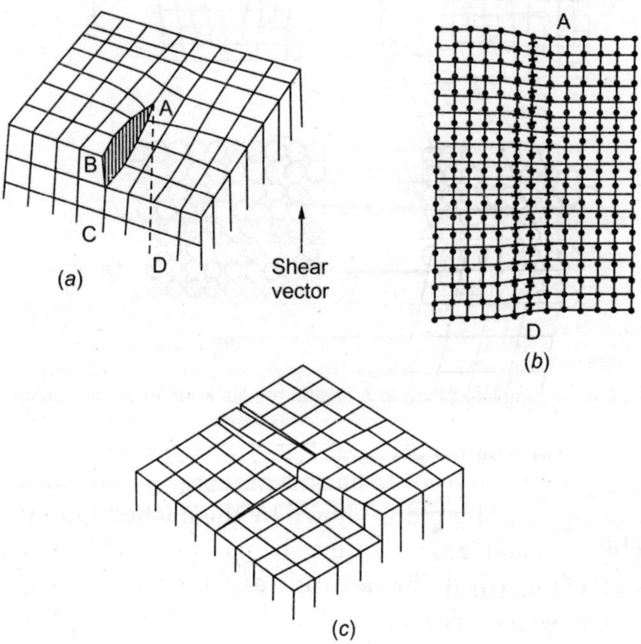

Fig. 1.26 Formation of screw dislocation

Burgers Vector

The Burgers vector of a dislocation is an important property of a dislocation because, if the Burgers vector and the orientation of the dislocation line are known, the dislocation is completely described. This indicates how much and in what direction the lattice above the slip plane appears to have been shifted with respect to the lattice below the slip plane. Figure 1.27 shows a method of determining the *Burgers vector* applied to an edge dislocation. It is first necessary to choose arbitrarily a positive direction for the dislocation and then to find out the vector which closes a circuit (line of arrows). In the present case, a clock-wise circuit of atom-to-atom steps is considered round a dislocation. It is seen that *A* fails to close the circuit when repeated in a perfect lattice *B* unless completely by a closure vector equal to the *Burgers vector b*, which is found normal to the dislocation.

In Fig. 1.28 for a screw dislocation, the Burgers circuit fails to close by the amount indicated by the Burgers vector *b*, which is parallel to the dislocation line.

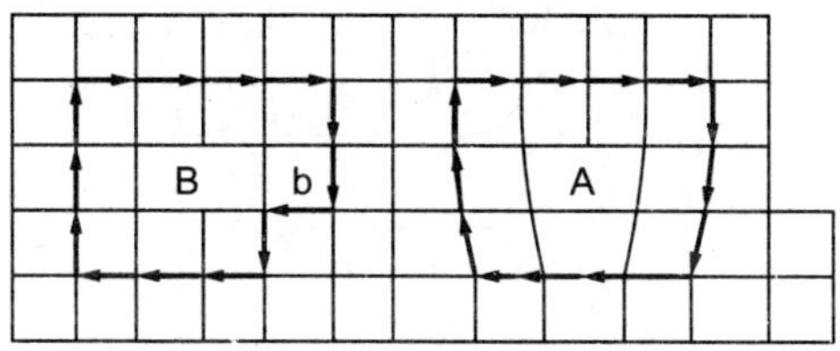

Fig. 1.27 Burgers vector for edge dislocation

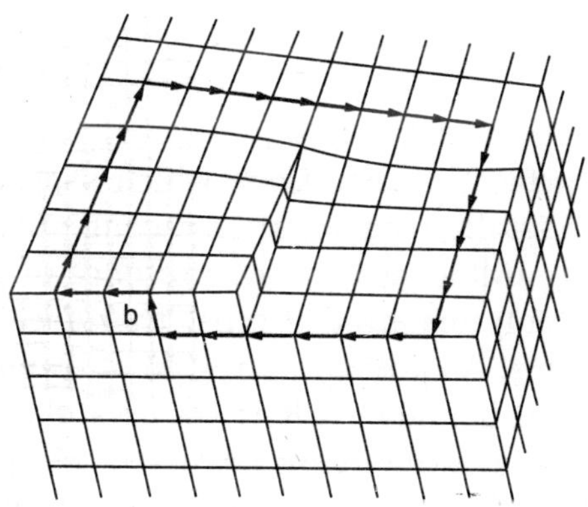

Fig. 1.28 Burgers vector for screw dislocation

Surface Imperfections

Surface imperfections of a structural natural arise from a change in the stacking of atomic planes on or across a boundary. The change may be one of the orientations or of the stacking sequence of the planes. In geometric concept, surface imperfections are two-dimensional. They are of two types: External and internal.

External Surface Imperfections

The external type is just what its name implies, the imperfections represented by a boundary. The most obvious boundary is the external surface. Although we may visualise a surface as simply a terminus of the crystal structure, we should appreciate the fact that the atoms on the surface cannot be compared with the atoms within a crystal. The surface atoms have neighbours on one side only, while atoms inside the crystal have neighbours on either side of them. This is shown in Fig. 1.29. Since these surface atoms are not entirely surrounded by others, they possess higher energy than that of internal atoms. This energy of the surface atom, for most metals, is of the order 1 J/m^2.

Internal Surface Imperfections

Internal surface imperfections are manifested by such defects as grain boundaries, tilt boundaries, twin boundaries, and stacking faults.

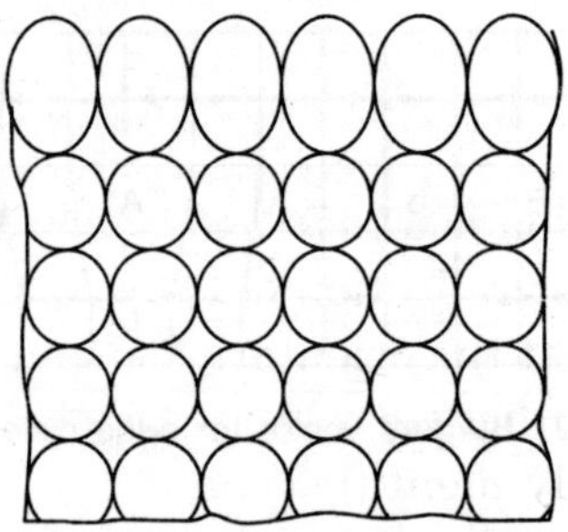

Fig. 1.29 External surface defects

Grain Boundaries

Grain boundaries are those surface imperfections which separate crystals or grains of different orientation in a polycrystalline aggregation during nucleation or crystallisation. The shape of a grain is usually influenced by the presence of surrounding grains. The boundary atoms in two randomly oriented grains, therefore, cannot have a perfect complement of surrounding atoms. As a result, a region of transition exists in which the atomic packing is imperfect. In the boundary where the crystals or grains change abruptly orientation difference between neighbouring grains is more than 10–15°, and the boundaries are known as high-angle grain boundaries. These imperfections in the grain boundary enable the microscopist to see them, for in a transparent crystalline material they may scatter light and in opaque material they can be etched chemically. Although a grain boundary is two-dimensional, it has a definite thickness of 2 to 10 or more atomic distances. The mismatch of the orientation of adjacent grain produces a less efficient packing of the atoms at the boundary. Thus, the atoms along the boundary have a higher energy than those within the grains. The boundary between two crystals which have different crystalline arrangements or different compositions, is called an interphase boundary or commonly an interface.

Tilt Boundaries

Tilt boundary is another surface imperfection. This is called *low-angle boundary* as the orientation difference between two neighbouring crystals is less than 10°. This is why the disruption in the boundary is not so drastic as in the high-angle boundary. In general, low-angle boundaries can be described by suitable arrays of dislocation. A low-angle tilt boundary is composed of edge dislocation lying one above the other in boundary. The angle or tilt will be $q = \frac{b}{D}$; where, b is the magnitude of the Burgers vector and D is the average vertical distance between dislocations (see Fig. 1.30).

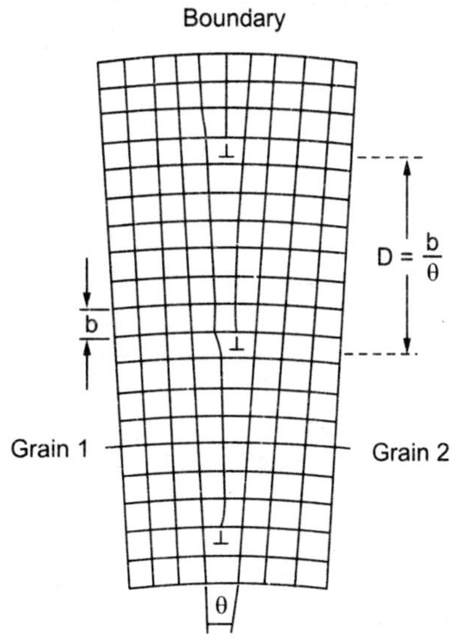

Fig. 1.30 Tilt boundary

Twin Boundaries

Another planar surface imperfection is a *twin boundary*. The atomic arrangement on one side of a twin boundary is a mirror reflection of the arrangement on the other side. Twin boundaries occur in pairs, such that the orientation change introduced by one boundary is restored by the other. The region between the pair of boundaries is called the *twinned region*. Twin boundaries are easily identified under an optical microscope. Twins which form during the process of recrystallization are called *annealing twins* and those which form during plastic deformation of the material are called *deformation twins* (see Fig. 1.31).

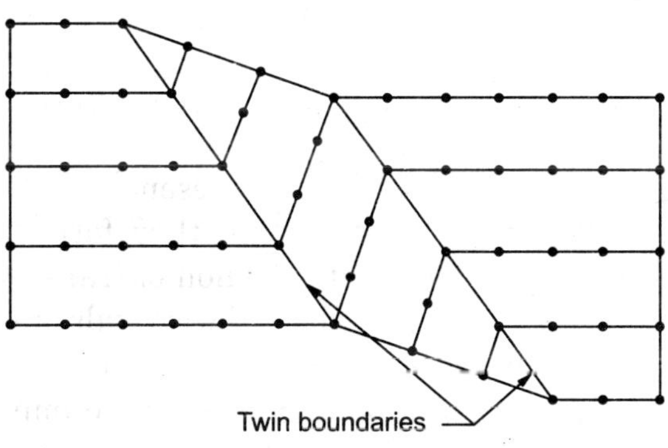

Fig. 1.31 Twin boundary

Stacking Defect

A stacking fault is a surface imperfection that arise from the stacking of one atomic plane out of sequence on another while the lattice on either side of the fault is perfect. For example, the stacking sequence in an ideal fcc crystal may be described as *ABC ABC*

ABC .... But the stacking fault might change the sequence to *ABC AC ABC* .... Faults in the stacking sequences of fcc crystals are shown in Fig. 1.32.

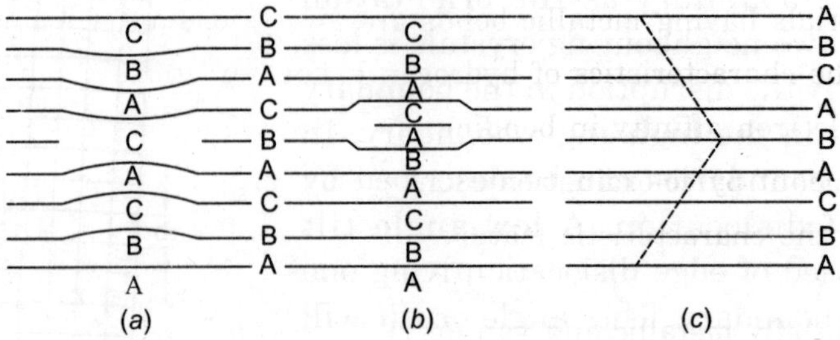

Fig. 1.32 Faults in stacking sequences (*a*) Intrinsic fault (*b*) Extrinsic fault (*c*) Twin fault

Volume Defects

Volume defects such as cracks may arise when there is only small electrostatic dissimilarity between the stacking sequences of close packed planes in metals. Further when clusters of atoms are missing, a large vacancy or void is got which is also a volume imperfection. Foreign particle inclusions, large voids or non-crystalline regions which have the dimensions of the order of 0.20 nm are also called volume imperfections.

SHORT QUESTIONS

1.1.1 Differentiate primary bonds and secondary bonds.

1.1.2 Discuss the interaction between the atoms in a diatomic molecule with suitable graphs. *(Anna. U., BE and Part-time BE, November 1993)*

1.1.3 Why high external energy is required to melt ionic crystals?

1.1.4 Reaction between ionic compounds in solution state is high. Give reasons.

1.1.5* Show that the Madelung constant for a linear ionic solid having 2N ions of alternate charge $\pm e$ is 2 ln 2. *(Madras U., BE, November 1992)*

1.1.6 Explain electric dipole. Give an example.

1.1.7 Explain the features of metallic bond with illustrations. *(Anna. U., BE April 1999)*

1.1.8 In crystals there is an attractive interaction of one kind or other which is quite appreciable for nearest neighbours. Explain why they do not collapse?

1.1.9* If 440×10^3 kJ of energy is required to break the H — H bonds in a kmol, compute the energy required in eV to break one H — H bond. *(M.K.U., BE, April 1988)*

1.1.10 Explain covalent bonds with two examples.

1.1.11 Describe the formation of hydrogen bonded crystals.

1.1.12 Explain van der Waals forces and molecular crystals. *(Anna. U., BE, April 1990)*

1.1.13 List out the different characteristics of hydrogen bonded crystals.

1.1.14 Experiments suggest that CCl_4 has no net dipole moment. Give the reasons.

1.1.15 The materials having metallic bonds are always opaque. Give one line answer.

1.1.16* Discuss the characteristics of hydrogen bonded crystals.

1.1.17 Explain electron affinity in bonding. *(Anna. U., BE, April 1983)*

1.1.18* What is meant by co-ordination number in solids?

1.1.19 What are the characteristic features of covalent bonded crystals? *(Anna. U., BE, May 1990)*

1.1.20 Describe briefly metallic and van der Waals bonds. *(Anna. U., BE, November 1989)*

1.1.21 Compare and contrast the characteristics of ionic and covalent bonds. *(Anna. U., BE, November 1986)*

1.1.22* Give reasons:

(*i*) magnesium oxide is bonded ionically

(*ii*) van der Waals bond is weak

(*iii*) liquid carbon tetrachloride cannot conduct electricity

(Question 1.1.22 has been taken from the B. Tech. question paper of Imperial College of Science & Technology, London).

1.2.1 Define crystal structure, crystal lattice and Bravais space lattice.

1.2.2 Find the radius of the interstitial sphere that just fit into the void at the centre between the body centred atoms of the bcc structure. *(Anna. U., BE, November 1995)*

1.2.3 Distinguish crystalline and amorphous substances.

1.2.4 Explain the terms basis, lattice and crystal structure.

1.2.5* Distinguish space lattice and crystal structure.

1.2.6 Explain symmetry elements and symmetry operations in crystals. *(REC, Jaipur, December 2001)*

1.2.7 Get the relation for the radius of the atom in terms of lattice parameter in fcc structure. *(Anna. U., BE/B. Tech, November 1996)*

1.2.8 Describe the seven crystal systems.

1.2.9* Show that for a cubic lattice, the lattice constant a is given by

$$a = \left[\frac{nM_A}{N_A\rho}\right]^{1/3}$$

where n is the number of atoms in a unit cell, M_A is the atomic weight, N_A is Avogador's number and ρ is the density of the material *(Anna. U., BE/B. Tech, November 1998)*

1.2.10 How many atoms are there in the unit cell of diamond?

1.2.11 Distinguish primitive cell and unit cell.

1.2.12 List out the different kinds of symmetry elements and their total number in a cube. *(Anna. U., BE, May 1997)*

1.2.13 Show that $\frac{a}{c}$ ratio for an ideal hexagonal crystal is $\sqrt{\frac{3}{8}}$. *(M. U., B. Tech/BE May 1984)*

1.2.14 If the height of the unit cell of zinc (hcp) is 0.495 nm, compute the radius of zinc atom.

1.2.15 Explain rotation-inversion axis in a cubical crystal.

1.2.16 Explain co-ordination number, nearest neighbour distance and packing factor with the help of a highly crystalline solid. *(M. U., B. Tech/BE, 1979)*

1.2.17 What is meant by atomic radii in a crystal? How many atoms are there in the unit cell of diamond?

1.2.18* Give the general explanation of closest packing with suitable figures.

1.2.19 What are Miller indices? Explain Burgers vectors. *(Anna. U., BE/B.Tech., May 2002)*

1.2.20 Silicon crystallizes in dc structure. If the lattice constant at 25°C is 0.5 nm. How many unit cells will be there in a cubical crystal of side 0.5 cm.

1.2.21 What is Bravais lattice? Draw the planes (1 1 0) and (1 1 1) in a cubical unit cell. *(Anna. U., BE, April 2004)*

1.2.22 The theoretical value of *c/a* ratio for zinc is 1.63. But the actual value of *c/a* for zinc in hcp structure is 1.856. Explain.

1.2.23 Define atomic packing factor. What are Schottky defects. *(Anna. U., BE./B. Tech., May 1996)*

1.2.24* What are Miller indices? How are they obtained? *(JNTU, 2002)*

1.2.25 Give the crystal structure of the following: (*a*) gold, (*b*) germanium, (*c*) barium, (*d*) zinc. (JNTU; BE April 2001).

1.2.26 What are Schottky defects? How are they formed? *(Anna. U., BE, November 2003)*

1.2.27 Explain the terms *i*) basis, *ii*) space lattice, *iii*) unit cell. *(JNTU, 2002)*

1.2.28 Copper crystallizes in fcc structure. If the radius of the atom is *r*, find the distance between the parallel planes (1 1 1) and (2 2 2). *(Bangalore University, May 2001)*

1.2.29 Draw the planes (1 1 0) and (1 1 1) in a cubic crystal.

1.2.30 Name different kinds of symmetry elements that are present in cubic crystal giving the total number. *(University of Groningen, Netherlands, U.G., 2002)*

1.2.31 Discuss dc structure and get its packing factor. *(Anna. U., BE, May 1990)*

1.2.32 If 0.56 nm is the edge of the unit cell of NaCl, compute the shortest distance between Na^+ ions. *(IIT Mumbai, B. Tech. 1998)*

1.2.33* Obtain the Miller indices of a plane which intercepts at *a*, *b* and *c* in fcc structure. Also get the density of atoms in this plane. *(M.K. University, B. Tech/BE, 1989)*

1.2.34 Show that the Miller indices are the same for a family of crystallographic planes which are parallel to each other. *(Anna. U., BE/B. Tech., May 1989)*

Answers to the starred questions

1.1.5* Let us take Na^+ at the point 0. The interaction between this and the nearest chlorine ions at r_0 is

$$-\frac{e^2}{4\pi\epsilon_0 r_0}+\left[-\frac{e^2}{4\pi\epsilon_0 r_0}=-\frac{2e^2}{4\pi\epsilon_0 r_0}\right]$$

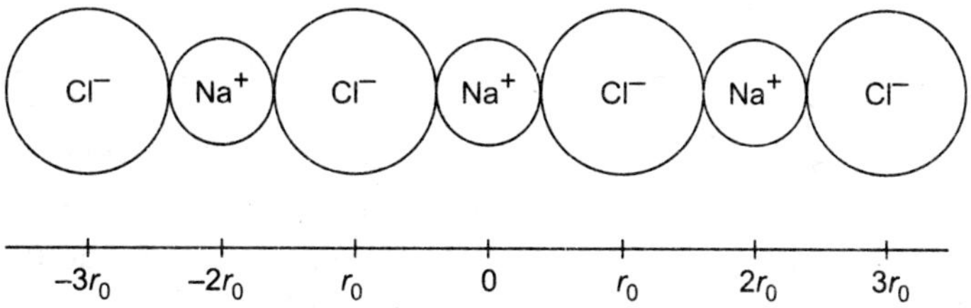

Similarly the repulsive energy due to the two positive ions at a distance $2r_0$ is $\frac{2e^2}{4\pi\epsilon_0 2r_0}$

and that due to the next set of Na^+ ions is $-\frac{2e^2}{4\pi\epsilon_0 3r_0}$. Thus the total energy due to all the ions in the linear array is

$$-\frac{e^2}{4\pi\epsilon_0 r_0}\left[2\left(1-\frac{1}{2}+\frac{1}{3}-\frac{1}{4}+\ldots\right)\right]$$

$$=-\frac{e^2}{4\pi\epsilon_0 r_0}(2\log_e 2)$$

i.e., $\alpha = 2\ ln\ 2 = 1.38$

1.1.9* The energy required to break 6.02×10^{26} bonds is 440×10^6 joule. Hence the energy required to break one H — H bond is

$$\frac{440\times10^6}{6.02\times10^{26}\times1.6\times10^{-19}} = 4.6 \text{ eV}$$

$$E_d = 4.6 \text{ eV}$$

1.1.16* (*i*) they have the tendency to form groups of many molecules.

(*ii*) they have increased binding energy in comparison with similar molecules without hydrogen bonded ones.

(*iii*) they have weak binding.

(*iv*) they have low electric conductivity and poor thermal conductivity.

(*v*) they are transparent.

1.1.18* It is the number of atoms around an atom or the number of atoms which are just in the neighbourhood of atoms.

1.1.22* (*i*) because magnesium and oxygen are both divalent

$$2Mg + O_2 \rightarrow Mg^{++} + 2O^{--} \rightarrow 2MgO$$

(*ii*) because it bonds only molecules together

(*iii*) because it does not contains C^{4+} and Cl^- ions

1.2.5* Space lattice is nothing but the geometry of a set of points in space whereas the crystal structure denotes the real ordering or arrangement of the constitutent atoms or ions.

1.2.9* Let M_A, ρ and a be the atomic weight, density of the element and a is the lattice parameter of the unit cell and n – be the number of atoms in the unit cell. N_A is Avogadro's number.

Thus

$\frac{M_A}{\rho}$ (molar volume) contains N_A atom

a^3 will have $\frac{\rho N_A a^3}{M_A}$ atoms. This must be equal to the number of atoms in the unit cell say n

$$i.e., \quad \left[\frac{\rho N_A a^3}{M_A}\right] = n; \; a^3 = \frac{M_A n}{\rho N_A}$$

$$a = \left[\frac{M_A n}{\rho N_A}\right]^{1/3} \quad \textbf{\textit{Answer}}$$

1.2.18* Closest packing is a way of arranging equidimensional objects such that the available space is more efficiently filled. Under this each object will be in contact with maximum number of like objects.

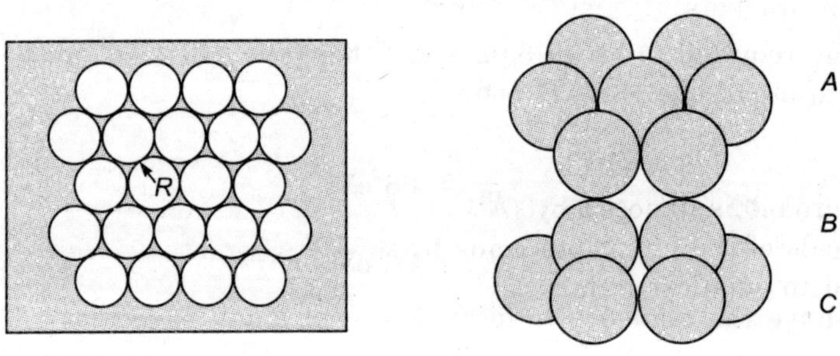

(a) (b)

Closest packing of spheres in two-dimension and in three-dimension

1.2.24.* Let a plane cut the axes at 2*a*, 3*b* and *c* on *x*, *y* and *z* axes.

(*a*) Determine the co-ordinates of the intercepts made by the plane along the three crystallographic axes (*x*, *y*, *z*).

x	y	z
$2a$	$3b$	c
pq	qb	rc

(*b*) Express the intercepts as multiples of the unit cell dimensions, or lattice parameters along the axes, i.e.,

$$\frac{2a}{a} \qquad \frac{3b}{b} \qquad \frac{c}{c}$$

$$2 \qquad 3 \qquad 1$$

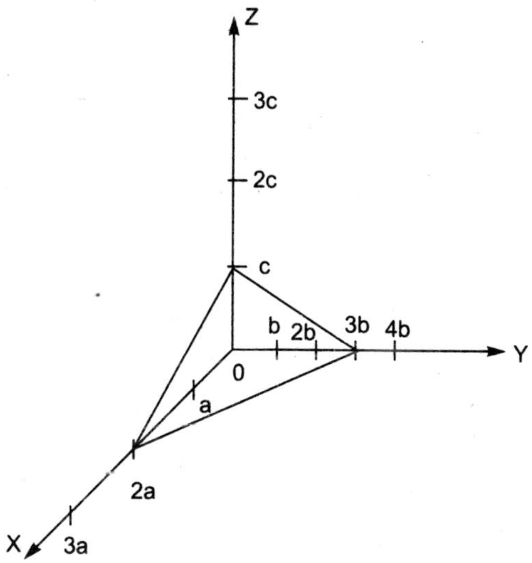

(*c*) Determine the reciprocals of these numbers

$$\frac{1}{2} \qquad \frac{1}{3} \qquad 1$$

(*d*) Reduce these reciprocals to the smallest set of integral numbers and enclose them in brackets.

$$6\times\frac{1}{2} \qquad 6\times\frac{1}{3} \qquad 6 \times 1$$

$$(3 \qquad 2 \qquad 6)$$

In general it is denoted by $(h\ k\ l)$. Thus Miller indices may be defined as the reciprocals of the intercepts made by the plane on the crystallographic axes when reduced to smallest numbers.

1.2.33*

$$a \qquad a \qquad a$$

$$\frac{a}{a} \qquad \frac{a}{a} \qquad \frac{a}{a}$$

$$1 \qquad 1 \qquad 1$$

$$(1 \qquad 1 \qquad 1)$$

This plane is shown below.

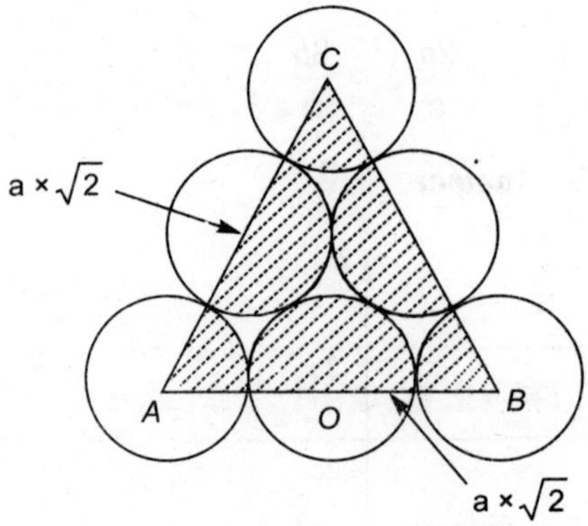

The face diagonal is $\sqrt{2} \times a$

Base calculation :

$$\cos 60° = \frac{AO}{a\sqrt{2}}$$

$$AO = \alpha\sqrt{2}\cos 60°$$

$$2 \times AO = a\sqrt{2} \times 2 \times \frac{1}{2} = a \times \sqrt{2}$$

The number of atoms in the plane is,

$$n = \frac{1}{6} \times 3 + 1\frac{1}{2} = 2 \text{ atoms}$$

Height of the triangle,

$$\cos 30° = \frac{\sqrt{3}}{2} = \frac{H}{a\sqrt{2}}$$

$$\text{or} \quad H = a\left[\frac{3}{2}\right]^{1/2}$$

$$\text{Area} = \frac{1}{2} \times 2 \times AO \times H = \frac{1}{2}\sqrt{2}a^2\left[\frac{3}{2}\right]^{1/2}$$

$$= \frac{\sqrt{3}}{2}a^2 \text{ with } a = \frac{4r}{\sqrt{2}}$$

$$\text{Thus} \quad \text{Area} = \frac{\sqrt{3}}{2}\left(\frac{16r^2}{2}\right) = 4r^2\sqrt{3}$$

The planar density of (1 1 1) plane is

$$\frac{2}{4r^2\sqrt{3}} = \frac{0.288}{r^2}$$

$$\boxed{p.d = \frac{0.288}{r^2}}$$ ***Answer***

REVIEW QUESTIONS

1.1.1 The potential energy of a diatomic molecule is given in terms of the interatomic distance r by the equation

$$U(r) = -\frac{a}{r^m} + \frac{b}{r^n}$$

Obtain the relation for the equilibrium spacing of the atoms and hence get the expression for dissociation energy. *(Anna. U., M.Sc., November 1983)*

1.1.2 Describe the formation of NaCl lattice from neutral sodium chlorine atoms with suitable figures and equations. *(Anna. U., BE, November 1984)*

1.1.3 What is meant by bond energy of NaCl molecule? Give the important steps to calculate the lattice energy of ionic crystals and hence obtain the final expression for the cohesive energy per kmol of the crystal.

1.1.4 Explain the difference between ionic and covalent types of binding. What do you understand by Madelung constant? How madelung constant can be computed?

1.2.1 Explain the terms: lattice, basis, structure and unit cell. Discuss the seven systems of crystals and the fourteen types of Bravais lattices with suitable figures.

1.2.2 What is meant by symmetry elements and symmetry operations? Discuss the various types of symmetry elements in a cubic crystal. *(Calicut. U., BE, May 1992)*

1.2.3 Estimate the fraction of total volume occupied by atoms in body centred and face centred cubic structures. Give two examples in each structure. *(M.U., BE,/B. Tech. November 1990)*

1.2.4 Discuss the various symmetry elements associated with a crystal. Show that an actual crystal cannot possess a five-fold rotational symmetry.

1.2.5 Compute the packing factor of bcc structure; why this structure is a closely packed one and not a closest one?

1.2.6 Show that an ideal hexagonal – close packed structure $\sqrt{\frac{c}{a}}$ is 1.663 and the density of atomic packing equal to that of the face-centred cubic structure. *(JNTU, BE, May 2001)*

1.2.7 Define Miller indices of a lattice plane and write the equation to a family of planes represented by the indices (h k l). Show that for cubic crystal the distance $d_{h,\,k,\,l}$ between adjacent $(h\ k\ l)$ planes is given in terms of the cubic edge 'a', by

$$d_{hkl} = \frac{a}{\sqrt{h^2 + k^2 + l^2}}$$

(M.K.U., P.G. April, 1999, Anna U., BE, May 1986)

1.2.8 Explain how Miller indices of a set of planes in a crystal are determined. Draw the figures of a simple cubic crystal and indicate (1 0 0), (1 1 0) and (1 1 1) planes in it. *(Metallurgy division, University of Manchester, UK, B. Tech, April 1982, IIT Delhi, B. Tech, May 1992).*

1.2.9 Explain with neat sketches the different imperfections in a crystal. Calculate the ratio of the number of vacancies in equilibrium at 300 K in aluminium to that produced by rapid quenching from 800 K. Enthalpy of formation of vacancies in aluminium is 63 kJ/mol. *(M.U., BE, November, 1992)*

1.2.10 Define space lattice. Name the seven crystal systems and give the relation of lengths of axes and the relation of angles between the axes of a unit cell in each type. Obtain the co-ordination number for bcc and fcc lattices. *(M.U., BE, April, 1994)*

1.2.11 What are Miller indices? Draw neat diagrams to indicate Miller indices of some important plane systems in a simple cubic crystal. Find the Miller indices of a crystal plane that makes from some origin an intercept of a on the a-axis, $2b$ on the b-axis and $3c$ on the c-axis. *(JNTU, BE, May 2001, M.U., BE, April 1994)*

1.2.12 Explain the dimensions of a unit cell. Derive an expression for the edge element of a cubic crystal in terms of the properties of the crystal material. Zinc has hcp structure. The radius of the atom is 0.153 nm. Compute the distance between the base and middle plane containing the three atoms. *(M.U., BE, May 2002)*

1.2.13 What are point, line and surface imperfections formed in solid materials? Illustrate these imperfections with suitable sketches. *(Anna.U., BE, April 1994)*

1.2.14 Show that for bcc and fcc crystal structure, the lattice constants are given by $a = \frac{4r}{\sqrt{3}}$ and $a = \frac{4r}{\sqrt{2}}$ respectively where r is the atomic radius. Explain the following terms used in crystallography: (1) primitive cell, (2) unit cell, (3) co-ordination number and (4) relative density of packing. *(Anna U., BE, May 1990)*

1.2.15 E..plain how x-rays are used for determining the crystal structure. A sodium chloride crystal of lattice constant 5.63 A° is used to measure the wavelength of x-rays. The diffraction angle is 5.2° for the d_{111} spacing of the chlorine ions. What is the wavelength? *(M.U., BE, November 1992)*

1.2.16 What is Bravais lattice? What are the different space lattices in the cubic system? How many lattice points per unit cell are there in each of these lattices? Deduce an expression for planes of the $(h\ k\ l)$ type in the case of a cubic structure. *(Anna U., BE, April 1985)*

1.2.17 Discuss in brief the type of bonding in a molecule of water. Sketch a unit cell of fcc lattice. What are Miller indices? In a simple cubic lattice with lattice parameter a, what is the interplaner distance between (1 1 1) type planes.

1.2.18 What is packing factor? Prove that the packing fraction of hcp is 0.74. Bring out the differences between edge dislocation and screw dislocation.

(Anna U., BE, 2004)

1.2.19 Show that the atomic packing factor of fcc and hcp structures are the same. What are Miller indices? Explain. How they are determined?

(Anna U., BE, December 2003)

1.2.20 Explain primitive cell and unit cell. Prove that five fold rotation axis is not compatible with a lattice. Classify the important types of crystal imperfections.

(Anna U., BE, December 2003, VTU, May 2003)

1.2.21 Explain the significance of Miller indices. Derive an expression for the number of Schottky defects in equilibrium at a temperature *T*. The fraction of vacancy sites in a metal is 1×10^{-10} at 500°C. What will be the fraction of vacancy sites at 1000°C?

(JNTU, B. Tech., 2002)

1.2.22 What are Miller indices? How are they obtained? Explain Schottky and Frenkel defects with the help of suitable figures. *(JNTU, B. Tech., 2000)*

1.2.23 Calculate APF value for fcc structure. Copper has fcc structure and atomic weight of copper is 63.54.

(Anna U., BE, May 1989. N.B: Some data are missing in the second part of the question*).*

1.2.24 Draw the following planes in a cubical unit cell (0 1 1), (1 0 2), (1 3 2) and (1 1 2). In an x-ray diffraction study, a radiation of wavelength. 1.71 A° was directed at a cubic crystalline sample. It was found that the first two Bragg reflections occur at angles of 30°, 35° 17′ respectively. Determine whether the crystal is bcc or fcc structure.

(VTU, BE, July 2004)

1.2.25 Discuss the crystal structure of diamond. Show that the packing fraction of fcc is 0.74. Discuss Bragg's spectrometer and explain how it is used to verify the Bragg's law. *(VTU, BE, July 2004)*

1.2.26 Discuss Bragg's law of x-ray diffraction. Describe the powder method to determine crystal structure. *(Anna U., BE, November 1995)*

PROBLEMS AND SOLUTIONS

1.1.1 The force of attraction between ions of Na and Cl is 3.02×10^{-9} newton when the two ions just touch each other. If the ionic radius of Na is 0.095 nm, what is the radius of Cl^-? *(M.U., Five year BE 1980, AMIE, Winter 1999)*

Solution:

$$F = 3.02 \times 10^{-9} = \frac{Z_1 Z_2 e^2}{4\pi \epsilon_0 r^2}$$

In this case $Z_1 = Z_2 = 1$

Hence

$$3.02 \times 10^{-9} = \frac{(1.6 \times 10^{-19})^2}{4\pi \times 8.85 \times 10^{-12} \times r^2}$$

i.e., $$r^2 = \frac{2.56\times10^{-38}}{4\times3.14\times8.85\times10^{-12}\times3.02\times10^{-9}}$$

$$= 0.0076 \times 10^{-17}$$

or $$r = 0.276 \text{ nm}$$

Thus

$$r = r(\text{Na}^+) + r(\text{Cl}^-)$$

Hence

$$r(Cl^-) = 0.276 - 0.095$$

i.e., $$r(Cl^-) = 0.181 \ nm$$ ***Answer***

1.1.2 Compute the net potential energy of a simple Na^+ Cl^- ion pair. The equilibrium distance between ions is 0.28 nm. The potential energy due to repulsion between electrons is given by $U_r = \frac{\beta}{r^8}$.

(Lincoln Laboratory, Massachusetts Institute of Technology, US, B. Tech-Mid semester, 1989).

Solution:

If F_1 is the force of attraction between the ions, then

$$U_a = \int F_1 dr = \int \frac{-Z_1Z_2e^2}{4\pi\epsilon_0 r^2} dr$$

In the case of Na^+ and Cl^- ions, $Z_1 = 1$, $Z_2 = -1$.

Hence $$U_a = \frac{Z_1Z_2e^2}{4\pi\epsilon_0 r} + C$$

When $r = \infty$, $U_a = 0$ and hence $C = 0$

$$U_a = \frac{Z_1Z_2e^2}{4\pi\epsilon_0 r}$$

$$U_a = \frac{Z_1Z_2e^2}{4\pi\epsilon_0 r}$$

$$U = U_a + U_r = \frac{Z_1Z_2e^2}{4\pi\epsilon_0 r} + \frac{b}{r^8}$$

$$F_r = \frac{d}{dr}\left(\frac{b}{r^8}\right) = -\frac{8b}{r^9}$$

But $F_a = -F_r$ at the equilibrium position with $F_a = 3.02 \times 10^{-9}$ N (taken from the previous problem).

Thus $$F_r = -3.02 \times 10^{-9} = -\frac{8b}{r^9}$$

$$b = \frac{r^9 \times 3.02 \times 10^{-9}}{8}$$

But $r = r_0 = 0.28 \times 10^{-9}$ metre

$$b = \frac{(0.28 \times 10^{-9})^9 \times 3.02 \times 10^{-9}}{8}$$

$$= \frac{0.28^9 \times 10^{-81} \times 3.02 \times 10^{-9}}{8} = \frac{0.28^9 \times 10^{-90} \times 3.02}{8}$$

$$b = 4 \times 10^{-96}$$

Thus $$U = \frac{1 \times (-1) \times (1.6 \times 10^{-19})^2}{4\pi \epsilon_0 \, r_0} + \frac{b}{r_0^8}$$

$$= \frac{-2.56 \times 10^{-38}}{4\pi \times 8.85 \times 10^{-12} \times (0.28 \times 10^{-9})} + \frac{4 \times 10^{-96}}{(0.28)^8 \times 10^{-72}}$$

$$= -8.3 \times 10^{-19} + 1 \times 10^{-19}$$

$$= -7.3 \times 10^{-19} \text{ joule}$$

$$\boxed{U = -7.3 \times 10^{-19}\, J}$$ ***Answer***

1.1.3 If the repulsive energy is of the Mott-Gurney form $A_0 \exp\left[\frac{(r_a + r_c - r)}{\rho}\right]$ where r_a and r_c are the basic radii of anions and cations respectively, then determine A_0 and ρ for NaCl if the cohesive energy per ion pair is 6.61 eV and the interatomic separation is 0.282 nm. The ionisation energy and the basic radii of Na are 5.14 eV and 0.875 Å respectively where as the electron affinity and the basic radii of Cl are 3.61 eV and 0.1475 nm respectively. *(UPSC May 2000, B. Tech., IIT-Mumbai, May 1992)*

Solution:

Cohesive energy per ion pair is

$$U = I - E - C + R \qquad ...(1)$$

where I, E, C and R are ionisation energy of Na, electron affinity of Cl, coulomb energy and repulsive respectively.

$$C = \frac{1.75e}{4\pi \epsilon_0 \, r} = \frac{2.52}{r} \text{ eV with } r \text{ in nm}$$

Now $$-6.61 = (5.14 - 3.61) - \frac{2.52}{r} + A_0 \exp\left[\frac{(r_a + r_c - r)}{\rho}\right]$$

i.e., $$-6.61 = 1.53 - \frac{2.52}{r} + A_0 \exp\left[\frac{(0.235 - r)}{\rho}\right] \quad \text{...(2)}$$

as

$$r_a + r_c = (0.1475 + 0.0875) \text{ nm} = 0.235 \text{ nm}$$

$$\left[\frac{dU}{dr}\right]_{r=r_0} = 0 = 0 + \frac{2.52}{r_0^2} - \frac{A_0}{\rho}\exp\left[\frac{(0.235 - r_0)}{\rho}\right] \text{ with } r_0 = 0.282 \text{ nm}$$

or $$A_0 \exp\left[\frac{(0.047)}{\rho}\right] = 31.69\,\rho \quad \text{...(3)}$$

Substituting this in Eqn. (2), one gets

$$\boxed{\rho = \mathbf{0.025} \text{ nm}}$$ ***Answer***

Substituting ρ in Eqn. (3), we get

$$\boxed{A_0 = 5.146 \text{ eV}}$$ ***Answer***

1.1.4 Calculate the ratio of ionic radii for an atomic co-ordination number of three as shown in figure. *(V.U.T., Karnataka, BE May 2002)*

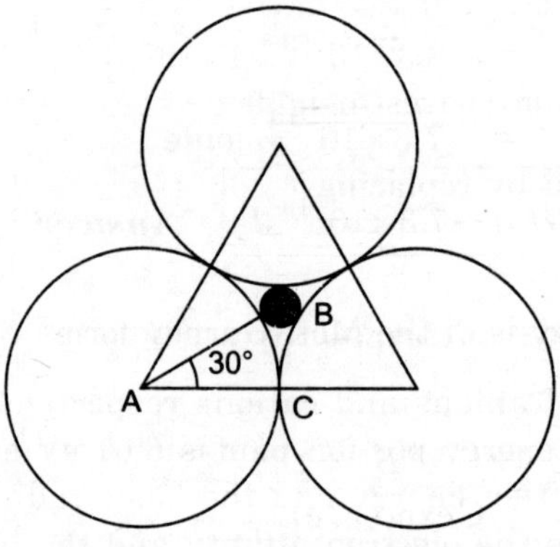

Solution:

$$\cos 30° = \frac{R}{R+r} = 0.866$$

$$\frac{R+r}{R} = \frac{1}{0.866} = 1.154$$

$$1 + \frac{r}{R} = 1.154; \quad \frac{r}{R} = 1.154 - 1$$

$$\boxed{\frac{r}{R} = 0.154}$$ ***Answer***

1.1.5 Mr. Rudden used to add 500 cc of water to 500 cc of milk every time to prepare tea in his shop. Can you suggest a method to estimate the number of water molecules added each time?

Solution:

Yes sir. I can suggest a small formula to calculate the number of molecules present in 500 cc of water.

If M and ρ are the molecular weight and density of water, then $\frac{M}{\rho}$ is the molar volume of water. Thus $\frac{18}{1}$ cc of water contains 6.02×10^{23} water molecules;

Hence 500 cc of water will have $\frac{500 \times 6.02 \times 10^{23}}{18} = 1.67 \times 10^{25}$ water molecules.

$$\boxed{1.67 \times 10^{25}}$$ ***Answer***

1.1.6 The total energy per kmol of a crystal is given by the equation

$$U(r) = N_A\left[\frac{B}{r^n} - \frac{\alpha e^2}{r}\right]$$

What is the equilibrium nearest-neighbour separation r_0 at which $U(r_0)$ in the above equation is unchanged by replacing $\frac{B}{r^n}$ by $C \exp(-r/\rho)$?

Solution:

$$B = r^n C \exp(-r/\rho) \text{ with } r = r_0$$

i.e.,

$$r^n = \frac{B}{C \exp(-r/\rho)}$$

Differentiating this,

$$nr^{n-1} = \frac{B}{\rho C}\exp(r/\rho)$$

Substituting for B

$$nr^{n-1} = \frac{1}{\rho C}\exp(r/\rho)\left[\frac{Cr^n}{\exp(r/\rho)}\right]$$

$$\frac{nr^n}{r} = \frac{r^n}{\rho} \text{ with } r = r_0$$

Thus $$\boxed{r_0 = n\rho}$$ ***Answer***

1.1.7 Calculate the value of the Madelung constant for the structure in the following figure. All bond lengths are equal and all bond angles are 90°. Assume that there are no ions other than those shown in figure and that the charges on the cations and anions are + 1 and – 1.

(Department of Chemistry, B. Tech. programme 1997. The Open University, U.K.)

Solution:

There are seven ions in the structure; six cations and an anion. First we shall calculate the contribution to the potential energy of interactions of the six cations with the central anion. Each cation is distance r_o from the central ion.

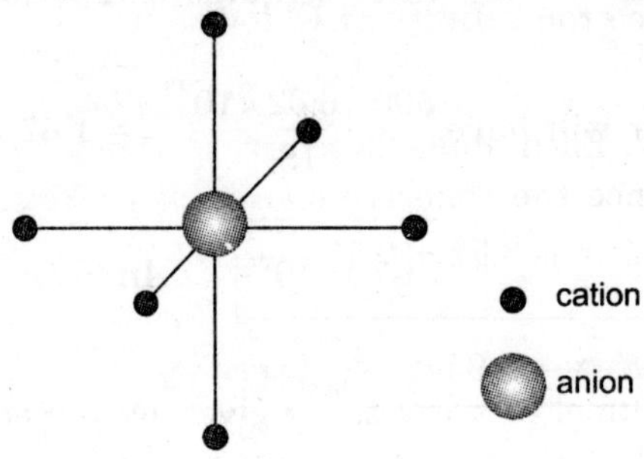

Thus $$E_1 = -\frac{6e^2}{4\pi\epsilon_0 r_0}$$

Each cation also interacts with diametrically opposite cations (distance $2r_o$). There are three such interactions, so

$$E_2 = +\frac{3e^2}{4\pi\epsilon_0 2r_0}$$

Finally, E_3 is calculated from interactions between adjacent cations [distance $(\sqrt{2})r_0$] of which there are twelve.

Now

$$E_3 = +\frac{12e^2}{4\pi\epsilon_0 (\sqrt{2})r_0}$$

Thus

$$E = E_1 + E_2 + E_3 = -\frac{e^2}{4\pi\epsilon_0 r_0}\left[6 - \frac{3}{2} - \frac{12}{\sqrt{2}}\right]$$

$$= -\frac{Ae^2}{4\pi\varepsilon_0 r_0}$$

with $$A = 6 - 1.5 - 8.5 = -3.99$$

$$\boxed{A = -3.99}$$ ***Answer***

1.1.8 Calculate the cohesive energy of NaCl from the following data: r_0 = 0.281 *nm*, A = 1.748, n = 9, ionisation energy of Na = 5.1 eV, electron affinity of Cl is 3.61 eV. *(JNTU, BE May 2002)*

Solution:

The P.E. of an ion pair is $V = -\dfrac{Ae^2}{4\pi\epsilon_0 r_0}\left[\dfrac{n-1}{n}\right] = -7.44$ eV

To calculate the cohesive energy, we must add to this, the energy required to make an ion pair from Na and Cl atoms. This energy is the difference between ionisation energy of Na and electron affinity of Chlorine.

i.e.,

5.14 – 3.61 = 1.53 eV. Each atom therefore contributes half of 1.53 i.e., 0.77 eV to the cohesive energy. Hence the cohesive energy of NaCl/atom is thus

$$E_c = -3.72 + 0.77 = -2.95 \text{ eV}.$$

$$\boxed{E_c = -2.95 eV}$$ ***Answer***

1.2.1 Explain allotropic forms of iron with ranges of temperatures. Whether any change in volume is noticed when a metal exists in more than one form? *(Metallurgy division, University of Manchester, U.K, B. Tech., September 1982)*

Solution:

When an element or a compound can exist in more than one crystalline forms at different temperatures, the phenomenon is termed allotropy or isomorphism. The most common example is that of iron which shows different allotropic forms at different temperatures. Between – 273 and 912°C iron is bcc crystal also called α – iron. Between 912 and 1394°C iron is in fcc form, also known as γ – iron. Between 1394°C and its melting point, iron is in bcc form called δ – iron. These are shown in the following figure.

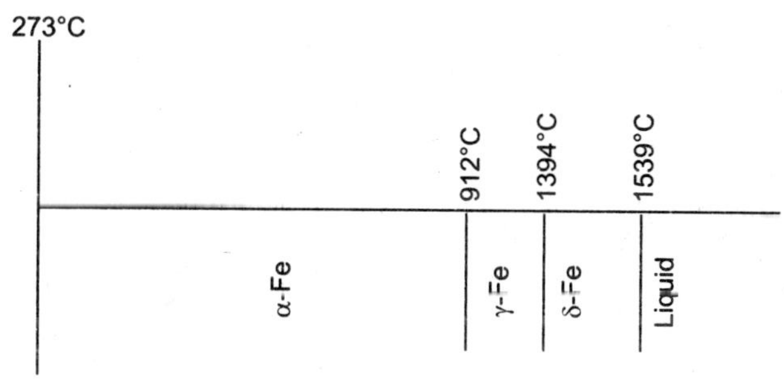

For bcc structure, $a = \dfrac{4r}{\sqrt{3}}$ with 2 atoms in an unit cell

For fcc structure, $a = \dfrac{4r}{\sqrt{2}}$ with 4 atoms in an unit cell

Volume of bcc unit cell per atom

$$V_b = \frac{a^3}{2} = \left[\frac{4}{\sqrt{3}}\right]^3 \frac{r^3}{2} = 6.158r^3$$

Volume of fcc unit cell per atom

$$V_f = \frac{a^3}{4} = \frac{1}{4}\left[\frac{4r}{\sqrt{2}}\right]^3 = \frac{16r^3}{2\sqrt{2}} = 5.658r^3$$

Now it is established that when bcc structure changes to fcc, there is a reduction of volume for a given mass. the change of volume per atom is,

$$\Delta V = 6.158\ r^3 - 5.657r^3$$

$$\Delta V = 0.501\ r^3$$

Answer

1.2.2 Calculate the lattice constant of iron. Given: density of iron 7.86 kg/cm^3; atomic weight of iron 55.85 and Avogadro's number 6.023 × 10^{23}/mol. *(Anna U. May 2004)*

Solution:

This problem carries some errors. The density is 7.86 gm/cm^3 or 7860 kg/m^3 and it is suggested to use 6.023 × 10^{26}/kmol for Avogadro's number instead of 6.023 × 10^{23} /gmol as SI system is followed almost everywhere.

$\frac{M_A}{\rho} m^3$ will contain N_A atom.

Thus a^3 will have $\frac{\rho N_A a^3}{M_A}$ atom. This must be equal to n = 2

Thus $$n = \frac{\rho N_A a^3}{M_A}$$

or $$a = \left[\frac{M_A n}{\rho N_A}\right]^{1/3}$$

with

(i) ρ = 7860 kg/m^3

(ii) $n = 2$

(iii) $M_A = 55.85$

(iv) $N_A = 6.02 \times 10^{26}$/kmol

$$a = \left[\frac{55.85 \times 2}{7860 \times 6.02 \times 10^{26}}\right]^{1/3} = 0.287 \times 10^{-9} \text{ metre}$$

a = 2.87 nm ***Answer***

1.2.3 The lattice constant of aluminium (fcc) structure is 1.4 times the lattice constant of chromium (bcc) structure. Compare the radii of the atoms of these metals. *(Department of Electronic and Electrical Engineering, Loughborough University of Technology, U.K., November 1979).*

Solution:

$$a_A = \frac{4r_A}{\sqrt{2}}$$

$$a_C = \frac{4r_C}{\sqrt{3}}$$

$$\frac{a_A}{a_C} = 1.4 = \frac{r_A}{r_C}\sqrt{\frac{3}{2}}$$

or $$r_A = \frac{1.4 \times 1.414}{1.732} r_C = 1.14 r_C$$

$$\boxed{r_A = 1.14 r_C}$$ ***Answer***

1.2.4 In a crystal whose primitives are 0.12 nm, 0.18 nm and 0.2 nm, a plane (2 3 1)′ has an intercept of 0.12 nm on the x-axis. Find the intercepts on the y and z axes.

(Bangalore U., BE, May 1982)

Solution:

Let p, q, r be the three intercepts, then

$$p : q : r = \frac{0.12}{2} : \frac{0.18}{3} : \frac{0.2}{1}$$

Thus $$\frac{p}{q} = \frac{0.12 \times 3}{2 \times 0.18} = 1$$

i.e., $q = p = 0.12 nm$ as p = 0.12 nm

Similarly,

$$\frac{q}{r} = \frac{0.18}{3 \times 0.2} = \frac{0.18}{0.6} = 0.3$$

$$r = \frac{q}{0.3} = \frac{0.12}{0.3} = 0.4 nm$$

$$\boxed{q = 0.12\ nm,\ r = 0.4\ nm}$$ ***Answer***

1.2.5 The toy-car of Mr. Ajay was heavily damaged; he threw all the parts except the chase. It is of α-iron with bcc structure at 300 K and a rectangular plate with dimensions 10 cm, 8 cm and 1 cm. Compute the number of unit cells in this specimen. The nearest neighbour distance of this structure is 0.244 nm.

(University of Maryland, US, April 1982)

Solution:

$$a = \frac{4r}{\sqrt{3}} = \frac{2 \times 2r}{\sqrt{3}} = \frac{2 \times 0.244}{1.732}$$

$$a = 0.282\ nm$$

Volume considered 80 cm^3

Thus

$(0.282 \times 10^{-7})^3$ cm^3 has one unit cell

80 cm^3 therefore will have $\dfrac{80}{0.282^3 \times 10^{-21}} = \dfrac{80 \times 10^{21}}{0.282^3} = 3.57 \times 10^{24}$ cell

$\boxed{3.57 \times 10^{24}}$ ***Answer***

1.2.6 Aluminium has density of 2698 kg/m^3. Its structure is fcc and atomic weight is 26.98. Calculate

(*a*) How many atoms are contained in 1 m^3 of this metal?

(*b*) Get the size of the unit cube for aluminium.

(*c*) Calculate the atomic radius aluminium.

(*d*) Find the radius of the interstitial sphere that can just fit into the void at the body centre of this fcc structure coordinated by the facial atoms.

(Anna. U. BE, November 1984, JNTU, BE, December 2001)

Solution:

(*a*) $\dfrac{26.98}{2698}$ m^3 will contain 6.02×10^{26} atom

1 m^3 may contain $\dfrac{2698 \times 6.02 \times 10^{26}}{26.98}$

$\boxed{n = 6.02 \times 10^{28}}$ ***Answer***

(*b*) $a = \left[\dfrac{M_A n}{\rho N_A}\right]^{1/3} = \left[\dfrac{26.98 \times 4}{2698 \times 6.02 \times 10^{26}}\right]^{1/3} = 0.406 \times 10^{-9}$ m

$\boxed{a = 0.406\ nm}$ ***Answer***

$a = 0.406$ nm

$V = a^3 = 6.69 \times 10^{-29}$ m^3

$\boxed{V = 6.69 \times 10^{-29} \text{m}}$ ***Answer***

(*c*) $a = \dfrac{4r}{\sqrt{3}}; r = \dfrac{a \times 1.732}{4} = \dfrac{0.406 \times 1.732}{4} = 0.176$ nm

$\boxed{r = 0.176 \text{ nm}}$ ***Answer***

(*d*) Let r be the radius of the atom and R be the radius of the sphere which can just fit.

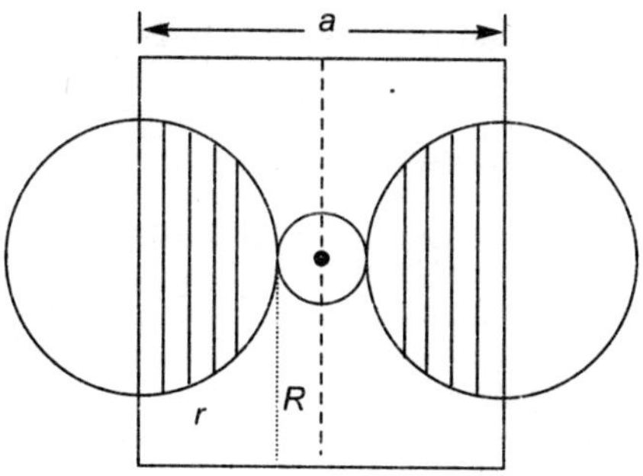

$2(r+R)=a;\ r+R=a/2$

or $$R=\frac{a}{2}-r=\frac{4r}{2\sqrt{2}}-r$$

$$R=r(\sqrt{2}-1)=0.414r=0.414\times 0.176$$

$$\boxed{R = 0.0729 \text{ nm}}$$ ***Answer***

1.2.7 The fraction of vacancy sites in a metal is 1×10^{-10} at 500°C. What will be the fraction of vacancy sites at 1000°C? *(Anna U., M.Sc., September 1992)*

Solution:

$$\frac{n}{N}=1\times 10^{-10}=\exp\left(-E_v/k_BT\right)$$

$$\frac{E_v}{k_BT}=10 ln 10$$

$$E_v=773\times 10\times 2.3031\times k_B$$

$$\frac{n_1}{N_1}=\exp\left(-\frac{7730\times 2.3031\times k_B}{k_B\times 1273}\right)$$

$$\boxed{\frac{n_1}{N_1}=8.44\times 10^{-7}}$$ ***Answer***

1.2.8 NaCl has an fcc structure. The density of NaCl = 2.18 gm/cm^3. Calculate the distance between two adjacent cations. Given: molecular weight of NaCl is 58.5.

(University of Minnesota, US, Mid-Semester B. Tech., 1994)

Solution:

Mass of one molecule $\dfrac{58.5}{6.02\times 10^{23}}=9.708\times 10^{-23}$ gm

$$\rho=\frac{M}{V}=\frac{n\times m}{1}\text{ with } V=1$$

Hence the number of molecules in 1 cm^3 of NaCl is

$$n = \frac{\rho}{\text{mass of one molecule}} = \frac{2.18}{9.708\times10^{-23}}$$

i.e., $n = 22.46 \times 10^{21}$ molecules/cc

Hence number of ions in one cm^3 = 2 × 22.46 × 10^{21}

$= 44.92 \times 10^{21}$

One unit cell contains 8 ions; therefore number of unit cells in one cm^3 is

$$\frac{44.92\times10^{21}}{8} = 5.62\times10^{21}$$

i.e.,

1 cm^3 contains 5.62 × 10^{21} unit cell

a^3 will contain $\frac{5.62\times10^{21}\times a^3}{1}$ = 1 unit cell

or $$a^3 = \frac{1}{5.62\times10^{21}} = 0.178 \times 10^{-21}$$

or $$a = 5.62 \times 10^{-8} \text{ cm}$$

Cation-cation distance is the face diagonal of the primitive cell

$$D^2 = \left(\frac{a}{2}\right)^2 + \left(\frac{a}{2}\right)^2 = \frac{a^2}{2}$$

$$D = \frac{a}{\sqrt{2}} = \frac{5.626\times10^{-8}}{1.414} = 3.98\times10^{-8}\text{ cm}$$

or $\boxed{D = 0.398 \text{ nm}}$ ***Answer***

1.2.9 A diffraction pattern of a cubic crystal of lattice parameter a = 0.316 nm is obtained with a monochromatic x-ray beam of wavelength 0.154 nm. The first four lines on this pattern were observed to have the following values:

Line:	1	2	3	4
θ°	20.3	29.2	36.7	43.6

Determine the interplanar spacings and Miller indices of the reflecting planes.

Solution:

$$d = \frac{\lambda}{\sin\theta}; \sin\theta = \frac{\lambda}{d}$$

Line	θ	$\sin\theta$	d (nm)	$\frac{a^2}{d^2} = h^2 + k^2 + l^2$
1	20.3	0.3460	0.2240	2
2	29.2	0.4886	0.1570	4
3	36.7	0.5980	0.1290	6
4	43.0	0.6900	0.1150	8

For determining (h, k, l) one has to resort to trial and error method.

$h^2 + k^2 + l^2 = 2 \quad = 1^2 + 1^2 + 0; \quad h\,k\,l \quad = (1\;1\;0)$

$h^2 + k^2 + l^2 = 4 \quad = 2^2 + 0 + 0; \quad h\,k\,l \quad = (2\;0\;0)$

$h^2 + k^2 + l^2 = 6 \quad = 2^2 + 1^2 + 1^2; \quad h\,k\,l \quad = (2\;1\;1)$

$h^2 + k^2 + l^2 = 8 \quad = 2^2 + 2^2 + 0; \quad h\,k\,l \quad = (2\;2\;0)$

EXERCISE

1.1 The potential energy of a diatomic molecule is given in terms of the interatomic distance r by the expression $U(r) = -\frac{a}{r^2} + \frac{b}{r^{10}}$. Calculate the equilibrium spacing of the two atoms and the dissociation energy. Given: $a = 1.44 \times 10^{-39}$ J m^2 and $b = 2.19 \times 10^{-115}$ J m^{10}.
(M.K. U., BE May 1990)
(***Ans***: 0.047 nm, 4.33×10^{-2} eV)

1.2 How many kg-atom of potassium will be required to treat with 709.1 kg of chlorine to form KCl?
(***Ans***: 20 kg-atom)

1.3 (*a*) How much energy is required to form a Li^+ and Br^- ion pair from a pair of Li and Br pair of atoms?

(*b*) What must be the separation between a Li^+ and Br^- ion pair so that their total energy is to be zero. Given: the ionisation energy of Li = 5.4 eV and electron affinity of Br = 3.36 eV.
(Anna. U. Part-time BE, May 1992)
(***Ans***: 2.04 eV, 0.71 nm)

1.4 The ionic radii of Cs and Cl are 0.165 nm and 1.81 nm and their atomic weights are 133 and 35.5 respectively. Calculate the density of CsCl.
(JNTU May 1997)
(***Ans***: 4.37×10^3 kg/m^3)

1.5 Assuming an overlap interaction between nearest neighbours of the type $\phi(r) = B \exp(-r/\rho)$ where B and ρ are constants, calculate the equilibrium spacing r_o in terms of B and ρ.
(IIT Madras, BE May 1990)
(***Ans***: $\rho \ln B$)

1.2.1 Calculate the atomic density (i.e., the number of atoms per unit area of a lattice plane) in (1 0 0) and (1 1 0) planes of a bcc crystal having a as its lattice constant.

(AMIE, September 1988)

(Ans: $1/a^2$, $\sqrt{2}/a^2$)

1.2.2 How is the distance between two planes of same indices calculated? Calculate the ratio of the distances of the planes with Miller indices (1 1 1) and that with Miller indices (1 0 2) in lead crystallizes in fcc structure. Given radius of lead atom 0.174 nm. *(Anna U., B.E. November 1992)*

(Ans: $d_1/d_2 = 1.3$)

1.2.3 Magnesium has hcp structure. The radius of magnesium atom is 0.1605 nm. Calculate the volume of unit cell of magnesium. ***(Ans***: 1.4×10^{-28} m^3)

1.2.4 Miss. Mithila a seventh standard student geometrically prepared a cube of common salt as her project work in the junior school level. When it was shown to me I found that it is a cube of NaCl with 1 cm edge. Compute the number of sodium ions and the number of unit cells if the lattice constant is 0.56 nm. ***(Ans***: 22.8×10^{21}, 5.7×10^{21})

1.2.5 Copper has fcc structure and its atomic radius is 0.128 nm. Calculate the density of copper if its atomic weight is 63.5. (Anna U., B.E., May 1989 and JNTU June 2004)

(Ans: 8.9×10^3 kg/m^3)

1.2.6 Given the following data; determine whether a gaseous molecule $A^+ B^-$ will be stable with respect to the separated A and B gaseous atoms:

First ionisation energy of A = 500 kJ/mol

Electron affinity for B atom = – 335 kJ/mol

Interionic separation = 0.3 nm ***(Ans***: – 297 kJ/mol)

1.2.7 The lattice constant of the unit cell of α- iron is 0.287 nm. Find the number of atoms/ mm^2 of the planes (1 1 0) and (1 1 1). α-iron has bcc structure.

(Ans: 1.72×10^{13}, 2.11×10^{13})

(The Open University, UK)

1.2.8 Calculate the ratio of the number of vacancies in equilibrium at 300 K in aluminium to that produced by rapid quenching at 800 K. Enthalpy of formation of vacancies in aluminium is 68 kJ/mol. ***(Ans***: 3.75×10^{-8})

(M.U., BE, November 1992)

1.2.9 A gold smith used to prepare samples by adding 20 cc of copper in 80 cc of pure gold every time for making ornaments. Once he had a wire of 200 cm with a uniform diameter 0.5 cm prepared as per the above ratio. Find out the number of unit cells in copper that is added to draw the wire. Also compute the number of copper atoms present. Given : Copper takes fcc structure with a lattice constant of 0.36 nm.

(Ans: 1.7×10^{23}m, 6.8×10^{23})

(GATE 1995)

Chapters 2 & 3

UNIT 3

Elements of Statistical Mechanics and Principles of Quantum Mechanics

2.1 ELEMENTS OF STATISTICAL MECHANICS

2.1.1 INTRODUCTION

Just as the failure of classical mechanics made it necessary to establish quantum mechanics for the description of nature on the atomic scale, the failure of Maxwell-Boltzmann statistics to provide correct solutions for many problems involving fundamental particles led to the development of new systems of quantum statistics. In the present work we will reexamine the statistical bases underlying the Maxwell-Boltzmann statistics and introduce modifications which lead to the Fermi-Dirac statistics, applicable to all fundamental particle with spin one-half, and to Bose-Einstein statistics, applicable to particles which have zero or integer spin.

2.1.2 METHODS OF STATISTICAL MECHANICS

There are three kinds of identical particles:

(*i*) Identical particles of any spin which are so much separated in the assembly that they can be distinguished from one another. The molecules of a gas are particles of this kind. The Maxwell-Boltzmann distribution holds good.

(*ii*) Identical particles of 0 or integral spin which cannot be distinguished from one another. These are called Bose particles (or Bosons) and do not obey Pauli's exclusion principle. Photons, phonons and a-particles are of this kind. The Bose-Einstein distribution holds good for them.

(*iii*) Identical particles of odd half-integral spin which cannot be distinguished from one another. These are called Fermi particles (or Fermions) and do obey Pauli's exclusion principle. Electrons, protons and neutrons are particles of this kind. The Fermi-Dirac distribution holds for them.

The energy distribution law for particles of kind (*i*) can be derived by methods of classical statistics as well as of quantum statistics, but that for particles of kind (*ii*) and (*iii*) can be derived by methods of quantum statistics only.

Maxwell-Boltzmann (Classical) Distribution

Let us consider a system composed of a very large number (N) of distinguishable identical particles. Suppose the energies of the particles are limited to the values E_1, E_2 ....,E_r. This means n_1 of N has an energy E_1, n_2 of N has an energy E_2 and so on.

Thus

$$N = n_1 + n_2 + \cdots + n_r = \text{a constant}$$

or

$$\delta N = \delta n_1 + \delta n_2 + \cdots + \delta n_r = 0$$

i.e.,

$$\sum \delta n_i = 0 \qquad ...(2.1.1)$$

Similarly

$$E = E_1 n_1 + E_2 n_2 + \cdots + E_r n_r \qquad ...(2.1.2)$$

i.e.,

$$\delta E = E_1 \delta n_1 + E_2 \delta n_2 + \cdots + E_r \delta n_r = 0$$

i.e.,

$$\sum_i E_i \delta n_i = 0 \qquad ...(2.1.3)$$

When we consider the gas in equilibrium, the probability is maximum. Hence δP will be zero. Again when P is maximum, $\log P$ is maximum.

or

$$\delta(\log P) = 0 \qquad ...(2.1.4)$$

But

$$P = \Pi \left[\frac{(g_i)^{n_i}}{n_i!} \right] \qquad ...(2.1.5)$$

Here Π is the product sign and g_i is the priori probability for a molecule to have the energy E_i.

Taking logarithm of Eqn. 2.1.4,

$$\log_e P = \sum n_i \log_e g_i - \log_e n_i!$$

By Stirling's theorem,

$$\log n! = n \log n - n$$

Hence

$$\log P = \sum n_i \log g_i - (n_i \log n_i - n_i)$$

$$= \sum n_i - \sum (n_i \log n_i - n_i \log g_i)$$

$$\log P = N - \sum n_i \log (n_i / g_i)$$

$$d(\log P) = -\sum dn_i \log\frac{n_i}{g_i} - \sum n_i \frac{dn_i}{n_i}$$

$$= -\sum dn_i \log\frac{n_i}{g_i} - \sum dn_i$$

Equations 2.1.4 and 2.1.1, $d(\log P) = 0$ and $\sum dn_i = 0$.
Therefore

$$\sum \log\left(\frac{n_i}{g_i}\right) dn_i = 0 \qquad ...(2.1.6)$$

Let us incorporate into Eqn. (2.1.6) by making use of Lagrange's method of undertermined multipliers. Multiplying Eqn. 2.1.1 by a factor α and Eqn. (2.1.3) by β and adding to Eqn. (2.1.6), we Sct

$$\sum\left[\log\left(\frac{n_i}{g_i}\right) + \alpha + \beta E_i\right] = 0$$

Here $e^{-\alpha} = \frac{N}{Z}$, where Z is called partition function. $\beta = \frac{1}{kT}$

i.e., $\log\frac{n_i}{g_i} = -\alpha - \beta E_i$

Therefore $\frac{n_i}{g_i} = e^{-\alpha} e^{-\beta E_i}$ ...(2.1.7)

This equation gives the distribution of molecules over the various energy states for the most probable configuration. This relation is known as Maxwell-Boltzmann distribution.

Study of Bose-Einstein Distribution

The Maxwell-Boltzmann distribution governs identical particles which can be distinguished from one another in some way (as molecules in a gas) and as such they can be given names or numbers. On the other hand Bose-Einstein distribution governs identical particles which can never be distinguished from one another (as radiation photons in an enclosure). As such there is no way to number them or give them names.

The basic assumption of the *B-E* distribution is that any number of particles can be in any quantum state and that all the quantum states are equally probable. Suppose each quantum state corresponds to an elementary cell in the phase space. The number of different ways in which n_i indistinguishable particles can be distributed in g_i phase cells is

$$\frac{(n_i + g_i - 1)!}{n_i!(g_i - 1)!}$$

When the number of cells is sufficiently large, then the previous term can be written as

$$\frac{(n_i + g_i)!}{n_i! g_i!}$$

The probability W of the entire contribution of N particles is the product of the numbers of different arrangements of particles among the states having each the same energy. Thus

$$W = \Pi \frac{(n_i + g_i)!}{n_i! g_i!}$$

For the equilibrium (most probable) condition, we must have

$$\delta W = 0$$

Subject to the limitation

$$\sum \delta n_i = 0$$

and $$\sum E_i \, \delta n_i = 0$$

This leads to the result

$$n_i = \frac{g_i}{[e^{\alpha} e^{E_i/kT} - 1]} \quad ...(2.1.8)$$

where e^{α} is a constant, k is Boltzmann constant and T is Kelvin temperature. This is Bose-Einstein distribution law.

Photon Gas

Thermal radiation from at absolute temperature T and in thermal equilibrium consists of light quanta. It is considered as photon gas. Each photon has an energy hv and momentum $\frac{h}{\lambda}$. Photons are indistinguishable from one another. We can derive Planck's formula for black body from Bose-Einstein statistics. The number of photons in a system is not necessarily constant. Hence the condition $\sum dn_i = 0$ is no longer valid. Due to this reason $\alpha = 0$ in the B.E. distribution law. Further $\beta = \frac{1}{kT}$. Therefore Eqn. 2.1.8 becomes

$$n_i = \frac{g_i}{[e^{E_i/kT} - 1]} \quad ...(2.1.9)$$

The energy spectrum of the photons is taken to be continuous. Replacing g_i by $g\,(E)\,dE$ and n_i by dn, we get

$$dn = \frac{g(E)dE}{[e^{hv/kT} - 1]}$$

But $E = hv$, $g\,(E)\,dE$ will be $g\,(v)\,dv$.

The total number of Eign states between frequencies υ and $\nu + d\nu$ is given by

$$g(\upsilon)\, d\upsilon = 4\pi V\left(\frac{\upsilon^2}{c^3}\right) d\upsilon$$

Taking into account the doubling of the states due to polarization of the photons, the number of states available for the photons in the frequency range

$$\upsilon \text{ and } \upsilon + d\upsilon = 8\pi V\frac{\upsilon^2}{c^3} d\upsilon$$

Thus

$$dn = \left[\frac{8\pi V\upsilon^2 d\upsilon}{c^3}\right] \times \frac{1}{\left[\left(e^{h\upsilon/kT}\right)-1\right]}$$

i.e.,

$$\frac{dn}{V} = \frac{8\pi\upsilon^2}{c^3}\frac{1}{\left[e^{h\upsilon/kT}\right]-1} \qquad \text{...(2.1.10)}$$

This equation represents the number of photons per unit volume lying in the frequency range υ and $\upsilon + d\upsilon$.

Thus the energy density of radiation ($E\upsilon\, d\upsilon$) with frequencies between υ and $\upsilon + d\upsilon$ is

$$E_\upsilon d\upsilon = \frac{8\pi h}{c^3}\frac{\upsilon^3 d\upsilon}{\left[e^{(h\upsilon/kT)}-1\right]} \qquad \text{...(2.1.11)}$$

This is Planck's law of radiation.

Fermi-Dirac (Quantum) Distribution

Fermi-Dirac distribution applies to indistinguishable particles, like electrons, which are governed by Pauli's exclusion principle. The general method of deriving the Fermi-Dirac distribution law is similar to that for the Bose-Einstcin statistics except that now each phase cell, that is, each quantum state, cannot be occupied by more than one particle. This leads to the following function:

$$n_i = \frac{g_i}{\left[(e^{\alpha}e^{E_i/kT})-1\right]} \qquad \text{...(2.1.12)}$$

Again n_i is the number of particles whose energy is E_i, and g_i is the number of states that have the same energy E_i.

Comparison of the Three Distribution Laws

For this purpose, let us define a quantity

$$f(E_i) = \frac{n_i}{g_i}$$

$f(E_i)$ is called occupation index of a state of energy E_i. Sometime $F(E_i)$ is used in literature. It represents the average number of particles in each of the states of that energy. Thus

$$f_{MB} = \frac{1}{e^{\alpha}\, e^{E_i/kT}}$$

$$f_{BE} = \frac{1}{[(e^{\alpha}\, e^{E_i/kT}) - 1]}$$

$$f_{FD} = \frac{1}{[(e^{\alpha}\, e^{E_i/kT}) + 1]}$$

In the following diagrams we plot each occupation index $f(E_i)$ against energy E_i for two different values of T and α.

In the M-B distribution (Fig. 2.1) the occupation index $f(E_i)$ falls purely exponentially with increase in energy E_i. In fact, it falls by a factor $\frac{1}{e}$ for each increase of kT in the energy E_i.

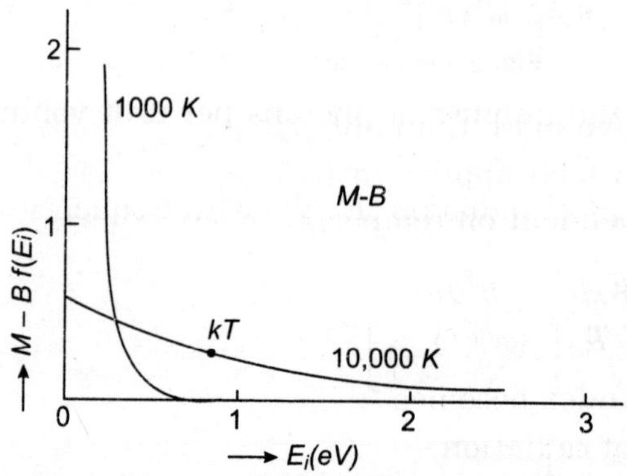

Fig. 2.1 Maxwell-Boltzmann Distribution

The B-E occupation index against energy is plotted in Fig. 2.2 for temperatures 1000 K and 10000 K, in each case for $\alpha = 0$ (corresponding to a gas of photons). For $E_i >> kT$, the $B.E.$ distribution approaches the exponential form characteristic of the M-B

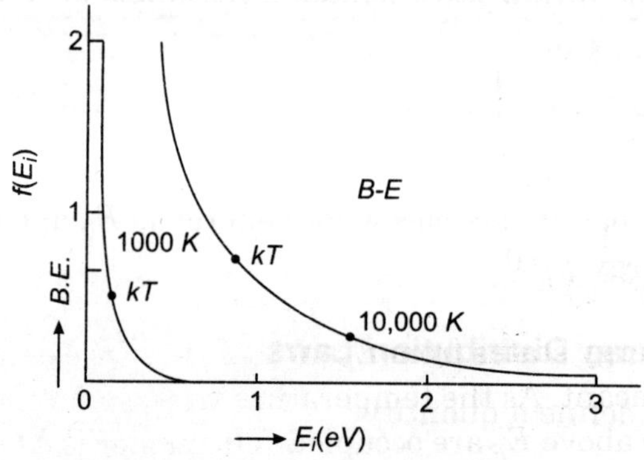

Fig. 2.2 Bose-Einstein Distribution

distribution. In this region the average number of particles per quantum state is much less than one. However, for $E_i << kT$, the -1 term in the denominator causes the occupation index much greater compared to that of the *M-B* distribution. This means that for energies small compared to kT, the number of particles per quantum state is greater for the *B.E.* distribution than for the *M-B* distribution.

The *F-D* distribution is plotted in Fig. 2.3 for three different values of T and α. This

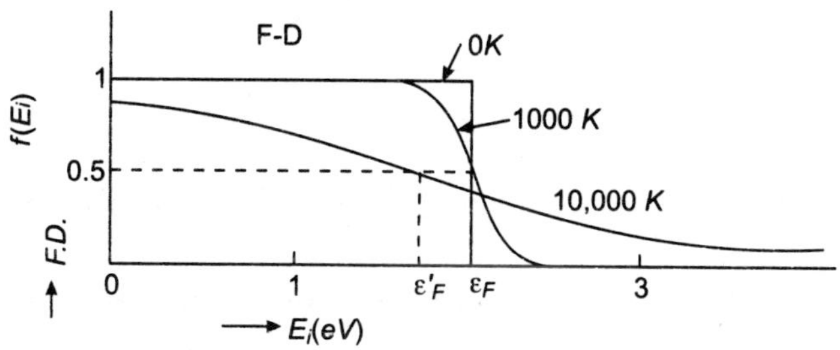

Fig. 2.3 Fermi-Dirac Distribution

signi fies that we cannot have more than one particle per quantum state as required by Pauli's exclusion principle which applies in this case. Further, in this distribution, the parameter α is strongly dependent on temperature, and we write

$$\alpha = -\frac{E_F}{kT}$$

so that the *F-D* occupation index becomes

$$f_{FD}(E_i) = \frac{1}{e^{(E_i - E_F)/kT} + 1} \qquad ...(2.1.13)$$

where E_F is called the Fermi energy.

Let us consider the situation at the absolute zero of temperature. At $T = 0$, $(E_i - E_F)/kT = -\infty$ (for $E_i < E_F$ and $+\infty$ (for $E_i > E_F$), so that

for $E_i < E_F$,

$$f_{FD}(E_i) = \frac{1}{e^{-\infty} + 1} = 1$$

and for $E_i > E_F$,

$$f_{FD}(E_i) = \frac{1}{e^{\infty} + 1} = 0$$

Thus, at $T = 0$, all energy states from $E_i = 0$ to $E_i = E_F$ are occupied because $f_{FD}(E_i) = 1$; while all above E_F are vacant. As the temperature rises, some of the states just below E_F become vacant while some above E_F are occupied. The higher the temperature, the more is the spreading in $f_{FD}(E_i)$.

At $E_i = E_F$, we have

$$f_{FD} = \frac{1}{(e^0 + 1)} = \frac{1}{2}, \text{ at all temperatures.}$$

This means that the probability of finding an electron with energy equal to the Fermi energy in a metal is $\frac{1}{2}$ at any temperature.

In literature sometimes the above equation is written as

$$F(E) = \frac{1}{1 + e^{(E-E_F)/kT}}$$

where $F(E)$ is the probability of occupation of the energy state with E as the energy and E_F is Fermi energy.

Thus Fermi energy is defined as the highest energy state for the particle to occupy at 0K. or it is the energy state at which the probability of occupation is $\frac{1}{2}$ for other temperatures.

2.1.3 BLACK BODY RADIATION AND PLANCK'S LAW

An elementary study of heat radiation helps us to understand that when a body is heated it emits radiation. This radiation emitted by hot bodies is usually known as thermal radiation or emmission of heat. It is interesting to note that even at ordinary temperatures a body emits over a range of frequencies. The relative brightness of the different frequencies depends on the temperature of the body.

Here we shall describe a black devised by Ferry assuming that no actual body is a perfect black body. It consists of a hollow thick-walled sphere painted lamp black internally and provided a small circular opening to enter the radiation. In front of the opening, there is a projection to prevent the direct reflection of radiation from inner surface. When any radiation enters the opening, it suffers multiple reflections inside the sphere and then absorbed. Similarly, when the walls of such a cavity are heated to a temperature T, the radiation emitted fills the cavity and then leaps from the opening. These are called black body radiation and is a characteristic of its temperature. It must be remembered that thermal radiation and light radiation are identical. The only difference is the wavelength of visible light is smaller than that of thermal radiation.

Measurements on the spectral distribution of black body radiation have been carried out by a number of workers. Figure. 2.4 shows the results of Lummer and Pringsheim. Here a graph is emissive power or intensity and wavelength for various temperatures.

Following are the important findings of Lummer and Pringsheim:

(*i*) The energy is not uniformly distributed in the radiation spectrum of black-body.

(*ii*) At a given temperature, the intensity of radiation increases with increase in wavelength and becomes maximum at a particular wavelength. With further increase in wavelength, the intensity of radiation decreases.

(*iii*) The wavelength corresponding to the maximum energy represented by the peak of the curve shifts towards shorter wavelengths as the temperature increases.

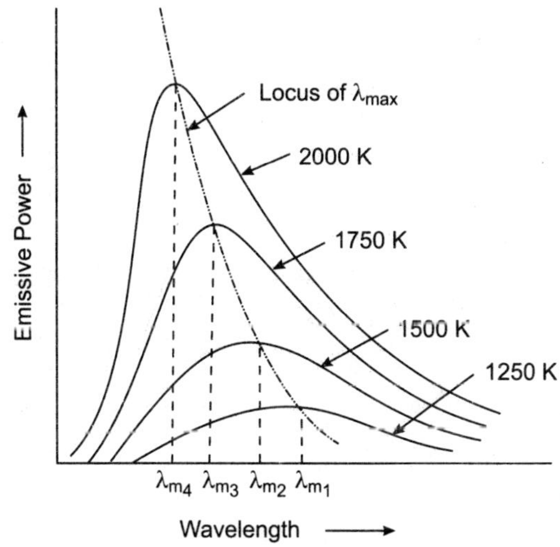

Fig. 2.4 Studies of black-body radiation by Lummer and Pringsheim.

Planck's Radiation Formula

Planck assumed the presence of number of resonators (harmonic osciullators) inside the cavity of the black-body and used the quantum hypothesis to derive the formula. According to Planck's quantum hypothesis, an oscillator can have only discrete energies given by

$$E_n = nh\nu = n\varepsilon \qquad ...(2.1.4)$$

with $n = 0, 1, 2$ ...
where ν is the frequency of the oscillator, h is Planck's costant n is (an integer including 0) known as quantum number. Thus an oscillator can have energies 0, $h\nu$, $2h\nu$,... and the oscillator cannot emit or absorb energy continuously but in quanta of energy $h\nu$. We shall first calculate average energy of the oscillator of frequency ν. If N is the total number of oscillators and E their total energy, then average energy per Planck's oscillator is given by

$$\overline{E} = \frac{E}{N} \qquad ...(2.1.15)$$

Let N_0, N_1, ..., N_r, ... be the number of oscillators having energies 0, $h\nu$, $2h\nu$, ..., $rh\nu$, ... respectively. The relative probability that an oscillator will have the energy $h\nu$ at temperature T is $\exp\left(-\frac{h\nu}{kT}\right)$ where k is Boltzmann constant. In other words, according to Maxwell's law of molecular motions if ε is a certain amount of energy, the probabilities that a system will have energies 0, ε, 2ε ... $r\varepsilon$ are in the ratio

$$1 : e^{-\varepsilon/kT} : e^{-2\varepsilon/kT} : \cdots e^{-r\varepsilon/kT} \cdots$$

As said earlier if N_0 is the number of resonators having energy zero, then the number of resonators N_1 having energy ε will be $N_0\, e^{-\varepsilon/kT}$, the number of resonators N_2 having energy ε will be $N_0\, e^{-2\varepsilon/kT}$ and so on.

Thus, we have

$$N_r = N_0 \exp\left(-\frac{rh\upsilon}{kT}\right)$$

and

$$N = N_0 + N_1 + N_2 + \cdots + N_r + \cdots$$

$$= N_0 + N_0 \exp\left(-\frac{h\upsilon}{kT}\right) + N_0 \exp\left(-\frac{2\,h\upsilon}{kT}\right) + \cdots + N_0 \exp\left(-\frac{rh\upsilon}{kT}\right) + \cdots$$

$$N = \frac{N_0}{\left[1 - \exp\left(-\frac{h\upsilon}{kT}\right)\right]} \qquad \ldots(2.1.16)$$

as $(1-x)^{-1} = 1 + x + x^2 + \cdots$

Now the total energy E is given by

$$E = N_0 \times 0 + h\upsilon N_0 \exp\left[-\frac{h\upsilon}{kT}\right] + 2\,h\upsilon N_0 \exp\left[-\frac{2\,h\upsilon}{kT}\right]$$

$$+ \cdots rh\upsilon\, N_0 \exp\left[-\frac{rh\upsilon}{kT}\right] + \cdots$$

or,

$$E = N_0\, h\upsilon \exp\left(-\frac{h\upsilon}{kT}\right)\left[1 + 2\exp\left(-\frac{h\upsilon}{kT}\right) + 3\exp\left(-\frac{2h\upsilon}{kT}\right) + \cdots\right]$$

$$= N_0\, h\upsilon \exp\left(-\frac{h\upsilon}{kT}\right)\frac{1}{\left\{1 - \exp\left(-\frac{h\upsilon}{kT}\right)\right\}^2} \qquad \ldots(2.1.17)$$

as

$$1 + 2x + 3x^2 + \cdots rx^{r-1} + \cdots = \frac{1}{(1-x)^2}$$

Now the average energy of the oscillator is given by

$$\overline{E} = \frac{E}{N} = \frac{N_0\, h\upsilon \exp\left(-\frac{h\upsilon}{kT}\right)\left\{1 - \exp\left(-\frac{h\upsilon}{kT}\right)\right\}^{-2}}{N_0 \Big/ \left[1 - \exp\left(-\frac{h\upsilon}{kT}\right)\right]}$$

$$\overline{E} = \frac{h\upsilon \exp\left(-\frac{E}{kT}\right)}{\left\{1-\exp\left(-\frac{E}{kT}\right)\right\}} \quad \text{with } E = h\upsilon$$

$$\overline{E} = \frac{h\upsilon}{\left\{\exp\left(\frac{h\upsilon}{kT}\right)-1\right\}} \qquad ...(2.1.18)$$

Thus we see that the average energy of oscillator given by Eqn. (2.1.17) is different from the energy kT of a classical oscillator.

We know that the number of oscillators per unit volume in frequency range υ and $\upsilon + d\upsilon$ is given by

$$N = \frac{8\,\pi\upsilon^3 d\upsilon}{c^3} \qquad ...(2.1.19)$$

Multiplying Eqn.(2.1.19) with the average energy of oscillators, given by Eqn. (2.1.18), we get the total energy per unit volume belonging $d\upsilon$ or the energy density belonging to the range $d\upsilon$ as

$$E_\upsilon d\upsilon = \left(\frac{8\,\pi\upsilon^3}{c^3}\right)\left[\frac{h\upsilon}{\exp\left(\frac{h\upsilon}{kT}\right)-1}\right]d\upsilon \qquad ...(2.1.20)$$

This is the famous Planck's radiation law.

But $\upsilon = \frac{c}{\lambda}$ and $|\,d\upsilon\,| = -\frac{c}{\lambda^2}\,d\lambda$

i.e;

$$E_\lambda d\lambda = \frac{8\,\pi h}{c^3}\left(\frac{c^3}{\lambda^3}\right)\frac{1}{\left\{\exp\left(\frac{ch}{\lambda kT}\right)-1\right\}}\left(\frac{c}{\lambda^2}d\lambda\right)$$

Therefore

$$E_\lambda d\lambda = \left[\frac{8\,\pi hc}{\lambda^5}\right]\frac{1}{\left\{\exp\left(\frac{ch}{\lambda kT}\right)-1\right\}}d\lambda \qquad ...(2.1.21)$$

This equation of Planck well agrees with the experimental curves throughout the whole range of wavelengths.

With the help of Planck's radiation law, Wien's displacement law and Rayleigh-Jeans law can be derived in the following way:

Wien's Formula

For small temperatures, λT is small. Now exp $(ch/\lambda kT) >> 1$ and 1 can be neglected in the denominator or Eqn. (2.1.20)

$$E_\lambda d\lambda = \frac{8\pi ch}{\lambda^5}\left\{\exp\left(-\frac{ch}{\lambda kT}\right)\right\} d\lambda \quad ...(2.1.22)$$

which is Wien's law

Rayleigh-Jeans Law

For large temperatures λT is large

$$\text{Now} \quad \exp\left(\frac{ch}{\lambda kT}\right) = 1 + \frac{ch}{\lambda kT} + \frac{1}{\lfloor 2}\left(\frac{ch}{\lambda kT}\right)^2 + \cdots$$

$$= 1 + \frac{ch}{\lambda kT}$$

$$\text{Therefore} \quad E_\lambda d\lambda = 1 + \frac{ch}{\lambda kT}$$

$$\text{Thus} \quad E_\lambda d\lambda = \left(\frac{8\pi ch}{\lambda^5}\right)\frac{1}{\left\{\left(1+\frac{ch}{\lambda kT}\right)-1\right\}} d\lambda$$

$$\text{i.e.,} \quad E_\lambda d\lambda = \left(\frac{8\pi ch}{\lambda^5}\right)\times\frac{\lambda kT}{ch} d\lambda = \frac{8\pi kT}{\lambda^4} d\lambda \quad ...(2.1.23)$$

This is Rayleigh-Jeans law.

Here it should be clearly understood that the quantization of energy in the oscillators is not only confined to the atoms of black body but it is a general concept. Due to the smallness of Planck's constant, the possibility of the observation of this concept in oscillators, macroscopic dimensions such as simple pendulum, oscillatory spring, LCR circular etc is excluded.

Concept of Electron Gas and Fermi Energy

Let us discuss this small topic by considering the metallic crystal copper. Each copper atom contributes roughly one electron to the crystal as a whole. These free electrons, leaving the metal atoms in place ions, serve as a pervasive 'glue' moving among the ions and binding them together. Excellent thermal conductivity comes also from the free electron. Since they are used to behave as gas molecules, classical laws were applied and may results were obtained. Pity that the conclusion arrived at did not agree with the experimental results.

Hence it was later assumed as free electrons obeying Fermi-Dirac statistics; hence they are called Fermions. Now we disscuss the behaviour of the Fermions and arrive at the following results.

The free electron gas in a metal obeys Fermi-Dirac statistics. Suppose in an assemblage of fermions, there are $M(E)$ allowed quantum states in an energy range between E and $E + dE$ and $N(E)$ is the number of particles in the same range, then $N(E)$ states are filled and $M(E) - N(E)$ are vacant. The F-D distribution function $F(E)$. is defined as

$$F(E) = \frac{N(E)}{M(E)} = \frac{1}{[1+e^{(E-E_F)/KT}]}$$

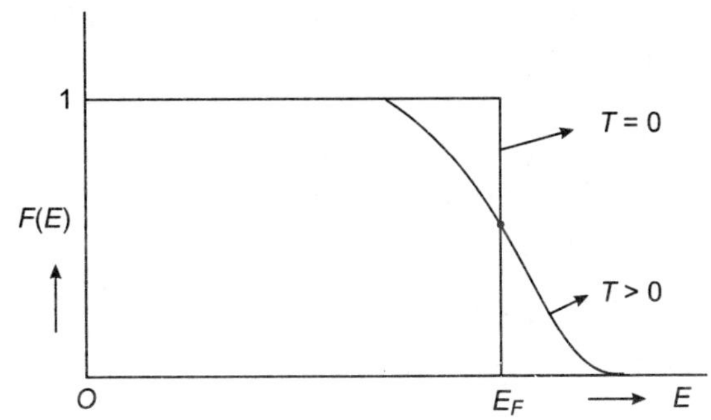

Fig. 2.5

$\frac{N(E)}{M(E)}$ represents the fraction of the possible quantum states which is occupied.

The distribution of electrons among the levels is usually described by the distribution function $F(E)$. It is defined as the probability that the level E is occupied by an electron. Thus if the level is certainly empty, then $F(E) = 0$. If the level is certainly full, then $F(E) = 1$. In general $F(E)$ has a value between zero and unity.

The distribution function for electrons at $T = 0$K has the form

$$F(E) = 1 \text{ when } E < E_F$$

$$F(E) = 0 \text{ when } E > E_F$$

That is, all levels below E_F are completely filled and all levels above E_F are completely empty. This function is plotted in Fig. 2. which show's the discontinuity at the Fermi-energy. As the temperature rises, $F(E)$ changes, from 1 to 0 more and more gradually as shown in the graph.

For $E = E_F$, $E_F = \frac{1}{1+e^0} = \frac{1}{2}$, at all temperatures. Thus the probability of finding an electron with energy equal to the Fermi energy in a metal is $\frac{1}{2}$ at any temperature. Since the electrons are confined inside the crystal, their wave properties will limit the energy values which they may have. Let *g(E) dF* be the number of quantum states available to electrons with energies between *E* and *dE*. It can be shown to be

$$g(E)\,dE = \frac{\pi}{2}\left[\frac{8m}{h^2}\right]^{3/2} V\,E^{1/2}dE \qquad ...(2.1.24)$$

Now

$$N = \int g(E)\,dE\,F(E) = \frac{\pi}{2}\left[\frac{8m}{h^2}\right]^{3/2} V\int_0^{E_F} E^{1/2}dE$$

with $E = E_F$ at 0K

g (E) dE is called the density of energy states. *i.e.*, number of energy states in a volume *V* lying between *Ed E + dE*

$$N = \frac{\pi}{2}\left(\frac{8m}{h^2}\right)^{3/2} V\frac{E_F^{3/2}}{3/2}$$

$$N = \frac{\pi}{3}\left(\frac{8m}{h^2}\right)^{3/2} VE_F^{3/2}$$

or,

$$E_F^{3/2} = \frac{\pi}{3}\left[\frac{8m}{h^2}\right]^{3/2}\frac{V}{N}$$

Since $\frac{V}{N} = \frac{1}{n}$ the density of free electrons, we write

$$E_F^{3/2} = \frac{\pi}{3n}\left[\frac{h^2}{8m}\right]^{3/2}$$

or,

$$E_F = \left(\frac{3n}{\pi}\right)^{2/3}\left[\frac{h^2}{8m}\right] \qquad ...(2.1.25)$$

An effective temperature of the electron gas, known as the Fermi temperature, defined by

$$T_F = \frac{E_F}{k}.$$

3.1 PRINCIPLES OF QUANTUM MECHANICS

3.1.1 INTRODUCTION

The experiment such as those in which e/m of the material particles (electrons, protons) is measured, indicate that the matter is corpuscular in nature. Up to 1924 the matter was considered to be completely corpuscular in nature, but in that year Louis de-Broglie proposed that a material particle such as an electron, proton, atom etc. might have a dual nature as light does.

It has just been argued that the conflict between wave and particle models for the electron is more apparent than real and that the two modes are equivalent. We therefore require some method of translating the terms of one model into the other. This is provided by just two equations. These are postulated rather than derived but they are found to relate correctly the wave characteristics to the particle characteristics under a given set of conditions.

3.1.2 WAVES AND PARTICLES AND PLANCK'S QUANTUM THEORY

According to corpuscular theory of light, light consists of minute fast moving elastic particles known as corpuscles. However, phenomena of interference, diffraction, pholarization etc. could not be explained on the basis of corpuscular theory. To explain these phenomena Huygen proposed the wave theory of light. According to this theory light travels in the form of waves. There are certain other experimentally observed phenomenon such as Compton effect, photoelectric effect, emission and absorption of light which could not be explined on the basis of above mentioned theories. These phenomena gave brith to the quantum theory of light.

In order to explain the shape of the black body radiation, Max Planck in 1900 proposed quantum theory. According to this theory, matter is composed of large number of oscillating particles which vibrate with different frequencies while according to classical theory the particles can have any value of frequency (or vibrational energy). Moreover, in quantum theory, the energy of oscillating particles is quantized. The oscillating particle cannot have any energy but only those energies given by

$$E = nh\nu$$

where Planck's constant h - 6.625×10^{34} J – s, ν the oscillator frequency and n is an integer such as 1, 2, 3, ⋯ . In this theory, it is lso assumed that the vibrating particle does not radiate energy continuously but discontinuously in terms of discrete quanta or photon.

3.1.3 DE BROGLIE THEORY OF MATTER WAVES

The first equation, due to de Broglie, relates wave length λ to momentum p:

$$\boxed{\lambda = \frac{h}{p}} \qquad \text{...(3.1.1)}$$

where h is Planck's constant. The second equation, referred to as the Einstein equation, relates energy E to frequency υ:

$$\boxed{\upsilon = \frac{E}{h}} \qquad ...(3.1.2)$$

If we take the *classical kinetic energy*

$$E = \frac{1}{2}mv^2$$

or $$p = mv = \sqrt{2mE}$$

From Eqn. (3.1.1)

$$\lambda = \frac{h}{p} = \frac{h}{\sqrt{2mE}} \text{ with } E = \text{eV}$$

$$\lambda = \frac{h}{\sqrt{2\text{meV}}} = \frac{6.62\times10^{-34}}{\sqrt{2\times9.1\times10^{-31}\times1.6\times10^{-19}\times V}}$$

$$= \frac{1.227}{\sqrt{V}}\times10^{-9}\text{ m}$$

$$\boxed{\lambda = \frac{1.227}{\sqrt{V}}\text{ nm}} \qquad ...(3.1.3)$$

If the electron is accelerated through very high potential difference, it is necessary to take into account of the *relativistic increase in mass of the electron*. The equation used in such a case is

$$\lambda = \frac{1.227}{\sqrt{V}}\left[1 - \frac{\text{eV}}{4m_0c^2}\right] \qquad ...(3.1.3a)$$

Equations (3.1.3) and (3.1.3*a*) give the relations for de Broglie's wave length in the two cases.

Davisson and Germer's Experiment

The electron beam from an electron gun is accelerated and collimated to strike a nickel crystal. C is Faraday ionisation chamber for receiving the electron after they have been scattered by a nickel crystal. The ionisation chamber can be moved along a graduated circular scale so that it is able to receive the scattered electrons at all angles between 20° and 90° and their intensity is measured by the galvanometer current. The whole assembly is placed in a very high vacuum. Graphs are drawn at various voltages and the pronounced maximum is obtained for 54 volt at $\phi = 50°$. See Fig. 3.1.

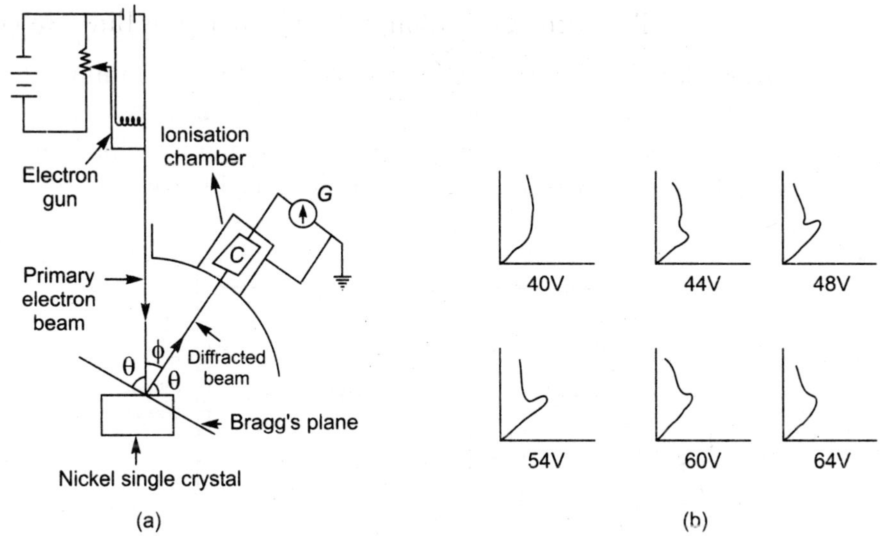

Fig. 3.1 Davisson-Germer experiment

$$\phi + \theta + \theta = 180°$$

$$2\theta = 180° - \phi = 180 - 50 = 130°$$

$$\theta = 65°$$

The *interplanar distance* for nickel is 0.1 nm.

Thus

$$2d \sin\theta = \lambda$$

$$\lambda = 2 \times 0.1 \times \sin 65° = 1.067 \text{ nm}$$

By de Broglie's hypothesis

$$\lambda = \frac{0.1227}{\sqrt{54}} \text{nm} = 0.167 \text{ nm}$$

Thus we see 100% agreement in the value.

G. P. Thomson's Experiments

In 1928 G.P. Thomson made investigations with high speed electrons, accelerated by a potential difference ranging from 10000 to 50000 volt, using a method analogom to Debye-Scherrer powder method of X-ray analysis of crystals. The high energy electron beam is produced by the cathode *C* which can be excited up to 50000 volt. The rays are passed through a diaphram A to obtain a fine beam of electrons which are then allowed to fall upon a very thin metallic film *F* of gold or aluminium foil. The thickness of the foil is of the order 1 Å. The photograph of the incoming beam from the foil is recorded by the photographic plate *P*. The whole apparatus is exhausted to a high vacuum so that the electrons may not loss their energy in collisions with the molecules of the gas. In the actual experiment instead of continuous patch, the distinct circular sharp rings are observed as in X-ray diffraction experiment. The crystals of the metal (gold or aluminium) which are at correct angle, scatter the electrons in accordance with the Bragg's law.

This experiment clearly demonstrates in a very striking manner that the electrons behave as waves since diffraction patterns can only be produced by waves.

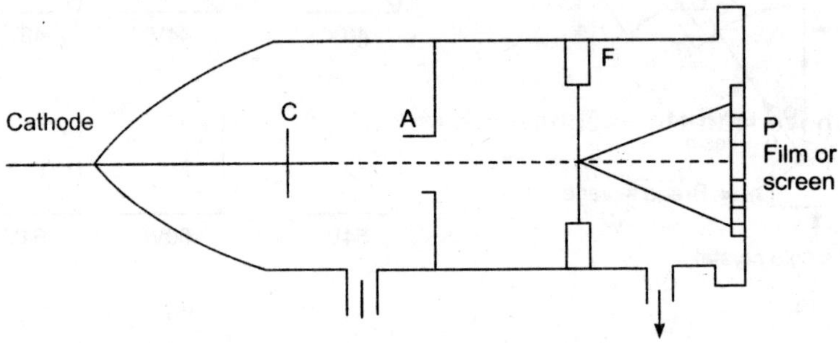

Fig. 3.2 G.P. Thomson's experiment

3.1.4 UNCERTAINTY PRINCIPLE

The *phase velocity* of a wave associated with a particle comes out to be greater than the velocity of light. This difficulty is resolved by postulating that the moving particle is associated with a 'wave packet' or 'wave group', rather than a single wave-train. Electrons possessing particle nature (deflected by magnetic and electric fields) can behave as a wave and the wavelength is given by de Broglie's equation. When such a wave extends through space and it is difficult to locate the exact position of the electron behaving as wave at any given time, because although the particle is somewhere within the wave packet moving with the group velocity, it is impossible to know where exactly the particle is and what exactly is its velocity at that moment. By making the wave packet very small, the position of the particle can be more or less fixed, but since it can be shown that the velocity spread of such a packet is large, the velocity of the particle becomes indeterminate. On the other hand if we consider large wave packet with many crests, so that the velocity spread is very small and the velocity can be determined fairly accurately, but the position of the particle becomes very uncertain. Extending these results to the limiting cases of infinitely large and infinitely small wave packets it is readily seen that certainly about velocity involves complete uncertainty about position and vice-versa. *Werner Heisenberg*, who was the first to release these consequences of wave mechanics, has expressed them mathematically by means of the following equation known as the *uncertainty principle* $\Delta x . \Delta p \geq h$ where Δx and Δp are the uncertainties involved in the measurement position and momentum of the particle and h is the Planck's constant. According to Heisenberg, it is impossible to measure the momentum and the position of the particle simultaneously with the wave packet.

We know, *group velocity* of the wave packet is given by

$$u = \frac{\Delta \omega}{\Delta k} \qquad \text{...(3.1.4)}$$

But $\boxed{\omega = 2\pi\upsilon}$ ...(3.1.5)

or $\Delta\omega = 2\pi\Delta\upsilon$

and also $k = \dfrac{2\pi}{\lambda}$

Since $\lambda = \dfrac{h}{p};\ k = \dfrac{2\pi\, p}{h}$

Thus $\boxed{\Delta k = \dfrac{2\pi}{h}\Delta p}$ ...(3.1.6)

Using the above said three equations, we get

$$u = \frac{2\pi\,\Delta\upsilon \times h}{2\pi\,\Delta p}$$

$$\boxed{u = \frac{h\,\Delta\upsilon}{\Delta p}} \quad ...(3.1.7)$$

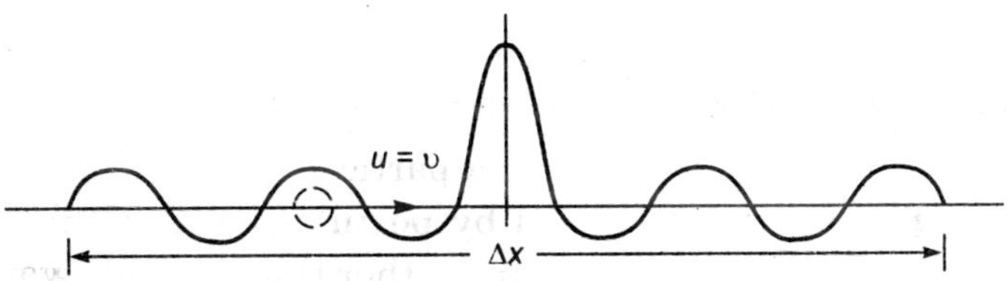

Fig. 3.3 A wide wave group–wavelength can be measured accurately but not the position.

Again $\boxed{u = \dfrac{\Delta x}{\Delta t}}$ ...(3.1.8)

Here Δt is the least time required for one complete wave packet to cross the reference point. The frequency $\Delta\upsilon$ is related to Δt by relation.

$$\boxed{\Delta t \geq \frac{1}{\Delta\upsilon}} \quad ...(3.1.9)$$

From Eqns. (3.1.7) and (3.1.9), we get,

$$\frac{\Delta x}{\Delta t} = \frac{h\Delta\upsilon}{\Delta p}$$

$$\boxed{\Delta x.\Delta p = h(\Delta t)(\Delta\upsilon)} \quad ...(3.1.10)$$

But from Eqn. (3.1.9) $\Delta\upsilon.\Delta t \geq 1$

Substituting for Δt from Eqn. (3.1.9) and Eqn. (3.1.10),

$$\boxed{\Delta x.\Delta p = h} \quad ...(3.1.11)$$

A more sophisticated derivation gives

$$\boxed{\Delta x . \Delta p = \frac{h}{2\pi}} \quad ...(3.1.12)$$

One should know the uncertainty principle and related equations have no relevance in macroscopic particles as h is extremely small.

The Planck's equation relating energy and frequency is

$$E = h\upsilon; \ \upsilon = \frac{E}{h}$$

Equation 3.1.9 is

$$\Delta\upsilon . \Delta t \geq 1$$

$$\Delta\left(\frac{E}{h}\right) \cdot \Delta t \geq 1$$

or

$$\boxed{\Delta E . \ \Delta t \geq \frac{h}{2\pi}} \quad ...(3.1.13)$$

This is the uncertainty relation between energy and time.

One Illustration

(Study of position of electron)

Suppose we try to measure the position and linear momentum of an electron using an imaginary microscope with a very high resolving power. The electron can be observed if at least one photon is scattered by it into the microscope lens. The *resolving power of the microscope* is given by the relation.

$$\Delta x = \frac{\lambda}{2 \sin \theta}$$

where Δx is the distance between two points which can be just resolved by microscope. This is the range in which the electron would be visible when disturbed by the photon. Hence Δx is the uncertainty involved in the position of the electron.

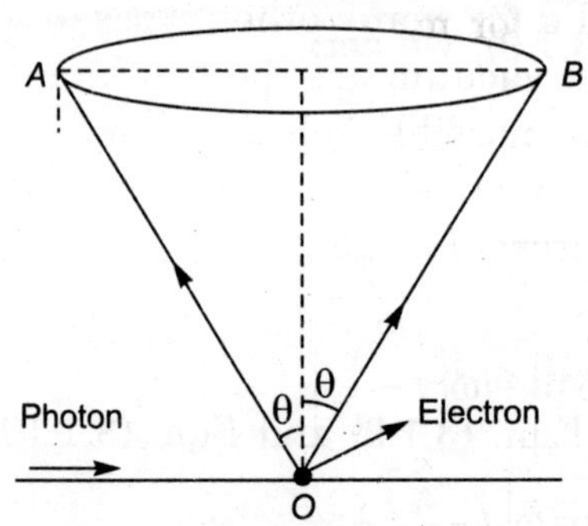

Fig. 3.4 Position of electron – uncertainty principle

However, the incoming photon will interact with the electron through the compton effect. To be able to see this electron, the scattered photon should enter the microscope within the angle 2θ. The momentum imparted by the photon to the electron during the impact is of the order of $\frac{h}{\lambda}$. The component of this momentum along OA is $-\frac{h}{\lambda}\sin\theta$ and along OB is $\frac{h}{\lambda}\sin\theta$.

Hence, the uncertainty in the momentum measurement in the x-direction is

$$\Delta p_x = \frac{h}{\lambda}\sin\theta - \left(-\frac{h}{\lambda}\sin\theta\right) = \frac{2h}{\lambda}\sin\theta$$

$$\therefore \quad \Delta x \times \Delta p_x = \frac{\lambda}{2\sin\theta}\left[\frac{2h\sin\theta}{\lambda}\right] = h$$

$$\boxed{\Delta x \times \Delta p_x = h}$$

3.1.5 WAVE MECHANICS–SCHRÖDINGER WAVE EQUATION

While the quantum theory of Planck, as extended to atomic structure by Bohr and others, have proved to be of utmost importance in the analysis of atomic spectra, some other methods of presenting atomic processes have been developed which are still in better agreement with the experimental facts. In this regard, E. Schrödinger, in 1926, directly started with Louis de Broglie's concept of matter waves and developed into a rigorous mathematical theory known as the 'wave mechanics'. The essential feature of this theory is the incorporation of expression for the de Broglie's wavelength into the general classical wave equation. As such a wave equation for a moving particle is derived which is called Schrödinger wave equation.

The partial differential equation of a wave motion is

Now $$\boxed{\frac{\partial^2 y}{\partial x^2} = \frac{1}{v^2}\frac{\partial^2 y}{\partial t^2}} \quad \text{...(3.1.14)}$$

where v is the velocity of the wave.

This wave equation is valid for many other types of waves also. In practice, sound waves in air and light waves in vacuum are the two important types of waves that obey Eqn. (3.1.14). Of the numerous possible solutions for the above equation, the solution corresponding to the case of undamped, monochromatic harmonic waves in the positive x-direction, is of particular interest. The general solution in this case is of the form

$$\boxed{\begin{aligned} y &= A\exp\left[-i\omega\left(t-\frac{x}{v}\right)\right] \\ y &= A\cos\omega\left[t-\frac{x}{v}\right] - iA\sin\omega\left[t-\frac{x}{v}\right] \end{aligned}} \quad \text{...(3.1.15)}$$

Only the real part of Eqn. (3.1.15) has some significance in the case of waves in stretched string, where y represents the displacement of the string from its normal position; in this case the imaginary part is discarded as irrelevant. This, however, cannot be directly followed in the case of material particles, because the wave function Ψ of de Broglie, unlike y, is not itself a measurable quantity and may, therefore, be complex.

Thus $$\Psi(x,t) = A\exp\left[-i\omega\left\{t-\frac{x}{v}\right\}\right]$$

$$= A\exp\left[-2\pi i\left\{\upsilon t-\frac{\upsilon x}{v}\right\}\right]$$

Using $E = mv^2 = h\upsilon$ or $\frac{E}{h} = \upsilon$ and $\frac{\upsilon}{v} = \frac{p}{h}$

Let $\hbar = 2\pi h$. The above equation becomes

$$\boxed{\Psi(x,t) = A\exp\left[-\frac{i}{\hbar}(Et-px)\right]} \qquad ...(3.1.16)$$

Now $$\frac{\partial\Psi}{\partial x} = A\exp\left[-\frac{i}{\hbar}(Et-px)\right]\left(\frac{ip}{\hbar}\right)$$

$$\frac{\partial^2\Psi}{\partial x^2} = A\exp\left[-\frac{i}{\hbar}(Et-px)\right]\left(\frac{i^2p^2}{\hbar^2}\right) = -\frac{p^2\Psi}{\hbar^2}$$

or $$-\hbar^2\frac{\partial^2\Psi}{\partial x^2} = p^2\Psi$$

$$\frac{\partial\Psi}{\partial t} = A\exp\left[-\frac{i}{\hbar}(Et-px)\right]\left(-\frac{iE}{\hbar}\right) = -\frac{iE\Psi}{\hbar}$$

$$-\frac{\hbar}{i}\frac{\partial\Psi}{\partial t} = E\Psi \qquad ...(3.1.16a)$$

The general equation for total energy is

$$E = \frac{1}{2}mv^2 + V(x);\ \text{or}\ E\Psi = \frac{p^2\Psi}{2m} + V\Psi$$

Substituting the above said values in Eqn. 3.1.16a, we get

$$-\frac{\hbar}{i}\frac{\partial\Psi}{\partial t} = -\frac{\hbar^2}{2m}\frac{\partial^2\Psi}{\partial x^2} + V\Psi$$

or $$\boxed{\frac{\hbar}{i}\frac{\partial\Psi}{\partial t} = \left[\frac{\hbar^2}{2m}\right]\left[\frac{\partial^2\Psi}{\partial x^2}\right] - V\Psi} \qquad ...(3.1.17)$$

This is the one dimensional Schrödinger time-dependent wave equation.

3.1.6 PHYSICAL SIGNIFICANCE OF THE WAVE FUNCTION

(1) The wave function $\Psi(x, t)$ is used to describe the behaviour of the system and, mathematically, $\Psi(x, t)$ can be a complex quantity.

(2) Max born postulated in 1926 that function $|\Psi(x,t)|^2 dx$ is the probability of finding the particle between x and $x + dx$ at a given time, or that $|\Psi(x,t)|^2$ is a probability density function. This is the function that has a physical meaning. The probability density $P = |\Psi|^2 = \Psi\Psi^*$, where ψ^* is the complex conjugate of Ψ.

(3) Any wave function can be represented as

$$\Psi(x,t) = \psi(x)\phi(t) = \psi\exp(i\omega t)$$

where $\psi(x)$ in independent of time and $\phi(t)$ is independent of x, the probability density can be expressed as

$$|\Psi(x,t)|^2 = |\psi(x)|^2 \exp(i\omega t)\exp(-i\omega t)$$

and is therefore independent of time. The time independent Schrödinger equation can be written as $\frac{d^2\psi}{dx^2} + \frac{2m}{\hbar^2}(E-V)\psi = 0$. Since the function $|\psi(x)|^2$ represents the probability density function, then for a single particle, we must have that

$$\boxed{\int_{-\infty}^{\infty} |\psi(x)|^2 dx = 1} \qquad \text{...(3.1.18)}$$

This relation is called normalization condition.

3.1.7 SCHRÖDINGER'S EQUATION (Steady-state Equation)

In a great many situations the potential energy of a particle does not depend upon time explicity. In other words the Schrödinger equation may be simplified by removing all reference to time t.

Look back Eqn. (3.1.16)

$$\psi = A\exp\left[-\frac{i}{\hbar}(Et - px)\right]$$

$$\psi = A\exp\left[-\frac{iEt}{\hbar}\right]\exp\left(\frac{ipx}{\hbar}\right)$$

$$\psi = A\psi\exp\left[-\frac{iEt}{\hbar}\right]$$

Now $$\frac{\partial\Psi}{\partial t} = A\psi\exp\left[-\frac{iEt}{\hbar}\right]\left(-\frac{iE}{\hbar}\right) = -\frac{AiE\psi}{\hbar}\exp\left(-\frac{iEt}{\hbar}\right)$$

and $$\frac{\partial^2\Psi}{\partial x^2} = A\frac{d^2\psi}{dx^2}\exp\left(-\frac{iEt}{\hbar}\right)$$

Substituting the values $\frac{\partial^2\Psi}{\partial x^2}$ and $\frac{\partial\Psi}{\partial t}$ in Eqn. (3.1.17) and simplifying, we get

$$-\left(\frac{\hbar}{i}\right)\left(\frac{iE\psi}{\hbar}\right) = \frac{\hbar^2}{2m}\frac{d^2\psi}{dx^2} - V\psi$$

$$-E\psi = \frac{\hbar^2}{2m}\frac{d^2\psi}{dx^2} - V\psi$$

$$\boxed{\frac{d^2\psi}{dx^2} + \left[\frac{2m}{\hbar^2}\right](E - V)\psi = 0} \qquad ...(3.1.19)$$

Here ψ is a function of x alone and is independent of time. Equation (3.1.19) is the steady-state form of Schrödinger equation. In three dimensional case, the related equation is:

$$\frac{\partial^2\psi}{\partial x^2} + \frac{\partial^2\psi}{\partial y^2} + \frac{\partial^2\psi}{\partial z^2} + \left[\frac{2m}{\hbar^2}\right](E - V)\psi = 0$$

$$\Delta^2\psi + \frac{2m}{\hbar^2}(E - V)\psi = 0 \qquad ...(3.1.20)$$

Here ψ is a function of (x, y, z), but independent of time.

Particle in a one-dimensional box

To apply Schrödinger equation to a particular problem, we substitute appropriate value of potential energy V and then solve the equation under given boundary conditions. Let us consider the problem of a particle in a one-dimensional box. If a particle is within a one dimensional box extending from $x = 0$ to $x = a$, then it is assumed that $V = 0$ everywhere inside the box and ∞ elsewhere.

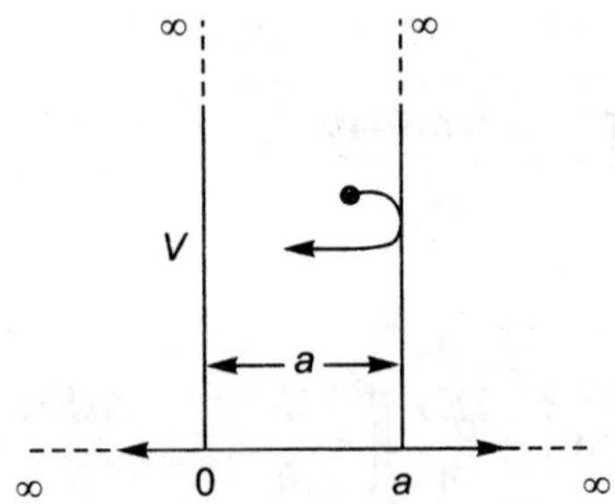

Fig. 3.5 Particle confined to a one dimensional box of length, a

Now the Schrödinger wave equation (see Eqn. 3.1.19) for this particle in the said box of parameter a is

$$\frac{d^2\psi}{dx^2} + \frac{2m}{\hbar^2} E\psi = 0$$

Substituting $2\pi\hbar = h$, then

$$\frac{d^2\psi}{dx^2} + \left[\frac{8\pi^2 mE}{h^2}\right]\psi = 0 \qquad ...(3.1.21)$$

or
$$\frac{d^2\psi}{dx^2} + k^2\psi = 0 \text{ with } k^2 = \frac{8\pi^2 mE}{h^2}$$

The general solution of this differential equation is of the form

$$\psi(x) = A\sin kx + B\cos kx \qquad ...(3.1.22)$$

This we can verify by substituting back into the differential equation.

To eliminate the arbitrary constants A & B, we shall apply the boundary conditions:

(*i*) when $x = 0$, $\psi = 0$ and hence $B = 0$

(*ii*) when $x = a$, $\psi = 0 = A \sin ka$

There will be no solution if $A = 0$ and hence ka must be equal to $n\pi$; or $\frac{n\pi}{a} = k$

Let us substitute this in Eqn. 3.1.22:

$$\psi_n(x) = A\sin\frac{n\pi x}{a}$$

If $dP = |\psi|^2 dx$ is the probability of finding the particle between x and $x + dx$, then the probability of finding the particle within the box is

$$P = \int_0^a |\psi|^2 dx = 1$$

i.e.,
$$A^2\int_0^a \sin^2\frac{n\pi x}{a} dx = 1$$

$$A^2\int_0^a \frac{1}{2}\left[1 - \cos\frac{2n\pi x}{a}\right]dx = 1$$

$$\frac{A^2}{2}\left[x - \frac{a}{2\pi n}\sin\frac{2n\pi x}{a}\right]_0^a = 1$$

$$\frac{A^2 a}{2} = 1; \; A = \sqrt{\frac{2}{a}}$$

Thus the wave function,

$$\boxed{\psi_n = \left[\sqrt{\frac{2}{a}}\right] \sin \frac{n\pi x}{a}} \quad \text{...(3.1.23)}$$

Refer again Eqn. 3.1.21

$$k^2 = \frac{8\pi^2 \text{mE}}{h^2}; \; E = \frac{k^2 h^2}{8\pi^2 \text{m}} \quad \text{with } k = \frac{n\pi}{a}$$

Thus

$$\boxed{E = \frac{n^2 \pi^2 h^2}{8\pi^2 \text{ma}^2}}$$

or

$$E_n = \frac{n^2 h^2}{8\text{ma}^2} \quad \text{...(3.1.24)}$$

with $n = 1, 2, 3, ...$

The above equation indicates that the particle can possess only a discrete set of values of energy, or the energy of the particle is quantized. The permitted values of energy are called eigen values.

The first three eigen functions ψ_1, ψ_2, ψ_3 together with the probability densities $|\psi_1|^2$ ... are shown in the following graphs for the benefit of the readers.

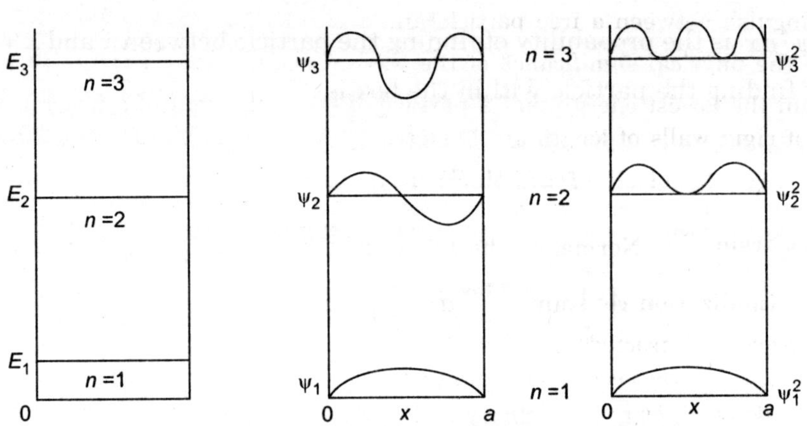

Fig. 3.5 Eigen functions and probability densities

SHORT QUESTIONS

3.1.1 What are matter waves? Write down the equation for the de Broglie wave length of an electron accelerated by a p.d of V volt.

3.1.2* A proton and an α-particle has the same kinetic energy. If the mass of the α-particle is four times that of a proton, how do their de Broglie wavelengths compare?

(R E C, Jaipur, B. Tech. December 2000)

3.1.3 Get the expression for the wave length of thermal neutron at T degree kelvin.

(M.U., BE, November 1990)

3.1.4 If $\psi = (A + i\beta)$, get $\psi\psi^*$. Also give the physical significance of the wave function ψ.

(University of Surrey, UK, May 1997)

3.1.5 Show that the spacing between the nth energy level and the next higher level in a one dimensional potential box increases by $(2n + 1)$. *(Anna. U., BE, May 2000)*

3.1.6 Let λ_1 be the de Broglie's wavelength associated with electrons accelerated by a potential of 36 volt. What is the p.d required to have a wavelength $2\lambda_1$?

(M.K.U., BE, May 1997)

3.1.7* Can matter waves travel faster than light? Justify your answer.

(IIT Bombay, B.Tech, 1989)

3.1.8 Explain why the waves nature of matter is not apparent in our daily life.

3.1.9 Give the basic principle of Davission-Germer experiment. At what voltage, they operated the cathode rays?

3.1.10 Explain Heisenberg's principle of uncertainty.

3.1.11 Do electrons exist inside the nucleus? Explain.

3.1.12 Distinguish between a free particle and a particle in a box.

3.1.13 Give the physical significance of the wave function.

3.1.14 Obtain the lowest energy that a particle of mass half that of a neutron can have in a box of rigid walls of length 10^{-14} metre. *(JNTU., B. Tech, November, 1998)*

3.1.15 The wave function of an electron in a one dimensional box of length a is

$\psi(x) = A\sin\frac{n\pi x}{a}$. Normalize the function and get the magnitude and dimension of

the normalization constant.

3.1.16* A particle is diffracted at a certain angle and the uncertainty in the measurement of the angle is one second. Estimate the error in the measurement of its angular momentum. *(M. U., BE, April 1988)*

3.1.17* Show that the kinetic energy associated with a plane electron wave is $\frac{h^2k^2}{8\pi^2 m}$.

Answers to the starred questions

3.1.2* $\frac{1}{2}mv^2 = E$; $v = \frac{2E}{m}$; $\lambda = \frac{h}{mv} = \frac{h\sqrt{m}}{m\sqrt{2E}}$

$$\lambda = \frac{h}{\sqrt{2Em}} \text{ or } \lambda_p = \frac{h}{\sqrt{2Em_p}}$$

Hence $\lambda_\alpha = \frac{h}{\sqrt{2E4m_p}}$

$$\frac{\lambda_\alpha}{\lambda_p} = \frac{1}{2}$$

$$\boxed{\lambda_p = 2\lambda_\alpha}$$ ***Answer***

3.1.7* Yes. Consider a body of rest mass m_0 moving with a velocity v. The angular frequency of the de Broglie wave associated with this body is

$$\omega = 2\pi\upsilon = 2\pi\frac{mc^2}{h} = \frac{2\pi}{h}\left[\frac{m_0c^2}{\sqrt{1-(v^2/c^2)}}\right]$$

The wave number is given by

$$k = \frac{2\pi}{\lambda} = \frac{2\pi mv}{h} = \frac{2\pi v m_0}{h\sqrt{1-(v^2/c^2)}}$$

Both ω and k are functions of velocity v,
The phase velocity w is given by

$$w = \frac{\omega}{k} = \frac{2\pi m_0c^2h}{h2\pi v m_0}$$

$$\boxed{w = \frac{\omega}{k} = \frac{c^2}{v}}$$ ***Answer***

The phase velocity thus exceeds both the velocity of the body and the velocity of light.

3.1.16*

180 × 60 × 60 sec corresponds to π radian

1 sec corresponds $\frac{\pi}{3600 \times 180}$ radian.

$$(\Delta L)(\Delta\Phi) = \frac{h}{2\pi}$$

$$\Delta L = \frac{h}{2\pi}\frac{1}{\Delta\Phi} = \frac{6.62\times10^{-34}\times3600\times180}{2\pi\times\pi}$$

$$\Delta L = 2.18\times10^{-29}\ \text{J s}$$

$$\boxed{\Delta L = 2.18\times10^{-29}\ J\,s}$$ ***Answer***

3.1.17* $\lambda = \frac{h}{mv}$ or $v = \frac{h}{m\lambda}$ and

$$\frac{1}{2}mv^2 = \frac{1}{2}m\frac{h^2}{m^2\lambda^2} = \frac{h^2}{2m\lambda^2}$$

But $\lambda = \frac{2\pi}{k}$

$$\frac{1}{2}mv^2 = \frac{h^2k^2}{2m4\pi^2}$$

$$\boxed{E = \frac{h^2k^2}{8\pi^2 m}}$$ ***Answer***

REVIEW QUESTIONS

2.1.1 State the basic assumptions on which Maxwell-Boltzmann, Bose-Finsteim and Fermi-Dirac statistics are developed. Give a comparison of the three statistics with suitable graphs. To which type of particles are they applicable?

2.1.2 Derive an expression for Fermi energy discussing the density of energy states.

2.1.3 Discuss black body radiation. Obtain Planck's radiation formula.

2.1.4 Describe the energy distribution in black-body spectrum. Show that the average energy of a Planck's oscillator is given by

$$\overline{E} = \frac{h\upsilon}{[\exp(h\upsilon/KT)-1]}$$

2.1.5 Derive the Bose-Einstein distribution law. How is could be used to obtain the Planck's formula for the black-body radiation.

2.1.6 Discuss Maxwell Boltzmann distribution and obtain the relation giving the distribution and obtain the relation giving the distribution of molecules over the varions energy states.

3.1.1 Give an account of Heisenberg's uncertainty principle. Outline an idealized experiment to bring out its significance. *(M.K.U., BE, April 1989)*

3.1.2 Describe Davisson and Germer experiments for the study of electron diffraction. Discuss the results obtained.

3.1.3 Show that the de Broglie wavelength of a beam of electrons accelerated by V volt is given by $\dfrac{h}{\sqrt{2mVe}}$. Discuss the salient features of the results of de Broglie's work. *(JNTU, BE, April 2001)*

3.1.4 Set up the Schrödinger equation in the time dependent form. Use it to arrive the eigen values of ψ and E for particles in a cubical box of side a. *(Anna. U., BE, May 2004)*

3.1.5 Discuss Schrödinger's time-independent wave equation for particles in a rectangular box. Discuss the results obtained. *(Anna. U., BE, May 2003)*

3.1.6 Formulate Schrödinger's time dependent and time independent wave equation. Briefly discuss the postulates made to get the above equation. *(NIT, Calicut April 2004)*

3.1.7 Derive time independent Schrödinger wave equation. An electron is bound in a one dimensional box of parameter 0.4 nm. What is its minimum energy? *(JNTU., BE, April, 2004)*

3.1.8 Obtain the time independent Schrödinger wave equation for a particle in a one dimensional potential well and infinite height. Discuss the solutions *(VTU April 2003)*

3.1.9 Write down the Schrödinger wave equation for a particle of mass m in an infinitely deep potential well with $V(x) = 0$ for $0 \le x \le a$ and $V(x) = \infty$ otherwise. Hence obtain the time independent equation *(Anna U., M. Sc, April, 1997)*

3.1.10 Solve the Schrödinger wave equation for a particle confined to move in a box of sides a, b, c along x, y and z directions respectively. Obtain the eigen function and eigen values of the system. Discuss the cases in which the energy levels become degenerate. What do you mean by degree of degeneracy? *(Anna U., Five Year BE, November, 1980)*

3.1.11 Derive time independent Schrödinger wave equation for a free particle in a one dimensional box. Explain the physical significance of the wave function.

An electron is bound in a one dimensional infinite well of width 0.1 nm. Find the difference in the energy values when the electron is in the ground state and third excited state. *(JNTU, BE, November 2002)*

PROBLEMS AND SOLUTIONS

2.1.1 Deduce Wien's displacement law from Planck's formula.

Solution: Planck's radiation relation is

$$E_\lambda d\lambda = \left[\frac{8\pi hc}{\lambda^5}\right]\frac{d\lambda}{[e^{ch/\lambda kT} - 1]}$$

To find the wavelength at which the spectral radiation is maximum, we put

$$\frac{dE_\lambda}{d\lambda} = 0$$

Thus, $E_\lambda = 8\pi ch\,(\lambda^{-5})\left(e^{ch/\lambda kT} - 1\right)^{-1}$

muct be differentiated.

i.e
$$8\pi ch\left[\left(-5\lambda^{-6}\right)\left(e^{ch/\lambda kT} - 1\right)^{-1} + \left(-\lambda^{-5}\right)\left(e^{\frac{ch}{\lambda kT}} - 1\right)^{-2}\left(-\frac{ch}{\lambda^2 kT}\right)\right] = 0$$

$$\frac{5}{\lambda} = \frac{ch}{\lambda^2 kT}\frac{e^{ch/\lambda kT}}{\left(e^{ch/\lambda kT} - 1\right)}$$

Putting $\frac{hc}{\lambda kT} = x$, we get

$$5 = \frac{x\,e^x}{e^x - 1} = \frac{x}{(1 - e^{-x})}$$

$$5 - 5e^{-x} = x$$

$$\frac{x}{5} + e^{-x} = 1$$

This equation has a single root given by x = 4·965, and therefore x must be a constant

i.e $\frac{hc}{\lambda kT} = 4 \cdot 965$

Therefore, the wavelength λ_m at which the spectral radiancy per unit range of wavelength has its maximum value is given by

$$\lambda_m T = \frac{ch}{4 \cdot 965} = b$$ ***Answer***

This is wien's displacement law.

2.1.2 Evaluate the surface temperature of the star which is radiating 1· 6 × 10^6 times more energy per unit area per second compared to the sun. Surface temperature of the sun is 6000 K.

Solution:

Let T be the surface temperature of the star. By Stefan's law

$$1 \cdot 6 \times 10^6 = \frac{T^4}{(6000)^4}$$

T = 120000 K. ***Answer***

2.1.3 Calculate the number of photons emitted in 5 hours by a 60 W sodium lamp. (λ = 589·3 mm)

Solution:

Energy of a photon

$$= h\upsilon = \frac{6\cdot62\times10^{-34}\times3\times10^{8}}{589\cdot3\times10^{-9}} \text{ Joule}$$

$$= 3\cdot374\times10^{-19} J$$

Number of photons emitted by the sodium lamp in one sec

$$= \frac{60}{3\cdot374\times10^{-19}}$$

$$= \frac{h\upsilon}{\left\{\exp\left(\frac{h\upsilon}{kT}-1\right)-1\right\}}$$

2.1.4 Calculate the average energy of an oscillator of frequency 5 · 6 × 10 [12] per second at 330 K rating in as (i) classical oscillator (ii) Planck's oscillator.

Solution:

(*i*) Average energy $\overline{E}$ of a classical oscillator is given by

$$\overline{E} = kT = 1\cdot38\times10^{-23}\times330$$

$$\overline{E} = 4\cdot554\times10^{-21} J. \quad \textbf{\textit{Answer}}$$

(*ii*)

$$\overline{E} = \frac{h\upsilon}{\left\{\exp\left(\frac{h\upsilon}{kT}-1\right)-1\right\}}$$

$$= \frac{6\cdot62\times10^{-34}\times5\cdot6\times10^{12}}{\exp\left(\frac{6\cdot62\times10^{-34}\times5\cdot6\times10^{12}}{1\cdot38\times10^{-23}\times330}-1\right)}$$

$$\overline{E} = 2\cdot945\times10^{-12} J. \quad \textbf{\textit{Answer}}$$

3.1.1 (*a*) Two low speed particles with masses m_1 and m_2 have the same kinetic energy. What is the ratio of their wavelengths?

(*b*) An electron and a proton have the same de Broglie wavelength. Which particle is moving faster?

(*c*) Calculate the de Broglie wavelength of electron which has kinetic energy equal to 15 eV.

(Meerut U., BE, April 2000)

Solution:

(*a*) $\lambda_1 = \frac{h}{\sqrt{2m_1E}}$ and $\lambda_2 = \frac{h}{\sqrt{2m_2E}}$

$$\boxed{\frac{\lambda_1}{\lambda_2} = \sqrt{\frac{m_2}{m_1}}}$$ ***Answer***

(b) $\lambda_e = \dfrac{h}{m_1 v_1} = \dfrac{h}{m_2 v_2} = \lambda_p$

Electron is faster ***Answer***

(c) $\lambda = \dfrac{h}{\sqrt{2mE}} = \dfrac{6.62\times10^{-34}}{\sqrt{2\times9.1\times10^{-31}\times15\times1.6\times10^{-19}}}$

$$\boxed{\lambda = 0.316 \text{ nm}}$$ ***Answer***

3.1.2 Compare the uncertainties in the velocities of an electron and proton confined in 1 nm box. *(M.K.U. BE, April 1991)*

Solution:

$$(\Delta x)_0 = (\Delta x)_p = 1\times10^{-9} \text{ m}$$

Uncertainties in the velocity of electron and proton are

$$m_e(\Delta v)_e = \frac{h}{(\Delta x)_e} \text{ and } m_p(\Delta v)_p = \frac{h}{(\Delta x)_p}$$

$$\frac{m_e(\Delta v)_e}{m_p(\Delta v)_p} = \frac{h\times(\Delta x)_p}{(\Delta x)_e\times h} = 1 \text{ as } (\Delta x)_p = (\Delta x)_e$$

$$\frac{(\Delta v)_e}{(\Delta v)_p} = \frac{m_p}{m_e} = \frac{1.67\times10^{-27}}{9.1\times10^{-31}} = 1.77\times10^3$$

i.e., $$\boxed{\frac{(\Delta v)_e}{(\Delta v)_p} = 1.77\times10^3}$$ ***Answer***

3.1.3 A particle is moving in a one-dimensional potential box (of infinite height) of width 1 nm. Calculate the probability of finding the electron within an interval of 0.1 nm at the centre of the box when it is in its state of the least energy. *(The Open University, UK, B. Tech. programme 2000)*

Solution:

The wave function of a particle of mass m in a potential box of width a is given by

$$\psi = \left[\sqrt{\frac{2}{a}}\right]\sin\frac{n\pi x}{a}$$

i.e., $$\psi = \left[\sqrt{\frac{2}{a}}\right]\sin\frac{\pi x}{a} \text{ with } n = 1$$

At the centre of the box we have $x = \frac{a}{2}$. The probability of finding the particle in unit interval at the centre of the box is given by

$$W = \psi^2 = \left(\sqrt{\frac{2}{a}}\sin\frac{\pi a/2}{a}\right)^2$$

$$= \frac{2}{a}\sin^2\frac{\pi}{2} = \frac{2}{a}$$

The probability of finding the particle in an interval of Δx at the centre of the box is

$$W = \frac{2}{a}\Delta x$$

$$W = \frac{2}{1}\times 0.1 = 0.2$$

$$\boxed{W = 0.2}$$ ***Answer***

3.1.4 The wave function ψ of a particle is given by

$\psi = A\exp(-kx)$ for $0 < x < \infty$ and $\psi = 0$ for $-\infty < x < \infty$

Obtain the value of A and the probability of finding the particle in the region $\frac{2}{k} < x < \frac{3}{k}$. *(Anna U., BE, May 1989)*

Solution:

$$\int_0^\infty |\psi|^2 dx = 1 \text{ or } \int_0^\infty A^2\exp(-2kx)dx = 1$$

$$A^2\left[\frac{\exp(-2kx)}{(-2kx)}\right]_0^\infty = 1$$

i.e.,

$$-\frac{A^2}{2k}(0-1) = 1$$

$$\boxed{A = \sqrt{2k}}$$ ***Answer***

Finding the probability in the given limits

$$\int_{2/k}^{3/k} A^2\exp(-2kx)dx = 2k\left[\frac{\exp(-2kx)}{(-2k)}\right]_{2/k}^{3/k}$$

$$= -[\exp(-6) - \exp(-4)]$$

$$\boxed{P = \left(\frac{1}{e^4} - \frac{1}{e^6}\right) = 0.0158}$$ ***Answer***

3.1.5 The wave function of a particle is given by $\psi(x)\,Ce^{-\alpha^2x^2}$ with $-\infty < x < \infty$ where C and α are constants. Calculate the probability of finding the particle in the region $0 < x < \infty$.

(Massachusetts Institute of Technology, US, B. Tech, 1990).

Solution:

The probability of finding the particle in the region $0 < x < \infty$ is

$$P = \int_0^\infty |\psi(x)|^2 dx = \int_0^\infty C^2 e^{-2\alpha^2x^2} dx$$

Also from standard integral

$$\int e^{-2\alpha^2x^2} dx \text{, the solution is } \frac{1}{2}\sqrt{\frac{\pi}{2\alpha^2}}$$

Now applying normalizing condition,

$$\int_{-\infty}^{\infty} |\psi(x)|^2 dx = 2C^2 \int_0^\infty e^{-2\alpha^2x^2} dx = 1$$

i.e., $$2C^2\left[\frac{1}{2}\left\{\frac{\pi}{2\alpha^2}\right\}^{1/2}\right] = 1$$

or $$C^2 \sqrt{\frac{\pi}{2\alpha^2}} = 1$$

$$C^2 = \sqrt{\frac{2\alpha^2}{\pi}}$$

Thus $$P = \left\{\sqrt{\frac{2\alpha^2}{\pi}}\right\}\frac{1}{2}\sqrt{\frac{\pi}{2\alpha^2}} = \frac{1}{2}$$

$$\boxed{P = \frac{1}{2}}$$ ***Answer***

3.1.6 A particle is confined to a one dimensional infinite potential well of width 0.2 nm. It is found that when the energy of the particle is 230 eV, its eigen function has 5 antinodes. Find the mass of the particle and show that it can never have energy equal to 1 keV. *(IIT Delhi, B.Tech, 2000)*

Solution:

(*a*) It is obvious from the problem that the particle is in a quantum state with $n = 5$. If E_5 and E_1 represent the energies in state 5 and state 1 respectively, then

$$E_5 = n^2E_1 \text{ with } n = 5$$

$$E_5 = 25E_1 = 230\,eV$$

$$E_1 = \frac{230 \times 1.6 \times 10^{-19}}{25}$$

$$E_1 = 14.7 \times 10^{-19} J$$

Again

$$E_1 = \frac{n^2h^2}{8ma^2} = 14.7 \times 10^{-19} J$$

or

$$m = \frac{(6.62 \times 10^{-34})^2}{8 \times (0.2 \times 10^{-9})^2 \times 14.7 \times 10^{-19}}$$

$$m = 9.3 \times 10^{-31} kg$$

(b) $E_n = 1 keV = n^2 E_1$

$$n^2 = \frac{E_n}{E_1} = \frac{1 \times 10^3 \times 1.6 \times 10^{-19}}{14.7 \times 10^{-19}}$$

$$\boxed{n = 10.4} \quad \textbf{Answer}$$

Since n is not an integer, E_n = 1 keV is not a permitted energy state.

3.1.7 Get the probability of finding an electron in a cubical box of parameter a from $0.45a$ to $0.55a$. *(Anna. U., M.Sc., May 1998)*

Solution:

$$P = \int_{x_1}^{x_2} |\psi_n^2| \, dx$$

$$= \frac{2}{a}\int_{x_1}^{x_2} \sin^2 \frac{\pi x}{a} \, dx$$

$$= \frac{2}{a}\int_{x_1}^{x_2} \frac{1}{2}\left(1 - \cos\frac{2\pi x}{a}\right) dx$$

$$= \frac{1}{a}\left[x - \frac{a}{2\pi}\sin\frac{2\pi x}{a}\right]_{x_1=0.45a}^{x_2=0.55a}$$

$$= \frac{1}{a}[0.55a - \frac{a}{2\pi}\sin 1.10\pi] - [(0.45a - \frac{a}{2\pi}\sin 0.9\pi]$$

$$= [0.55 - \frac{1}{2\pi}\sin 198°] - [0.45 - \frac{1}{2\pi}\sin 162°]$$

$$= (0.55 - 0.45) - \frac{1}{\pi}(\cos 180° \sin 18°)$$

$$= 0.10 + \frac{0.30}{\pi} = 0.198$$

$$\boxed{P = 19.8\%} \quad \textbf{Answer}$$

EXERCISE

3.1.1 A free particle is moving along the x-axis in a one dimensional box with boundaries at $x = 0$ and $x = 2.5$ nm. Calculate the probability of finding the particle within the range from 0.5 nm to 2.5 nm in its ground state. *(JNTU, B.E. November 2000)* *(**Ans**:1.25)*

3.1.2 Calculate the value of the difference between the energy of a hydrogen molecule bouncing back and forth in a one dimensional box of 1 cm long when $n = 2$ and $n = 3$ at 298 K.

*(**Ans**: 0.8 × 10^{-23} eV)*

3.1.3 Find the lowest energy of a neutron confined to a nucleus of size 10^{-14} metre. Given: mass of neutron 1.6 × 10^{-27} kg. *(**Ans**: 2.1 MeV)*

3.1.4 If the electron having a de Broglie wave length as 1.21 × 10^{-10}m is confined in a one dimensional box, how far apart must be the walls of the box when five loops of the de Broglie wave span are formed from one wall to the other.

(University of Groningen, Netherland, Winter 1999) *(**Ans**: 0.303 nm)*

3.1.5 Consider a grain dust ($m = 1$ μ gm) confined to move between two rigid walls separated by 0.1 mm. It requires 100 sec to cross the gap. What quantum number describes this motion? *(Anna U., B.E. April 1997)* *(**Ans**: 3 × 10^{14})*

Chapter 4

UNIT 4

Electron Theory of Metals and Band Theory of Solids

4.1 ELECTRON THEORY OF METALS

4.1.1 INTRODUCTION

The free electron theory of metals forms the basis of a simple explanation of several properties of metals, including the characteristic electrical and thermal conductivities. Proposed by *Drude* and *Lorentz*, soon after the discovery of the electron, it does this by considering a crystal in terms of a '*gas of free electrons*' and then treating this gas for the study of the properties of the crystal.

Electrical conductivity is the measure of the ease with which electrons move among atoms. It is the ease of movement of electrical charge from one position to another in a material. The charge is carried either by ions or electrons varies from material to material. Where the mobility is high, the material is called conductor and where mobility is low the material is called insulator. The in-between materials are called semiconductors. The number of conducting electrons per atom is determined by the atomic structure. Looking at Table 4.1 we can see that copper, silver and gold have only one such electron per atom; zinc and cadmium have two, while aluminium has three. The *unit valent metals* sodium and potassium are, of course, too reactive to be of engineering importance while if there are more than three valance electrons the character of the bonding changes and with it the electrical properties.

Table 4.1 Resistivity and valance electrons of some metals

Chemical symbol	*Resistivity (Ω m) at 300 K*	*Number of valance electrons per atom*
Cu	1.8×10^{-8}	1
Ag	1.6×10^{-8}	1
Au	2.4×10^{-8}	1
Cd	7.5×10^{-8}	2
Zn	6×10^{-8}	2
Al	2.7×10^{-8}	3

4.1.2 ELECTRONS IN A FIELD FREE CRYSTAL

The behaviour of the valance electrons when there is no potential gradient in the metal is quite different from that in a field. In the first case no current flows while in the second case a current flows in a direction opposite to the applied field. In the absence of an electric field the free electrons in a metal will be moving about at random in all directions and will be in temperature equilibrium with it. If the mass of an electron is m, then at the absolute temperature T it will possess an average random velocity (due to the *zig-zag motion*) given by the kinetic theory of gases.

Thus $$\frac{1}{2}m\bar{c}^2 = \frac{3}{2}k_BT \quad \text{...(4.1.1)}$$

where k_B is the Boltzmann constant

Now $$\bar{c} = \sqrt{\frac{3k_BT}{m}} \quad \text{...(4.1.2)}$$

$\bar{c}$ is sometimes called *root mean square velocity*. Its value for the free electron at 300 K is

$$\bar{c} = \sqrt{\frac{3\times1.38\times10^{-23}\times300}{9.1\times10^{-31}}}, \quad \bar{c} = 1.16 \times 10^5 \text{ m/s}$$

4.1.3 ELECTRON MOTION IN APPLIED ELECTRIC FIELDS

Having established the behaviour of electrons in the absence of an electric field, we now turn to the effects produced by applying a field. Since the electrons act between collisions as if they were in free space they are readily accelerated by a potential difference across the crystal, and a large current flows. In metals, the valence electrons are not fixed firmly to any particular atom, they move in all directions among the atoms with equal velocity. If electric field is applied across the material, the electrons which are moving towards positive pole get additional energy and are accelerated. While moving towards positive pole, they encounter with *local electric fields* around atoms and they get deflected or reflected. The movement towards negative electrodes requires extra energy and hence the velocity in that direction is decreased. Hence the net movement of electrons is towards the positive pole. By convention the direction of current is considered to be opposite to that of electrons.

The *mobility of electrons* is reduced because of the number of deflections and reflections which occur. The longer mean free paths results in *more acceleration*, *more drift velocity* and *more electron mobility* and greater conduction. Let us now imagine a cylinder of unit area of cross-section whose axis is also parallel to the x-direction.

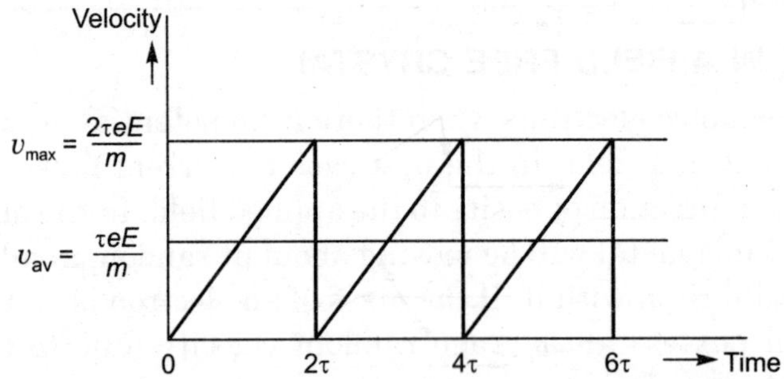

Fig. 4.1 Velocity versus time for free electrons in metals

An applied electric field (say $\vec{E}$) results in a force $e\vec{E}$ which accelerates an electron according to the equation

$$a = \frac{e\vec{E}}{m}$$

The electron has finite acceleration and therefore, we can consider that it loses all its kinetic energy at regular intervals by collisions with phonons, impurities or imperfections. This is shown in Fig. 4.1. Let the average time between collisions be 2τ where τ is known as the *relaxation time*. The average velocity of the electron is $v_{av} = \dfrac{0 + \dfrac{eE2\tau}{m}}{2}$. This velocity is also called *average drift velocity*. Thus

$$\boxed{v_d = \frac{eE\tau}{m}} \quad \text{...(4.1.3)}$$

The number of electrons crossing the plane at X for one sec is nv_d where n is the number of electrons per m^3. Hence the charge flowing through the layer per sec is nev_d. This is nothing but the *current density* as charge flowing through 1 sqm for 1 sec is current density by definition.

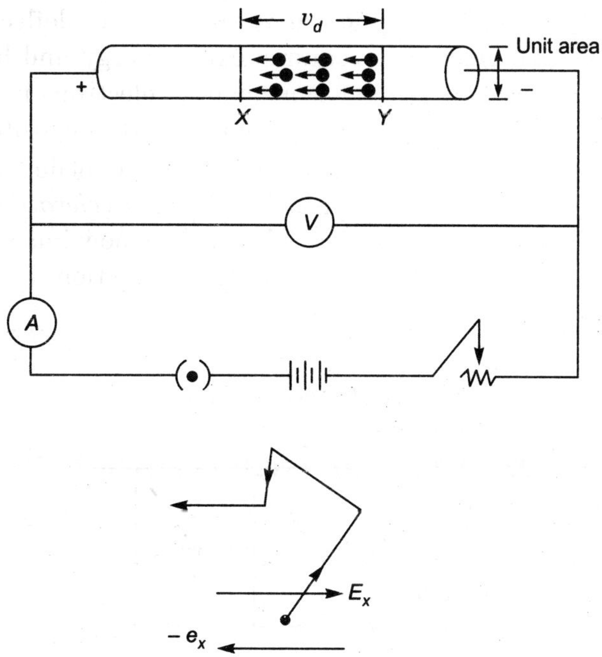

Fig. 4.2 Drift of electron in a field

i.e., $j = nev_d$

Substituting for v_d from Eqn. (4.1.3), one gets

$$j = \frac{neeE\tau}{m}$$

$$\boxed{j = \frac{ne^2E\tau}{m}} \quad \text{...(4.1.4)}$$

Hence $$\boxed{\sigma = \frac{j}{E} = \frac{ne^2\tau}{m}} \quad \text{...(4.1.5)}$$

or $$\boxed{\sigma = ne\mu} \quad \text{...(4.1.6)}$$

with $\mu = \frac{e\tau}{m}$ is called *mobility of charge carriers*. If the electrons have a *mean free path* λ, then the average time 2τ given by classical statistics is

$$2\tau = \frac{\lambda}{\bar{c}} \text{ where } \bar{c} = \sqrt{\frac{3k_BT}{m}}$$

Thus $$\sigma = \frac{ne^2\lambda}{2m\bar{c}} = \frac{ne^2\lambda}{2m}\sqrt{\frac{m}{3k_BT}}$$

$$\sigma = \frac{ne^2\lambda}{\sqrt{12mk_BT}}; \rho = \frac{\sqrt{12mk_BT}}{ne^2\lambda} \quad \text{...(4.1.7)}$$

i.e., $$\sigma \propto \lambda T^{-1/2}$$

This relation met with limited success only when compared with experiment. Also, in this classical treatment, since λ is almost independent of temperature T, therefore the result $\sigma \propto \frac{1}{T}$ is at variance with $\sigma \propto T^{-1}$ which is the experimental observation. This variance can be explained if one considers the *Fermi-Dirac statistics* instead of *classical* one. In that case $\tau = \frac{\lambda}{v_F}$ where v_F is the *Fermi velocity.*

4.1.4 SHORTCOMINGS OF CLASSICAL FREE ELECTRON THEORY OF METALS

The failure of classical free electron theory discussed in the previous section can be well understood by considering and discussing the following important propositions:

(*i*) Molar electronic specific heat

(*ii*) Computation of mean free path

(*iii*) Wiedemann-Franz law and Lorentz number

(*i*) *Molar electronic specific heat*

In many metals the most important types of internal energy are (*a*) the *vibrational energy* of the atoms about their mean *lattice positions* and (*b*) the kinetic energy of the free electrons. The kinetic energy associated with a free electron is $\frac{1}{2}m\bar{c}^2$. The kinetic energy with one kmol of a *monovalent metal* is:

$$U = \frac{1}{2}m\bar{c}^2 N_A = \frac{mN_A}{2}\left(\frac{3k_BT}{m}\right)$$

$$U = \frac{3}{2}N_A k_B T = \frac{3}{2}R_u T \quad \text{...(4.1.8)}$$

The *molar electronic specific heat* is:

$$c_{ve} = \frac{dU}{dT} = 1.5R_u$$

i.e., $$c_{ve} = 1.5 \times 1.38 \times 10^{-23} \times 6.02 \times 10^{26}$$

i.e., $$c_{ve} = 1.5\ R_u = 12.5 \times 10^3 \text{ J kmol}^{-1} \text{ K}^{-1}$$

$$c_{ve} = 1.5R_u = 12.3 \times 10^3 \text{ J kmol}^{-1} \text{ K}^{-1} \quad \text{...(4.1.9)}$$

This value of 1.5 R_u is about hundred times greater than the experimentally predicted value. Our assumption that all the free electrons contribute to the specific heat may be wrong and hence to be corrected.

(ii) Computation of mean free path

The microscopic relation for the resistivity of a metal is,

$$\rho = \frac{m}{ne^2\tau}; \; \tau = \frac{m}{ne^2\rho}$$

For copper, $n = 8.5 \times 10^{28}/\text{m}^3$ and $\rho = 1.69 \times 10^{-8}\ \Omega$ m.

Thus $$\tau = \frac{9.1\times10^{-31}}{8.5\times10^{28}(1.6\times10^{-19})^2\times1.69\times10^{-8}}$$

$$\tau = 2.47\times10^{-14} \text{ sec}$$

$$\lambda = \tau\bar{c} = 2.47\times10^{-14}\times1.154\times10^{5} = 2.85\times10^{-9} \text{ m}$$

$$\boxed{\lambda = 2.85\times10^{-9}\ m} \qquad ...(4.1.10)$$

The experimentally found value for λ is about ten times greater than this value of 2.85 nm. Probably the application of Fermi-Dirac statistics (instead of classical statistics) to the free electrons may help us to explain this difference in the values of λ.

(iii) Relation between thermal conductivity and electrical conductivity (Wiedemann-Franz law)

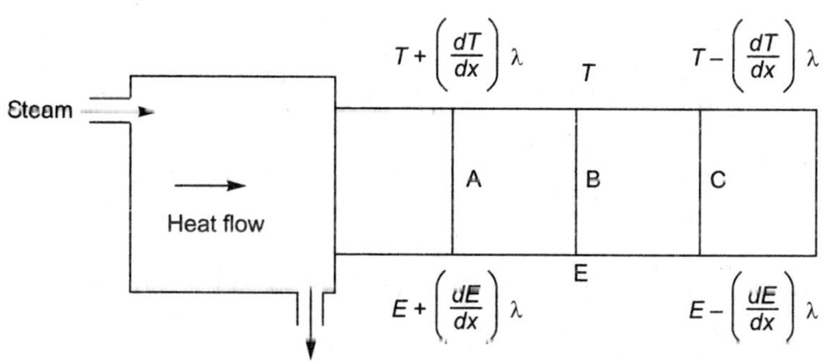

Fig. 4.3 Flow of heat through a copper rod at the steady state

Figure 4.3 shows the view of a copper rod of appreciable length with unit area of cross-section in the steady state.

Let $\lambda = AB = BC$ be the mean free path of the electron.

The excess of energy carried by an electron from A to B is $\left[\frac{dE}{dx}\right]\lambda$. Hence the excess of energy transported by the process of conduction through unit area in unit time at the middle layer B is $\frac{n\bar{c}\lambda}{6}\left[\frac{dE}{dx}\right]$. Similarly the deficit of energy transported through B in the opposite direction is $\frac{1}{6}n\bar{c}\lambda\left[\frac{dE}{dx}\right]$ assuming $\frac{1}{6}n\bar{c}$ is the number of free electrons flowing in a given direction through unit area in unit time.

Thus the net energy transported through unit area in unit time from A to B is:

$$\frac{1}{6}n\bar{c}\lambda\left[\frac{dE}{dx}\right]-\left[-\frac{1}{6}n\bar{c}\lambda\left[\frac{dE}{dx}\right]=\frac{n\bar{c}\lambda}{3}\left(\frac{dE}{dT}\right)\left(\frac{dT}{dx}\right)\right]$$

The general expression for the quantity of heat energy transported through unit area for unit time is $\sigma_T\left(\frac{dT}{dx}\right)$. Equating the two equations, one gets

$$\sigma_T = \frac{n\bar{c}\lambda}{3}\left(\frac{dE}{dT}\right)$$

But $\frac{dE}{dT}$ is the energy required to raise the temperature by one degree and hence it is $[c_v]_{el}$.

Now $$\boxed{\sigma_T = \frac{n\bar{c}\lambda}{3}[c_v]_{el} = \frac{n\lambda[c_v]_{el}}{3}\sqrt{\frac{3k_BT}{m}}} \quad ...(4.1.10a)$$

Refer Eqns. 4.1.1 and 4.1.8

$$[c_v]_{el} = \frac{3}{2}k_B \quad \text{with } n = 1 \text{ electron}$$

Thus $$\sigma_T = \frac{n\lambda}{3}\left(\frac{3}{2}k_B\right)\sqrt{\frac{3k_BT}{m}} = \frac{n\lambda k_B}{2}\sqrt{\frac{3k_BT}{m}}$$

The general expression for electrical conductivity by referring Eqn. (4.1.7)

$$\sigma = \frac{ne^2\lambda}{\sqrt{12mk_BT}}$$

Thus $$\frac{\sigma_T}{\sigma} = \frac{n\lambda k_B}{2}\sqrt{\frac{3k_BT}{m}}\left\{\frac{\sqrt{12mk_BT}}{ne^2\lambda}\right\}$$

$$\boxed{\frac{\sigma_T}{\sigma} = 3\left[\frac{k_B}{e}\right]^2 T} \quad ...(4.1.11)$$

This is known as *Wiedemann-Franz law* and the multiplying constant $3\left[\frac{k_B}{e}\right]^2$ is called *Lorentz number*

Example: For copper at 20°C, the thermal conductivity and electrical resistivity are respectively 386 W m^{-1} K^{-1} and 1.7×10^{-8} Ω m. Computation of Lorentz number is done as follows:

$$L = \frac{\sigma_T}{\sigma \times T} = \frac{\sigma_T \rho}{T}$$

or $$L = \frac{386 \times 1.7 \times 10^{-8}}{293}$$

i.e., $$\boxed{L = 2.26 \times 10^{-8}\,\mathrm{W\Omega K^{-2}}} \qquad ...(4.1.12)$$

This value of L does not match with the one calculated using RHS of Eqn. (4.1.11). This is another important drawback of the classical theory.

4.1.5 SOMMERFELD QUANTUM THEORY OF FREE ELECTRON

In order to account for the unexpectedly low value of specific heat and the other shortcomings of classical theory, Sommerfeld investigated the behaviour of a free electron gas taking into account the quantum theory and Pauli exclusion principle.

The other important shortcoming of classical theory is its inability to explain why some substances are conductors and others are not. Before discussing the quantum free electron theory, it is necessary to determine what restrictions are imposed by the laws of quantum mechanics on the energy values that an electron can have inside the crystal. It is assumed that the valence electrons travel in a constant potential (which is assumed to be zero) inside the metal and they are prevented from escaping outside by a high potential barrier. Quantum mechanics imposes some *restrictions* on their energy values trapped inside the crystal. The other one is: what are the possible energy states in which the large number of free electrons in thermal equilibrium are distributed? The first one has been arrived in the last chapter by giving a solution to Schrödinger equation and solving it. The second one is now answered using Fermi-Dirac statistics.

Fermi-Dirac Distribution and Fermi Energy

The Fermi-Dirac distribution function gives the most probable distribution of identical and indistinguishable particles (*fermions*) which obey the Pauli exclusion principle. In an ensemble of fermions in equilibrium at the kelvin temperature T, the probability that an electron has an energy E is given by

$$\boxed{F(E) = \frac{1}{[\exp(E - E_F)/k_B T]}} \qquad ...(4.1.13)$$

Here $F(E)$ is called the *Fermi function*, E is the energy of the level whose occupancy is being considered, E_F is the Fermi level and it is constant for a particular system. At absolute zero $F(E) = 0$ for $E > E_F$ and $F(E) = 1$ for $E < E_F$. Thus at absolute zero, the Fermi level divides the occupied states from the unoccupied states; *i.e.*, it is the highest energy state for the electrons to occupy at absolute zero.

To know the actual number of electrons with a given energy, one must know the number of states in the system which have the energy under consideration. Then by multiplying the number of states, by the *probability of occupation* we get the actual number of electrons. If $N(E)$ is the number of electrons in a system that have energy E, and $Z(E)$ is the number of states at that energy, then

$$\boxed{N(E)dE = Z(E)dE\ F(E)} \qquad ...(4.1.14)$$

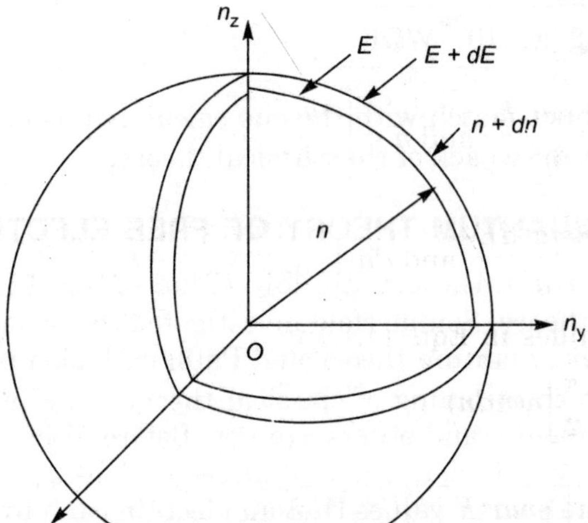

Fig. 4.4 The positive octant of *n*-space

The number of energy states, with a particular value of E, depends on how many combinations of the quantum numbers result in the same value of n. Since we are dealing with almost a continuum of energy levels, we may construct a space of points represented by the values n_x, n_y, n_z and let each point with integer values of the co-ordinates represent an energy state.

A radius vector n from the origin may be drawn to a point on the surface of a sphere of radius n such that all the points on the surface of this sphere will have the same energy. Thus n represents a vector to a point n_x, n_y, n_z ($n^2 = n_x^2 + n_y^2 + n_z^2$) in three dimensional space. In this space every integer specifies a state; *i.e.*, a unit cube contains exactly one state. Hence the number of states in any volume expressed in units of cubes of lattice parameters. Thus the number of available states within a sphere of radius n is $\frac{1}{8}\left[\frac{4\pi n^3}{3}\right]$.

The factor $\frac{1}{8}$ accounts for the fact that only positive integers are allowable and thus only one octant of the sphere alone be considered. Again the number of states within a sphere of radius $(n + dn)$ is, therefore, $\frac{1}{8}\left(\frac{4\pi}{3}\right)(n+dn)^3$. Thus the number of energy states having energy values lying between E and $E + dE$ or in the energy interval dE is

$$Z(E)dE = \frac{1}{8}\left(\frac{4\pi}{3}\right)\left[(n+dn)^3 - n^3\right]$$

$$\boxed{Z(E)dE = \frac{\pi}{2}\, n^2 dn} \qquad \text{...(4.1.15)}$$

The general expression for energy of the free electron is

$$E = \frac{n^2h^2}{8ma^2}$$

$$n^2 = \frac{8ma^2E}{h^2} \text{ and } n = \left[\frac{8ma^2}{h^2}\right]^{1/2} E^{1/2}$$

and $$2n\,dn = \frac{8ma^2dE}{h^2} \text{ and } dn = \frac{1}{2n}\left[\frac{8ma^2}{h^2}\right]dE$$

Substituting these values in Eqn. (4.1.15)

$$Z(E)dE = \left[\frac{\pi}{2}\right]n(ndn)$$

$$= \frac{\pi}{2}\left[\frac{8ma^2E}{h^2}\right]^{1/2}\left[n\frac{1}{2n}\left[\frac{8ma^2}{h^2}\right]dE\right]$$

$$Z(E)dE = \frac{\pi}{4}\left[\frac{8ma^2}{h^2}\right]^{3/2} E^{1/2}dE$$

It should be now remembered that Pauli's exclusion principle permits two electrons in each state, so that the number of energy levels actually available are

$$\boxed{Z(E)dE = 2\frac{\pi}{4}\left[\frac{8ma^2}{h^2}\right]^{3/2} E^{1/2}dE} \qquad \text{...(4.1.16)}$$

Thus the number of energy states in unit volume having energy values lying between E and $E + dE$ (called density of energy states) is

$$\boxed{Z(E)dE = \frac{\pi}{2}\left[\frac{8m}{h^2}\right]^{3/2} E^{1/2}dE} \qquad \text{...(4.1.17)}$$

The actual number of electrons $N(E)dE$ in a given energy interval dE will now be obtained by multiplying the number of energy states in the interval by the Fermi distribution function.

$$N(E)dE = Z(E)dE\,F(E)$$

$$= \frac{\pi}{2}\left[\frac{8ma^2}{h^2}\right]^{3/2} E^{1/2}dE\,F(E)$$

When $T = 0$, $F(E) = 1$ since $E < E_F$

$$\int N(E)dE = \left[\frac{\pi}{2}\right]\left[\frac{8ma^2}{h^2}\right]^{3/2} \int_0^{E_F} E^{1/2}dE$$

$$N = \left[\frac{\pi}{3}\right]\left[\frac{8m}{h^2}\right]^{3/2} a^3 E_F^{3/2} \quad ...(4.1.18)$$

Therefore the number of electrons per unit volume is

$$n = \frac{N}{a^3} = \frac{\pi}{3}\left[\frac{8m}{h^2}\right]^{3/2} E_F^{3/2}$$

$$E_F^{3/2} = \left[\frac{3n}{\pi}\right]\left[\frac{h^2}{8m}\right]^{3/2}$$

$$E_F = \left[\frac{h^2}{8m}\right]\left[\frac{3n}{\pi}\right]^{2/3}$$

$$E_F = \left[\frac{h^2}{2m}\right]\left[\frac{3n}{8\pi}\right]^{2/3} \text{ in joule; (or) } \left[\frac{h^2}{2me}\right]\left[\frac{3n}{8\pi}\right]^{2/3} \text{ eV} \quad ...(4.1.19)$$

$$E_F = 3.65 \times 10^{-19} n^{2/3} \text{(in electron volt)}$$

4.1.6 ELECTRON DISTRIBUTION IN METALS AT DIFFERENT TEMPERATURES AND OTHER CONCLUSIONS

The Fermi function is given by:

$$F(E) = \frac{1}{[1 + \exp(E - E_F)/k_BT]}$$

At 0 K for $E < E_F$, the function attains the value 1.

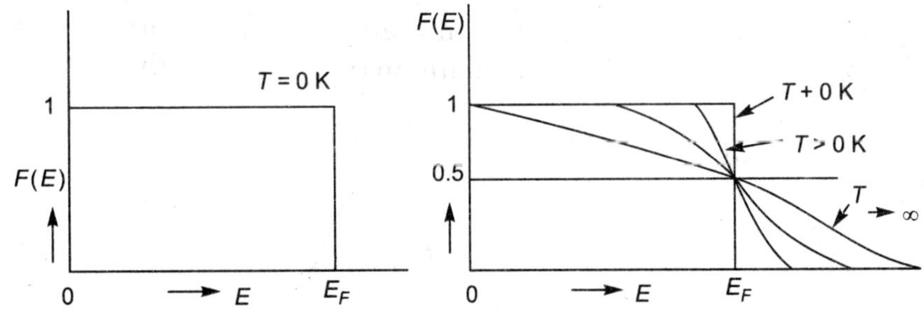

Fig. 4.5 The Fermi distribution function at different temperatures

When $E > E_F$, $F(E)$ will be zero. This means that at absolute zero, all the quantum states below E_F are occupied while all the quantum states having energies greater than E_F are unoccupied. So E_F is the maximum energy of the filled state. The Fermi function plot at a temperature greater than zero shows very little change from the Fermi plot at $T = 0$ K. Only very near E_F, $F(E)$ begins to fall and at $E = E_F$, $F(E) = \frac{1}{2}$. So, Fermi level is that level at which the probability occupation is $\frac{1}{2}$.

Sometimes in the literature, a temperature, T_F called *Fermi temperature*, and a velocity, v_F known as *Fermi velocity* are used. They are computed as under:

For sodium $E_F = 3.2$ eV

i.e., $$\frac{1}{2} m v_F^2 = E_F = 3.2 \times 1.6 \times 10^{-19} \text{ joule}$$

$$v_F^2 = \frac{2 \times 3.2 \times 1.6 \times 10^{-19}}{9.11 \times 10^{-31}} = \frac{3.2^2 \times 10^{12}}{9.11}$$

$$\boxed{v_F = 1.1 \times 10^6 \text{ m/s}} \qquad \ldots(4.1.20)$$

and $$k_B T_F = E_F$$

$$T_F = \frac{E_F}{k_B} = \frac{3.2 \times 1.6 \times 10^{-19}}{1.38 \times 10^{-23}}$$

$$\boxed{T_F = 37100 \text{ K}} \qquad \ldots(4.1.21)$$

Figure 4.6 shows the Fermi function at a finite temperature say 300 K. The Fermi curve at this temperature differs from the Fermi plot at 0 K in the smearing of the drop from 1 to 0. In the actual case, the extent of the smearing is very small except at very high temperatures. This can be looked into by considering the following example:

Let $F(E) = 0.2$ at 300 K

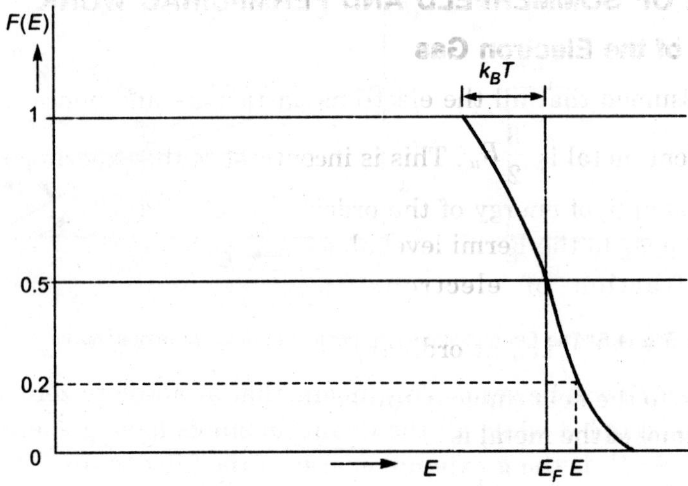

Fig. 4.6 Fermi function at 300 K

Now $$F(E) = 0.2 = \frac{1}{[1+\exp(E-E_F)/k_BT]}$$

i.e., $$0.2 = \frac{1}{1+e^x}$$

or $$0.2 + 0.2e^x = 1$$

$$0.2e^x = 0.8$$

$$e^x = 4;\ x = 1.39$$

i.e., $$\frac{E-E_F}{k_BT} = 1.39$$

$$E - E_F = 0.025 \times 1.39 = 0.035 \text{ eV}$$

Thus the extent of the smearing (0.035 eV) is very small compared with the distribution of 5 eV.

Inference

(i) At absolute zero, $F(E) = 1$ for $E < E_F$
and $F(E) = 0$ for $E > E_F$

i.e., at zero degree kelvin E_F is the maximum energy of the occupied state

(ii) As temperature increases, an electron may get energy of the order of k_BT and go to higher vacant state, and so Fermi function falls. But for energies below E_F, such that $(E_F - E) >> k_BT$, the value of $F(E)$ will be unity; i.e., only in the region of E_F, $F(E)$ begins to fall. When $E = E_F$, $F(E)$ will be $\frac{1}{2}$.

4.1.7 OUTCOME OF SOMMERFELD AND FERMI-DIRAC WORK

1. Heat Capacity of the Electron Gas

Classical theory assumed that all the electrons participate and hence the molar specific heat for a monovalent metal is $\frac{3}{2}R_u$. This is incontrast to the experimental value.

Instead for a supply of energy of the order of 0.025 eV (k_BT) at room temperature, only electrons very near to the Fermi level alone are excited to the unoccupied levels.

Hence the fraction of electrons that contribute to heat capacity is $\frac{k_BT}{E_F} = \frac{0.025}{5} = 0.005 = 0.5\%$ *i.e.*, at ordinary temperatures, less than 1% of the valence electrons contribute to the heat capacity unlike that in classical theory. Hence the energy associated with a kmol of the metal is

$$U = N_A\left[\frac{k_BT}{E_F}\right]\left(\frac{3}{2}k_BT\right)$$

$$U = \frac{3}{2}\frac{R_uk_BT^2}{E_F}$$

$$[c_v]_{el} = 3R_u\left[\frac{k_BT}{E_F}\right] \quad \text{...(4.1.22)}$$

$$[c_v]_{el} = 3R_u\left(\frac{0.025}{5}\right) = 0.015R_u$$

This value well agrees with experimental value. However, at very low temperatures, the lattice contribution is small and falls as T^3 and the electronic contribution is significant.

2. Mean Energy of Electron Gas at Absolute Zero

Refer Eqn. (4.1.19):

$$E_F = \left[\frac{h^2}{2m}\right]\left[\frac{3n}{8\pi}\right]^{2/3}$$

Now the energy associated with the electrons in the interval dE is

$$Z(E)dE\ F(E)\ E \text{ with } F(E) = 1$$

Hence the total energy,

$$U_0 = \int_0^{E_F} Z(E)dE\,E$$

$$U_0 = \int_0^{E_F}\left[\frac{\pi}{2}\right]\left[\frac{8ma^2}{h^2}\right]^{3/2} E^{1/2}EdE$$

$$U_0 = \frac{\pi}{2}\left[\frac{8ma^2}{h^2}\right]^{3/2}\int_0^{E_F} E^{3/2}\,dE$$

$$U_0 = \frac{\pi}{2}\left[\frac{8ma^2}{h^2}\right]^{3/2}\left(\frac{2}{5}E_F^{5/2}\right)$$

Thus the *mean energy* of the electron at 0 K is

$$\overline{E}_0 = \frac{U_0}{N}$$

Substituting the value of N from Eqn. (4.1.18), we get

$$\overline{E}_0 = \frac{\pi}{5}\left[\frac{8ma^2}{h^2}\right]^{3/2} E_F^{5/2}\,\frac{1}{\dfrac{\pi}{3}\left[\dfrac{8ma^2}{h^2}\right]^{3/2}(E_F)^{3/2}}$$

$$\boxed{\overline{E}_0 = \frac{3}{5}E_F} \qquad \ldots(4.1.23)$$

3. Effect of Temperature on Fermi Distribution

At higher temperatures, the number of electrons having energy values lying between E and $E + dE$ is given by:

$$dN = Z(E)dE\,f(E)$$

$$N = \int_0^{\infty} Z(E)dE\,f(E)$$

$$\overline{E} = \frac{1}{N}\int_0^{\infty} Z(E)\,E\,dE\,f(E)$$

$$\overline{E} = \frac{1}{N}\int_0^{\infty}\frac{Z(E)\,E\,dE}{[1+\exp(E-E_f)/k_BT]}$$

A rigorous mathematical treatment of the above equation yields two important results: one is Fermi energy of the electron at T K and the other one is mean energy of the electron at T K. The respective equations are given below

$$\boxed{\begin{aligned} E_f &= E_F\left[1-\frac{\pi^2}{12}\left[\frac{k_BT}{E_F}\right]^2\right] \\ \overline{E} &= \overline{E}_0\left[1+\frac{5\pi^2}{12}\left[\frac{k_BT}{E_F}\right]^2\right] \end{aligned}} \qquad \ldots(4.1.24)$$

Since $\frac{k_B T}{E_F}$ is very small, E_f does not vary much from E_F

The significance of the above two equations can be studied by considering the following problems:

(i) Heat capacity

$$[c_v]_{el} = \frac{5\pi^2 \overline{E}_0}{12}\left(\frac{k_B^2 2T}{E_F^2}\right) = \frac{5}{6}\frac{\pi^2 k_B^2}{E_F^2}\left(\frac{3}{5}E_F T\right)$$

$$= \frac{\pi^2}{2}\left(\frac{k_B T}{E_F}\right)k_B$$

Thus for a monovalent metal, the heat capacity per kmol of the metal due to the electrons is

$$[c_v]_{el} = \frac{\pi^2}{2}k_B N_A\left(\frac{k_B T}{E_F}\right) = \frac{\pi^2}{2}\left(\frac{k_B T}{E_F}\right)R_u \quad ...(4.1.25)$$

$$[c_v]_{el} = \frac{\pi^2}{2}\left(\frac{k_B T}{k_B T_F}\right)R_u = \frac{\pi^2}{2}\left(\frac{T}{T_F}\right)R_u = AT \quad ...(4.1.26)$$

with $$A = \left[\frac{\pi^2}{2}\right]\left[\frac{k_B^2 N_A}{E_F}\right]$$

For $E_F = 5$ eV and $T = 300$

$$[c_v]_{el} = \frac{\pi^2}{2}\left[\frac{0.025}{5}\right]R_u$$

$$[c_v]_{el} = 0.025 R_u \quad ...(4.1.27)$$

This value is much higher than that at 0 K (Refer Eqn. 4.1.22). However, this value of 0.025 R_u fairly agrees with the experimental values. The ratio of the electronic specific heat quantum mechanically and classically is

$$\frac{[c_v]_{el}\text{ (Q.Mech)}}{[c_v]_{el}\text{(Classical)}} = \left[\frac{\pi^2}{2}\right]\left[\frac{k_B T}{E_F}\right]R_u\frac{1}{\frac{3}{2}R_u} = \left[\frac{\pi^2}{3}\right]\left[\frac{k_B T}{E_F}\right] \quad ...(4.1.28)$$

The total specific heat is

$$c_v\text{ (Total)} = [c_v]_{el} + [c_v]_{la}$$

$$[c_v]_T = AT + BT^3 \quad ...(4.1.29)$$

(ii) Quantum treatment of electric conductivity and Ohm's law

The wave number

$$k = \frac{2\pi}{\lambda} \text{ or } \lambda = \frac{2\pi}{k}$$

But $$\lambda = \frac{h}{p} \text{ or } p = \frac{hk}{2\pi}$$

or $$\vec{p} = \hbar\vec{k}, \text{ i.e., } \vec{v} = \frac{\hbar\vec{k}}{m} \quad ...(4.1.30)$$

In *thermal equilibrium*, the valence electrons obeying classical free electron theory do random motion with no preferential velocity in any direction. It is now conveniently plotted the velocities of these free electrons in velocity space with v_F as the maximum velocity that an electron can assume *i.e.*, v_F is the actual value of the velocity of the electron at the Fermi level. The shape having v_F as a radius represents, the *Fermi surface*. Under normal conditions the origin and the centre of the sphere coincide so that the net momentum is zero. When the electrons are accelerated by an electric field, motion in the direction of the force is favoured. The whole *Fermi surface* is then moved from the origin and move steadily further as time proceeds. See Fig. 4.7.

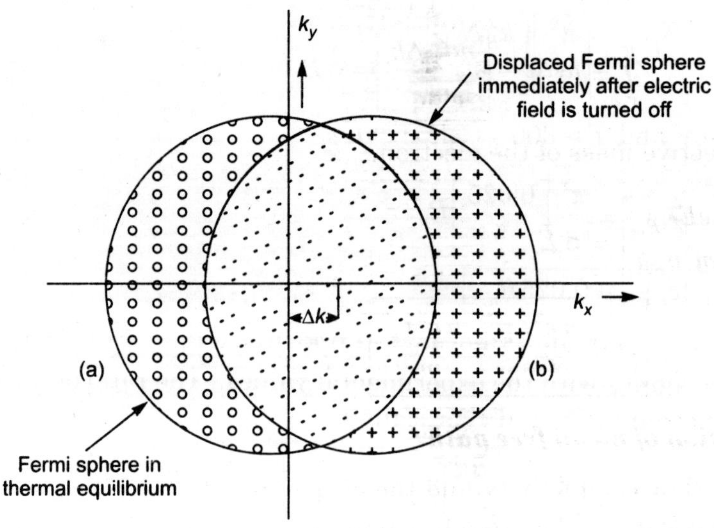

Fig. 4.7 Fermi sphere in (a) absence of a field, (b) presence of electric field.

$$\vec{v} = \left[\frac{\hbar\vec{k}}{m}\right]; m\vec{v} = \hbar\vec{k} \quad ...(4.1.31)$$

Now $$m\frac{d\vec{v}}{dt} = \hbar\frac{d\vec{k}}{dt} = e\vec{E}$$

This is the equation of motion of each electron in the *Fermi sphere* under the influence of the field $\vec{E}$

$$d\vec{k} = \frac{e\vec{E}\,dt}{\hbar}$$

Integrating

$$\vec{k}(t) - \vec{k}(0) = \frac{e\vec{E}\,t}{\hbar}$$

Thus if the electric field at a time $t = 0$ to a filled Fermi sphere centred at the origin of $\vec{k}$-space in the characteristic time,

$\tau = \tau_F = \dfrac{\lambda_F}{v_F}$, the sphere might have moved to a new centre at

$$\Delta\vec{k} = \frac{e\vec{E}\lambda_F}{\hbar v_F} \quad \text{with } \tau_F = \frac{\lambda_F}{v_F}$$

A steady current density is ultimately reached with,

$$\vec{J} = ne\Delta\vec{v} = \left[\frac{ne\hbar\Delta\vec{k}}{m^*}\right] = \sigma\vec{E}$$

with m^* the effective mass of the electron.

$$ne\hbar\left[\frac{e\vec{E}\lambda_F}{m^* v_F \hbar}\right] = \sigma\vec{E}$$

$$\boxed{\sigma = \frac{ne^2\lambda_F}{m^* v_F} = \frac{ne^2\tau_F}{m^*}} \quad \text{...(4.1.32)}$$

(iii) Computation of mean free path

Let the electric conductivity and the *effective mass* of copper be $5.76 \times 10^7\ \Omega^{-1}\ \text{m}^{-1}$ and 1.01 m respectively. The mean free path can be easily calculated assuming $E_F = 7$ eV.

Formula used:

$$\sigma = \frac{ne^2\tau_F}{m^*} \quad \text{or} \quad \tau_F = \frac{\sigma m^*}{ne^2}$$

$$\tau_F = \frac{1.01 \times 9.1 \times 10^{-31} \times 5.76 \times 10^7}{8.5 \times 10^{28} (1.6 \times 10^{-19})^2}$$

$$\tau_F = 2.4 \times 10^{-14}$$

$$\frac{1}{2} m v_F^2 = E_F = 7 \times 1.6 \times 10^{-19}$$

$$v_F^2 = \frac{14 \times 1.6 \times 10^{-19}}{9.1 \times 10^{-31}}$$

$$v_F = 1.55 \times 10^6 \text{ m/s}$$

$$\tau_F = \frac{\lambda_F}{v_F}$$

$$\lambda_F = \tau_F v_F = 2.4 \times 10^{-14} \times 1.55 \times 10^6$$

$$\boxed{\lambda_F = 37.8 \text{ nm}} \quad \text{...(4.1.33)}$$

This agrees with the experimental value fairly well.

(iv) ***Quantum mechanical treatment of Wiedemann-Franz law***

Refer Eqn. 4.1.10a and 4.1.25

$$\sigma_T = \frac{n \bar{c} \lambda}{3} [c_v]_{el}$$

As per quantum theory (by introducing v_F and λ_F for $\bar{c}$ and λ)

$$\sigma_T = \frac{n v_F \lambda_F}{3} \left[\frac{\pi^2}{2} \left\{ \frac{k_B T}{E_F} \right\} k_B \right] \text{ with } E_F = \frac{1}{2} m v_F^2$$

$$\sigma_T = \frac{\pi^2 n v_F \lambda_F}{6} \left[\frac{2 k_B^2 T}{m v_F^2} \right]$$

$$\sigma_T = \frac{\pi^2 n \lambda_F k_B^2 T}{3 m v_F}$$

The expression for electrical conductivity is $\sigma = \dfrac{n e^2 \lambda_F}{m v_F}$

$$\boxed{\frac{\sigma_T}{\sigma} = \frac{\pi^2}{3} \left(\frac{k_B}{e} \right)^2 T} \quad \text{...(4.1.34)}$$

$$\boxed{L = \frac{\sigma_T}{\sigma \times T} = \frac{\pi^2}{3} \left(\frac{k_B}{e} \right)^2 = 2.45 \times 10^{-8} \text{ W}\Omega\text{K}^{-2}} \quad \text{...(4.1.35)}$$

4.1.8 ELECTRON SCATTERING AND SOURCES OF RESISTANCES

When a metal contains impurities, the field in the vicinity of the impurities is in general different from that near the host atoms. The impurities act as *scattering centres* of electrons. Thus electrons in an impure metal are scattered by impurity atoms as well as by the thermal vibrations of the atoms. *i.e.*, in a pure metal, the electrical resistivity must go to zero as the temperature approaches absolute zero. However, experiments reveal that the resistivity of non-superconducting metals decreases to a small but finite value and remains practically constant and independent of temperature below 10 K down to nearly absolute zero as shown in Fig. 4.8. This is known as the *residual resistance* of the metal, which is attributed to the presence of impurity atoms in small concentrations and also geometrical imperfections like, *point defects, grain boundaries, stacking faults* etc. in the metal. These impurities and imperfections scatter electrons and relaxation time $\tau_1, \tau_2,...$ can be assigned to each type of impurity or imperfection in the metal. If we assume that the *scattering probabilities* per unit time are independent of each other, then the total scattering probability becomes the sum of all the individual probabilities. Since the scattering probability is inversely proportional to the *relaxation time*, therefore the relaxation time due to imperfection and impurities is

$$\frac{1}{\tau_i} = \frac{1}{\tau_1} + \frac{1}{\tau_2} + ...$$

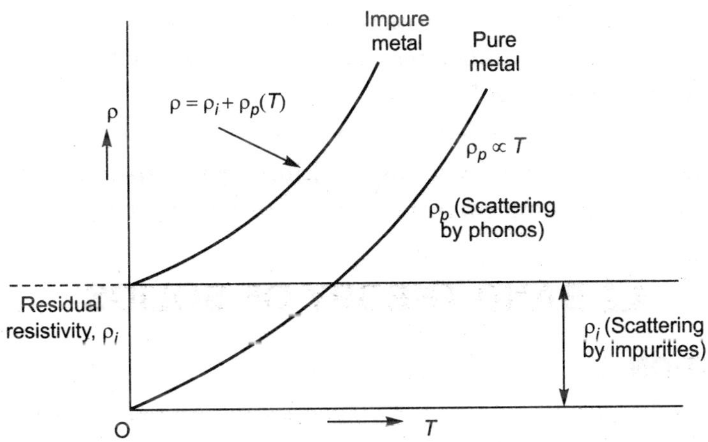

Fig. 4.8 Variation of electrical resistivity with temperature

If τ_p is the relaxation time due to phonons, then

$$\boxed{\frac{1}{\tau} = \frac{1}{\tau_p} + \frac{1}{\tau_i}} \quad ...(4.1.36)$$

Since the resistivity is proportional to $\frac{1}{\tau}$, we will have

$$\boxed{\rho = \rho_p(T) + \rho_i} \quad ...(4.1.37)$$

The resistivities arising from independent scattering processes are additive which is *Matthiesson's rule.*

The special features of metals with regard to their conductivities are summarized below:

1. The current density in the steady state is proportional to the electric field strength (Ohm's law)
2. For pure specimens, the electric conductivity (σ) and the thermal conductivity (σ_T) vary with temperature as follows:
   $\sigma \propto T^{-1}$ and σ_T = constant for $T > \theta_D$

   so that $\dfrac{\sigma_T}{\sigma \times T}$ is independent of temperature (Wiedemann-Franz law); and

   $\sigma \propto T^{-5}$ and $\sigma \propto T^{-2}$ for $T < \theta_D$ where θ_D is the characteristic *Debye temperature*. The relation $\rho \propto T^5$ is known as Bloch-Gruneisen law.
3. The resistivity of some metals becomes zero at very low temperatures near absolute zero, (e.g.) mercury aluminium, lead, etc. These metals are called *superconductors*.
4. The resistivity of metals having trace of impurities, the total electrical resistivity is given by
   $$\rho = \rho_p(T) + \rho_i$$
   where ρ_i is a constant that increases with increasing impurity content $\rho_p(T)$ is the temperature dependent part of the resistivity. This is *Matthiessen's rule* as mentioned earlier.
5. For most metals electrical resistivity decreases with increase of pressure.
6. The resistivity of alloys that exhibit order-disorder transitions shows pronounced minima corresponding to order phases.

4.2 BAND THEORY OF SOLIDS

4.2.1 INTRODUCTION

Sommerfeld, with the invention of quantum mechanics and use of Fermi-Dirac statistics modified the classical theory of free electrons and explained many phenomena where classical theory failed. The existence of a *thermionic work function* for each metal, the existence of a contact potential difference between dissimilar metals and also certain other phenomena were nicely explained by quantum theory.

The assumption that the conduction electron moves in a crystal without a potential energy does not seem to be reasonable. Semiconductors do not find an explanation on the basis of this theory. The shortcomings are due to the over simplified potential. The *periodicity character* of the potential extends to infinity in all directions inside the crystal. According to *Kronig-Penny* model, the potential energy of an electron, when it moves in the form of rectangular walls as shown in Fig. 4.9. It is assumed that the potential energy is zero when x lies between 0 and a and potential energy is V_0 when x lies between $-b$ and 0. *i.e.*, $V(x) = 0$ for $0 < x < a$ and $V(x) = V_0$ for $-b < x < 0$.

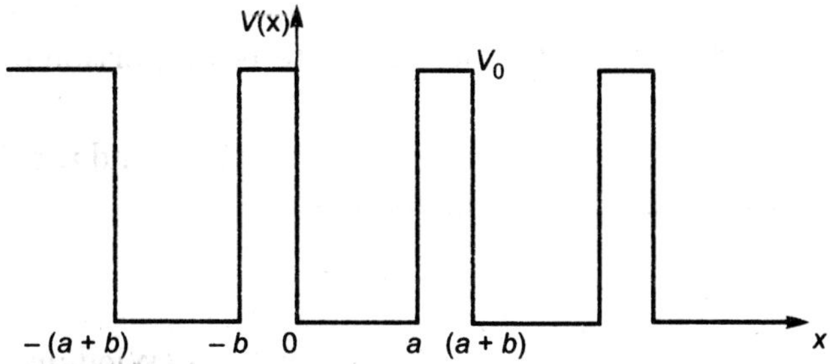

Fig. 4.9 One dimensional chain of potential wells (Kronig-Penny model)

Although the model is highly artificial, yet it illustrates many of the characteristic features of the behaviour of electrons in a *periodic lattice*. The wave functions related to this model may be obtained by solving Schrödinger equation for the following two regions.

$$\boxed{\frac{d^2\psi}{dx^2} + \left[\frac{2m}{\hbar^2}\right]E\psi = 0} \quad \text{...(4.2.1)}$$

for $0 < x < a$ with $V(x) = 0$

and
$$\frac{d^2\psi}{dx^2} + \left[\frac{2m}{\hbar^2}\right](E - V_0)\psi = 0$$

for $-b < x < a$ with $V(x) = V_0$

Let us now define the following real quantities

$$\alpha^2 = \frac{2mE}{\hbar^2} \text{ and } \beta^2 = \frac{2m}{\hbar^2}[V_0 - E]$$

Thus the two equations with the above assumption become

$$\boxed{\frac{d^2\psi}{dx^2} + \alpha^2\psi = 0} \quad \text{...(4.2.2)}$$

$$\boxed{\frac{d^2\psi}{dx^2} - \beta^2\psi = 0} \quad \text{...(4.2.3)}$$

Bloch has shown that the solutions are of the form

$$\boxed{\psi = u_k(x)\exp(ikx)} \quad \text{...(4.2.4)}$$

where $u_k(x)$ has the periodicity of the lattice, so that

$u_k(x) = u_k(x + a) = u_k(x + na)$ where n is an integer and k the wave number is:

$$\frac{2\pi}{\lambda} = \frac{p}{\hbar}$$

The common solution applicable to both the equations (4.2.2) and (4.2.3) suggested by Bloch is given by

$$\boxed{\psi = u_k(x)\exp(ikx)} \qquad ...(4.2.5)$$

where $u_k(x)$ is a *periodic function* in x such that $u_k(x) = u_k(x + a) = u_k(x + na)$) where a is an integer

From Eqn. (4.2.4)

$$\frac{d\psi}{dx} = \exp(ikx)\frac{du_k}{dx} + u_k(x)ik\exp(ikx)$$

$$\frac{d^2\psi}{dx^2} = \exp(ikx)\frac{d^2u_k}{dx^2} + \frac{du_k}{dx}[ik\exp(ikx)] + u_k(x)(i^2k^2)\exp(ikx) + ik\exp(ikx)\frac{du_k}{dx}$$

$$= -k^2u_k\exp(ikx) + 2ik\exp(ikx)\frac{du_k}{dx} + \exp(ikx)\frac{d^2u_k}{dx^2}$$

Substituting these values in Eqns (4.2.2) and (4.2.3), we get

$$\boxed{\frac{d^2u_1}{dx^2} + 2ik\frac{du_1}{dx} + (\alpha^2 - k^2)u_1 = 0} \qquad ...(4.2.6)$$

and

$$\boxed{\frac{d^2u_2}{dx^2} + 2ik\frac{du_2}{dx} - (\beta^2 + k^2)u_2 = 0} \qquad ...(4.2.7)$$

where u_1 represents the value of $u_k(x)$ in the interval $0 < x < a$ and u_2 that in the interval $-b < x < 0$. The solution of the Eqn. (4.2.6) is of the form

$$u_1 = \exp(mx)$$

Now

$$\frac{du_1}{dx} = m\exp(mx) \text{ and } \frac{d^2u_1}{dx^2} = m^2\exp(mx)$$

Substituting these values in Eqn. (4.2.6) and simplifying, we get,

$$m^2 + 2ikm + (\alpha^2 - k^2) = 0$$

$$m = \frac{-2ik \pm \sqrt{-4k^2 - 4(\alpha^2 - k^2)}}{2} = -ik \pm \sqrt{k^2 - (\alpha^2 - k^2)}$$

$$m_1 = -ik + i\alpha = i(\alpha - k)$$

$$m_2 = -ik - i\alpha = -i(\alpha + k)$$

Thus the general solution is

$$u_1 = A\exp[i(\alpha - k)x + B\exp[-i(\alpha + k)x]$$

Similarly, $\quad u_2 = C\exp[(\beta - ik)x] + D\exp[-(\beta + ik)x]$

where A, B, C and D are constants. Let us now apply the boundary conditions by seeing Fig. 4.9, we get

$$[u_1(x)]_{x=0} = [u_2(x)]_{x=0}; \quad \left[\frac{du_1(x)}{dx}\right]_{x=0} = \left[\frac{du_2(x)}{dx}\right]_{x=0}$$

Similarly,

$$[u_1(x)]_{x=a} = [u_2(x)]_{x=-b}; \quad \left[\frac{du_2(x)}{dx}\right]_{x=a} = \left[\frac{du_2(x)}{dx}\right]_{x=-b}$$

On applying these conditions, we may get four equations with $(A + B) = (C + D)$ as the first equation. These four equations will have non-vanishing solutions if and only if the determinant of the coefficients A, B, C, D vanishes. This requires

$$\begin{bmatrix} i(\alpha-k) & -i(\alpha+k) & (\beta-ik) & -(\beta+ik) \\ \exp[i(\alpha-k)a] & \exp[-i(\alpha+k)a] & \exp[-(\beta-ik)b] & \exp[(\beta+ik)b] \\ i(\alpha-k)\exp[i(\alpha-k)a] & -i(\alpha+k)\exp[-i(\alpha+k)a] & (\beta-ik)\exp[-b(\beta-ik)] & -(\beta+ik)\exp[b(\beta+ik)] \end{bmatrix} = 0$$

The final solution of this equation is

$$\boxed{\frac{(\beta^2-\alpha^2)}{2\alpha\beta}\sin\alpha a\sinh\beta b + \cosh\beta b\cos\alpha a = \cos k(a+b)} \qquad ...(4.2.8)$$

A fruitful discussion of this equation and arriving at useful results is not very easy. Hence Kronig Penney suggested that when V_0 tends to infinity, b approaches zero in such a way that the product $V_0\, b$ remains finite. Under such a situation, $\sinh\beta b \to \beta b$ and $\cosh\beta b \to 1$ as $b \to 0$. Now Eqn.(4.2.8) becomes

$$\beta b\left[\frac{(\beta^2-\alpha^2)}{2\alpha\beta}\right]\sin\alpha a + \cos\alpha a = \cos ka$$

But $$\beta^2 - \alpha^2 = \frac{2m}{\hbar^2}[V_0 - E + E]$$

$$\left[\frac{2mV_0}{2\hbar^2}\right](ab)\frac{\sin\alpha a}{\alpha a} + \cos\alpha a = \cos ka$$

$$\left[\frac{mV_0ab}{\hbar^2}\right]\left[\frac{\sin\alpha a}{\alpha a}\right] + \cos\alpha a = \cos ka$$

$$\boxed{P\left[\frac{\sin\alpha a}{\alpha a}\right] + \cos\alpha a = \cos ka} \qquad ...(4.2.9)$$

where $$P = \frac{mV_0ab}{\hbar^2}$$

Since V_0 tends to infinity, the quantity P is a measure of $V_0 b$ which is the area of the potential barrier. The physical significance of P is that as it increases, the area of the potential barrier increases. *i.e.*, the electron is bound more strongly to a particular potential well. When P tends to zero, the potential barrier becomes very weak. *i.e.*, the electron becomes free. Equation (4.2.9) will be satisfied only for those values of αa for which the left hand side lies between + 1 and – 1. Such values of αa alone will give a solution for the equation obtained by Bloch (Eqn. 4.2.9). In Fig. 4.10, $\frac{P \sin \alpha a}{\alpha a} + \cos \alpha a$ is plotted against αa for $P = \frac{3\pi}{2}$. The part of the vertical axis lying between horizontal lines represents the range acceptable to the left hand side $\left[\frac{P \sin \alpha a}{\alpha a} + \cos \alpha a\right]$. The abscissa αa is a measure of energy, because $\alpha^2 = \frac{2mE}{\hbar^2}$. By finding the value of αa at any point, the energy represented by the function at that point is calculated. The following interesting conclusions are being drawn:

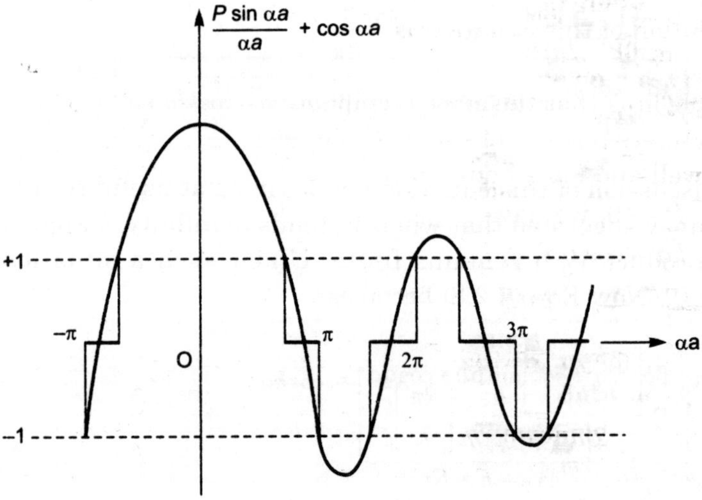

Fig. 4.10 Plot of $\frac{P \sin \alpha a}{\alpha a} + \cos \alpha a$

1. It is very clear from Fig. 4.10 that the energy spectrum of the electron consists of alternate regions of allowed energy which are drawn heavily and unallowed energy which are shown by white portion in between. These regions are called *allowed* and *forbidden energy bands*. The boundaries of allowed ranges of αa corresponds to the values of cos $ka = \pm 1$ or $ka = n\pi$, $k = \frac{n\pi}{a}$.
2. As the value of αa increases, the width of allowed energy band also increases while the width of forbidden band decreases.

3. It is known that P is a measure of the *potential barrier strength.* If P is large, the function described by the left hand side of the Equation (4.2.9) crosses + 1 and – 1 region at steeper angle; hence the allowed bands are narrower and the forbidden bands are wider. Also in equation (4.2.9), when $\alpha a = \pm n\pi$, $\cos \alpha a = \cos ka$ with again $ka = \pm n\pi$, or $k = \pm \frac{n\pi}{a}$. The values of k are *points of discontinuity* in the $(E - k)$ curve for electron in the crystal.
4. The energy spectrum in the two extreme cases using Eqn. (4.2.9) can now be discussed.

   (*i*) When $P \to \infty$, Eqn. (4.2.9) will have a solution only when $\sin \alpha a = 0$.

   or

   $\alpha a = \pm n\pi$ where n is an integer.

   Now $$\alpha = \pm\frac{n\pi}{a} \text{ or } \alpha^2 = \frac{n^2\pi^2}{a^2} = \frac{2mE}{\hbar^2}$$

   $$E = \frac{n^2\pi^2\hbar^2}{2ma^2}; \text{ or } \boxed{E = \frac{n^2\pi^2h^2}{8\pi^2ma^2} = \frac{n^2h^2}{8ma^2}} \quad (4.2.9a)$$

   The conclusion is that the energy is independent of k. The energy levels in this case are discrete and the electron is completely bound. It will be well within the potential well and moves only in the cell of width a.

   (*ii*) When $P \to 0$, we have

   $$\cos ka = \cos \alpha a$$

   $$a^2 = k^2 = \frac{2mE}{\hbar^2}$$

   $$\boxed{E = \frac{\hbar^2k^2}{2m} = \frac{h^2}{8\pi^2m}\left[\frac{2\pi}{\lambda}\right]^2} \quad ...(4.2.10)$$

   $$\boxed{E = \left[\frac{h^2}{2m}\right]\frac{1}{\lambda^2} = \frac{h^2}{2m}\left[\frac{p^2}{h^2}\right] = \frac{p^2}{2m} = \frac{1}{2}mv^2} \quad ...(4.2.11)$$

   This energy corresponds to a completely free particle. No energy level in this case exist. *i.e.*, all energies are allowed to the electrons.

(E – k) Curve and Brillouin Zones

Equation (4.2.10) gives an $(E - k)$ curve which is parabolic one, as shown by the dotted line in Fig. 4.11 while an electron acted on by the *lattice potential* of a crystal has an $(E - k)$ curve shown by the solid lines. Near the energy gaps at $ka = \pm n\pi$, the solid lines deviate markedly from the *parabola.*

when $k = \pm\frac{n\pi}{a}$, Eqn (4.2.10) gives the free electron energy.

$$E = \frac{\pi^2 n^2 \hbar^2}{2ma^2} = \frac{n^2 h^2}{8ma^2} \quad \text{...(4.2.12)}$$

which is exactly the same as the equation for a free particle in the potential well problem. If the size of the well is replaced by the distance between the atoms, say 0.3 nm, then the Eqn.(4.2.12) becomes $E = 4\ n^2$ eV.

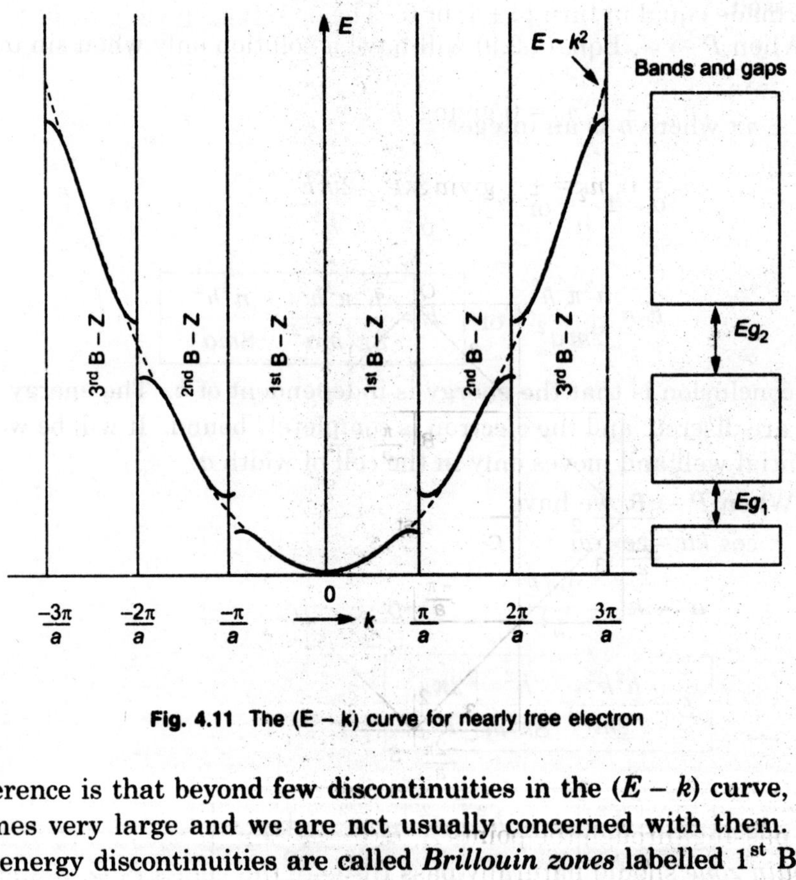

Fig. 4.11 The (E – k) curve for nearly free electron

The inference is that beyond few discontinuities in the $(E - k)$ curve, the electrons energy becomes very large and we are nct usually concerned with them. The regions between the energy discontinuities are called *Brillouin zones* labelled 1st Brillouin zone, 2nd Brillouin zone, etc. as shown in Fig. 4.12.

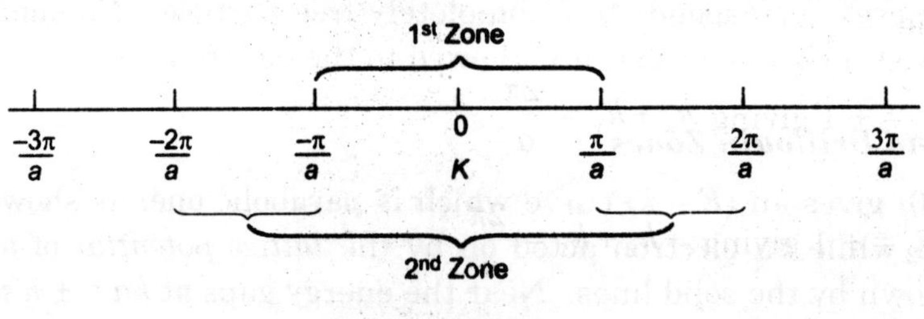

Fig. 4.12 Brillouin zones

Brillouin zones in two dimensions

In general, the condition for any energy discontinuity is

$$k = \pm \frac{n\pi}{a}$$

The corresponding condition for the two dimensional case is:

$$k_x n_1 + k_y n_2 = \frac{\pi}{a}(n_1^2 + n_2^2)$$

where n_1 and n_2 are integers referring to each of the axes. To sketch the first zone, n_1 and n_2 are made equal in turn to ± 1, or 0. The equations giving the borders of the first zone are:

$$n_1 = \pm 1,\ n_2 = 0 \text{ giving } k_x = \pm \frac{\pi}{a}$$

$$n_1 = 0,\ n_2 = \pm 1 \text{ giving } k_y = \pm \frac{\pi}{a}$$

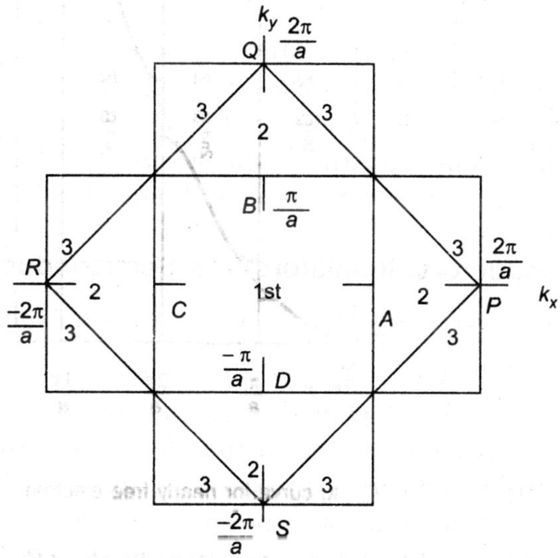

i.e., a square passing through the points *A*, *B*, *C* and *D* gives the *first Brillouin zone.* The *second Brillouin zone* should naturally pass through the points *P*, *Q*, *R* and *S*.

To get the sketch of the second zone, we must use the next set of integers above those for the first zone.

$$n_1 = +1,\ n_2 = +1 \text{ giving } k_x + k_y = \frac{2\pi}{a}$$

$$n_1 = -1,\ n_2 = +1 \text{ giving } -k_x + k_y = \frac{2\pi}{a}$$

$$n_1 = +1,\ n_2 = -1 \text{ giving } k_x - k_y = \frac{2\pi}{a}$$

$$n_1 = -1,\ n_2 = -1 \text{ giving } -k_x - k_y = \frac{2\pi}{a}$$

The above four equations describe a set of four lines at 45° to the k_x and k_y axes passing through a points P, Q, R and S. This means the second Brillouin zone is the area or region between P, Q, R, S and A, B, C, D. Similarly the third Brillouin zone may be obtained by giving n_1 and n_2 values of 0, ± 1 and ± 2.

Classification of Materials into Conductors, Semiconductors and Insulators

Since in a solid, many atoms are brought together the separation between the sublevels is very small. Consequently, the split energy levels are almost continuous and are said to form an energy band. The energy bands in a solid correspond to the energy levels in an atom. An electron in a solid can have only those diserete energies that lie within these energy bands. These bands are, therefore, called allowed energy bands. These (allowed) energy bands are, in general, separated by some gaps which have no allowed levels. These gaps (regions) are known as forbidden energy bands. Band corresponding to valence electrons is called valence band and the band beyond the forbidden band is called conduction band, into which, the electrons pass, and move freely.

Distinction between Conductors, Insulators and Semiconductors

(*i*) **Conductors (metals):** Let us consider the exaple of sodium ($Z = 11$) when has an electronic configuation $1^{s^2}, 2^{s^2}, 2p^6, 3s^1$. It shows that $1^{s^2}, 2^s$ and $2p$ bands in sodium are completely filled while the outermost (35) band has only one electron whereas it can accommodate one more; thesefore valence band is half filled. When an electric field is applied across a piece of sodium crystal, electron easily aquire additional energy, in the form of kinetic energy to move to conduction band where they can move freely, constituting and electric current sodium (and all other alkali metals) are, therefore, good conductors of electricity. Thus in conductors, bands either overlap or are only partially filled. Hence, the valence band itself is also the conduction band. Metals have a high electricl conductivity on account of the presence of large number of free electrons.

(*ii*) **Insulators:** The energy band diagram of insulators is shown in the second figure. A solid behaves as an insulated if it satisfies two conditions: (*a*) if has an even number of valence electrons per atom; (*b*) the valence and conduction bands are separated by an energy gap large compared with k_BT. Thus diamond with four valence electrons per atom and having a large gap of 6 eV between the valence band and the conduction band is an insulator. So the conduction band may have only little number of electrons even at high temperatures. Hence it is an insulator.

(*iii*) **Semiconductors:** Semiconductors have a forbidden band (energy gap) between filled valence band and unoccupied conduction band which is small compared to the gap in the case of insulators. This narrow gap is *Si* and *Ge* implies that some electrons from the completely filled valence band can be thermally excited to the empty conduction band where they are able to conduct small current on the application of electric field to the crystal.

Concept of Effective Mass of an Electron

Consider an isolated electron mass m and charge $-e$ in an electric field E. The force acting on it is $-eE$. It the acceleration produced in the electron is a such that $-eE = ma$. An electron in crystal may not be bound to any single atom. However, the neighbouring ions and electrons in the crystal do exert some force on an electron in a crystal. Hence by Newton's second law of motion we have.

$ma = -eE$ + force due to neighbouring ions and electrons. The latter force is not easily known quantitatively. We overcome this difficulty by simply transferring our ignorance to the left hand side of the equation by writing

$$m^* a = eE$$

Here m^* is called effective mass of the electron in the crystal.

SHORT QUESTIONS

4.1.1 If the mass of the neutron is 2025 times greater than that of electron, then show that the root mean square velocity of electron is 45 times that of the neutron. $T = 300$ K *(REC, Silchar, BE, May 2002)*

4.1.2* Discuss the motion of a bound electron from that of a free electron. *(Loughborough University of Technology, UK, BE April, 1997, Anna U., BE, November 1998)*

4.1.3 At what temperature, pressure remaining constant, the RMS velocity of a gas will be double its value at 0°C. *(University of Maryland, US, BE, 1985)*

4.1.4 What are Sommerfeld's assumptions to introduce wave mechanics to explain free electron theory of metals. *(Bangalore U., BE, May 1998)*

4.1.5 What are the shortcomings of classical free electron theory of metals? *(JNTU, BE, April 2000)*

4.1.6* Mention some of the important factors which reduce electron mobility in metals. *(Massachusetts, US, BE, May 2001)*

4.1.7 Explain density of energy states in a metal for the electrons to occupy. *(Anna U., M.Sc December 1997)*

4.1.8 If the Fermi energy of copper is 7 eV, the Fermi temperature is approximately 8×10^4 K. Explain. *(JNTU, BE, November 2000)*

4.1.9* At a temperature T K what is the energy of an electron? If the electron has an energy of V electron volt, then what is the temperature of the electron?

(U. of Groningen, Netherlands, UG, 2001)

4.1.10 Distinguish between degenerate and non-degenerate energy states.

(Loughborough University of Technology, UK, B. Tech., April 1989)

4.1.11* Show that the following two definitions of the resistivity ρ are equivalent

(*a*) resistance of a bar = $\dfrac{\rho \times \text{length of bar}}{\text{area of cross-section}}$

(*b*) $\rho = \dfrac{\text{electric field}}{\text{current density}}$ (Council of Engineering Institutions, New Jersey, 1995)

4.1.12 What is quantum physics? How does it differ from classical physics?

(Anna U., BE, May 1989)

4.1.13 What is meant by Fermi-Dirac distribution function? What are the kinds of particles to which it can be applied? *(JNTU, BE, November 2002)*

4.1.14 What are degenerate and non-degenerate energy states? Represent graphically the energy state with quantum numbers (1 2 3). Name it.

4.1.15* Explain anomalous expansion of water.

(Institute of Science and Technology, UK, B. Tech., Winter 1988, Anna U., BE, December 1990)

4.1.16 Discuss Brillouin zones in two dimensions.

(Physical metallurgy division, University of Florida, Mid-summer course, June 1980, Anna U., BE, May 1990)

4.1.17 We like to have a small discussion of the shortcomings of Sommerfeld's free electron theory of metals. Use lucid language in answering this short question.

(Stanford University, US, UG, B. Tech., Winter 1990, Anna U., Part time BE, November 1996)

4.1.18 Explain the terms gauge factor and Bridgeman constant.

4.1.19* Write a note on effective mass approximation of electrons.

(I.I.T. Bombay, BE, May 1992)

4.1.20 Give the salient features of high resistivity materials.

4.1.21 Distinguish between classical and quantum statistics.

Answers to the starred questions

4.1.2* The essential difference in the motion of a bound electron (e.g. in an atom) and a free electron consists in the fact that the potential energy of the latter is zero where as the bound electron is moving in a field of force and hence possesses a potential energy in addition to the kinetic energy.

4.1.6* (*i*) Thermal agitation: increase in temperature increases the resistance.
(*ii*) Plastic deformation: internal stresses are caused by cold working which distort the energy levels at which electron transfer takes place.
(*iii*) Impurities: they cause lattice distortion and decrease in mean free path.
These imperfections cause irregularities and hence reduce the mean free path of the electrons and thus reduce the mobility of the electron.

4.1.9*
$$E = k_B T \text{ joule} = \frac{k_B T}{e} \text{ electron volt}$$

$$E = \frac{1.38 \times 10^{-23} \times T}{1.6 \times 10^{-19}}$$

$$E = \frac{T}{11600} \text{eV}; \qquad E = \frac{eT}{11600} \text{ joule}$$

If an electron is under V volt, then the energy is eV joule

$$\text{eV} = \frac{eT}{11600}$$

$$T = 11600 \text{ V kelvin}$$

4.1.11* (*a*) resistance of a bar $= \dfrac{\rho \times \text{length of bar}}{\text{area of cross-section}}$

(*b*)
$$\rho = \frac{\text{electric field}}{\text{current density}}$$

Ohm's law is

$$V = IR$$

$$R = \frac{V}{I} = \frac{\rho l}{a}$$

But
$$E = \frac{V}{l} = \frac{I\rho}{a} = j\rho$$

$$\rho = \frac{E}{j}$$

Thus both are equivalent.

4.1.15* In majority of liquids, the increase in temperature produces an increase in volume of the liquids but water is a *notable* exception. The expansion of water is so markedly irregular that even ordinary laboratory methods can detect the anomaly. In the case of water, from 0°C to 4°C, instead of expanding actually, contracts, from 4°C upwards it expands, far from uniformity. In the range 5°C to 10°C, the *coefficient of expansion* is only 0.53×10^{-4}, where as in the range of 20°C to 40°C it is 3.02×10^{-4} and between 60°C and 80°C it is 5.87×10^{-4}.

Hence for a given mass of water, the volume is minimum or density is maximum at

4°C. In other words, in the expression for the volume coefficient, $\frac{dV}{dT}$ is negative for temperature less than 4°C and positive for temperature greater than 4°C. At 4°C $\frac{dV}{dT} = 0$. This means that the specific volume decreases first, reaching a minimum value at 4°C and then increases; it is evident that the density curve will follow an inverse path, increasing from 0°C to 4°C where it reaches a maximum and then decreases. It may be noted that the curve is not linear and hence the expansion is not uniform even beyond 4°C. This *anomalous behaviour* of water has been explained on the assumption that the three types of water molecules, H_2O, $(H_2O)_2$ and $(H_2O)_3$ which have different specific volumes are mixed in different temperatures, so that the maximum density at 4°C is probably due to greater proportion of the higher types of molecules. The result is, water at 4°C sinks to the bottom. A close *lookism* of this wonderful behaviour of water on either side of 4°C reveals that Nature has bestowed this special property to water to enable aquatic life to survive even during severe winter.

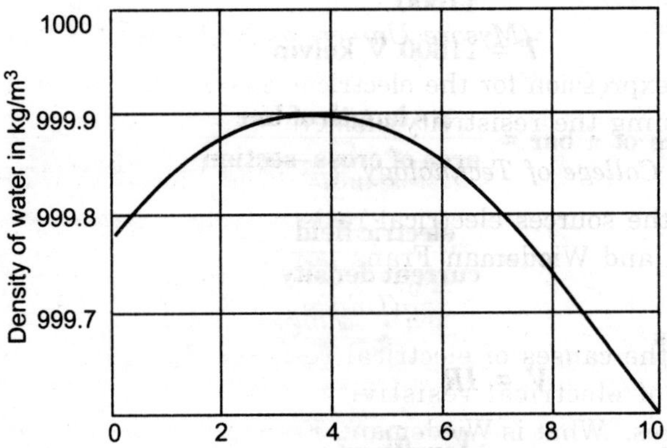

Fig. Anomalous expansion of water

4.1.19* For a completely free electron, the kinetic energy K.E = $\frac{p^2}{2m}$ with $p = \frac{h}{\lambda}$ or $\hbar k$. The $(E - k)$ curve is the parabola with $E = \frac{\hbar^2 k^2}{2m}$. Obviously the latter equation does not apply to the heavy curve of Fig. 4.11.3. This problem may be discussed introducing the concept of effective mass, m^*. The concept of effective mass is freely and widely used in physics of condensed materials. When a free electron is subjected to an electric field $\vec{E}$, it experiences an acceleration $\vec{a} = -\frac{e\vec{E}}{m}$. However, when the external electric field is applied to solids say a metallic crystal, only completely free electrons, if any have an acceleration $-\frac{e\vec{E}}{m}$. For example, a copper K electron is so tightly bound to its atom that it is not accelerated at all; its *effective mass* is infinite. For an electron which is not bound to any single atom, Newton's second law gives $m\vec{a} = eE +$ force due to neighbouring ion cores and electrons. These latter forces we do not know quantitatively, but we can transfer our ignorance to the left side of the equation by writing

$$m^*\bar{a} = -e\bar{E}$$

where m^* is the effective mass of the electron.

REVIEW QUESTIONS

4.1.1 What is electronic conductivity? Discuss the factors which are responsible for reducing its value. Discuss classical free electron theory of metals and obtain the expression for electrical conductivity. *(REC, Kurukshetra, BE, November 1998)*

4.1.2 Outline the classical free electron theory of metals. How is the conductivity of conductors related to the relaxation time?

(Mysore University BE, April 1997, JNTU, BE, November 2000)

4.1.3 Derive an expression for the electrical conductivity of a metal. Explain Matthiessen's rule governing the resistivity of metals.

(Imperial College of Technology, UK, B. Tech., 1992, Anna U., BE, November 1998)

4.1.4 What are the sources electrical resistivity in a metal? Give the electronic theory of conduction and Wiedeman Franz law.

(University of Surrey, UK, 1992, MKU, BE, May 1996)

4.1.5 What are the causes of electrical resistivity in a metal? Describe the temperature variation of electrical resistivity of metals (*i*) high temperatures and (*ii*) low temperatures. What is Wiedemann-Franz relation and how well it is obeyed in metals? *(Meerut University, M.Sc., April 1998)*

4.1.6 What is Fermi level? Derive an expression for the Fermi energy in terms of the number of electrons per unit volume. *(Anna U., BE, November 1998)*

4.1.7 What are the drawbacks of classical free electron theory of metals? How are these removed by the application of quantum states?

(JNTU, BE, May 1999, Anna U., BE, April 1997)

4.1.8 Explain Fermi-Dirac distribution for electrons in a metal. Obtain the expression for Fermi energy E_F at temperature $T = 0$ and relate it to Fermi energy at non-zero temperature T, where T is such that $k_BT << E_F$. *(VTU, BE, July 2002)*

4.1.9 Explain electric conductivity on the basis of Sommerfeld's model. Explain how this quantum theory is successful in explaining molar electronic specific heat and Wiedemann-Franz law. *(REC Jaipur, BE, December 2003)*

4.1.10 Write down the quantum mechanical expression for electrical conductivity and distinguish it from that of classical one. What are the sources of resistance in metals? How the resisitvity of alloys varies with temperature and impurity concentration? *(Anna U., BE, December 2003, JNTU, BE, May 2004)*

4.1.11 Obtain the mean energy of the electron at 0 K. If the mean energy at T K is given by the equation

$$\overline{E} = E_0\left[1+\frac{5\pi^2}{12}\left(\frac{k_BT}{E_F}\right)^2\right]$$

Show that the molar electronic specific heat is $\frac{\pi^2}{2}\left[\frac{T}{T_F}\right]R_u$.

(Anna U., M.Sc. May 1996)

4.1.12 Discuss the quantum treatment of electric conductivity and obtain the final expression for electric conductivity.

4.1.13 Show that as long as the Fermi energy E_F of the free electrons in a metal is very large compared to k_BT, the molar heat capacity of these electrons is

$$[C_v]_{el} = \frac{\pi^2}{2}\left[\frac{k_BT}{E_F}\right]R_u$$

The Fermi energy of copper is 7 eV, compute the value of Fermi velocity.

4.1.14 Give the physical meaning of wave function. What is Fermi energy? Using the expression for density of states, obtain the expression for Fermi energy.

(MKU, BE, March 1992, JNTU, BE, December 2000)

4.1.15 What are the factors that affect the resistivity of electrical materials? Describe the temperature variation of electrical conductivity of metals at (*i*) high temperatures, and (*ii*) low temperatures. *(REC, Calicut, BE, April 2001)*

PROBLEMS AND SOLUTIONS

4.1.1 The resistance of an aluminium wire of length 10 cm and diameter 1mm is 0.0034 Ω. If the Lorentz number of aluminium at room temperature is 2.01 × 10⁻⁸ W Ω K⁻² at 300 K, calculate the thermal conductivity of aluminium.

(University of Groningen, UK, UG winter 1989)

Solution:

$$R = \rho\frac{l}{a}$$

$$\rho = \frac{aR}{l} = \frac{\pi r^2 R}{10} = \frac{\pi\times(0.5\times10^{-3})^2\times0.0034}{10}$$

$$\rho = 2.7\times10^{-8}\ \Omega\ \text{m}$$

$$L = \frac{\sigma_T}{\sigma T} = 2.01\times10^{-8}$$

$$\sigma_T = L\sigma T = \frac{L \times T}{\rho}$$

$$\boxed{\sigma_T = 223 W m^{-1} K^{-1}}$$ ***Answer***

4.1.2 The Fermi energy and interatomic distances in several metals are given below: check in each case that the electrons of highest energy have a wavelength greater than twice the distance between atomic planes, and hence not diffracted by the lattice.

Metal	Na	K	Cu	Ag
Fermi energy (eV)	3.12	2.14	7.04	5.51
Inter planar spacing (Å)	3.02	3.88	2.09	2.35

(*University of Groningen, B. Tech., April 1990*)

Solution:

Let us consider Na

$$\frac{1}{2} m v_F^2 = 3.12 \times 1.6 \times 10^{-19} \text{ joule}$$

$$v_F^2 = \frac{6.24 \times 1.6 \times 10^{-19}}{9.1 \times 10^{-31}}$$

$$v_F = 1.048 \times 10^6 \text{ m/s}$$

$$\lambda_F = \frac{h}{m v_F} = \frac{6.62 \times 10^{-34}}{9.1 \times 10^{-31} \times 1 \times 10^6}$$

$$\boxed{\lambda_F = 7.2 \text{ Å}}$$ ***Answer***

This is much larger than the interplanar distance and so no diffraction is possible. Similarly try the other cases.

4.1.3 The Fermi energy of copper at 0 K is 7.04 eV. What is the percentage of change of Fermi energy at 1000 K ? (*IIT Bombay, B. Tech., 1990, Anna U., BE, April 2000*)

Solution:

$$E_f = E_F \left[1 - \frac{\pi^2}{12} \left(\frac{1.38 \times 10^{-23} \times 1000}{7.04 \times 1.6 \times 10^{-19}} \right)^2 \right]$$

$$= 7.04 \left[1 - \frac{\pi^2}{12} \left(\frac{1.38 \times 10^{-20}}{7.04 \times 1.6 \times 10^{-19}} \right)^2 \right]$$

$$= 7.04 [1 - 0.00074]$$

$$E_f = 7.03 \text{ eV}$$

$$E_F - E_f = 7.04 - 7.03 = 0.01$$

$$\frac{0.01 \times 100}{7.04} = 0.142\%$$

$$\boxed{0.142\%}$$ ***Answer***

4.1.4 Compute the Fermi kinetic energy for aluminium assuming that there are three free electrons per atom. Over what temperature range is this energy greater than 10 k_BT? The density of aluminium is 2750 kg/m^3.

(California University of Technology, US, UG November 1989, VTU, BE, July 2004)

Solution:

$$\frac{27}{2750}\text{ m}^3 \text{ contain } 3 \times 6.02 \times 10^{26} \text{ free electrons}$$

$$1\text{ m}^3 \text{ will have } \frac{2750\times3\times6.02\times10^{26}}{27}$$

$$= 1.8 \times 10^{29}$$

$$E_F = 3.65\times10^{-19}\times(1.8\times10^{29})^{2/3} = 11.6 \times 1.6 \times 10^{-19}$$

$$10k_BT = 11.6\times1.6\times10^{-19} \text{ joule}$$

$$T = \frac{11.6\times1.6\times10^{-19}}{10k_B} = \frac{11.6\times1.6\times10^{-19}}{10\times1.38\times10^{-23}}$$

$$\boxed{T = 1.34\times10^4 K} \quad \textbf{\textit{Answer}}$$

Thus E_F will be greater than 10 k_BT above 1.34 × 10^4 K.

4.1.5 Show that the wavelength associated with an electron having an energy equal to the Fermi energy is given by

$$\left[\frac{8\pi}{3n}\right]^{1/2}$$

If this wavelength is 5 Å, get the Fermi temperature.

(University of Sheffield, US, UG May 1988, Anna U., BE, April 1998)

Solution:

$$\frac{1}{2}mv_F^2 = E_F$$

$$m^2v_F^2 = 2m\,E_F$$

$$p_F = \sqrt{2m\,E_F}$$

$$\lambda_F = \frac{h}{p_F} = \frac{h}{\sqrt{2mE_F}}$$

Substituting for E_F

$$\lambda_F = \frac{h}{\sqrt{2m}} = \left[\frac{2m}{h^2}\right]^{1/2}\left[\frac{8\pi}{3n}\right]^{1/3}$$

$$\boxed{\lambda_F = \left[\frac{8\pi}{3n}\right]^{1/3}} \quad \textbf{\textit{Answer}}$$

$$\lambda_F^3 = \left[\frac{8\pi}{3n}\right]$$

$$n = \frac{8\pi}{3}\left[\frac{1}{\lambda_F^3}\right]$$

$$= \frac{8\pi}{3}\left[\frac{1}{(0.5\times10^{-9})^3}\right] = 6.7 \times 10^{28}$$

$$E_F = 3.65\times10^{-19}\times(6.7\times10^{28})^{2/3} \text{ eV}$$

$$E_F = 6.0 \text{ eV}$$

$$k_B T_F = 6\times1.6\times10^{-19}$$

$$T_F = \frac{9.6\times10^{-19}}{1.38\times10^{-23}} = 7\times10^4 \text{ K}$$

$$\boxed{T_F = 7\times10^4 \text{ K}}$$ ***Answer***

4.1.6 In the Kronig-Penny model determine the energy of the lowest band for P << 1.

Solution:

The energy of the lowest band corresponds to $k = \pm\frac{n\pi}{a}$ with n = 1 and

$$\frac{P}{a\alpha}\sin\alpha a + \cos\alpha a = \pm1$$

Considering only the magnitude ± 1, we can write

$$\frac{P}{a\alpha}\sin\alpha a = (1-\cos\alpha a)$$

$$\frac{P}{a\alpha}\left[2\sin\frac{\alpha a}{2}\times\cos\frac{\alpha a}{2}\right] = (1-\cos\alpha a) = 2\sin^2\frac{\alpha a}{2}$$

$$\frac{P}{a\alpha} = \frac{2\sin^2\frac{\alpha a}{2}}{\left(2\sin\frac{\alpha a}{2}\times\cos\frac{\alpha a}{2}\right)}$$

$$\tan\frac{\alpha a}{2} = \frac{P}{\alpha a}$$

when 1 >> P, then

$$\frac{P}{a\alpha} = \frac{\alpha a}{2}; \alpha^2 = \frac{2P}{a^2} = \frac{2mE}{\hbar^2} \quad \text{(Refer Eqn. 4.1.1)}$$

$$E = \frac{\hbar^2}{ma^2}P = \frac{h^2 P}{4\pi^2 ma^2}$$

$$\boxed{E = \frac{\hbar^2}{ma^2}P}$$ ***Answer***

EXERCISE

4.1.1 Copper has electrical resistivity of 1.5×10^{-8} Ω m at 290 K. If the density of copper is 8.9×10^3 kg/m^3, determine the relaxation time.

(Metallurgy Division, University of Manchester, April 1985, MKU December 1990)
***(Ans**: 2.7×10^{-14} s)*

4.1.2 If 8.5×10^{28} is the number of free electron in one cubic metre, get the contribution of these electrons to the specific heat taking $k_B = 1.38 \times 10^{23}$ J sec.

(Royal Military College, Canada, UG March 1985, Anna U., BE, December 1990)
***(Ans**: 12.75×10^5 J kg^{-1} K^{-1})*

4.1.3 If the probability of an electron occupying an energy state is 0.3 at 300 K, compute the interval between this state and the Fermi state.

(Imperial College, London, winter 1998, Anna U., May 2000)
***(Ans**: 0.021 eV)*

4.1.4 Assuming free electron theory, calculate the Fermi energy of fcc aluminium with 0.404 nm as the dimensions.

(Department of Chemistry, The Open University, UK, May 2000)
***(Ans**: 11.7 eV)*

4.1.5 Given the Fermi energy of sodium to be 3.1 eV and the observed conductivity to be 2.1×10^{17} esu at 0 K. Calculate the relaxation time of the electron. Also calculate the electron mean free path.

(University of Maryland, UK, UG winter 1975, Anna U., May 1999)
***(Ans**: 3.2×10^{-14} sec, 3.3 nm)*

4.1.6 The Fermi energy of silver is 5.5 eV (*a*) What is the average energy of the free electron in silver at 0 K? (*b*) At what temperature a classical free (such as ideal gas) molecule will have the same kinetic energy? *(JNTU, BE, May 2000)*
***(Ans**: 3.3 eV, 2.25×10^4 K)*

4.1.7 Compute the electronic specific heat of copper at 600 K. Take the Fermi energy at 7.04 eV. *(University of Rajasthan, Jaipur, B.Sc., April 2004)*
***(Ans.** 3×10^3 J Kmol^{-1}K^{-1})*

4.1.8 Copper has electrical conductivity at 300 K as 6.4×10^7 Ω^{-1} m^{-1}. If Wiedemann-Franz law is valid, then determine the electronic contribution to the thermal conductivity of copper. *(MKU, BE, April 1992)*
***(Ans**: 470 Ω m^{-1} K^{-1})*

4.1.9 An alloy of a metal is found to have a resistivity of 10^{-6} Ω m at 300 K. When it is heated to a temperature of 1000 K, the resistance increases by 7%: Using Matthiessen's rule, determine the resistivity of the alloy. *(Anna U., BE, April 1995)*
***(Ans**: 0.97×10^{-10} Ω m)*

4.1.10 If the Fermi energy of a metal at 0 K is 0.36x5 eV with a concentration of 10^{27} atom/m^3, compute the constant of proportionality. ***(Ans**: 3.65×10^{-19})*

Chapter 5

UNIT 5

Semiconductor Physics and Physics of Semiconductor Devices

5.1 SEMICONDUCTOR PHYSICS

5.1.1 INTRODUCTION

In a covalent crystal each bond contains two electrons with opposite spin directions. One must not think, however, that these electrons are immobile, condemned for ever to stay in the same bond. The probability distributions are smeared out so that those of neighbouring bonds over lap slightly, and this make it possible for a pair of electrons in adjacent bonds to change places with one another. Indeed, all the electrons are continually doing so, moving through the net work of bonds at greater velocity keeping Pauli's principle intact. However, the *charge density* at each point remains unchanged, and Pauli's principle is satisfied. We thus decide that no net flow of charge in any direction is possible *i.e.*, a covalent crystal cannot carry electric current. The above argument suggests that a crystal with covalent bonding should be a good insulator, and in some cases this is substantially correct. Our prototype covalent solid, diamond, has a resistivity of 10^{12} Ω m at room temperature, and *polymers* such as *polyethylene* and *polyvinyl chloride* have resistivities of about 10^{15} Ω m in their pure state. However, two elements, silicon and germanium, both of which are of great technological importance, have much lower resistivities. Why is this? If we compare the bond strengths of diamond, silicon, and germanium, we find that they decrease as the atomic weight increases. It is thus easier to break the bonds in germanium than in silicon, while it is hardest of all to break them in diamond. If the bond is sufficiently weak, the *thermal vibration* of the atoms may occasionally be sufficient to sever a bond, even when the temperature is not very high. Naturally, the higher the temperature the more likely it is that any one bond will be broken, and hence more bonds will be broken at a given time.

Now breaking a bond does not affect the atoms which it joins together-each still has three bonds to keep it in place. But breaking a bond does entail releasing one of the electrons which form it. Thus at any temperature a small proportion of the large number of bonding electrons will be set free. Being no longer confined in the bonds, these electrons may wander freely through the lattice and conduct an electric current.

$10^{16}/m^3$ is the order of density of charge carriers in pure silicon at 300 K. This is about 10^{12} times smaller than the number of silicon atoms/m^3. It is also interesting to note the number of charge carriers in metals (say copper) is of the order of $10^{28}/m^3$. Hence for more current to flow greater voltage must be applied in a semiconductor. However, by doping (incorporating) trace of suitable *foreign atoms* in regular lattices of silicon or germanium may increase the charge carriers many fold just like pluming the cake with small quantity of pieces of nuts at regular intervals enhances the taste substantially. *One general information is that the statistical study of these doped semiconductors is not a cake walk but a walk on the cake.*

5.1.2 ENERGY BANDS IN A SEMICONDUCTOR

The present problem is to study the bands arising from the 3*s* and 3*p* states of silicon as a function of interatomic distance.

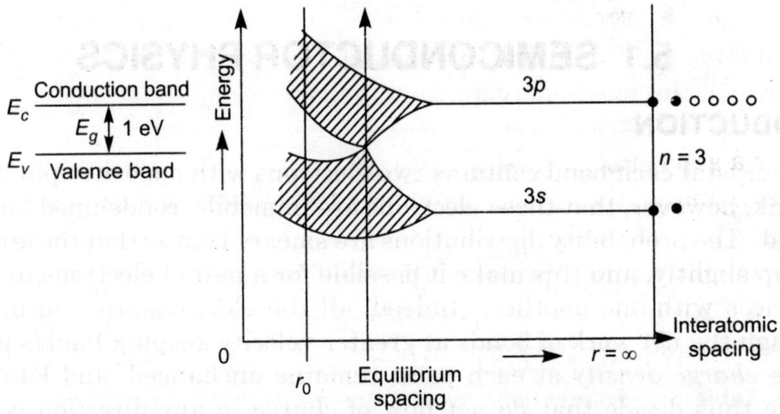

Fig. 5.1 The bands arising from the 3*s* and 3*p* states of silicon as a function of interatomic distance

The electronic configuration of silicon (14) is $1s^2$, $2s^2\ 2p^6$, $3s^2\ 3p^2\ 3d^0$. The outermost shell contains 4 electrons of which 2 in the 3*s* level and the other 2 in the 3*p* level though it can accommodate 6 in total. When the silicon atoms come close together strong interaction starts and the sharp 3*s* and 3*p* levels widen into bands. At one particular distance of separation (called *cross over point*), the bands merge and split again. What is actually happening at the *equilibrium distance* r_0 where stability is reached? The *unoccupied states* put together form the *upper conduction* band separated by the occupied states put together form the lower *valence band*, by a *forbidden energy* gap. This *band gap* (E_g) is 0.67 eV for Ge and 1.106 eV for Si at 300 K.

The energies of the levels in the valence band must all be below those of the conduction band, for we have seen that an electron has to acquire energy from the *thermal vibrations* in the lattice in order to leave a bond and become free. This situation is shown on the energy band diagram in Fig. 5.2.

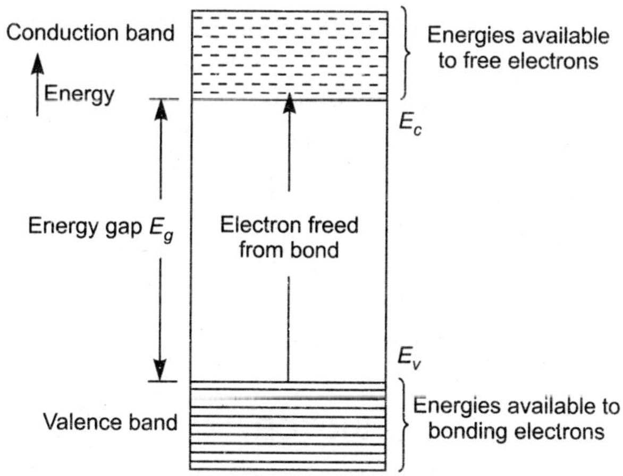

Fig. 5.2 The energy diagram for a semiconductor, showing the energy gap

There is, in fact, a minimum value for the amount of additional energy which a bonding electron must acquire in order to leave the bond. The absolute minimum for this energy is that which the most energetic bonding electron needs in order just to become free, with no kinetic energy left over to carry it away. This is represented in the energy level diagram by a gap between the highest level in the band of energies occupied by the bonding electrons, that is, the valence band, and the lowest level in the conduction band as shown in Fig. 5.2.

The arrow in this figure represents the transition of an electron from the valence band to a state in the conduction band, *i.e.*, from a bond to freedom. This leaves vacant a state in the valence band, *i.e.*, a 'hole'.

Table 5.1. Energy gaps, electron densities and conductivities

Material	E_g (eV)	*Calculated value of n at 300 K* (m^{-3})	*Measured value of* (σ) *at 300 K* ($\Omega^{-1}m^{-1}$)
Copper	–	8.5×10^{28}	5×10^{7}
Aluminium	–	1.8×10^{29}	3.3×10^{7}
Gray tin	0.08	5.4×10^{24}	10^{6}
Germanium	0.65	9.3×10^{19}	2.2
Silicon	1.15	6×10^{15}	5×10^{-4}
Diamond	5.3	1.1×10^{-20}	10^{-12}
Li F	11	2×10^{-70}	10^{-11}

5.1.3 THEORY OF INTRINSIC SEMICONDUCTORS

When an electron near the top of the valence band (*i.e.*, in a bond) collides with a vibrating atom, it can absorb some of the atoms' energy and jump into the conduction band. This is just another way of describing how *thermal vibrations* can break a bond and create an *electron-hole pair*. Referring Eqn. (5.1.1), one can write the number of energy states in the conduction band having energy values lying between E and $E + dE$ (considering a volume of one metre^3) as

$$Z(E)dE = \frac{\pi}{2}\left[\frac{8m}{h^2}\right]^{3/2} (E - E_c)^{1/2} dE$$

Hence the number of electrons in this interval dE is

$$N(E)\, dE = Z(E)dE\, F(E)$$

$$= \frac{\pi}{2}\left[\frac{8m}{h^2}\right]^{3/2} (E - E_c)^{1/2} \frac{dE}{[1 + \exp(E - E_F)/k_B T]}$$

Thus

$$\int N(E)dE = \frac{\pi}{2}\left[\frac{8m}{h^2}\right]^{3/2} \int_{E_c}^{\infty} \frac{(E - E_c)^{1/2}\, dE}{[1 + \exp(E - E_F)/k_B T]} \qquad ...(5.1.1)$$

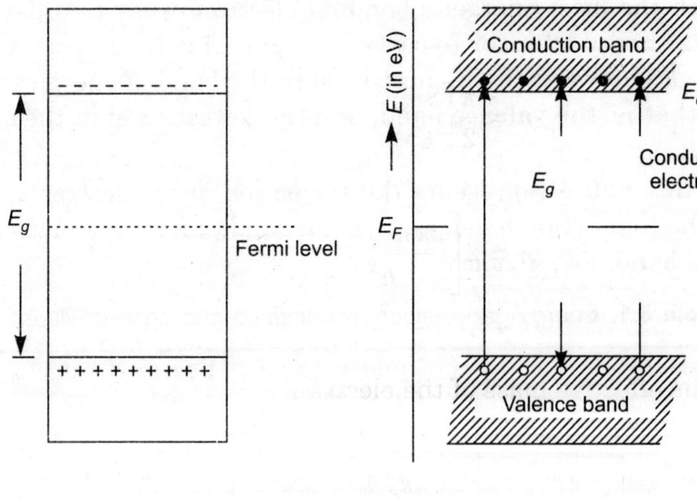

Fig. 5.3 Thermal excitation of electrons

At 300 K, $(E_c - E_F)$ as well as $(E - E_F)$ will be at least 20 times greater than k_BT. Thus Eqn. (5.1.1) becomes

$$n = \frac{\pi}{2}\left[\frac{8m}{h^2}\right]^{3/2} \int_{E_c}^{\infty} (E - E_c)^{1/2} \exp[(E_F - E)/k_B T]dE$$

Let

$$x = E - E_c \text{ and therefore } dx = dE$$

and

$$E = x + E_c$$

Thus

$$n = \frac{\pi}{2}\left[\frac{8m}{h^2}\right]^{3/2} \int_0^\infty x^{1/2} \exp[(E_F - x - E_c)/k_BT]dx$$

or

$$n = \frac{\pi}{2}\left[\frac{8m}{h^2}\right]^{3/2} e^{\frac{E_F-E_c}{k_BT}} \int_0^\infty x^{1/2} e^{-x/k_BT} dx$$

Using $\alpha = \frac{1}{k_BT}$,

$$n = \frac{\pi}{2}\left[\frac{8m}{h^2}\right]^{3/2} e^{\frac{E_F-E_c}{k_BT}} \int_0^\infty x^{1/2} e^{-\alpha x}\, dx$$

The solution of this integral using gamma function is

$$\int_0^\infty x^{1/2} e^{-\alpha x}\, dx = \frac{\sqrt{\pi}}{2}\frac{1}{\alpha^{3/2}} = \frac{\sqrt{\pi}}{2}(k_BT)^{3/2}$$

Thus

$$n = \frac{\pi}{2}\left[\frac{8m}{h^2}\right]^{3/2} e^{\frac{E_F-E_c}{k_BT}} \quad \frac{\sqrt{\pi}}{2}(k_BT)^{3/2}$$

$$n = 2\left[\frac{2\pi m_e^* k_BT}{h^2}\right]^{3/2} e^{\frac{E_F-E_c}{k_BT}}$$

with m_e^* — the effective mass of the electron

i.e.,

$$\boxed{n = N_e\, e^{\frac{(E_F-E_c)}{k_BT}}} \quad \text{...(5.1.2)}$$

Calculation of Holes

$[1 - F(E)]$ is the probability of missing the electron in the valence band which gives the probability of finding a hole there.

i.e.,

$$1 - F(E) = 1 - \frac{1}{1 + e^{\frac{(E-E_F)}{k_BT}}}$$

$$= \frac{1 + e^{\frac{(E-E_F)}{k_BT}} - 1}{1 + e^{\frac{-(E_F-E)}{k_BT}}}$$

Since $\frac{-(E_F-E)}{k_BT}$ is a large negative quantity,

$$1 - F(E) = e^{\frac{(E-E_F)}{k_BT}}$$

Referring Eqn. 5.1.1 and proceeding in the same line, one will get,

$$P(E)dE = Z(E)dE[1 - F(E)] = \frac{\pi}{2}\left[\frac{8m_p^* k_BT}{h^2}\right]^{3/2} (E_v - E)^{1/2} e^{\frac{(E-E_F)}{k_BT}} dE$$

A simplification as in the case of electrons yields

$$p = 2\left[\frac{2\pi m_p^* k_BT}{h^2}\right]^{3/2} e^{\frac{(E_v-E_F)}{k_BT}}$$

or

$$\boxed{p = N_p e^{\frac{(E_v-E_F)}{k_BT}}} \qquad ...(5.1.3)$$

5.1.4 FAVOUR OF THESE EQUATIONS FOR DENSITY OF CHARGE CARRIERS

The vanity of these equations is discussed below systematically with physical significance and conclusions: (Favour is features in Shakespearean language).

(i) Position of Fermi level

In an intrinsic semiconductor, the number of electrons and the number of holes in a given volume are the same.

i.e.,

$$[m_e^*]^{3/2} e^{\frac{(E_F-E_c)}{k_BT}} = [m_p^*]^{3/2} e^{\frac{(E_v-E_F)}{k_BT}}$$

or

$$e^{\frac{2E_F}{k_BT}} = \left[\frac{m_p^*}{m_e^*}\right]^{3/2} e^{\frac{(E_c+E_v)}{k_BT}}$$

i.e.,

$$\frac{2E_F}{k_BT} = \frac{(E_c + E_v)}{k_BT} + \frac{3}{2} ln\left[\frac{m_p^*}{m_e^*}\right]$$

i.e.,

$$E_F = \frac{(E_c + E_v)}{2} + \frac{3k_BT}{4}\ln\left[\frac{m_p^*}{m_e^*}\right] \quad ...(5.1.4)$$

The Fermi level lies exactly at the middle of the forbidden gap if $m_p^* = m_e^*$

(ii) Calculation of intrinsic carriers

1. If $m_e^* = m_p^*$, then *Fermi level* lies exactly mid-way between the conduction band and valence band. Usually $m_p^* > m_e^*$. In that case, the Fermi level is slightly shifted from this position towards the conduction band.
2. Computation of intrinsic charge carriers: This term is usually represented as n_i. Thus

$$n_i^2 = np = (N_eN_p)\ e^{\frac{E_v - E_c}{k_BT}}$$

$$n_i^2 = np = (N_eN_p)\ e^{\frac{-(E_c - E_v)}{k_BT}}$$

or

$$n_i^2 = 4\left[\frac{2\pi k_BT}{h^2}\right]^3 (m_e^*m_p^*)^{3/2}\ e^{-\frac{E_g}{k_BT}}$$

$$n_i = 2\left[\frac{2\pi k_BT}{h^2}\right]^{3/2} (m_e^*m_p^*)^{3/4} e^{-\frac{E_g}{2k_BT}} \quad ...(5.1.5)$$

If $m_e^* = m_p^*$ is equal to rest mass m, then

$$n_i = 2\left[\frac{2\pi k_BT}{h^2}\right]^{3/2} (m^{3/2})\exp(\ E_g/2k_BT)$$

$$n_i = 2\left[\frac{2\pi mk_BT}{h^2}\right]^{3/2} T^{3/2}\ e^{-\frac{E_g}{2k_BT}} \quad ...(5.1.6)$$

$$n_i = 4.83 \times 10^{21} \times T^{3/2} \times e^{-\frac{Eg}{2k_BT}}$$

$$n_i = C \times T^{3/2}\ e^{-\frac{Eg}{2k_BT}} \quad ...(5.1.7)$$

with $C = 4.83 \times 10^{21}$

(iii) Conclusions

1. For a given semiconductor, we see an exponential addition of charge carriers as the temperature increases.
2. The magnitude of C is 4.83×10^{21} assuming $m_e^* = m_p^* = m = 9.1 \times 10^{-31}$ kg.
3. n_i is now calculated for $E_g = 1$ eV at 300 K as given below:

$$n_i = 4.83 \times 10^{21} \times 300^{3/2}\ e^{-\frac{1}{2k_B T}}$$

where

$$2k_B T = \frac{2 \times 1.38 \times 10^{-23} \times 300}{1.6 \times 10^{-19}} = 0.0518 \text{ eV}$$

Thus

$$n_i = 4.83 \times 10^{21} \times 300^{3/2}\ e^{-\frac{1}{2k_B T}}$$

$$\boxed{n_i = 1.03 \times 10^{17} \text{ /m}^3} \qquad ...(5.1.8)$$

4. *Study of electric conductivity of intrinsic semiconductors*
The general expression for conductivity in the case of metal is $\sigma = ne\mu$. For intrinsic semiconductors, it is

$$\sigma_i = \sigma_n + \sigma_p = ne\,\mu_e + pe\,\mu_p$$

i.e.,

$$\boxed{\sigma_i = n_i e(\mu_e + \mu_p)} \qquad ...(5.1.9)$$

Knowing, the values of the mobilities μ_e and μ_p it is possible to compute σ_i.

5. *Evaluation of energy gap of a semiconductor*
This is a very interesting proposition to work out. The mobility of charge carriers is inversely proportional to $T^{3/2}$
i.e.,

$$\mu_e \propto \frac{1}{T^{3/2}};\ \mu_e = \frac{\alpha}{T^{3/2}}$$

Similarly

$$\mu_p = \frac{\beta}{T^{3/2}}$$

Hence

$$(\mu_e + \mu_p) = \frac{(\alpha + \beta)}{T^{3/2}}$$

From equations (5.1.6) and (5.1.9), one gets

$$\sigma_i = n_i\, e\,(\mu_e + \mu_p) = C\,T^{3/2}\; e^{-\frac{E_g}{2k_BT}}\; \frac{(\alpha+\beta)}{T^{3/2}}$$

i.e.,

$$\sigma_i = C\,(\alpha+\beta)\; e^{-\frac{E_g}{2k_BT}} = B\; e^{-\frac{E_g}{2k_BT}}$$

or

$$\boxed{\rho_i = A\; e^{\frac{E_g}{2k_BT}}} \qquad \ldots(5.1.10)$$

Experiment to determine ρ_i

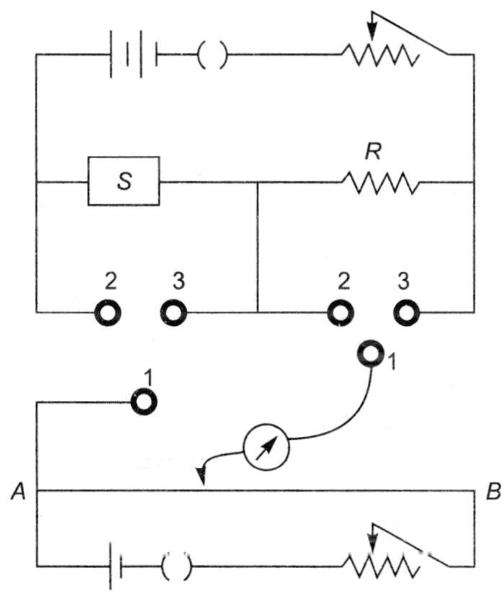

Fig. 5.4 Experimental arrangement for measuring ρ_i at different temperatures

A small crystalline specimen of the semiconductor is taken. Its opposite faces are copper plated. Two leads soldered to each face. One pair of leads is used to measure the p.d using a potentiometer. The other pair is used to measure the current. The specimen is inserted in a pyrex glass tube. The tube is immersed in an oil bath to vary the temperature. Alternatively the resistance of the specimen may be compared with a standard resistance using a potentiometer as shown in Fig. 5.4. S is the specimen and R is a resistance box. The resistance in the box is adjusted such that the balancing length at room temperature is at about the middle of the potentiometer wire AB. In the two way switches if 1 and 2 are connected the p.d across the specimen S gets balanced on the potentiometer. If 1 and 3 are connected the p.d across R gets balanced. The ratio of the two balancing lengths gives the ratio of the resistance of the specimen and R. From this the resistance of the specimen can be calculated. The tube containing the specimen is then placed in the oil bath. Using a thermometer the resistance at varies temperatures can be measured and a tabulation may be made as follows:

The resistivity at various temperatures may be obtained using the dimensions of the crystal.

T (K)	$\frac{1}{T}$ (K^{-1})	R Ω	ρ_i Ω m	$ln\ \rho_i$
T_1	$\frac{1}{T_1}$	R_1	ρ_1	–
T_2	$\frac{1}{T_2}$	R_2	ρ_2	–
T_3	$\frac{1}{T_3}$	R_3	ρ_3	–
T_4	$\frac{1}{T_4}$	R_4	ρ_4	–
T_5	$\frac{1}{T_5}$	R_5	ρ_5	–
T_6	$\frac{1}{T_6}$	R_6	ρ_6	–

Equation 5.1.10 now follows

$$ln\ \rho_i = \frac{E_g}{2k_B T} + ln\ A$$

This is the equation of a straight line. Draw a graph between $ln\ \rho_i$ and $\frac{1}{T}$. The slope will be $\frac{E_g}{2k_B}$ from which E_g can be determined.

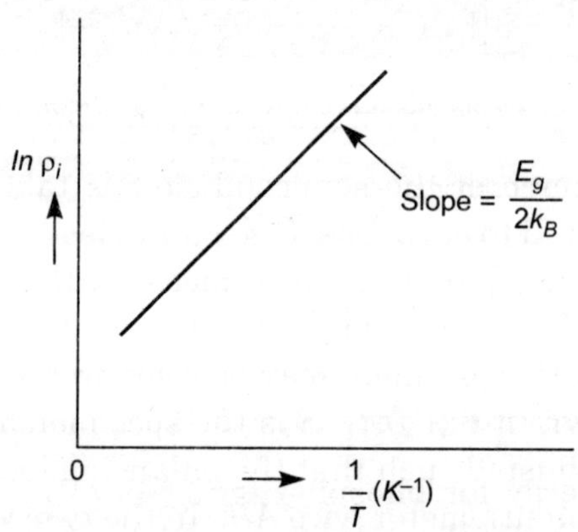

Fig. 5.5 $ln\rho_i$ versus $\frac{1}{T}$

5.1.5 IMPURITY SEMICONDUCTORS (*Theory and Practical*)

Historically the intrinsic semiconductors that have been studied extensively are germanium and silicon. In modern appliances silicon is used more than germanium because it is more abundantly available. Intrinsic semiconductors, however, are not much practical use because in them only 1 atom in 10^{10} contributes to electrical conduction. But it is found that the presence of suitable impurities even to the extent of 1 in 10^8 alters significantly the conduction of germanium and silicon. These impurities can be deliberately introduced in carefully controlled proportions. As a result we get a semiconductor of any *predetermined conductivity*. This process of adding an impurity to an intrinsic semiconductor is called *dopants*. The impurity that is added is called the dopant. The resulting material is called an *extrinsic semiconductor*.

Germanium and silicon belong to the fourth column in the periodic table. Consider the elements in the adjacent third and fifth columns. We here decide the effect of adding phosphorus (V^{th} column) and then Boron from the third column to silicon which is present in the IVth column.

Since at 0K, the electronic system is in its lowest energy state, all valence electrons will be in the valence band and all the phosphorus atoms will be unionised. The picture is similar to that for an intrinsic semiconductor in that the valence band is filled and the conduction band is empty. However, four of the valence electrons of the pentavalent impurity atom form covalent bonds with adjacent silicon atoms, but the fifth one occupies a state which lies 0.045 eV below the conduction band.

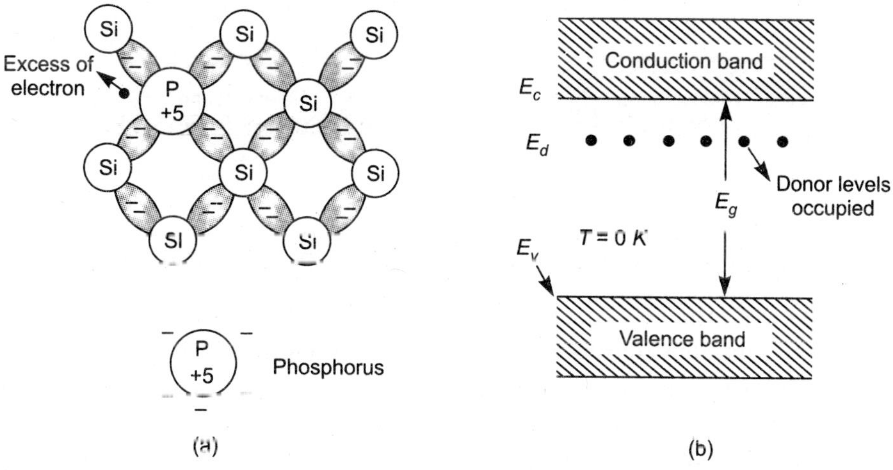

Fig. 5.6 (*a*) Representation of *n*-type silicon (*b*) Energy band diagram at 0 K

Since this ionization energy for phosphorus (0.045 eV) is much smaller than the energy E_g, required to break one of the valence bonds, the phosphorous atoms will donate their fifth electrons to the conduction band even at much lower temperature than the valence band will. An impurity of this kind is called a *donor*, since it donates conduction electrons without producing holes in the valence band. The donor level represents the ground state of the fifth electron of the impurity atom. The small gap between the donor level and the conduction band emphasizes that many of the extra electrons are thermally excited into the conduction band at comparatively low temperatures. Even at room temperature nearly all the donor atoms will have lost their electrons to the conduction band in this way.

The ionized donors remain fixed, positively charged centres incorporated in the lattice, but the additional electrons are free to move in the conduction band and they substantially increase the conductivity of the crystal. Thus, if pentavalent impurities exist in an otherwise intrinsic semiconductor, electrons are added to the conduction band, out numbering the holes in the valence band. The electrons then become the majority carriers and the holes the minority carriers. This type of material is now called *n-type semiconductor*.

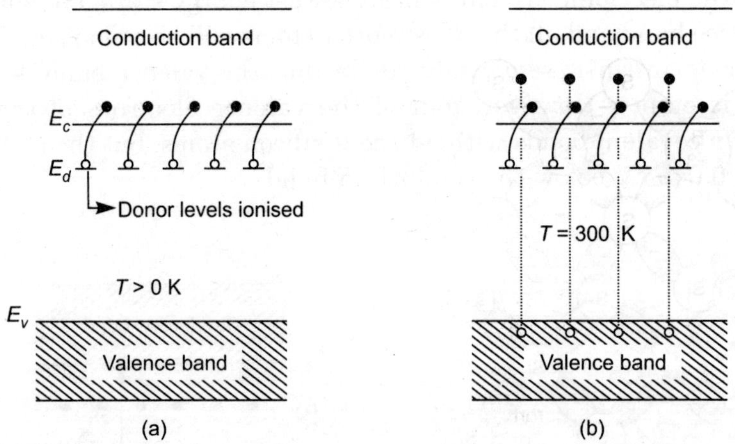

Fig. 5.7 (*a*) Energy band diagram of *n*-type semiconductor at $T > 0$ (*b*) At $T >> 0$ K (say 300 K)

The electrons *are majority carriers* and thermally generated holes are the minority carriers in n-type semiconductors. In an n-type semiconductor if n and p represent respectively the electron and hole concentrations, it can be shown that $np = n_i\, p_i = n_i^2$ where n_i and p_i are the intrinsic values. Here $n > n_i$ but $p < p_i$.

Let us now see the effect of adding a trivalent impurity say boron into silicon-silicon lattices.

Table 5.2. Ionization energy from donor and accepter states in silicon and germanium: Values are quoted in eV

Doping element	*Number of valence electrons*	*Silicon* (14) ($1s^2\ 2s^2\ 2p^6\ 3s^2\ 3p^2$)		*Germanium* (32) ($1s^2\ 2s^2\ 2p^6\ 3s^2\ 3p^2\ 3d^{10}\ 4s^2\ 4p^2$)	
		Donor ($E_c - E_d$)	*Accepter* ($E_a - E_v$)	*Donor* ($E_c - E_d$)	*Accepter* ($E_a - E_v$)
Phosphorous (15)	5	0.045	–	0.0120	–
Arsenic (33)	5	0.049 – 0.056	–	0.0127	–
Antimony (51)	5	0.039	–	0.0096	
Born (5)	3	–	0.045	–	0.0104
Aluminium (13)	3	–	0.057 – 0.067	–	0.0102
Gallium (31)	3	–	0.065 – 0.071	–	0.0108
Indium (49)	3	–	0.16	–	0.0112

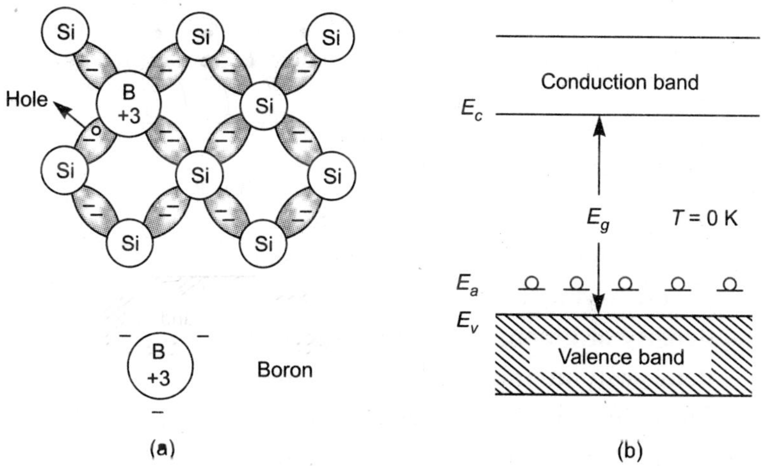

Fig. 5.8 (*a*) Representation of *p*-type silicon (*b*) Energy band diagram at 0 K

The three valence electrons enter into covalent bonding, but a region of stress exists where the fourth bond ought to be, any stray electron in the vicinity can be trapped by the impurity centre which then becomes negatively charged.

The regions from which the electrons came in are positively charged. In effect the trivalent impurity has injected a hole into the crystal. Such an impurity is called an *accepter impurity* since it accepts electrons into the impurity centre. Silicon or germanium doped in this way has many holes as charge carriers and is therefore termed a *p*-type semiconductor. In *p*-type semiconductors $p > n_i$ and $n < n_i$ with $pn = p_i n_i = n_i^2$.

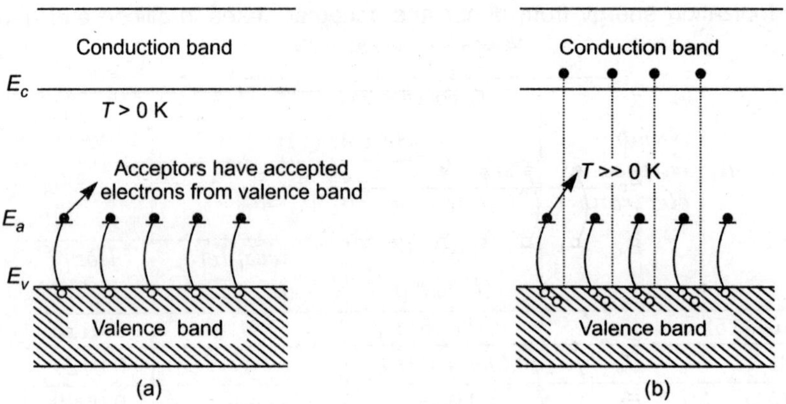

Fig. 5.9 (*a*) Energy band diagram of *p*-type semiconductor at $T > 0$ (*b*) at say 300 K

5.1.6 CARRIER DYNAMICS AND STATISTICS OF IONISED ATOMS

Let us now build up the same picture using the energy band model using some selected modifications. The salient feature of integrated-circuit technology is the control of the conductivity of silicon that is achieved by the controlled introduction of electrically active elements into the lattice. The following symbols are used while deriving the various relations:

N_d	– density of donor atoms
N_d^+	– density of ionized donor atoms
N_a	– density of acceptor atoms
N_a^-	– density of ionized accepter atoms
n	– usual symbol for density of electrons
p	– general symbol for density of holes
$(N_d - N_d^+)$	– density of unionized donor atoms
$(N_a - N_a^-)$	– density of unionized acceptor atoms
E_F	– Fermi level in intrinsic semiconductors
E_f	– Fermi level in impurity semiconductors

1. ***General equation for an impurity semiconductor***

You are now considering a pure semiconductor (say silicon) doped with both types of impurities. Let N_d and N_a be the concentration of the impurities.

One metre cube is the volume considered in almost all discussions. Let us now dope the original semiconductor with N_d and N_a impurity atoms in say unit volume of silicon. As the temperature increases, electrons are first excited from the donor levels to the conduction band and from the valence band to the acceptor levels. The electrons leaving the donor levels leave behind positively ionized donor atoms; similarly the electrons leaving the valence band make the acceptor atoms negatively charged and leave behind holes in the valence band. As the temperature is further increased, electrons are able to go over to the conduction band from the valance band.

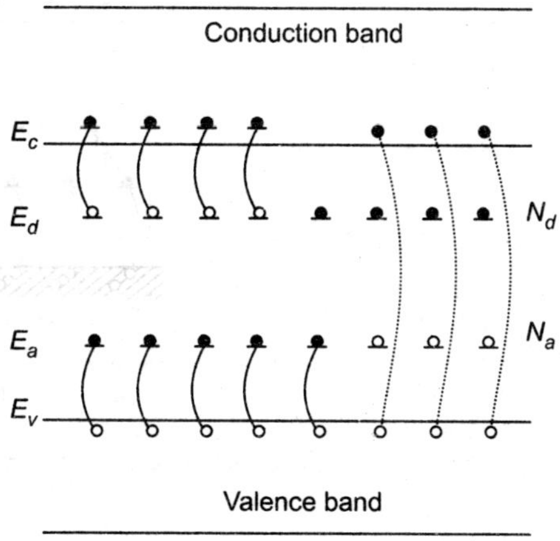

Fig. 5.10 A semiconductor with both type of impurities

For charge neutrality

$$(ne + N_a^- e) = (pe + N_d^+ e)$$

i.e.,

$$\boxed{n + N_a^- = p + N_d^+} \qquad \text{...(5.1.11)}$$

Apply Fermi-Dirac statistics and expand the above equation

$$N_a^- = N_a F(E_a) = \frac{N_a}{[1 + \exp(E_a - E_f)/k_B T]}$$

$$N_d^+ = N_d[1 - F(E_d)]$$

Now

$$N_d^+ = N_d \left[1 - \frac{1}{[1 + \exp(E_d - E_f)/k_B T]}\right]$$

$$= N_d \frac{[1 + \{\exp(E_d - E_f)/k_B T\} - 1]}{1 + [\exp(E_d - E_f)/k_B T]}$$

$$N_d^+ = \frac{N_d}{1 + [\exp(E_f - E_d)/k_B T]}$$

Substituting all these values in Eqn. 5.1.11, one gets

$$2\left[\frac{2\pi m_e^* k_B T}{h^2}\right]^{3/2} \exp[(E_f - E_c)/k_B T] + \frac{N_a}{[1+\exp(E_a - E_f)/k_B T]}$$

$$= 2\left[\frac{2\pi m_p{}^* k_B T}{h^2}\right] \exp[(E_v - E_f)/k_B T] + \frac{N_d}{1+[\exp(E_f - E_d)/k_B T]}$$

or

$$N_e \exp[(E_f - E_c)/k_B T] + \frac{N_a}{[1+\exp(E_a - E_f)/k_B T]}$$

$$= N_p \exp[(E_v - E_f)/k_B T] + \frac{N_d}{1+[\exp(E_f - E_d)/k_B T]} \qquad ...(5.1.12)$$

2. *Fermi level in an n-type semiconductor at very low temperatures*

In an n-type semiconductor $N_a = 0$ and at very low temperatures (near 0K or at 0K) all the donor levels are filled with electrons. When the temperature is slightly increased, a small fraction of donor atoms are ionised without disturbing the valence band.

Thus

$$n = N_d F(E_c) = \frac{N_d}{[1+\exp(E_c - E_f)/k_B T]}$$

i.e.,

$$n = N_d[\exp(E_f - E_c)/k_B T] \quad \text{as} \quad k_B T \to 0 \qquad ...(5.1.12a)$$

Similarly

$$N_d^+ = N_d[1 - F(E_d)]$$

$$N_d^+ = N_d\left[1 - \frac{1}{1+\exp(E_d - E_f)/k_B T}\right]$$

$$= \frac{[\exp(E_d - E_f)/k_B T]\, N_d}{1+\exp[-(E_f - E_d)/k_B T]}$$

At small temperatures, $-(E_f - E_d)$ is small and hence $\exp[-(E_f - E_d)/k_B T] \to 0$

Thus

$$N_d^+ = N_d\,[\exp(E_d - E_f)/k_B T] \qquad ...(5.1.12b)$$

But

$$n = N_d^+$$

$$\frac{E_f - E_c}{k_B T} = \frac{E_d - E_f}{k_B T} \quad \text{(Refer Eqn. 5.1.12}a\text{ and 5.1.12 }b\text{)}$$

Thus

$$\boxed{E_f = \frac{E_c + E_d}{2}} \qquad \ldots(5.1.13)$$

i.e., at 0 K or at temperatures near 0 K, E_f lies exactly mid way between E_c and E_d levels.

3. *Theory of n-type semiconductors*

Refer Eqn. (5.1.12); assume both N_a and holes are absent:
Under these conditions,

$$N_e \exp[(E_f - E_c)/k_B T] = \frac{N_d}{1 + [\exp(E_f - E_d)/k_B T]}$$

i.e.,

$$N_e \exp[(E_f - E_c)/k_B T] + N_e \exp[(E_f - E_c + E_f - E_d)/k_B T] - N_d = 0$$

i.e.,

$$\exp(2E_f / k_B T)\{N_e \exp[-(E_c + E_d)/k_B T]\} + \exp(E_f / k_B T)[N_e \exp(-E_c / k_B T)] - N_d = 0$$

Thus

$$\exp(E_f / k_B T) =$$

$$\frac{-N_e \exp(-E_c / k_B T) \pm \sqrt{N_e^2 \exp(-2E_c / k_B T) + 4N_e N_d \exp[-(E_c + E_d)/k_B T)]}}{2N_e \exp(-E_c / k_B T)\exp(-E_d / k_B T)}$$

$$= \frac{-1 \pm \sqrt{1 + 4\dfrac{N_d}{N_e} \exp[(E_c - E_d)/k_B T)]}}{2\exp(-E_d / k_B T)}$$

i.e.,

$$\boxed{\exp(E_f / k_B T) = \frac{-1 + \sqrt{1 + \dfrac{4N_d}{N_e} \exp[(E_c - E_d)/k_B T)]}}{2\exp(-E_d / k_B T)}} \qquad \ldots(5.1.14)$$

There are mainly two cases for discussions: either the first term under the radical is small or second term is small.

Case 1

The second term is small (*i.e.*, at higher temperatures and smaller N_d)
i.e.,

$$\frac{4N_d}{N_e}\exp[(E_c - E_d)/k_BT) << 1$$

This implies small N_d or higher value for T, which is a bad approximation, because we do not know how small or large they have to be and if either N_d is too small or temperature is too high, the assumption about the hole suppression becomes untrue. The only argument is:

Using $(1+x)^{1/2} = 1+\frac{x}{2}$, we have

$$e^{E_f/k_BT} = \frac{-1+1+\frac{4N_d}{2N_e}\exp[(E_c - E_d)/k_BT)]}{2\exp(-E_d/k_BT)}$$

$$\boxed{e^{E_f/k_BT} = (N_d/N_e)\exp(E_c/k_BT)} \qquad ...(5.1.15)$$

$$N_d = N_e[\exp(E_f - E_c)/k_BT]$$

This is nothing but the general equation for concentration of electrons say, n.
i.e.,

$$\boxed{N_d = n} \qquad ...(5.1.16)$$

This shows that all the donor atoms are ionised with the data that the holes are suppressed. Referring Eqn. (5.1.15) one can write:

$$\frac{E_f}{k_BT} = \frac{E_c}{k_BT} + ln\frac{N_d}{N_e}$$

$$\boxed{E_f = E_c - k_BT\, ln\frac{N_e}{N_d}} \qquad ...(5.1.17)$$

At 0 K, this equation is not valid, because we have assumed T is large. Similarly if $T \to \infty$, again it is not valid because of suppression of holes becomes untrue. The only valid information is that at 0 K, $E_f = \frac{E_c + E_d}{2}$ and E_f begins to fall as temperature increases and finally coincides with $\frac{E_g}{2}$.

An interesting illustration:

Data supplied:

(*i*) $T = 300$ K

(*ii*) $N_d = 10^{22}/\text{m}^3$

Refer Eqn. (5.1.17)

$$(E_c - E_f) = k_B T\, ln\left[\frac{N_e}{N_d}\right]$$

$$= 0.025 \times ln\left[\frac{4.83 \times 300^{3/2} \times 10^{21}}{10^{22}}\right]$$

$$(E_c - E_f) = 0.203 \text{ eV}$$

and now

$$(E_d - E_f) = (E_c - E_f) - (E_c - E_d)$$

$$(E_d - E_f) = 0.203 - 0.02 = 0.18 \text{ eV}$$

If N_d^+ and N_d' are the ionised and unionised atoms respectively, then

$$N_d' = (N_d - N_d^+) = N_d F(E_d) \text{ with } n = N_d^+$$

$$(N_d - N_d^+) = \frac{N_d}{1 + \exp[(E_d - E_f)/k_B T]}$$

i.e.,

$$N_d' = \frac{10^{22}}{1 + \exp(0.18/0.025)}$$

$$N_d' = 7.5 \times 10^{18}$$

But

$$N_d^+ + N_d' = N_d$$

$$n + N_d' = N_d\ ;\ n = N_d - N_d'$$

or

$$n = 10^{22} - 7.5 \times 10^{18}$$

$$\boxed{n = 10^{22}\left[1 - \frac{7.5}{10^4}\right] = 10^{22}} \quad ...(5.1.18)$$

But the magnitude of n_i in an intrinsic semiconductor at 300 K is approximately 10^{17}.

Case 2

$$\frac{4N_d}{N_e}\exp[(E_c - E_d)/k_BT) >> 1$$

This implies large N_d or low T. Here there are no limiting problems as in case 1, but there may be some practical difficulties. Now Eqn. (5.1.14) becomes

$$\exp(E_f / k_BT) = \frac{2(N_d / N_e)^{1/2}\exp[(E_c - E_d)/2k_BT)]}{2\exp(-E_d / k_BT)}$$

$$\boxed{\exp(E_f / k_BT) = (N_d / N_e)^{1/2}\exp[(E_c + E_d)/2k_BT)]} \qquad ...(5.1.19)$$

Now

$$\frac{E_f}{k_BT} = \frac{(E_c + E_d)}{2k_BT} + \frac{1}{2}ln\frac{N_d}{N_e}$$

$$E_f = \frac{(E_c + E_d)}{2} - \frac{k_BT}{2}ln\left(\frac{N_e}{N_d}\right)$$

At 0 K Fermi level lies exactly half way between the donor level and bottom of the conduction band. As T increases, the Fermi level drops. Also for a given temperature, the Fermi level shifts upward as the concentration increases. This is shown in Fig. 5.11.

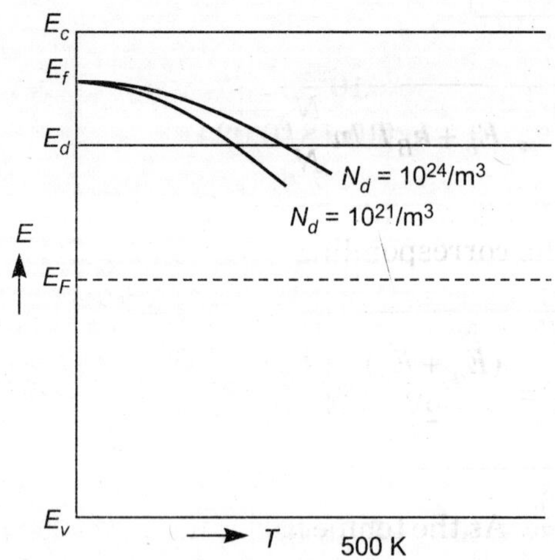

Fig. 5.11 Variation of Fermi level with temperature for different concentrations in an n-type semiconductor

Density of electrons

The general expression for the number of electrons in every cubic metre is

$$n = N_e e^{E_f / k_B T} e^{-E_c / k_B T}$$

Substituting the value of $e^{E_f / k_B T}$ from Eqn. (5.1.19)

$$n = N_e (N_d / N_e)^{1/2} \exp[(E_c + E_d)/2k_B T] \exp(-E_c / k_B T)$$

$$\boxed{n = (N_e N_d)^{1/2} \exp[-(E_c - E_d)/2k_B T)]} \qquad ...(5.1.20)$$

This equation indicates that the density of electrons in the conduction band is proportional to the square root of the donor concentration. Also as the temperature increases, the Fermi level moves downward crosses the donor level and approaches the intrinsic level at $\frac{E_g}{2}$.

p-type semiconductor

In a similar way the theory of *p*-type semiconductors can be discussed. The final results for selected physical parameters are listed below:

(*i*) $\exp(-E_f / k_B T) = \dfrac{-1 + \{1 + (4N_a / N_p)\exp[(E_a - E_v)/k_B T]^{1/2}\}}{2\exp(E_a / k_B T)}$

At high temperature and lower concentration of N_a

$$\boxed{N_a = p} \qquad ...(5.1.21)$$

and

$$\boxed{E_f = E_v + k_B T \, ln \frac{N_p}{N_a}} \qquad ...(5.1.22)$$

(*ii*) At large N_a or low T, the corresponding equation for E_f is

$$\boxed{E_f = \frac{(E_a + E_v)}{2} + \left[\frac{k_B T}{2}\right] ln \frac{N_p}{N_a}} \qquad ...(5.1.23)$$

If $T = 0$, $E_f = \dfrac{(E_a + E_v)}{2}$. As the temperature increases Fermi level shifts upward as in Fig. 5.12. N_p represents $\left[\dfrac{2\pi m_p^* k_B T}{h^2}\right]^{3/2}$ with m_p^*, the effective mass of the hole.

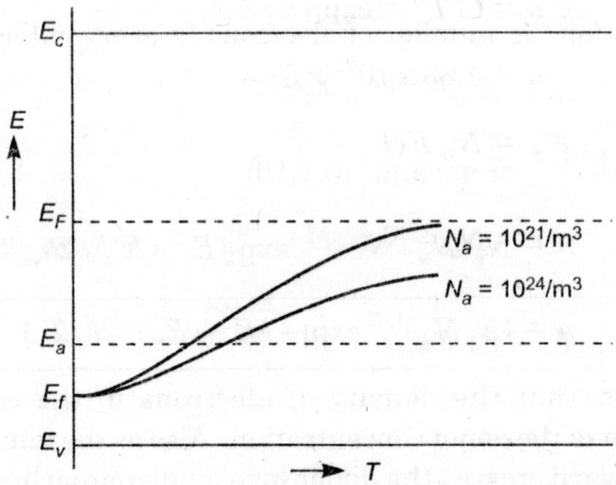

Fig. 5.12 The variation of Fermi level with temperature for different concentrations.

The final expression for density of holes is

$$p = (N_a\, N_p)^{1/2} \exp\left[-(E_a - E_v)/2k_BT\right] \quad \text{...(5.1.24)}$$

This equation indicates that the density of holes in the valence band is proportional to the square root of the acceptor concentration. Another important conclusion is that as the temperature increases, the Fermi level moves upward, crosses the acceptor level and then approaches the central portion of the forbidden gap which makes the substance an intrinsic one.

One chasing illustration

Given:

(*i*) Sample used: *n*-type germanium

(*ii*) Concentration of donor impurities $10^{22}/\text{m}^3 = N_d$

(*iii*) Temperature 27° C

(*iv*) Forbidden energy gap 0.67 eV = E_g

(*v*) N_d^+ ionized donor atoms/m^3

(*vi*) N_d' unionized donor atoms/m^3

(*vii*) $\mu_e = 0.38\text{m}^2/\text{V sec}$, $\mu_p = 0.18\text{ m}^2/\text{V sec}$

Way out to reach the other side of the bank with the above data:

Number of electrons reaching the conduction band from donor level is

$$n = N_d^+ = N_d - N_d'$$

or

$$n = N_d - N_d'$$

But

$$n = C\, T^{3/2}\, \exp[(E_f - E_c)/k_BT]$$

$$n = 4.83 \times 10^{21} \times 300^{3/2} \times \exp[-(E_c - E_f)/k_BT]$$

$$N'_d = N_d\, F(E_d)$$

Also

$$n = N_d - N_dF(E_d) = N_d[1 - F(E_d)]$$

or

$$24.96 \times 10^{24} \times \exp[-(E_c - E_f)/k_BT] = N_d[1 - F(E_d)]$$

i.e.,

$$24.96 \times 10^{24} \times \exp[-(E_c - E_f)/k_BT] = 10^{22}\left[1 - \frac{1}{1 + \exp[E_d - E_f)/k_BT]}\right] = 10^{22}$$

as $[-(E_f - E_d)/k_BT]$ is very small compared with 1.

or

$$\exp[(E_c - E_f)/k_BT] = \frac{24.96 \times 10^{24}}{10^{22}} = 2496$$

i.e.,

$$\left(\frac{E_c - E_f}{k_BT}\right) = ln\ 2496$$

$$E_c - E_f = k_BT\ ln\ 2496 = 0.025 \times ln\ 2496$$

$$E_c - E_f = 0.196\ \text{eV} \qquad ...(1)$$

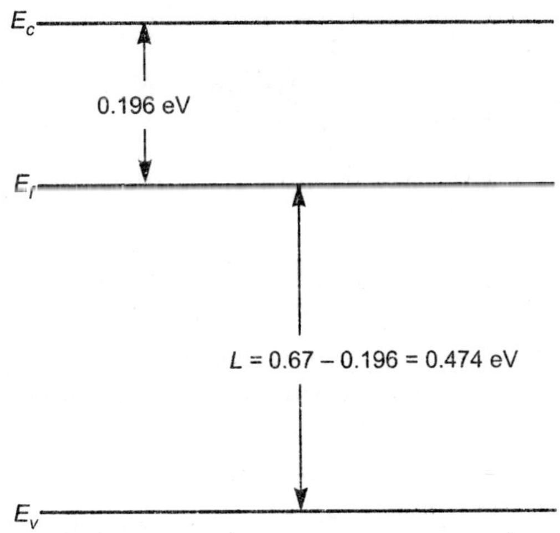

Fig. 5.13 Impurity and intrinsic carriers in an *n*-type semiconductor

Now the number of electrons jumping to the conduction band from the valence band

$$\left.\begin{aligned} n' = p_n &= N_e \exp(-L/k_BT) \\ &= 24.96\times10^{24}\times\exp(-0.474/0.025)\end{aligned}\right\}\text{ assuming } T = 300\text{ K everywhere}$$

$$n' = p_n = 1.5\times10^{17}$$

Thus

$$n\times p_n = 10^{22}\times1.5\times10^{17}$$

or

$$\boxed{n\,p_n = 1.5\times10^{39}} \qquad ...(2)$$

However, the intrinsic charge carriers if it is purely an intrinsic semiconductor is

$$n_i = N_e \exp(-E_g/2k_BT)$$

$$n_i = 24.96\times10^{24}\times\exp(-0.67/0.05)$$

$$n_i = 3.78\times10^{19}$$

$$\boxed{n_i^2 = 1.43\times10^{39}} \qquad ...(3)$$

The interesting conclusion is that the product of n p_n and n_i^2 is the same. The immediate question is: then how the conductivity is increased many fold in n-type semiconductor.

The answer is:

In intrinsic semiconductor

$$\sigma_i = n_i\,e(\mu_e+\mu_p) = 3.78\times10^{19}\times1.6\times10^{-19}(\mu_e+\mu_p)$$

$$\sigma_i = 6.08\,(\mu_e+\mu_p) = 6.08\times0.56 = 3.4\,\Omega^{-1}\,\text{m}^{-1}$$

In *n*-type, it is

$$\sigma_n = n\,e\mu_e + p_n\,e\mu_p$$

$$\sigma_n = (10^{22}\,e\,\mu_e + 1.5\times10^{17}\mu_p\,e)$$

$$\sigma_n = e\left(\mu_e\,10^{22} + \frac{1.5\times10^{22}}{10^5}\mu_p\right)$$

$$\sigma_n = 10^{22}\times1.6\times10^{-19}(\mu_e + 1.5\times10^{-5}\mu_p)$$

$$\sigma_n \approx 1.6\times0.38\times10^3 = 0.608\times10^3\,\Omega^{-1}\,\text{m}^{-1}$$

with $\mu_e = 3.8$ m^2/V-sec and neglecting the term $1.5\times10^{-5}\,\mu_p$.

Thus

$$\frac{\sigma_n}{\sigma_i} = \frac{0.608 \times 10^3}{3.4} = 0.179 \times 10^3;\ \sigma_n \approx 0.18 \times 10^3 \sigma_i$$

$$\sigma_n = 0.179 \times 10^3 \times 3.4 = 6.1 \times 10^2 = 610$$

$$\boxed{\sigma_n = 6.1 \times 10^2\ \Omega^{-1}\ \text{m}^{-1}}$$

i.e., the conductivity in an extrinsic semiconductor becomes a few hundred times greater than that of an intrinsic semiconductor. By adopting tricky mathematical methods, it is always viable and possible to simplify even complicated equations in any area-establishing that a hard-working teacher can not only take the horse to the pond but make it drink which may not be easily possible in other fields.

One interesting outcome is that the number of electrons from the valence band to the conduction band strictly depends on the position of the Fermi level whether it is a pure semiconductor or a doped one. This is elaborated below:

In an intrinsic semiconductor $(E_c - E_f) = (E_f - E_v)$; *i.e.*, $2E_F = (E_c + E_v)$; $E_F = \frac{(E_c + E_v)}{2}$ *i.e.* $\frac{E_g}{2}$ from the valance band. Now the number of electrons coming to the conduction band is $n = C\,T^{3/2} \exp(-E_g / 2k_BT)$. If it is an n-type semiconductor, then $(E_f - E_v) = L$ and not $\frac{E_g}{2}$.

Hence,

$$\boxed{n = C\,T^{3/2} \exp(-L / k_BT)} \qquad \text{...(5.1.25)}$$

5.1.7 VARIATION OF CARRIER CONCENTRATION CUM CONDUCTIVITY WITH TEMPERATURE

The temperature variation of n and p for an n-type semiconductor is shown in Fig. 5.14. There are three different regions corresponding to three different ranges of temperatures, very near absolute zero, the Fermi level is in between the donor level and the lower edge of the conduction band and the donor ionization is about to start. As the temperature is raised the donor atoms get ionized and the electrons go to the conduction band crossing the small impurity gap quite easily. The electron concentration in the conduction band steadily increases, and this region is called the *impurity range*. The Fermi level moves towards the donor level. At about room temperature all the donor atoms are ionized and so the concentration of the electrons in the conduction band remains constant over a certain temperature range. This region is called the *exhaustion range* and in this range Fermi level crosses the donor level. Hence the conductivity must be essentially constant over this region. In fact there will be a slight decrease in conductivity with increasing temperature in this region, because of the decrease in electron mobility. As the temperature is still further raised, the thermal ionization of the atoms in the valence band starts and electrons cross the forbidden gap thereby increasing the electron concentration in the conduction band considerably. As the valence band is an *inexhaustible source* of electrons unlike the donor level, more and more of the electrons from the valence band reach the conduction band and completely out number the donor electrons.

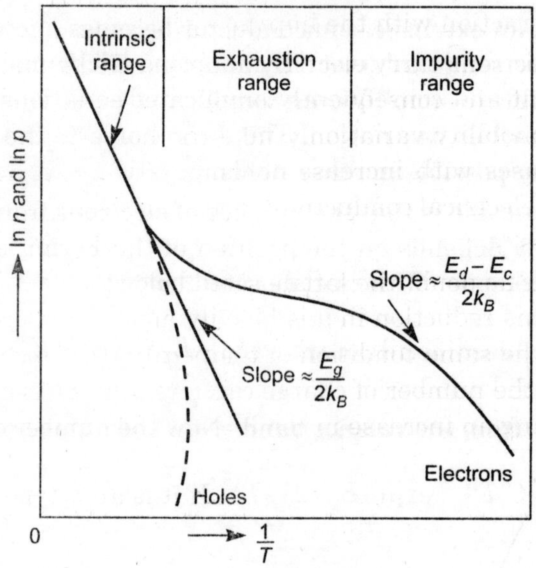

Fig. 5.14 Variation of the electron and hole concentration in an *n*-type semiconductor with the reciprocal of the absolute temperature

The Fermi level keeps on its downward trend until it comes to the middle of the forbidden gap the material practically becomes intrinsic one at this temperature. The hole concentration follows the dotted curve as the temperature increases and over the intrinsic range both the hole and electron concentration curves overlap, since they become equal in number. In practical devices it is necessary either in the impurity range or in the *exhaustion range* to take full advantage of the properties introduced by the impurities. Consequently the temperature at which the intrinsic conductivity becomes dominant, determine the highest possible temperature. For Ge (E_g = 0.67 eV), it is about 100°C and for Si (E_g = 1.1eV), it is about 200°C. Also the lower donor densities have larger flat region.

The resistivity of extrinsic semiconductors

The conductivity of semiconductors is more complicated than that of metals: besides the electron scattering processes being temperature dependent as in a metal, the actual number of current carriers *n* and their energy distribution also vary with temperature.

We know that

$$\rho = \frac{1}{\sigma} = \frac{1}{ne\mu} \qquad ...(5.1.26)$$

Mobility, the *measure of the ease with* which the carriers can move under the influence of an electric field is determined by the scattering of the current carrying particles. Lattice scattering occurs due to thermal vibration of the lattice atoms. Charges are also deflected or scattered by the interaction with the impurity scattering.

In a metal only electrons carry electric charges unlike semiconductors. This number is substantially constant and consequently any variation of conductivity in a metal with temperature is due to mobility variation. The main type of scattering in metals is *lattice scattering* which increases with increase of temperature and hence mobility decreases. Hence in a metal σ the electrical conductivity decreases as the temperature increases, or ρ increases as T increases.

In a semiconductor no doubt the lattice scattering reduces the mobility on increasing the temperature; but this reduction in μ is heavily masked by the exponential increase of charge carriers under the same condition of temperature. The other aspect is that as the temperature increases the number of charge carriers increase rapidly by the ionization of the donor atoms resulting an increase in conductivity which is quite different from that in a metal. See Fig. 5.15.

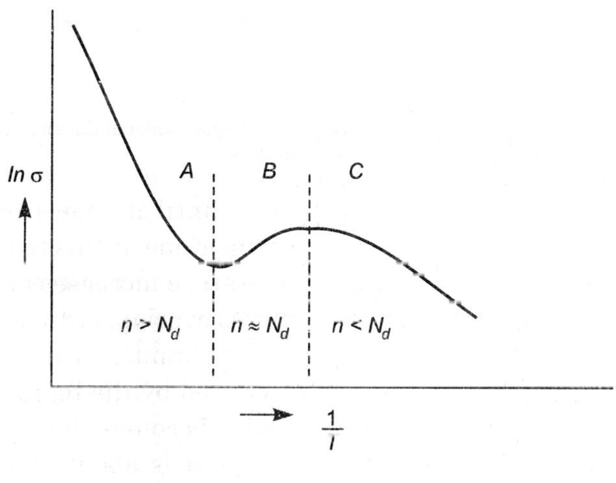

Fig. 5.15 Electrical conductivity $ln\ \sigma$ against $\frac{1}{T}$ for an n-type semiconductor (N_d is impurity atom)

In the region B, even though the charge carriers remain the same, a decrease in conductivity is noticed as the temperature increases. This is because of the decrease in μ when T increases. The strong dependence of the resistance of semiconductors on temperature is used in a device, called *thermistor*. These are bulk semiconductor resistors with a large temperature coefficient of resistivity and a *non-linear current-voltage characteristic*. They are used in measuring temperature and power of ultra high frequency radiation, for timing relays and for temperature compensation in various electric circuits.

5.1.8 MECHANISM OF CURRENT CONDUCTION IN SEMICONDUCTORS

When no external field is applied to a semiconductor each carrier (electron or hole) moves in a random way owing to its thermal energy. The frequent change in direction of the path of the carrier occurs due to scattering by the vibrations of the lattice atoms and by the coulomb field of the ionized donor and acceptor atoms. Application of an external field superimposes on the random motion of the carriers a drift velocity. In the steady state, the rate of momentum gained from the field equals the rate of loss of momentum due to scattering. A steady drift velocity is thus achieved giving rise to a steady flow of current. We shall now consider in detail the mechanism of current flow when a voltage is applied to (*i*) an *n*-type semiconductor and (*ii*) a *p*-type semiconductor.

Biased n-type Semiconductor

An *n*-type semiconductor has (*i*) electrons as majority carriers, (*ii*) an (almost) equal number of immobile positive ions, and (*iii*) holes as minority carriers. Assume that the hole contribution to current is negligible. Let us now consider an *n*-type semiconductor placed between a pair of electrodes across which a voltage is applied. Due to the field produced by the voltage, there will be a steady drift of the free electrons towards the positive electrode. The electrons reaching the positive terminal disappear at the metal of the electrode and the immobile positive ions in the vicinity of the negative electrode remain unneutralized due to the drift of the free electrons. These ions immediately attract electrons. Thus a continuous flow of electrons from one terminal of the voltage source to the other terminal via the semiconductor takes place. The rate of flow of electrons from the negative electrode to the semiconductor and from the semiconductor to the positive electrode is determined by the applied voltage and the conductivity of the semiconductor which accounts for the current flow.

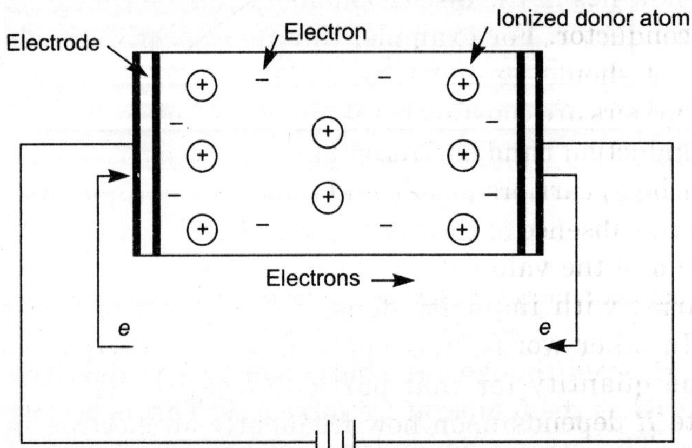

Fig. 5.16 Biased *n*-type semiconductor

5.1.9 MINORITY CARRIERS AND THEIR LIFE TIME

In intrinsic semiconductors, both electrons and holes (free charge carriers) simultaneously take part in the electric conductivity whereas in extrinsic case, the conductivity is mainly due to the charge carriers of one sign, to electrons in n-type semiconductors and to holes in the p-type semiconductors. Such carriers are known as majority carriers. In addition to majority carriers, an extrinsic semiconductor has as well minority carriers, holes in n-type semiconductors and electrons in p-type semiconductors.

In extrinsic semiconductors, the concentration of majority carriers is far greater than the minority carriers concentration. Consider an extrinsic semiconductor given a thermal energy by radiation such that electrons excite from valence band to conduction band. This effectively does not change the density of majority carriers but percentage basis, do appreciably change the density of minority carriers. When irradiation is stopped, the excess carriers (conduction electrons and holes) simply combine and thereby disappear. In other words, the excess conduction electrons returns to the valence band and combine with the excess holes. The phenomenon is called *recombination*.

Life Time of Minority Carriers

Under thermal-equilibrium conditions the product of electrons and holes in any semiconductor must be constant at any temperature. The value of this constant will depend on semiconductor parameters such as the energy gap. When carriers are injected via the energy of an electric field or light, equilibrium is disturbed and carriers in excess of the equilibrium *np* product are introduced. For reasons of electrical neutrality the numbers of electrons and holes so introduced must be equal; we shall also assume that their number is small compared to the equilibrium number so that quasi-thermal equilibrium occurs. This possibility of introducing excess carriers, yet maintaining electrical neutrality, distinguishes semiconductors from metals and results from the two-carrier nature of the semiconductor. This accounts for the fact that transistors utilizing two electric carriers-bipolar transistors are fabricated from semiconductor materials and not metals.

Our interest now lies in the discussion about the life time of the minority carriers in an extrinsic semiconductor. For example, the life of positive hole in an n-type material may be discussed. It should be understood that in thermal equilibrium generation and recombination of carriers are constantly taking place, but on the average for every electron excited into the conduction band one recombines or drops back into the valence band. Let the *generation rate* be g_T carriers/m^3-sec and the *recombination rate* be R. For a particular semiconductor (in the absence of the field), the thermal generation rate g_T is a function of temperature alone and the value of g_T is an empirical quantity of the specimen at hand and therefore varies with impurity density, imperfection present in the specimen, deformation etc. In other words, it is not a unique property of say germanium, but is certainly a unique quantity for that particular sample of germanium. However the recombination rate R depends upon how frequently an electron in the conduction band encounters a vacant state (hole) in the valence band. *i.e.*, the rate is proportional to the concentration of holes and electrons. The more the holes there are, the sooner they will encounter an electron.

i.e.,

$$R \propto n\,p$$

$$R = r\,np$$

where r the constant of proportionality is called *recombination coefficient.* The net rate of change of hole concentration in n-type semiconductors without minority carrier injection is

$$\boxed{\frac{dp}{dt} = g_T - r\,n\,p} \qquad ...(5.1.27)$$

Under equilibrium conditions $\frac{dp}{dt}$ is zero, and the rate of formation of electron-hole pairs is equal to the rate recombination.

Thus

$$\boxed{g_T = r\,n_0\,p_0} \qquad ...(5.1.28)$$

where p_0 and n_0 are the equilibrium concentration of electrons and holes respectively. If now the crystal is slightly disturbed from equilibrium by a few excess holes Δp and an equal number of electrons Δn, the net rate of excess hole carrier recombination (or loss) at any instant is given by

$$\boxed{-\frac{d(\Delta p)}{dt} = r(n_0 + \Delta n)(p_0 + \Delta p) - r\,n_0\,p_0} \qquad ...(5.1.29)$$

This assumes r is unchanged since the density of excess carriers introduced is small. Eqn. (5.1.29) becomes,

$$-\frac{d(\Delta p)}{dt} = r[n_0\,p_0 + n_0\,\Delta p + p_0 \Delta n + \Delta n\,\Delta p - n_0\,p_0]$$

$$\approx r\,[n_0\,\Delta p + p_0\,\Delta n] \text{ neglecting } (\Delta n)\,(\Delta p)$$

$$\boxed{\frac{d(\Delta p)}{dt} = -r\ \Delta p\,(n_0 + p_0)} \qquad ...(5.1.30)$$

The quantity $\frac{1}{r\,(n_0 + p_0)}$ is usually defined as the *minority carrier life time* τ, and Eqn. (5.1.30) is then written as

$$\frac{d(\Delta p)}{dt} = -\frac{\Delta p}{\tau} = \frac{(p - p_0)}{\tau}$$

or

$$\boxed{\frac{d(\Delta p)}{\Delta p} = -\frac{dt}{\tau}} \qquad ...(5.1.31)$$

Integrating,

$$ln\,\Delta p = -\frac{t}{\tau} + lnC$$

i.e.,

$$\boxed{\Delta p = C\ e^{-t/\tau}} \qquad ...(5.1.32)$$

When $t = 0$, $\Delta p = (\Delta p)_0 = C$

i.e.,

$$\boxed{\Delta p = (\Delta p)_0\ e^{-t/\tau}} \qquad ...(5.1.33)$$

where $(\Delta p)_0$ is the initially injected number of holes at $t = 0$ and Δp is the number of holes left at the instant t. It is apparent now that the excess of holes injected decay exponentially with time *i.e.*, if the initial hole concentration is $(\Delta p)_0$, the fraction of holes left after time t is $(\Delta p)_0\ e^{-t/\tau}$. If a time $t = \tau$ has elapsed since concentration $(\Delta p)_0$ existed, the concentration of holes will be $\frac{1}{e}(\Delta p)_0$ or $\frac{(\Delta p)_0}{2.718} = 37\%$ of $(\Delta p)_0$.

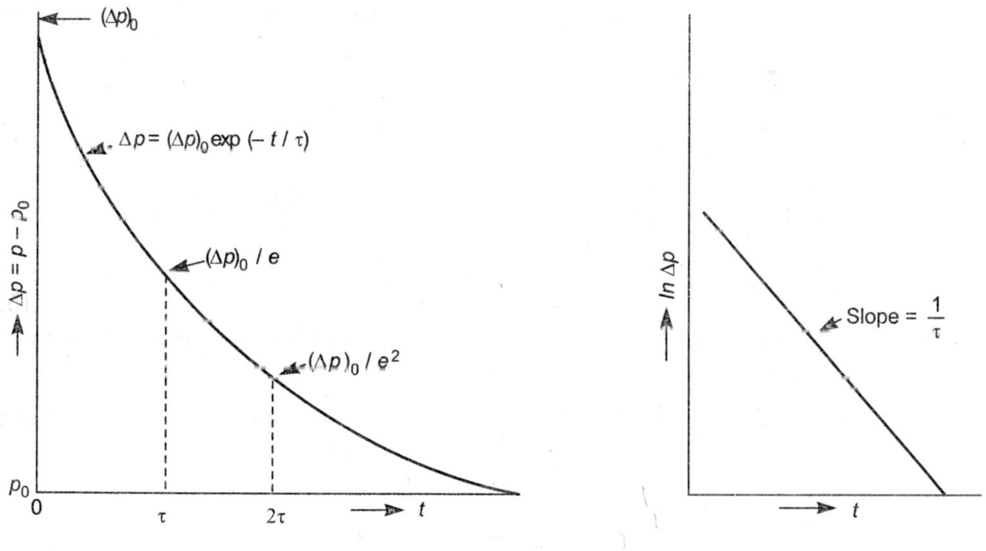

Fig. 5.17(*a*) The decay of excess hole concentration with time (*b*) Logarithm of the concentration *vs* time

5.1.10 TRANSPORT PROPERTIES OF SEMICONDUCTORS

Mobility: If an electric field is applied to a semiconductor the carriers in it develop a *drift velocity* of the carrier for unit electric field is called its mobility. It is nothing but the ratio of the drift velocity to the applied electric field. The symbol used is μ. The unit is m^2/V-sec. *i.e.*,

$$\mu = \frac{v_d}{E}$$

The current density for unit applied electric field is called *conductivity*.
i.e.,

$$\sigma = \frac{J}{E}$$

The charge flowing for unit area for unit time is current density and it is nev_d

$$j = nev_d$$

or

$$\sigma = \frac{j}{E} = \frac{nev_d}{E} = ne\mu$$

where μ is the mobility of charge carriers.
The total conductivity of the semiconductor is

$$\sigma = \sigma_n + \sigma_p = e(n\mu_n + p\mu_p)$$

For *n*-type

$$\sigma_n = ne\mu_n$$

and for *p*-type

$$\sigma_p = ne\mu_p$$

The actual calculation of mobility in any case may be of two components; one is μ_L due to scattering by phonons and is proportional to $\frac{1}{T^{3/2}}$. The other is mobility due to ionised impurities scattering μ_l is $T^{3/2}$ temperature dependence as shown below. The actual mobility value is given by

$$\frac{1}{\mu} = \frac{1}{\mu_L} + \frac{1}{\mu_l}$$

$$\boxed{\frac{1}{\mu} = \frac{a}{T^{3/2}} + bT^{3/2}} \quad ...(5.1.34)$$

Drift and Diffusion

The net current that flows across a semiconducting crystal has two components: 1. *Drift current*, 2. *Diffusion current*. A semiconductor may have a nonuniform concentration of the charge carriers. Suppose the concentration p of the holes varies with distance along the positive x-direction. If the concentration gradient $\frac{dp}{dx}$ is negative, the holes will diffuse in the positive x-direction. This results in a *diffusion current* in the positive x-direction.

The diffusion process is analogous to the heat conduction in a rod, we can write the diffusion of holes in semiconductors by the equation.

$$p \propto \frac{dp}{dx}$$

or

$$p = -D_p \frac{dp}{dx}$$

where p is the density of holes diffusing through unit area in unit time. $\frac{dp}{dx}$ is concentration gradient. Thus *diffusion current* density.

$$J_{pD} = -eD_p \frac{dp}{dx}$$

where e is the charge on each carrier and D_p is the *diffusion constant or diffusivity* of the holes. The minus sign is introduced because $\frac{dp}{dx}$ is negative.

The current due to the drift on the application of the field is

$$J_{pD} = pev_p = pe\mu_p E$$

Thus the total current due to both the applied electric field and diffusion is

$$\boxed{J_p = pe\mu_p E - eD_p \frac{dp}{dx}} \quad ...(5.1.35)$$

In a similar way the total current density due to the electrons can be obtained by replacing the minus sign with a plus sign and p by n.

Thus

$$\boxed{J_n = ne\mu_n E + eD_n \frac{dn}{dx}} \quad ...(5.1.35a)$$

Thus the total current density is

$$J = J_n + J_p = e\left\{\left(n\mu_e E + D_n \frac{dn}{dx}\right) + \left(p\mu_p E - D_p \frac{dp}{dx}\right)\right\} \quad ...(5.1.36)$$

Einstein Relation

There exists a definite relationship between the mobility and the diffusion constant of a particular type of charge carrier. The higher the value of mobility of charge carrier, the greater will be its tendency to diffuse. The electron current density due to diffusion is $D_n \frac{dn}{dx}$. When the steady state is reached the conduction current density J_c must be equal and opposite to the diffusion current density. *i.e.*,

$$J_c = -J_{nD}$$

$$\sigma E = -eD_n \frac{dn}{dx}$$

$$ne\mu_n E = -eD_n \frac{dn}{dx} \quad ...(5.1.37)$$

The presence of the electric field produces a potential difference $V(x)$ over a distance say x. The variation of n with x is given by the Boltzmann relation,

$$n = C\exp(-eV/k_BT) = C\exp(-eEx/k_BT)$$

$$\frac{dn}{dx} = C\exp(-Eex/k_BT)(-eE/k_BT)$$

$$\frac{dn}{dx} = -\frac{neE}{k_BT} \quad ...(5.1.38)$$

Substituting the value of $\left(\frac{dn}{dx}\right)$ in Eqn. (5.1.37)

$$ne\mu_n E = \left\{\frac{eD_n}{k_BT}\right\}(neE)$$

$$D_n = \frac{\mu_n k_B T}{e}$$

Thus

$$\frac{D}{\mu_n} = \frac{k_B T}{e}$$
$$\frac{D_p}{\mu_p} = \frac{k_B T}{e} \qquad ...(5.1.39)$$
$$\frac{D_p}{D_n} = \frac{\mu_p}{\mu_n}$$

Equation (5.1.39) is called *Einstein's diffusion equation.*

Life Time and Diffusion Length

In a semiconductor which is in *thermal equilibrium*, though the concentration of electrons and holes are steady, the equilibrium is a dynamic one. That means the generation of the electron-hole pairs is exactly equal to the recombination of such electron hole pairs at any instant. It is possible to change such equilibrium concentration of electrons and holes by several processes. These carriers do not remain indefinitely and take part in the diffusion process but recombine to form neutral particles. It is found that the excess carrier density decreases exponentially with time and the carrier density at any time t is found to be proportional to exp $(-t/\tau)$ where τ is the *lifetime of the carriers*. The life time τ can be defined as the time taken for the carrier density to fall to $\frac{1}{e}$ of its initial value. The average distance L of a carrier diffuses before recombination, is related to the diffusion coefficient and the life time by the equation

$$L_n = \sqrt{D_n \tau_n}$$
$$L_p = \sqrt{D_p \tau_p} \qquad ...(5.1.40)$$

5.1.11 THE HALL EFFECT

Suppose a material (metal or semiconductor) carrying an electric current is placed in a magnetic field. Then an electric field is produced inside the material in a direction which is at right angles to both the current and the magnetic field. This effect was discovered in 1879 by Edwin H. Hall who was working under Rowland in Johns Hopkins University. This phenomenon is known as the *Hall effect*, and the generated voltage is known as *Hall voltage*.

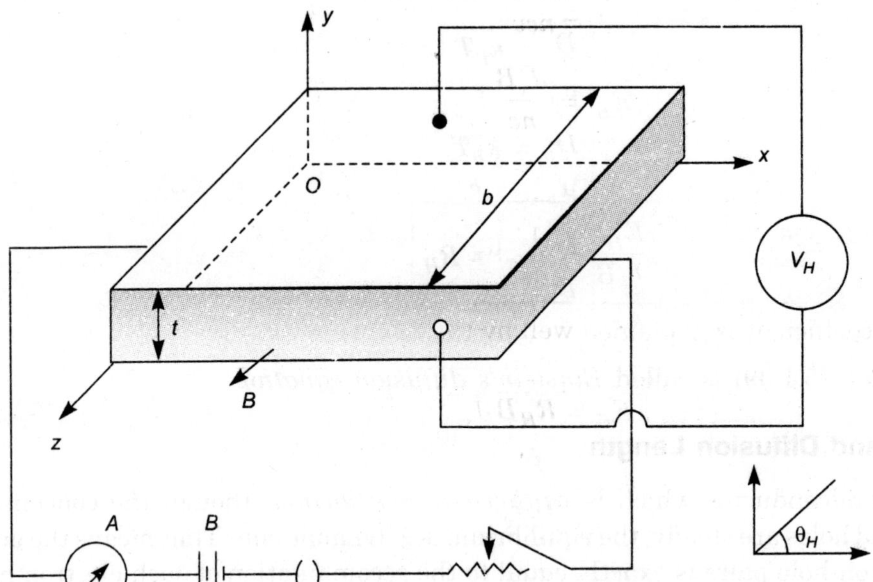

Fig. 5.18 Experimental set up for the measurement of Hall voltage

Theory and Experiment

Let us consider an n-type semiconductor. Let v be the velocity of the electron flowing right to left just opposite to the conventional flow of current. A downward force on each electron of magnitude (Bev) on account of the magnetic field perpendicular to the applied electric field is quite possible. This causes the electron to be deflected in a downward direction and causes a negative charge to accumulate on the bottom face of the slab. A potential difference is therefore established from top to bottom of the specimen with bottom face negative. This potential difference causes a field E_H in the negative y-direction. Equilibrium occurs when

$$eE_H = Bev$$

$$\boxed{E_H = Bv} \quad \text{...(5.1.41)}$$

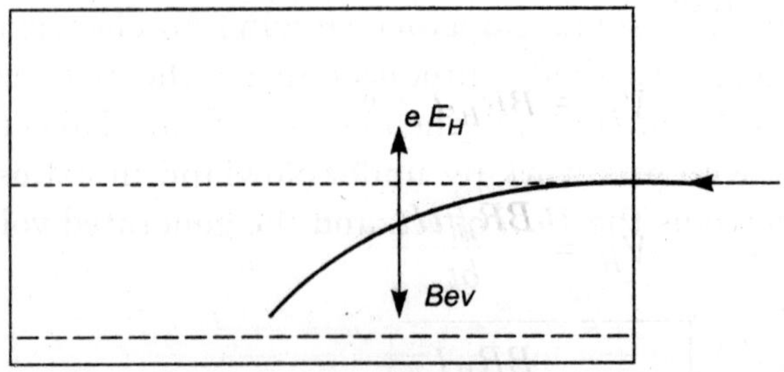

Fig. 5.19 Motion of electrons

If J_x is the current density in the x-direction, then

$$J_x = nev$$

$$E_H = \frac{J_x B}{ne}$$

or

$$\boxed{\frac{E_H}{J_x B} = \frac{1}{ne} = R_H} \quad ...(5.1.42)$$

Hall coefficient is explained well by the following equation

$$\boxed{\begin{aligned} E_H &= R_H B J_x \\ R_H &= \frac{E_H}{B J_x} \\ R_H &= \frac{1}{ne} \end{aligned}} \quad ...(5.1.43)$$

Negative sign is used because of the electronic field developed is in the negative y-direction.

i.e.,

$$\boxed{R_H = \frac{E_H}{B J_x} = -\frac{1}{ne}} \quad ...(5.1.44)$$

The experiment is conducted and the Hall voltage V_H is measured. The length, breadth and thickness l, b and t are also measured.

Now

$$E_H = \frac{V_H}{t}$$

or

$$V_H = E_H t$$

Referring Eqn. (5.1.44)

$$V_H = BR_H J_x t \text{ with } J_x = \frac{I_x}{bt}$$

$$V_H = \frac{BR_H t I_x}{bt}$$

$$\boxed{V_H = \frac{BR_H I_x}{b}} \quad ...(5.1.45)$$

The vital quantity measured in the experiment is the Hall voltage V_H. This applies to complete free particles. A computation of the average speed leads to the conclusion that

$$R_H = -\frac{3\pi}{8ne} = -\frac{1.18}{ne} \text{ instead of } -\frac{1}{ne}$$

Now conductivity,

$$\sigma_n = ne\,\mu_n$$

and

$$\mu_n = \frac{\sigma_n}{ne} = -\frac{\sigma_n R_H}{1.18}$$

$$\mu_p = \frac{\sigma_p}{pe} = \frac{\sigma_p R_H}{1.18} \quad ...(5.1.46)$$

There is another interesting quantity called *Hall angle* (θ_H) defined by the equation,

$$\tan\theta_H = \frac{E_H}{E_x}$$

But

$$E_H = R_H J_x B = \frac{ne v_x B}{ne}$$

or

$$E_H = \frac{v_x B E_x}{E_x} \quad ...(5.1.47)$$

Thus

$$\frac{E_H}{E_x} = \tan\theta_H = \frac{v_x B}{E_x} = \mu B \quad ...(5.1.48)$$

where μ is the mobility of charge carriers.

Applications of Hall Effect

(*i*) *Determination of type of semiconductor*: For an *n*-type semiconductor the Hall coefficient is negative whereas for *p*-type semiconductor it is positive. Thus, the sign of Hall coefficient can be used to determine whether a given semiconductor is *n*-type or *p*-type.

(*ii*) *Calculation of carrier concentration*: The Hall voltage V_H is measured as usual by placing the two probes at the centres of the top and bottom faces of the sample. If the magnetic flux density is B wb/m^2, then

$$n = \frac{1}{e R_H}$$

where,

$$R_H = \frac{1}{n e}$$

(*iii*) *Determination of mobility*: If the conduction is due to one type of carriers, e.g., electrons, we have

$$\sigma = ne\mu_n$$

$$\mu_n = \frac{\sigma}{ne} = \sigma R_H \quad \text{...(5.1.49)}$$

or

$$\mu_n = \sigma\left[\frac{V_H b}{I_x B}\right] \quad \text{...(5.1.50)}$$

i.e., knowing σ, the mobility μ_n can be determined.

(*iv*) *Measurement of magnetic flux density*: Since Hall voltage V_H is proportional to the magnetic flux density B for a given current I_x through a sample, the Hall effect can be used as the basis for the design of a magnetic flux density meter.

(*v*) *Measurement of power in an electromagnetic wave*: In an electromagnetic wave in free space the magnetic field H and the electric field E are at right angles. Thus, if a semiconductor sample is placed parallel to E, it will derive a current I in the semiconductor. The semiconductor is subjected simultaneously to a transverse magnetic field H producing a Hall voltage across the sample. The Hall voltage will be proportional to the product E and H *i.e.*, to the magnitude of the *poynting vector* of electromagnetic wave. Thus, the Hall effect can be used to determine the power flow in the electromagnetic wave.

(*vi*) *Hall effect multiplier*: If the magnetic flux density B is produced by passing a current I' through an air core coil, B is proportional to I'. The Hall voltage is thus proportional to the product of I and I'. This forms the basis of multiplier.

5.1.12 DIRECT AND INDIRECT GAP IN SEMICONDUCTORS

The condition for electron excitation by electromagnetic radiation is not merely $\hbar\omega \geq$ energy gap. There is an additional requirement that the wave vector must be conserved *i.e.*, momentum must also be conserved.

$$\vec{k} + \vec{k}_{ph} = \vec{k}'$$

where $\vec{k}$ = electron wave vector before excitation, $\vec{k}_{ph}$ = photon wave vector and $\vec{k}'$ = electron wave vector after excitation. Due to the high velocity of electromagnetic radiation, often the photon wave vector is very small compared with $\vec{k}$ and $\vec{k}'$ and hence the minimum of the conduction band has the same wave vector as the maximum of the valence band. The excitation under such conditions is called a direct transition and semiconductors having such a band structure are called direct gap semiconductors e.g., Insb and GaAs. On the other hand, important semiconductors like Si and Ge have more complicated band structure and excitation across the band gap often requires the additional interaction of phonons for wave vector conservation. These are called indirect transitions / processes and such semiconductors are termed as indirect gap semiconductors.

5.2 PHYSICS OF SEMICONDUCTOR DEVICES

5.2.1 ADVANTAGES OF SEMICONDUCTOR DEVICES

Most of the devices in electronics today are built from semiconductors. The vacuum devices like diode, triode, etc. have been replaced by semiconductor devices like *p-n* junction diodes, transistors, etc. The advantages of semiconductor devices over vacuum tubes are the following:

(*i*) As the name indicates, vacuum tubes require vacuum, but semiconductors do not require vacuum.

(*ii*) Vacuum tubes have filaments, and so require supply for heating the filament. Semiconductors do not have filaments.

(*iii*) Semiconductor devices are smaller in size, lighter in weight, and are mechanically very rugged.

(*iv*) Operation of semiconductor devices requires low-voltage power supply. The power consumption is also smaller than that of the corresponding vacuum tubes.

(*v*) The semiconductor devices require very small warm-up time and, therefore, operate immediately after the supply voltages are switched on

(*vi*) The performance of semiconductor devices is more reliable than that of vacuum tube devices.

(*vii*) The operating life of a semiconductor device is appreciably longer than that of its equivalent vacuum tube.

5.2.2 THE *p-n* JUNCTION

A semiconductor device can be defined as a unit which consists, partly or wholly, of semiconducting materials and which can perform useful functions in electronic apparatus and solid state research. Examples of semiconductor devices are *semiconductor diodes* (*p-n junction*), transistors, integrated circuits, etc. Almost all semiconductor devices are comprised of a single crystal semiconductor incorporating two or more semiconducting regions of different impurity density. The difference in the electric fields and carrier densities associated with differently doped regions, called junctions, permit a wide range of essentially non-linear conductivity effects in devices incorporating two, three or more distinct regions. Most semiconductor devices can be understood through the simplest of such functions, called *p-n* junction, which is a system of two semiconductors in physical contact, one with excess of electrons (*n*-type) and the other with excess of holes (*p*-type).

A *p*-type or *n*-type crystal when used separately conducts just like any poor conductor without asymmetry, and this phenomenon by itself is not very important. Almost all the semiconductor devices in use today are combinations of the two types of extrinsic crystals, the *p*- and *n*-types. The simplest semiconductor device is a combination of one *p*-type and one *n*-type crystal called the *p-n* junction.

It should be emphasised at this point that it is not possible to make a *p-n* junction by placing *p*-material in contact with *n*-material. One reason is that even the most smoothly polished surfaces pressed against one another have only relatively small areas of contact. Furthermore, it is impossible to get surfaces absolutely clean, and the effect of drift or other contaminating agents on the surface is such as to reduce the area of contact and cause other deleterious effects.

The fabrication of junctions between *p*- and *n*-type materials, which have become of great technical importance, will now be described. A *p-n* junction is, in a sense, an internal boundary within a single crystal. On one side donors and on the other side acceptors are in excess. Ordinarily, the transition is not completely sharp and the width over which the conductor changes from *n*-type to *p*-type is the most important parameter. Double junctions such as a *p-n-p* arrangement can also be formed on the same principle.

Experimentally there are three principal ways of preparing *p-n* junction devices; the *grown junction method,* the *alloying method* and the *diffusion method.* In the grown junction method, a crystal of semiconductor is grown, for example, from an initially *n*-type melt, which while the crystal is being formed, is counterdoped by adding enough acceptor impurities so that the subsequently grown portion of the crystal is *p*-type. The resulting junction is usually graded rather than abrupt. Although accurate control can be maintained over the impurity concentrations on either side of the junction by this method, the difficulty of locating the junction in the grown crystal and the difficulty of attaching leads to narrow grown junction regions has prevented widespread adoption of the technique in manufacturing.

In the alloying method an alloy pellet or foil is melted upon a semiconductor base crystal. The alloy pellet contains impurities which will give rise to the conductivity type opposite to that of the original crystal; after it has melted the temperature is raised to a value high enough to allow the molten alloy to dissolve away the underlying semiconductor crystal to some extent. The liquid phase is then a solution of the semiconductor material in the liquid alloy which is approximately saturated at the maximum fabrication temperature. Since the equilibrium solubility of the semiconductor in the liquid metal normal increases with temperature, when the device is slowly cooled the solution will become supersaturated with respect to the solute and some of the dissolved semiconductor atoms will recrystallise at the liquid-solid interface, continuing the single-crystal structure of the substrate. Since the liquid phase contains impurities associated with the conductivity type opposite to that of the original crystal, the regrown material will be of the opposite conductivity type, and there will be a *p-n* junction at the interface between original and regrown material. Upon further cooling the entire alloy phase solidifies; it is then a simple matter to make electrical connections to the original crystal and to the resolidified alloy pellet. The alloying process generally produces junctions that are quite abrupt. Although it is sometimes difficult to exercise the strictest dimensional control over devices made by the alloying process, the ease and simplicity of the technique has led to its widespread use as a production method for many different types of devices.

In the diffusion method, an impurity element is diffused at an elevated temperature into a base crystal whose conductivity type is opposite to that which is produced by the presence of the diffusing impurity in the crystal lattice. The depth of the junction below the surface of the crystal and the impurity concentration gradient at the junction can be controlled by controlling the diffusion time and temperature and the surface concentration of the diffusing impurity. Diffused junctions may be either graded or quite abrupt, depending upon the diffusion time, surface concentration and base crystal impurity density. Although the attachment of leads to diffused layers presents some technical problems, the ease and accuracy with which the thickness of diffused layers can be controlled has resulted in the widespread use of diffusion as a manufacturing technique.

Let us now begin the discussion of the characteristics of a *p-n* junction by assuming that a junction has just been made in a crystal-growing furnace. Figure 5.20 shows such a crystal, which we will assume is at room temperature.

The uncircled charges on each side of the junction represent the free carriers, *i.e.*, the plus signs on the *p* side represent free holes, and the minus signs on the *n* side represent free electrons. These holes and electrons are present because of the thermal ionisation of the donors and acceptors.

The charges shown in circles represent the ionised impurities, which cannot move. On the *p* side, the acceptors are shown with negative charge. This is because the thermal energy associated with room temperature causes electrons from the valence band to be activated into the empty bands of the acceptor atoms. The acceptor atoms are negatively charged because of the presence of the extra electrons.

On the other hand, the donor impurities on the *n* side are positively charged because they have supplied electrons to the conduction band.

On the *p*-side of the junction, a high concentration of free holes exists. On the *n*-side, a high concentration of free electrons is present. Because of thermal vibration, these concentrations on either side would like to spread out. The spreading-out process due to random velocity and scattering is known as *diffusion*. Its effect on the charge distributions is to spread or disperse it in a manner similar to the spreading or diffusion of ink by capillarity when a drop is put on a blotter. Thus holes from the *p*-side of the junction diffuse over to the *n*-side, and electrons from *n*-side diffuse into the *p*-region because of thermal agitation.

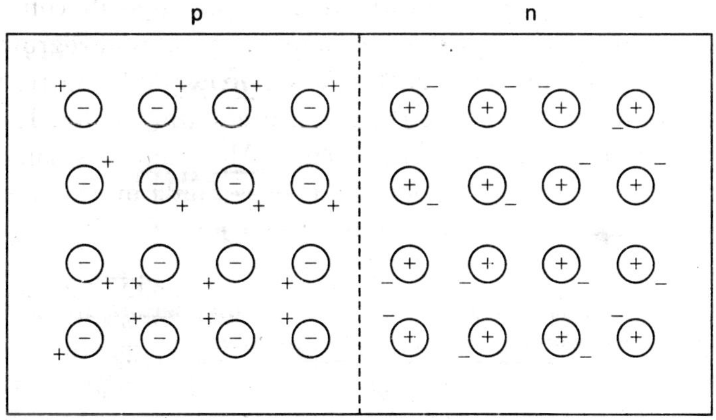

Fig. 5.20 Newly made *p-n* junction. Diffusion and recombination have not yet occurred

However, when either type of charge penetrates the boundary region into the other side, it finds itself surrounded by carriers of the opposite sign. Recombination is very likely under such conditions. For this reason, quite soon after a junction is made, the boundary region between the *p*- and the *n*-sides becomes devoid of holes and electrons because both readily find carriers of the opposite type with which to recombine.

Away from the boundary, toward the outside of either of these regions, free carriers still exist, and the charged impurity atoms are still effectively neutralised by the space charge of free carriers of opposite sign around them. But right near the junction region, the impurities have no neutralising space charge present, because the free carriers in this region have been annihilated. This leaves ionised acceptors (–) near the junction on the *p*-side and ionised donors (+) near the junction on the *n*-side. Both are without compensating free-carrier charge. When a sufficient number of donor and acceptor ions are uncovered, further diffusion is stopped. It is because now, a *barrier* is set up against further movement of charge carriers. This is called *potential barrier* or *junction barrier*. The potential barrier is of the order of 0.1 to 0.3 volt. Figure 5.21 shows the electrostatic potential difference across the *p-n* junction. How this potential barrier is developed? When a sufficient number of donor and acceptor ions is uncovered, further diffusion is prevented. It is because now positive charge on *n*-side repels holes to cross from *p*-type to *n*-type and negative charge on *p*-side repels free electrons to enter from *n*-type to *p*-type. Because of this a difference in potential exists between the two sections, which inhibits further electron-hole combinations at the junction, and the Fermi level of the two sides is in the same level as shown in Fig. 5.22. The region across the *p-n* junction in which the potential changes form positive to negative is called the depletion region. The width of this region is of the order 10^{-8} m. Since this region has immobile (fixed) ions which are electrically charged, it is called as the *space-charge region*. Outside this region on each side of the junction, the material is still neutral. The potential barrier for germanium *p-n* junction is about 0.3 volt. This potential barrier discourages the diffusion of majority carriers across the junction. However, the potential barrier helps minority carriers to drift across the junction. But electric current cannot flow as no external electric field is applied to the system. The voltage developed across the junction may be represented by a ficticious battery having a voltage V_B.

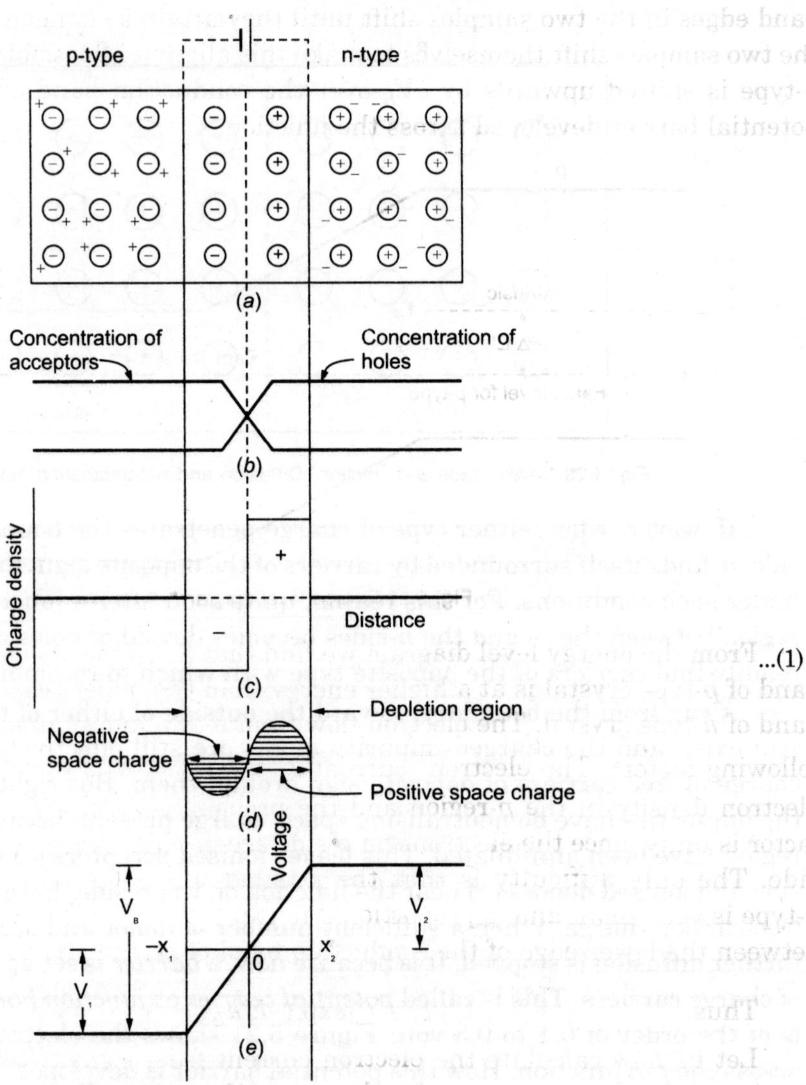

Fig. 5.21 (*a*) Diagram showing the position of carriers in a *p-n* junction
(*b*) Variation of concentration of immobile ions
(*c*) Charge density
(*d*) Space charge
(*e*) Potential hill caused by charge distribution

A question will arise whether this space charge equivalent battery will be able to deliver any current when connected in a circuit containing no other battery. In order to answer this, consider the energy level diagram of a (*p-n*) junction. The Fermi levels in the two samples are at different positions with respect to the band edges. Fermi level is something like potential and so it must be at the same level in a system in thermal equilibrium. At the instant *p-n* junction is formed, Fermi level in *n*-type is at a higher level than in *p*-type. Because of this difference the electrons will flow from material which has a higher Fermi level to a material which has a lower Fermi level until they attain a common level. The band edges in the two samples shift until they attain a common level. The band edges in the two samples shift themselves to make this alignment possible. The conduction band of p-type is shifted upwards by eV_B over the conduction band of n-type where V_B is the potential barrier developed across the junction.

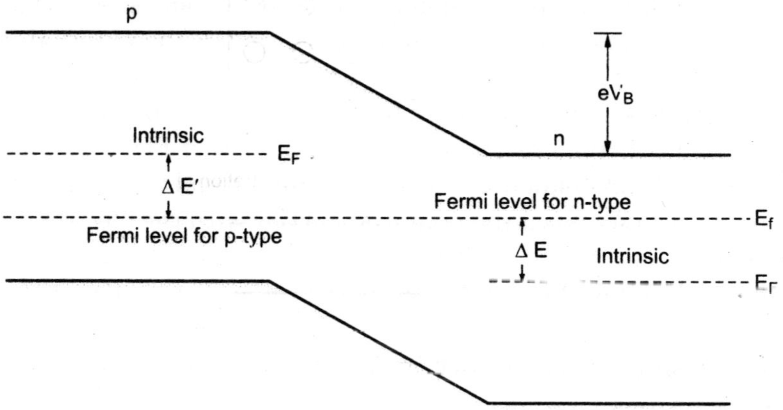

Fig. 5.22 Energy level diagram of *p-n* junction

From the energy level diagram we find that the minority electrons in the conduction band of p-type crystal is at a higher energy than the majority electrons in the conduction band of n-type crystal. The electron flow from p to n and from n to p are governed by the following factors. The electron current $\vec{I_e}$ form p to n is proportional to the product of electron density in the p-region and the probability factor. In this case the probability factor is unity since the electrons on p-side are having energy higher than electrons on n-side. The only difficulty is that the number of electrons in the conduction band of p-type is very small and is proportional to exp $(-E/k_BT)$ where E is the energy difference between the lower edge of the conduction band of p-type and Fermi level.

Thus, $$\vec{I_e} = C \exp(-E/k_BT) \qquad ...(5.2.1)$$

Let us now calculate the electron current from right to left (n-side to p-side). The density of electrons is proportional to exp $[(E_f - E_c)/k_BT]$ or exp $[-(E_c - E_f)/k_BT]$. The probability that the electron will have sufficient energy to climb up the energy ladder of height $[E - (E_c - E_f)]$ is proportional to exp $[-(E - E_c + E_f)/k_BT]$.

Thus, $$\overleftarrow{I_e} = C \exp[-(E_c - E_f)/k_BT] \exp[-(E - E_c + E_f)/k_BT]$$

$$\overleftarrow{I_e} = C \exp(-E/k_BT)$$

i.e., $$\overleftarrow{I_e} = \vec{I_e} \qquad ...(5.2.2)$$

Thus we see that no net current flows across the junction because of the space charge equivalent battery. The same argument holds good for holes also.

5.2.3 THE POTENTIAL BARRIER AT THE *P-N* JUNCTION

The potential difference (or the barrier potential V_B) due to accumulation of an excess of immobile charge in each layer near the junction, under the equilibrium condition at the temperature T degree kelvin may be obtained as follows: When no external voltage is applied to the *p-n* junction crystal, no current passes through the external circuit even when the *p* and the *n*-regions are shorted as shown in Fig. 5.23. Thus, the net current through the junction must be zero separately both for free electrons and positive holes. Considering only the free electron current at the *p-n* junction of Fig. 5.23, one finds that there are two components of current density flowing across the junction. J_n is the component of the current density that flows down the potential barrier with the electron flowing from *p*-region to *n*-region and it is the drift current density due to the free electrons under the influence of the field on account of the potential barrier and J_n may be written as

$$J_n = n\, e\, \mathrm{m}_n E$$

where E is the electric field intensity caused due to the potential barrier voltage V_B and n is the density of free electrons.

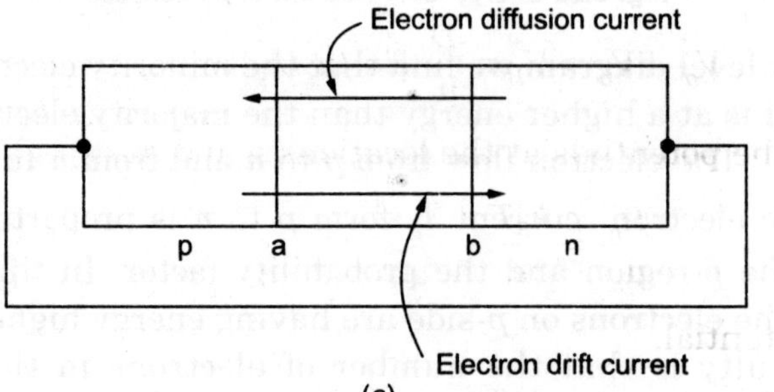

(*a*)

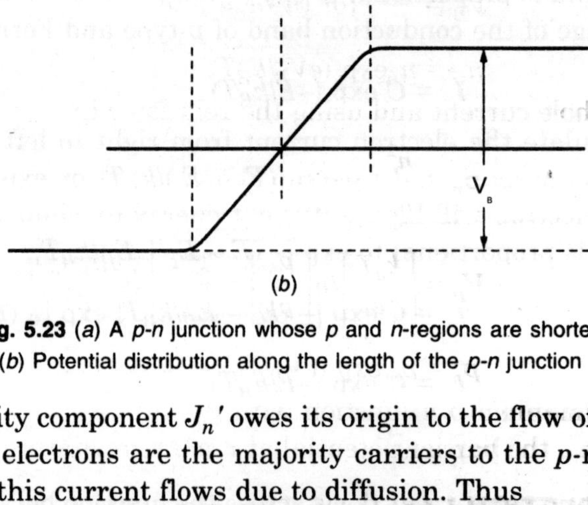

(*b*)

Fig. 5.23 (*a*) A *p-n* junction whose *p* and *n*-regions are shorted
(*b*) Potential distribution along the length of the *p-n* junction

The current density component J_n' owes its origin to the flow of electrons from the *n*-region where the free electrons are the majority carriers to the *p*-region where they are minority carriers and this current flows due to diffusion. Thus

$$J_n' = e\, D_n \frac{dn}{dx} \tag{5.2.3}$$

where D_n is the *diffusion constant* for electron and $\left(\frac{dn}{dx}\right)$ is the gradient of the electron density in the x-direction (length of the crystal is along the x-direction). When the externally applied voltage is zero, the sum of the components is equal to zero. Hence

$$J_n + J_n' = n e\mu_n E + e D_n\left[\frac{dn}{dx}\right] = 0 \qquad ...(5.2.4)$$

$$-E\,dx = \left[\frac{D_n}{\mu_n} \times \frac{dn}{n}\right]$$

$$-\int_a^b E\,dx = \left[\frac{D_n}{\mu_n}\right]\int_{n_p}^{n_n} \frac{dn}{n}$$

or
$$-\int_{V_a}^{V_b} dV = \frac{D_n}{\mu_n}\int_{n_p}^{n_n} \frac{dn}{n}$$

where n_n and n_p are the densities of the free electrons in the n-region and p-region respectively,

$$V_a - V_b = V_B = \frac{D_n}{\mu_n}[ln\ (n_n/n_p)]$$

where V_a and V_b are the potentials at the locations a and b respectively.

But
$$\frac{D_n}{\mu_n} = \frac{k_B T}{e}$$

Hence barrier potential,

$$V_B = \frac{k_B T}{e}\{ln\ (n_n/n_p)\} \qquad ...(5.2.4a)$$

$$n_n = n_p \exp\ (eV_B/k_B T) \qquad ...(5.2.5)$$

Considering tho hole current and using the relationship

$$p_n = \frac{n_i^2}{n_n}\text{; we can show that}$$

$$V_B = \left[\frac{k_B T}{e}\right] ln\left[\frac{p_p}{p_n}\right] = \frac{k_B T}{e}\left[ln\frac{p_p n_n}{n_i^2}\right]$$

or
$$p_p = p_n \exp\ (eV_B/k_B T) \qquad ...(5.2.6)$$

For a typical germanium p-n junction with p- and n-regions doped to the order of 10^{22} impurity atoms/m^3, the barrier potential at room temperature is about 0.3 volt.

5.2.4 WIDTH OF DEPLETION LAYER OF THE *p-n* JUNCTION

We have seen that the space charge on the n-side is positively charged and on the p-side it is negatively charged. Figure 5.21 shows the space charge region in the two sides of the junction and the consequent potential variation. Let X_1 and X_2 be the widths of the space charge region in the p and n sides respectively. The area of the depletion layer in each region depends on the concentration of the impurities in the regions. The effective area of the depletion layer can easily be calculated with the help of the *Poisson's equation* which states that the second derivative of the potential with respect to the distance is proportional to the charge density. In the simple one-dimensional case when the voltage varies only in the x-direction along the length of the crystal, the *Poisson's equation* may be written as

$$\frac{d^2V}{dx^2} = -\frac{\rho}{\varepsilon}$$

where V is the voltage and ρ is the volume density of charge in the material of permittivity ε. In the portion of the depletion layer situated in the p-region near the junction the charge density may be given by

$$\rho = -eN_a$$

where N_a is the density of the acceptor atoms and the negative sign is used since the acceptor atoms are negatively ionised. In order to determine the width of the depletion region X_1 in the p-region, we shall apply Poisson's equation in this region.

i.e.,
$$\frac{d^2V}{dx^2} = -\frac{\rho}{\varepsilon} = \frac{eN_a}{\varepsilon} \qquad ...(5.2.7)$$

On integration,
$$\frac{dV}{dx} = \left[\frac{eN_a}{\varepsilon}\right]x + A \qquad ...(5.2.8)$$

and
$$V = \left[\frac{eN_a}{\varepsilon}\right]\left[\frac{x^2}{2}\right] + Ax + B \qquad ...(5.2.9)$$

where A and B are arbitrary constants and may be determined by the boundary conditions. At the right hand side of this portion of the depletion layer in Fig. 5.21 we assume that $V = 0$ when $x = 0$ so that all voltages are measured with respect to the potential at the boundary between the p-type and n-type materials. Using this boundary condition in equation (5.2.8), one gets

$$B = 0 \qquad ...(5.2.10)$$

In the p-type semiconducting material the potential is constant at the end of the depletion region (Fig. 5.21). Thus, the depletion region may be assumed to end at a point $x = -X_1$ where $dV/dx = 0$, so that we can realise that at this point the field strength is zero. Using the second boundary condition in equation (5.2.8), one gets

$$A = \frac{eN_aX_1}{\varepsilon} \qquad ...(5.2.11)$$

On substitution of the values of the constants A and B in equation (5.2.9),

$$V = (1/2)\frac{eN_a x^2}{\varepsilon} + \frac{eN_a X_1 x}{\varepsilon}$$

If
$$V = -V_1 \text{ at } x = -X_1$$

$$-V_1 = (1/2)\left[\frac{eN_aX_1^2}{\varepsilon}\right] - \frac{eN_aX_1^2}{\varepsilon}$$

or $$-V_1 = -\frac{eN_aX_1^2}{2\varepsilon} \qquad \text{...(5.2.12)}$$

In a similar way the Poisson's equation may be applied for the portion of the depletion layer situated in the n-region near the junction. Considering that the charge density in this region is due to the positively ionised donor atoms, we may write the Poisson's equation for this portion of the depletion layer as

$$\frac{d^2V}{dx^2} = \frac{eN_d}{\varepsilon}$$

where N_d is the density of the donor atoms in the n-region of the p-n junction.

On integration, $$\frac{dV}{dx} = -\frac{eN_d}{\varepsilon}x + C \qquad \text{...(5.2.13)}$$

and $$V = (-1/2)\frac{eN_dx^2}{\varepsilon} + Cx + D$$

The boundary conditions are now applied as follows:

(*i*) When $x = 0$, $V = 0$, one gets $D = 0$

(*ii*) When $x = X_2$, $dV/dx = 0$, one gets

$$C = \frac{eN_dX_2}{\varepsilon} \qquad \text{...(5.2.14)}$$

Thus, $$V = (-1/2)\frac{e\,N_dx^2}{\varepsilon} + \frac{eN_dX_2x}{\varepsilon}$$

If $V = V_2$ at $x = X_2$, one gets

$$V_2 = (-1/2)\left[\frac{e\,N_dX_2^2}{\varepsilon}\right] + \frac{e\,N_dX_2^2}{\varepsilon}$$

i.e., $$V_2 = (1/2)\left[\frac{eN_dX_2^2}{\varepsilon}\right]$$

Here V_2 is the potential at the point $x = X_2$ where the depletion layer in the n-region ends.

Thus, the potential barrier at the junction,

$$V_B = V_2 - (-V_1) = V_1 + V_2$$

i.e., $$V_B = \frac{e}{2\varepsilon}(N_aX_1^2 + N_dX_2^2) \qquad \text{...(5.2.15)}$$

The equation of charge neutrality is

$$eN_aX_1 = N_dX_2e$$

or
$$X_2 = \frac{N_aX_1}{N_d} \qquad ...(5.2.16)$$

On substitution of the value of X_2 in equation (5.2.15), one gets

$$V_B = \frac{e}{2\varepsilon}\left[\frac{N_a^2X_1^2}{N_d} + N_aX_1^2\right]$$

$$= \frac{eN_aX_1^2}{2\varepsilon}\left[1 + \frac{N_a}{N_d}\right]$$

$$X_1 = \left[\frac{2\varepsilon V_B}{eN_a\left(1 + \frac{N_a}{N_d}\right)}\right]^{1/2} \qquad ...(5.2.17)$$

If this value of X_1 is substituted in equation (5.2.16), one gets,

$$X_2 = (N_a/N_d)\left[\frac{2\varepsilon V_B}{eN_a\left(1 + \frac{N_a}{N_d}\right)}\right]^{1/2}$$

$$= \left[\frac{2\varepsilon V_B}{(eN_d^2/N_a)\left(1 + \frac{N_a}{N_d}\right)}\right]^{1/2}$$

$$= \left[\frac{2\varepsilon V_B}{eN_d\left(1 + \frac{N_d}{N_a}\right)}\right]^{1/2} \qquad ...(5.2.18)$$

The total width of the depletion layer is given by

$$X = X_1 + X_2$$

$$X = \left[\frac{2\varepsilon V_B}{eN_a\left(1 + \frac{N_a}{N_d}\right)}\right]^{1/2} + \left[\frac{2\varepsilon V_B}{eN_d\left(1 + \frac{N_d}{N_a}\right)}\right]^{1/2}$$

$$= \left[\frac{2\,\varepsilon\, V_B}{(eN_a/N_d)(N_a+N_d)}\right]^{1/2} + \left[\frac{2\,\varepsilon\, V_B}{(eN_d/N_a)(N_a+N_d)}\right]^{1/2}$$

$$= \left[\frac{2\,\varepsilon\, V_B}{e(N_a+N_d)}\right]^{1/2} \{[(N_d/N_a)^{1/2} + (N_a/N_d)^{1/2}]\}$$

$$= \left[\frac{2\,\varepsilon\, V_B}{e(N_a+N_d)}\right]^{1/2} \left[\frac{(N_d+N_a)}{(N_aN_d)^{1/2}}\right]$$

i.e.,
$$\boxed{X = \left[\frac{2\,\varepsilon\, V_B(N_a+N_d)}{e\,N_aN_d}\right]^{1/2}} \quad \text{...(5.2.19)}$$

Thus, the width of the depletion layer can easily be determined if the value of the potential barrier V_B and the densities of the donor and the acceptor atoms in the n-type and the p-type regions of the p-n junction are known.

The portion of the depletion layer situated to the left of the junction contains a certain amount of negative charge, while the portion of the same situated to the right of the junction contains an amount of positive charge. The potential developed across these two portions is equal to the barrier potential V_B. Therefore, there exists a certain amount of capacitance at the junction. If Q is the quantity of charge contained in each portion of the depletion layer, then

$$Q = e\,N_aX_1S = e\,N_dX_2S \quad \text{(where } S \text{ is area of cross-section)}$$

Substituting for X_1 from equation (5.2.17)

$$Q = \left[\frac{2e^2N_a^2S^2\,\varepsilon\, V_B}{eN_a\left(1+\dfrac{N_a}{N_d}\right)}\right]^{1/2}$$

$$Q = \left[\frac{2e\,N_aN_dV_B\,\varepsilon\, S^2}{(N_a+N_d)}\right]^{1/2}$$

Thus the junction capacitance for unit area is

$$C_j = \frac{dQ}{dV_B} = [V_B]^{-1/2}\left[\frac{e\,\varepsilon\, N_aN_d}{2(N_a+N_d)}\right]^{1/2}$$

i.e.,
$$C_j = \left[\frac{e\,\varepsilon\, N_aN_d}{2V_B(N_a+N_d)}\right]^{1/2}$$

$$C_j = \varepsilon \left[\frac{eN_a N_d}{2\,\varepsilon\, V_B (N_a + N_d)} \right]^{1/2}$$

$$C_j = \frac{\varepsilon_0 \varepsilon_r}{X_1 + X_2} \text{ [farad per unit area of cross-section]} \quad (5.2.20)$$

5.2.5 VOLT-AMPERE CHARACTERISTICS OF THE *p-n* JUNCTION

The *p-n* junction possesses rectifying properties just like a vacuum tube diode and it offers an extremely low resistance to the flow of electric current in one particular direction and a very high resistance to the current flow in the opposite direction. It has already been pointed out that the barrier potential V_B developed across the junction is of such polarity that it tends to oppose the movement of the positive holes due to diffusion from the *p*-region to the *n*-region (see Fig. 5.22) and that of free electrons due to diffusion from the *n*-region to the *p*-region. Again, all the charge carriers in any semiconducting material do not possess the same amount of kinetic energy but have a range from a very high value to an extremely low value. It has already been mentioned that the distribution of kinetic energy or velocity to different charge carriers is decided by Fermi-Dirac statistics. Due to the diffusion of charge carriers, as the potential barrier across the *p-n* junction slowly develops, it becomes impossible for more and more charge carriers to cross the junction. The ones that have highest kinetic energy may be able to surmount the potential barrier. The process of crossing the junction may continue till an equilibrium condition is established and no free carrier, however large its kinetic energy, is able to cross the junction. After reaching equilibrium the only mechanism by which the free carriers can be made to cross the junction is to increase its kinetic energy by some external means, such as by the application of an external e.m.f. across the junction. The band diagram of an unbiased *p-n* junction is shown in Fig. 5.24.

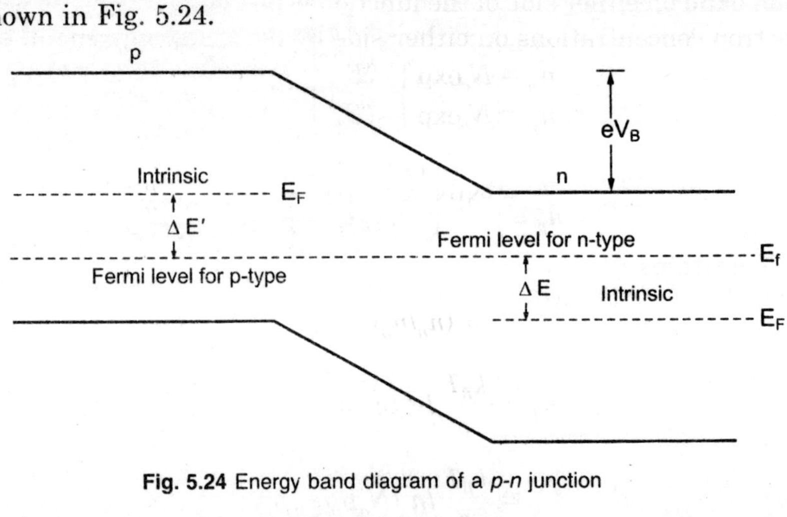

Fig. 5.24 Energy band diagram of a *p-n* junction

For an *n*-type semiconductor the Fermi level lies near the conduction band edge E_c and for *p*-type semiconductor it is near the valence band edge E_v. When a *p-n* junction is formed, under equilibrium condition the Fermi level attains a constant value throughout the sample as shown in Fig. 5.24. Clearly, then the conduction band edge E_{cp} of the p-type will not be at the same level as E_{cn} of the n-type. Similarly, E_{vp} will not be in line with E_{vn}. From Fig. 5.24, we find

$$E_B = E_{cp} - E_{cn} = E_{vp} - E_{vn} = eV_B \qquad ...(5.2.21)$$

where E_B is the *barrier energy*, e is the electronic charge and V_B is the barrier potential. Note that although a potential $V_B = \left(\frac{E_B}{e}\right)$ exists at the p-n junction a voltmeter connected across p-n junction will not read this potential. The reason for this is as follows:

If this is possible, let us assume that a current flows due to the barrier voltage in a short-circuit p-n junction diode. This current will heat the connecting metal wire. Since there is no external source of energy, the heating of the wire must take place with a simultaneous cooling of the p-n junction diode. But under thermal equilibrium this situation cannot exist. Hence we conclude that the current through the circuit is zero. This means that the barrier voltage must be balanced by the metal-to-semiconductor contact potentials at the ends of the diode. A voltmeter connected to the terminals of the diode shows zero voltage reading.

On the other hand when the positive terminal of a battery is connected to the p-type side and the negative terminal to the n-type side of a p-n junction, the junction allows a large current to flow through it. In this case, the p-n junction, is said to be forward-biased. When the terminals of the battery are reversed, *i.e.*, the positive terminal is connected to the n-type side, and the negative terminal to the p-type side, the junction allows a very small current to flow through it. Under this condition the p-n junction is called reverse-biased. This behaviour of a p-n junction makes it suitable for use as a rectifier.

The potential energy difference $(E_{cp} - E_{cn})$ between (see Fig. 5.24) free electrons in the conduction band on either side of the junction is just equal to eV_B, so we can now relate V_B to the electron concentrations on either side by the following general equations

$$n_n = N_c \exp\left[-(E_{cn} - E_f)/k_BT\right]$$

and

$$n_p = N_c \exp\left[-(E_{cp} - E_f)/k_BT\right]$$

Thus,

$$\frac{n_n}{n_p} = \exp\left[\frac{(E_{cp} - E_{cn})}{k_BT}\right] = \exp\left[\frac{eV_B}{k_BT}\right]$$

Taking logarithms

$$\frac{eV_B}{k_BT} = ln\ (n_n/n_p)$$

$$V_B = \frac{k_BT}{e} ln\ (n_n/n_p)$$

$$= \frac{k_BT}{e} ln\ (N_d p_p/n_p\, p_p)$$

i.e.,

$$= \frac{k_BT}{e} ln\ (N_d N_a/n_i^2) \qquad ...(5.2.22)$$

Typically, in silicon we may calculate V_B for $N_a = 10^{22}$ and $N_d = 10^{24}$ with $n_i = 1.4 \times 10^{16}/\text{m}^3$.

$$V_B = 0.82 \text{ volt at 300 K}$$

$$ne\mu_n E = -eD_n \frac{dn}{dx}$$

Putting, $E = -\frac{dV}{dx}$ [and rearranging]

$$\frac{dV}{dx} = (D_n/\mu_n)\left[(1/n)\frac{dn}{dx}\right] \quad ...(5.2.22a)$$

$$\int_0^{V_B} dV = (D_n/\mu_n)\int_{n_p}^{n_n} \frac{dn}{n}$$

$$V_B = (D_n/\mu_n)\ ln\ (n_n/n_p) \quad ...(5.2.23)$$

This looks very like equation (5.2.4*a*) indeed use of Einstein's relation $D_n/\mu_n = k_B T/e$ shows that they are identical.

This is a comforting result, as this proof of equation (5.2.23) is dubious, since the occurrence of diffusion within the junction region is not obvious: The mean free path is normally no smaller than the thickness of the depletion layer. The importance of equation (5.2.23) is that the same equations are needed for discussing the theory of the junction when a bias voltage is applied across it.

5.2.6 APPLICATION OF VOLTAGE ACROSS A *p-n* JUNCTION

In Fig. 5.25, the circles with negative sign in the depletion layer represent acceptor ions and the circles with positive sign represent donor ions. It may be noted that there are no free or mobile charge carriers in the depletion layer. But it contains only immobile positive and negative ions. It has been observed that an electric potential is developed across the *p-n* junction due to charge separation on its either side. This electric potential is known as barrier potential and represented by the symbol V_B. The value of this *barrier potential* at room temperature is about 0.3 volt for germanium and 0.7 volt for silicon. It has been observed that there is no further diffusion of electrons or holes across the junction due to the barrier potential unless an external field is established across the *p-n* junction.

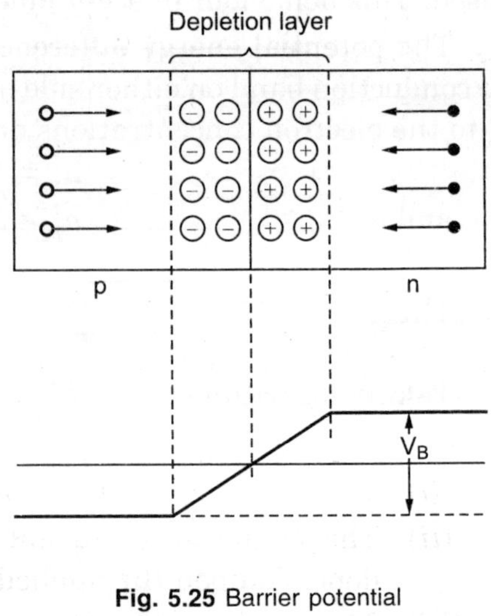

Fig. 5.25 Barrier potential

A voltage or p.d. across a *p-n* junction may be applied in any one of the following two methods:

(*i*) Forward biasing (*ii*) Reverse biasing.

(*i*) Forward Biasing

For a current to flow through the junction, a battery must be connected to the diode to overcome the potential barrier. And the polarity of the battery should be such that the majority carriers in both sections are driven toward the junction. When the battery is connected in this way, it provides forward bias causing forward or high current to flow, because it allows the majority carriers to provide the current.

To apply forward bias, positive terminals of the battery is connected to *p*-side and negative terminal to *n*-side as shown in Fig. 5.26 (*a*). The applied forward potential establishes an electric field which acts against the potential barrier field. Obviously, the resultant field is weakened and the barrier height is reduced at the *p-n* junction. Since the potential barrier height is very small, a small forward voltage is sufficient to completely eliminate the barrier. Obviously, at some forward voltage the potential barrier at the *p-n* junction can be eliminated altogether. Then the junction resistance will become almost zero, and a low resistance path is established for the entire circuit. Thus a large current is generated in the circuit even for the small potential difference applied. Such a circuit is called forward bias circuit and the current is called forward current. The salient features of the forward bias circuit are summarised below:

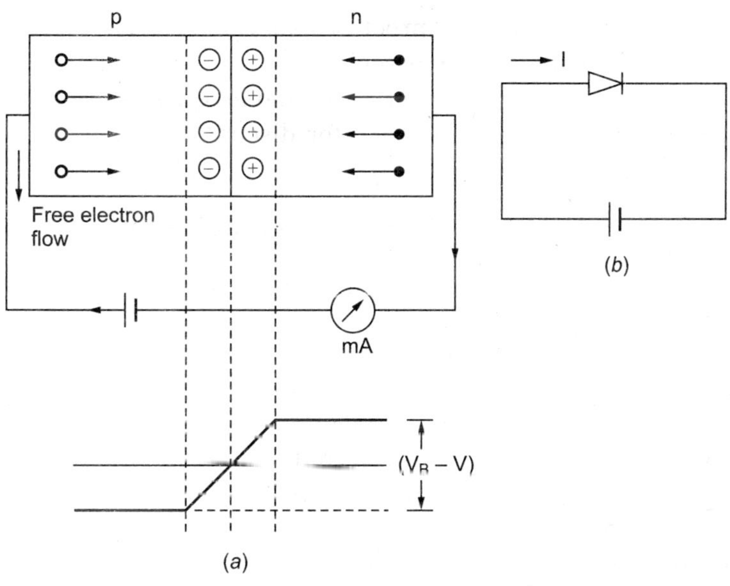

Fig. 5.26 (*a*) Forward biasing
(*b*) Symbolic representation of *p-n* junction

(*i*) The *p-n* junction offers low forward resistance to current flow.

(*ii*) The magnitude of current in the circuit due to the establishment of low resistance depends upon the applied forward bias.

(*iii*) At some forward voltage, the potential barrier is eliminated altogether.

The current flow may be understood as follows:

When the electron hole combination takes place near the junction, a covalent bond near the positive terminal of the battery breaks down. This causes the liberation of an electron which enters the positive terminal. This action creates a new hole which moves towards the junction. For each electron in n-region that combines with a hole from p-region, an electron enters the crystal from the negative terminal. The constant movement of the electrons towards the positive terminal and the holes towards the negative terminal produces a high forward current. As the applied voltage is increased further, the electrons and holes having lower energy will be able to cross the potential barrier and the current will increase further.

(*ii*) Reverse Biasing

Here, the positive terminal of the battery is connected to n-region and the negative terminal to p-region as shown in Fig. 5.27. In this case, the applied reverse voltage establishes an electric field in the same direction as that of the field due to barrier potential. As a result of this, the resultant field is strengthened and the barrier height (V_B) increases as shown in Fig. 5.27. It has been observed that the increased barrier potential prevents the flow of charge carriers across the junction.

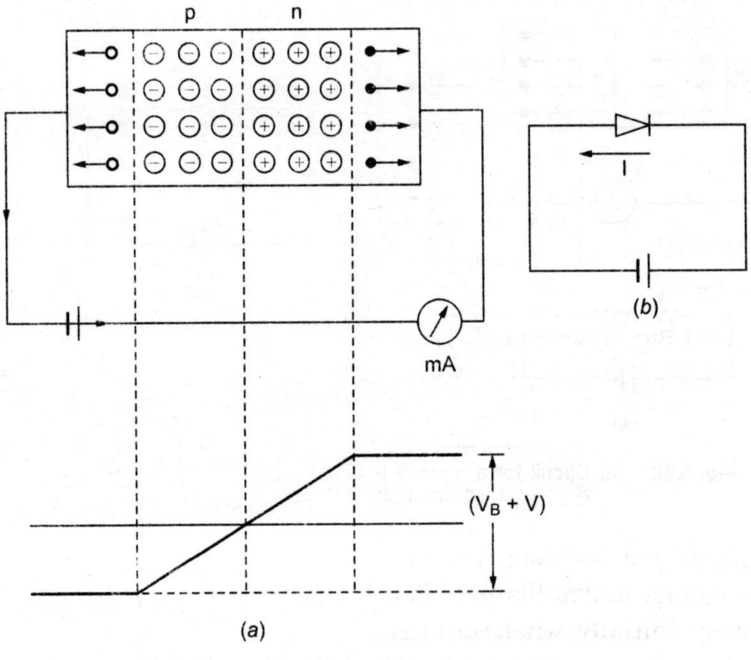

Fig. 5.27 (*a*) Reverse biasing
(*b*) Symbolic representation of *p-n* junction

Thus a high resistance path is established for the entire circuit, and the current does not flow through the junction, *i.e.*, the effect of reverse bias is to increase the potential barrier, thus allowing a very little current to flow. When the junction is reverse biased, the electrons in n-type semiconductor and holes in p-type semiconductor are attracted away from the junction under the action of the applied voltage. Since there is no recombination of electron-hole pairs, the current is negligibly small, *i.e.*, the junction has a high resistance.

V-I Characteristics of a *p-n* Junction

The volt-ampere characteristic of a *p-n* junction is a curve between the voltage across the current through it. Usually, the voltage is plotted along the horizontal axis and the current along the vertical axis.

In a forward biased *p-n* junction, the voltmeter and milliammeter are connected to record the values of voltage and current respectively. A resistance R is attached in the circuit to control the flow of current through the *p-n* junction as shown in Fig. 5.28 (*a*).

Initially, when the applied voltage is zero, no current flows through the circuit. But as the applied voltage is increased by adjusting the potential divider Rh, a small current starts flowing in the circuit. As the applied voltage is further increased, the current increases sharply. It is because of the fact, that initially when the applied voltage is small, the entire voltage is consumed in overcoming the barrier potential. However, once the applied voltage exceeds the barrier potential, the resistance of the *p-n* junction reduces and the current increases sharply. If we plot a graph with voltage along the *x*-axis and current in (mA) along the *y*-axis, we obtain a curve as shown in Fig. 5.28(*b*).

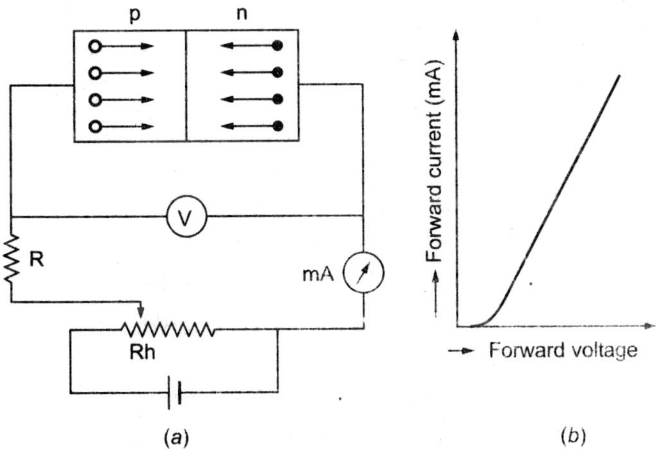

Fig. 5.28 (*a*) Circuit for a forward biased *p-n* junction
(*b*) V-I Characteristic for a forward biased *p-n* junction

In a reverse biased *p-n* junction, the battery connections are opposite to those in forward biasing. The voltage and milliammeter are connected to record the values of voltage and current respectively. Initially, when the applied voltage is zero, no current flows through the circuit. But as the voltage is increased a small current starts flowing in the reverse direction due to opposite battery connections, the barrier potential at the junction is increased. As a result of this, the junction resistance becomes very high, and practically no current flows through the circuit. However, a very small current (microampere) flows in the circuit. This current is due to the flow of minority carriers and known as reverse saturation current. As the voltage is further increased the value of reverse saturation current does not increase.

But it has been observed that at a particular voltage, the breakdown of the junction occurs and the reverse saturation current increases to a large value. The voltage, at which this sudden increase of current takes place, is known as *breakdown voltage* (V_{Br}). Now if we plot a graph with voltage along the horizontal axis and current in (mA) along the vertical axis, we obtain a curve as shown in Fig. 5.29 (*b*).

5.2.7 RECTIFIER EQUATION

The net current density across a *p-n* junction can be analysed in terms of the four components J_1, J_2, J_3 and J_4. The current density J_1 is due to the flow of minority electrons on the conduction band of *p*-region to conduction band of *n*-region. Since electrons carry a negative charge, the flow of free electrons

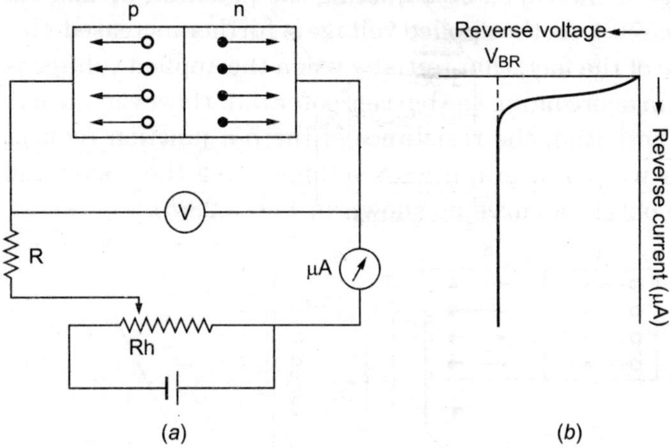

Fig. 5.29 (*a*) Circuit for a reverse biased *p-n* junction
(*b*) V-I Characteristic for a reverse biased *p-n* junction

from the *p*-region to the *n*-region is equivalent to a conventional current flow from the *n*-region to the *p*-region, *i.e.*, J_1 flows from right to left.

The majority electrons in the conduction band of the *n*-region can flow to the conduction band of the *p*-region giving rise to a current density J_2 flowing from left to right.

The majority holes in the valence band on the *p*-region can flow to the valence band on the *n*-region giving rise to a current density J_3 flowing from left to right.

The minority holes in the valence band on the *n*-region can flow to the valence band on the *p*-region giving rise to a current density J_4 from right to left (see Figs. 5.26 and 5.30). Hence the net current density J flowing from left to right is given by

$$J = (J_2 + J_3) - (J_1 + J_4) \quad ...(5.2.24)$$

We can now evaluate these four components. J_1 is the current density caused by the flow of electrons from the *p*-type down the energy hill to the *n*-region. In a typical *p-n* junction this current is very small and is also a constant for the following reasons. The minority electrons are scarce in the *p*-region and so the current carried by them also must be small. Their motion through the bulk material to the junction boundary is mainly by diffusion, since the applied electric field is only across the barrier region and not in the bulk region. Once the electrons is at the junction boundary it will be swept away through the junction, down the potential hill into the *n*-region. The diffusion of the electron from

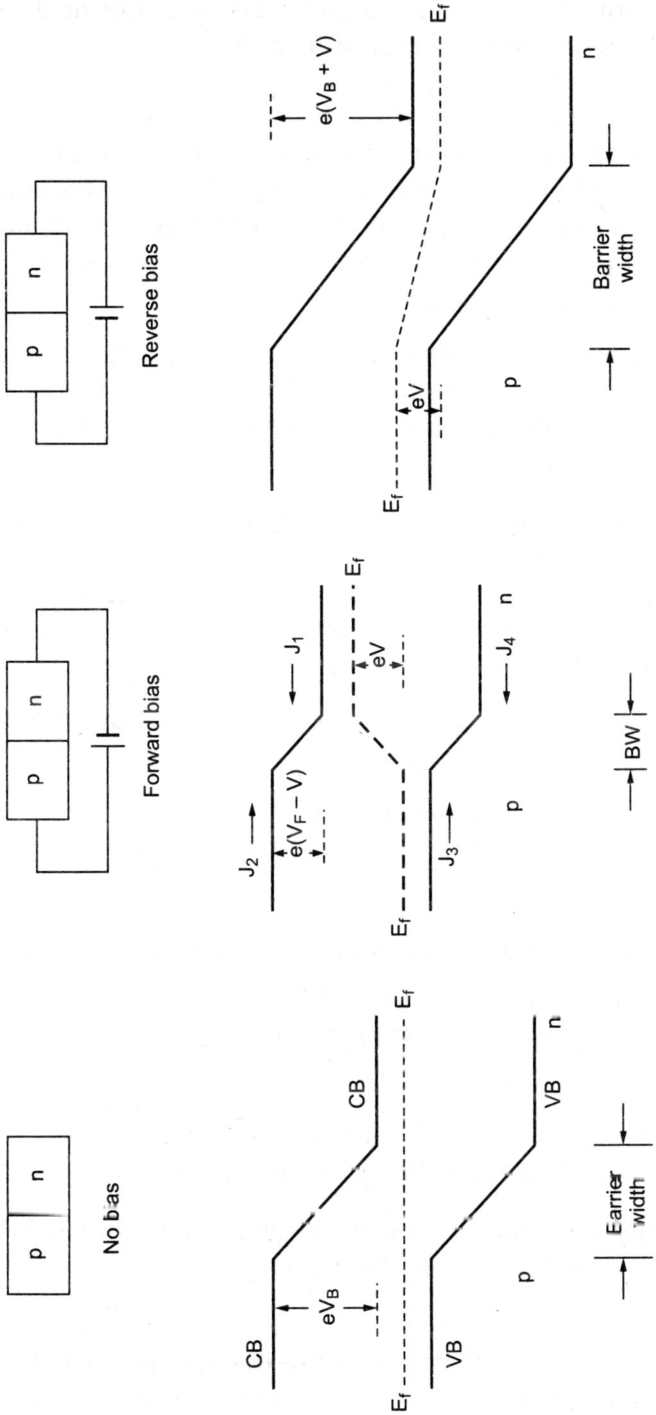

Fig. 5.30 Energy band diagram and barrier width

the bulk to the junction boundary is not affected by the applied voltage whether in the forward direction or in the reverse direction. Again the potential hill across the junction is always downward for the electron flowing from left to right whether the biasing is forward or reverse. Hence J_1 is practically constant and for this reason it is called the saturation current density. J_2 is caused by the flow of the majority electrons from the conduction band of the n-region to the conduction band of the p-region. The conduction band provides a copious supply of electrons, but the potential barrier $V_T = (V_B - V)$ is the only impediment. Only those electrons which have sufficient energy to overcome the potential barrier V_T will flow across the junction to the left. The current density J_2 is given by

$$J_2 = C_1 \exp(-eV_T/k_BT) \quad ...(5.2.25)$$

where C_1 is a constant. A similar argument can be used to find the flow of holes in the valence band. J_4 is the saturation current density which does not depend on the applied voltage. J_3 is the hole current which flows up the potential hill and is given by

$$J_3 = C_2 \exp(-eV_T/k_BT) \quad ...(5.2.26)$$

where C_2 is another constant. When there is no applied voltage ($V = 0$), the net current density in each band is separately zero. Current flow in one band is independent of the flow in the other band since there cannot be any accumulation of charge in either band under equilibrium conditions. Applying these conditions C_1 and C_2 can be evaluated.

When $V = 0$, $V_T = V_B$, then $J_2 = J_1$ and $J_3 = J_4$

Substituting these values of J_2 and J_3 in equations (5.2.25) and (5.2.26) we get

$$J_1 = C_1 \exp(-eV_B/k_BT)$$

and

$$J_4 = C_2 \exp(-eV_B/k_BT)$$

or

$$C_1 = J_1 \exp[eV_B/k_BT]$$

$$C_2 = J_4 \exp[eV_B/k_BT]$$

Substituting the values C_1 and C_2 in equations (5.2.25) and (5.2.26) one gets

$$J_2 = J_1 \exp[e(V_B - V_T)/k_BT] \quad ...(5.2.27)$$

and

$$J_3 = J_4 \exp[e(V_B - V_T)/k_BT] \quad ...(5.2.28)$$

$$J_2 + J_3 = \{(J_1 + J_4) \exp[e(V_B - V_T)/k_BT]\}$$

Thus equation for J is given by [Refer equation (5.2.24)]

$$J = (J_1 + J_4)\{\exp[e(V_B - V_T)/k_BT] - 1\}$$

But $(J_1 + J_4) = J_0$ represents the total saturation current density and $(V_B - V_T) = V$ is the applied voltage. The net current density becomes

$$J = J_0[\exp(eV/k_BT) - 1] \quad ...(5.2.29)$$

Equation (5.2.29) is known as the *rectifier equation.* V is positive when the applied voltage biases the junction in the forward direction and V is negative when the biasing is reverse. The experimental circuit for studying the diode characteristic and the graphical plot of equation (5.2.29) are shown in Figs. 5.31 and 5.32. In the positive quadrant when a forward bias voltage is applied of magnitude $V > \frac{k_B T}{e}$ exp $(eV/k_B T) >> 1$ and J becomes equal to J_0 exp $(eV/k_B T)$ *i.e.*, J increases exponentially with applied voltage. In the negative quadrant when a reverse bias voltage of magnitude $-V >> k_B T/e$ is applied, exp $(-eV/k_B T) << 1$ and J becomes practically equal to J_0. That is, the reverse current remains constant at the small value J_0, until the junction breaks down. It is found that the theoretical plot agrees with the experimental curve obtained for diodes.

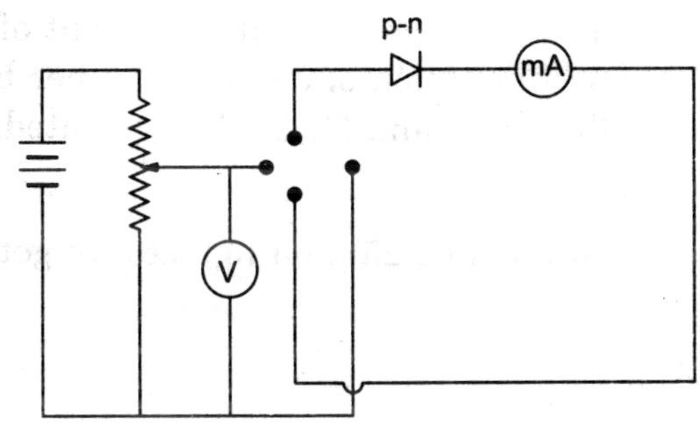

Fig. 5.31 Circuit for studying diode characteristic

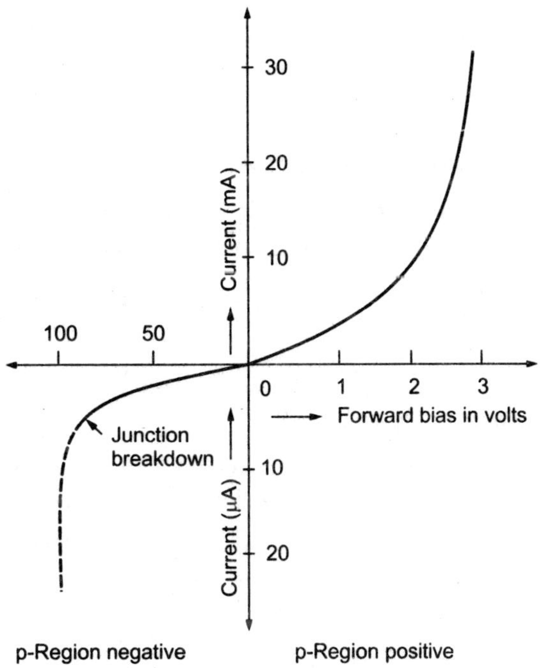

Fig. 5.32 Graphical plot of equation (5.2.29)

Zener Diode

We have seen that a junction diode, when subjected to a large reverse bias, breaks down. In this breakdown region the reverse current increases very rapidly. This breakdown is attributed to two main causes, namely, the avalanche breakdown and the *Zener breakdown*. In the *avalanche breakdown*, the conduction band electrons make ionising collisions with the atoms of the crystal. These collisions produce further electrons, which in turn collide with further atoms. The cumulative effect of such collisions results in the breakdown of the junction, called the *avalanche effect*. Zener breakdown occurs when the applied electric field is so high that the valence band electrons are pulled out to the conduction band in large numbers resulting in the breakdown. This phenomenon is also called the *field emission* of electrons. The voltage at which the Zener breakdown occurs can be carefully controlled by adjusting the doping concentrations of the junction materials. When the junction materials are heavily doped, the depletion region will be thin and even at low voltages a strong electric field of the order of 10^5 volt/cm appears, resulting in the Zener breakdown or field emission. When the junction materials are lightly doped the depletion region will be very wide and at a higher voltage (not strong enough to cause field emission) the carriers will be accelerated enough, to cause ionisation by collision resulting in avalanche breakdown. It is found that such breakdowns will be nondestructive provided, the excess heating is prevented by limiting the junction current. The current voltage characteristic in the breakdown region or the Zener region as it is called, is practically vertical and this property is very useful as voltage reference. A junction diode used in the Zener region is called the Zener diode and its action is similar to that of voltage regulator gas tubes. Zener diodes have the advantage of *lower dynamic resistance* and a *higher voltage and current range*.

5.2.8 CONTINUITY EQUATION

Since the action of semiconductor devices like the diode and the transistor depend on the generation and transport of excess minority carriers, it is necessary to have an equation governing their motion in a semiconducting crystal. This equation is known as the *continuity equation*. The continuity equation is nothing more than a mathematical way of expressing the principles of conservation of particles inside a given volume of semiconducting crystal. Such an analysis is made taking into account the following processes for holes and electrons separately.

1. The electrons and holes generated by thermal processes or otherwise have a finite lifetime. The generation and recombination of the electrons and holes must be considered. The rate of change of excess hole concentration in a given volume element is given by the ratio of the excess concentration to the carrier lifetime

   i.e.,
   $$\frac{d(\Delta p)}{dt} = -\frac{\Delta p}{\tau}$$

   putting $\Delta p = p - p_0$ where p_0 is the equilibrium density and p is the concentration at the instant t, we get

   $$\frac{d}{dt}(p - p_0) = -\frac{(p - p_0)}{\tau} \quad ...(5.2.30)$$

   i.e.,
   $$\frac{dp}{dt} = \frac{(p_0 - p)}{\tau_p}$$

   p_0/τ_p gives the generation rate and p/τ_p gives the recombination rate and so (dp/dt) gives the net rate of change of hole density due to generation and recombination.

2. Let the volume element considered have unit area in the *y*-*z* plane and a length dx along the *x*-axis as shown in Fig. 5.33. Let E be the applied electric field along the *x*-axis. The drift current density J_c in the positive *x*-direction is given by

   $$J_c^+ = ep\mu_p E \quad ...(5.2.30a)$$

where the superscript + indicates that the current is due to positive holes and subscript c indicates that the current is due to conduction or drift motion in the electric field.

The diffusion current density for positive holes is given by

$$J_d^+ = -eD_p \frac{dp}{dx} \quad \text{...(5.2.31)}$$

The sum of the drift and diffusion currents is given by

$$J = J_c^+ + J_d^+ = ep\mu_p E - eD_p\left[\frac{dp}{dx}\right]$$

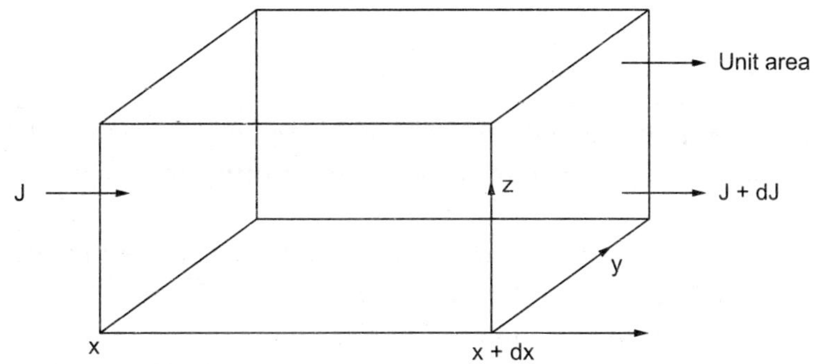

Fig. 5.33 Drift and diffusion current in a volume element

Considering only the drift and diffusion motions, the rate of change of hole density with time from the given volume element dx is given by $\left(\frac{dp}{dt}\right) dx$. If the corresponding current density that is responsible for this change in the hole density is dJ, then

$$dJ = -e\left[\frac{dp}{dt}\right] dx \quad \text{...(5.2.32)}$$

i.e.,

$$\frac{dp}{dt} = -(1/e)\frac{dJ}{dx} \quad \text{...(5.2.33)}$$

The negative sign indicates that the excess concentration will decrease with time: or $\frac{dp}{dt}$ is negative.

Thus,

$$\frac{dp}{dt} = -(1/e)\frac{dJ}{dx} = -(1/e)\left[e\mu_p E\frac{dp}{dx} - eD_p\frac{d^2p}{dx^2}\right]$$

$$\frac{dp}{dt} = -(\mu_p E)\frac{dp}{dx} + (D_p)\frac{d^2p}{dx^2} \quad \text{...(5.2.34)}$$

3. Lastly there may be sources and sinks for excess minority carriers in the volume element. A reverse biased p-n junction acts as a sink for minority carriers. Photo emission of minority carriers is also possible. Let the rate of change of hole density

due to such processes be given by G. Then the complete continuity equation becomes

$$\frac{dp}{dt} = \frac{p_0 - p}{\tau_p} - (\mu_p E)\frac{dp}{dx} + (D_p)\frac{d^2 p}{dx^2} + G \qquad \text{...(5.2.35)}$$

In three-dimensions the equation becomes

$$\frac{dp}{dt} = \frac{p_0 - p}{\tau_p} - \mu_p E \nabla p + D_p \nabla^2 p + G \qquad \text{...(5.2.36)}$$

From similar arguments we can write the continuity equation for electrons as

$$\frac{dn}{dt} = \frac{n_0 - n}{\tau_n} + \mu_n E\frac{dn}{dx} + D_n\frac{d^2 n}{dx^2} + G \qquad \text{...(5.2.37)}$$

In order to find the current density due to the contribution of the different carriers, the appropriate continuity equation has to be solved taking into account the suitable boundary conditions.

Let us now apply the continuity equation to calculate the current components J_1 and J_4 in a biased *p-n* junction. The current J_1 consists of minority electrons moving by diffusion on the *p*-side and falling down at the junction due to the potential well, *i.e.*, the junction acts like a sink for the electrons.

Far away from the junction boundary the electron density n_0 is the normal equilibrium value. At the junction boundary the electron density is zero. So there is a concentration gradient for the minority electrons in the *p*-region as shown in Fig. 5.34.

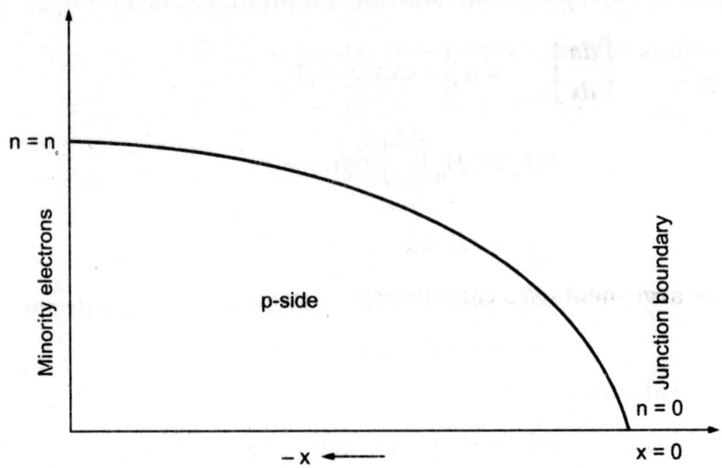

Fig. 5.34 Concentration gradient for the minority electrons in the *p*-region

Because of this concentration gradient the minority electrons diffuse to the junction boundary giving rise to the current J_1. The appropriate equation is given by

$$(D_n)\frac{d^2 n}{dx^2} - \frac{n - n_0}{\tau_n} = 0 \qquad \text{...(5.2.38)}$$

$\left[\frac{dn}{dt}\right.$ = 0 under steady state conditions. $\mu_n E\left(\frac{dn}{dx}\right) = 0$ since there is no electric field in the bulk region. $G = 0$, since there are no sources or sinks in the bulk material].

The solution of the equation (5.2.38) is of the form

$$n - n_0 = A \exp\left[\pm x/\sqrt{D_n \tau_n}\right] \quad ...(5.2.39)$$

since $(n - n_0)$ cannot increase with distance, the negative exponent is in admissible. So the admissible solution is

$$n - n_0 = A \exp\left[x/\sqrt{D_n \tau_n}\right] \quad ...(5.2.40)$$

Assuming the junction to be a sharp one at $x = 0$, we have at $x = 0$, $n = 0$ and so $A = -n_0$. Thus equation (5.2.40) becomes

$$n = n_0 - n_0 \{\exp\left[x/\sqrt{D_n \tau_n}\right]\} \quad ...(5.2.41)$$

$$= n_0\left[1 - \exp\left(x/\sqrt{D_n \tau_n}\right)\right] \quad ...(5.2.41a)$$

Since $\sqrt{D_n \tau_n}$ has the dimension of length, it is called the *diffusion length* for the electron (L_n). When $x = -L_n$, $n = n_0\left(1 - \frac{1}{e}\right)$. Thus diffusion length is defined as the distance at which the concentration increases to $\left(1 - \frac{1}{e}\right)$ of its final value. Thus equation (5.2.41) can be written as

$$n = n_0\left[1 - \exp\left(x/L_n\right)\right] \quad ...(5.2.42)$$

n_0 must be substituted as n_p^0 indicating the equilibrium density of electrons in p-type.

Thus, $$\left[\frac{dn}{dx}\right]_{x=0} = n_p^0\left[-\exp(x/L_n)\frac{1}{L_n}\right]_{x=0} = -n_p^0/L_n \quad ...(5.2.43)$$

Thus, $$J_1 - eD_n\left(\frac{dn}{dx}\right) \text{ at } x = 0 \quad ...(5.2.44)$$

or $$J_1 = eD_n n_p^0/L_n \quad ...(5.2.45)$$

From similar arguments we can show that

$$J_4 = -eD_p\, p_n^0/L_p \quad ...(5.2.46)$$

The total saturation current

$$J_0 = J_1 + J_4 = -\left[eD_n n_p^0/L_n + eD_p\, p_n^0/L_p\right] \quad ...(5.2.47)$$

The negative sign indicates that the current is opposite to the positive x-direction. Since p_n^0 and n_p^0 are small, the saturation current is small and is also constant. Substituting this value of J_0 in equation (5.2.29) one gets,

$$\boxed{J = \{(eD_n n_p^0/L_n) + (eD_p\, p_n^0/L_p)\}[\exp(eV/k_B T) - 1]} \quad ...(5.2.48)$$

Conclusion

From the above rectifier equation we see that in a forward biased *p-n* junction the currents are carried by the injected minority carriers. The component J_3 is due to the holes injected into the *n*-region where they become minority carriers. Similarly the component J_2 is due to the electrons injected into the *p* region where they become minority carriers. These injected carriers increase minority carrier densities close to the depletion region, and so begin to move by diffusion to regions of lower concentration away from the depletion region. These minority carriers suffer recombination and at a distance approximately equal to 4 times the diffusion length, the carrier concentrations come to the normal equilibrium values. In devices it is preferable to have efficient injection of minority carriers in one direction only. For example, in a *p-n-p* transistor the injected hole current J_3 must be appreciable compared to the electron current injected from *n* to *p*. In the *n-p-n* transistor the injected electron current from *n* to *p* must be appreciable compared to the hole current injected from *p* to *n*. This can be done by proper doping of the samples. We define a parameter called the minority carrier injecting efficiency or simply the emitter efficiency (g) to describe the characteristic of a good emitter. In a forward biased *p-n* junction the ratio of the hole current to the sum of the hole current and the electron current is called the *emitter efficiency* of the *p*-material. That is

$$\gamma = \frac{J_3}{(J_2 + J_3)}$$

$$= \frac{eD_p p_n^0/L_p}{[eD_n n_p^0/L_n + eD_p p_n^0/L_p]}$$

$$= \frac{1}{[eD_n n_p^0/L_n + eD_p p_n^0/L_p]/eD_p p_n^0/L_p}$$

$$= \frac{1}{1 + eD_n n_p^0 L_p/eD_p p_n^0 L_n}$$

i.e.,

$$\gamma = \frac{1}{1 + D_n n_p^0 L_p/D_p p_n^0 L_n} \quad ...(5.2.49)$$

$$\sigma_p = e\mu_p^0 p_p^0 = e\mu_p^0 n_i^2/n_p^0 \quad ...(5.2.50)$$

$$\sigma_n = e\mu_n^0 p_n^0 = e\mu_n^0 n_i^2/p_n^0 \quad ...(5.2.51)$$

Thus,

$$n_p^0 = \frac{e\mu_p^0 n_i^2}{\sigma_p} \quad \text{and} \quad p_n^0 = \frac{e\mu_n^0 n_i^2}{\sigma_n} \quad ...(5.2.52)$$

$$\frac{D_n n_p^0}{D_p p_n^0} = \left[\frac{\mu_n^0}{\mu_p^0}\right]\left[\frac{e\mu_p^0 n_i^2 \sigma_n}{\sigma_p e\mu_n^0 n_i^2}\right] = \sigma_n/\sigma_p$$

$$\gamma = \frac{1}{[1 + \sigma_n L_p / \sigma_p L_n]} » [1 - \sigma_n L_p/\sigma_p L_n]$$

So for greater emitter efficiency

$$\sigma_n L_p < \sigma_p L_n \qquad ...(5.2.53)$$

Smaller σ_n means, smaller conductivity, higher resistivity and hence smaller doping levels for the *n*-type material. Smaller L_p means, smaller diffusion length for the injected holes in the *n*-region and longer L_n means larger diffusion length for the injected electrons in the *p*-region. Greater σ_p means higher doping levels in the *p*-region. These factors will be important in the design of efficient transistors.

5.2.9 SOME SPECIAL *p-n* JUNCTION DIODES

We have already seen that the depletion region around a *p-n* junction contains immobile positive charges on the *n*-side and immobile negative charges on the *p*-side. Also we know that, the application of a reverse voltage to a *p-n* junction causes these uncovered charges to increase. This may be considered as a capacitive effect and the junction is assumed to exhibit a capacitance, called the junction capacitance. This capacitance is not a constant, but decreases with the increase in reverse voltage. This property of the *p-n* junction finds application in a special type of diodes, called *varactors* or *varicaps*. Such diodes, among others, are discussed below.

Varactor Diode

It is a device whose reactance can be varied in a controlled way with a bias voltage. We have seen that a *p-n* diode has a voltage-dependent junction capacitance. Therefore, *p-n* junction diodes find applications as varactors, also called varicaps.

For an abrupt *p-n* junction, the junction capacitance varies inversely as the square root of the reverse bias voltage. For a linearly graded junction, this capacitance varies inversely as the one-third of the reverse bias voltage. The voltage sensitivity of the junction capacitance is thus greater for an abrupt junction than for a linearly graded junction. By using special growth techniques, the junction capacitance may be made to vary inversely as the square of the reverse bias voltage. When such a junction capacitor is used with an inductor in a resonant circuit, the resonant frequency will vary linearly with the applied voltage. Such varactors are useful in frequency modulation.

Varactor diodes may be used to replace the bulky variable capacitors in the tuning stage of a radio receiver or in a TV tuner. This makes the circuit compact and improves its performance. Varactors are also used in self-balancing bridge circuits, in special type of amplifiers (called parametric amplifiers), in harmonic generators, in microwave frequency multipliers, and in active filters.

Breakdown Diode

When *p-n* junction diodes are operated in the saturation portion of the reverse-voltage characteristic (Fig. 5.32), they are called breakdown diodes.

The following two mechanisms are responsible for breakdown in a *p-n* junction diode:

(*i*) As the applied reverse bias in a *p-n* junction increases, the field across the junction increases correspondingly. At some value of the bias, the field becomes large so that a thermally generated carrier while traversing the junction acquires a large amount of energy from the field. This carrier is then able to disrupt covalent bond and create a new electron-hole pair while colliding with an immobile ion. The new carriers again pick up sufficient energy from the applied field and collide with other immobile ions generating further electron-hole pairs. This process being cumulative in nature, results in generation of an avalanche of carriers within a very short time. This mechanism of carrier generation is referred to as avalanche multiplication. The result of the process is the flow of a large amount of current at some value of the reverse bias.

(*ii*) Zener breakdown takes place when the reverse-bias field across a *p-n* junction is such that the field may exert a strong force on a bound electron to tear it out from a covalent bond. Thus, a large number of electron-hole pair will be generated through a direct rupture of the covalent bonds. The electron-hole pairs so generated increase the reverse current. Note that, contrary to avalanche multiplication, in Zener breakdown, the carrier generation is not due to the collisions of carriers with the immobile ions.

For breakdown voltages below 6 volt true Zener breakdown occurs. For higher voltages the avalanche multiplication is predominant. In spite of the two distinct mechanisms, breakdown diodes are ordinary called Zener diodes. The symbol for a Zener diode is shown in Fig. 5.35. The voltage-current characteristic of a Zener diode is given by the dashed portion of the characteristic curve in Fig. 5.32.

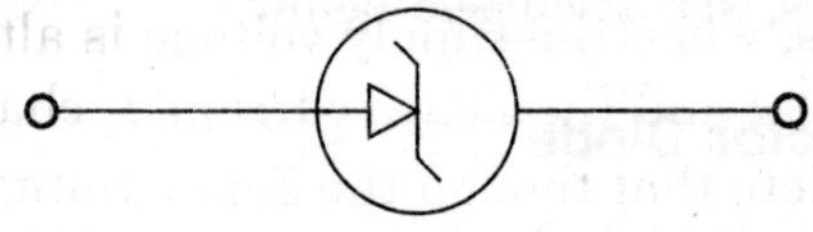

Fig. 5.35 Symbol used for a Zener diode

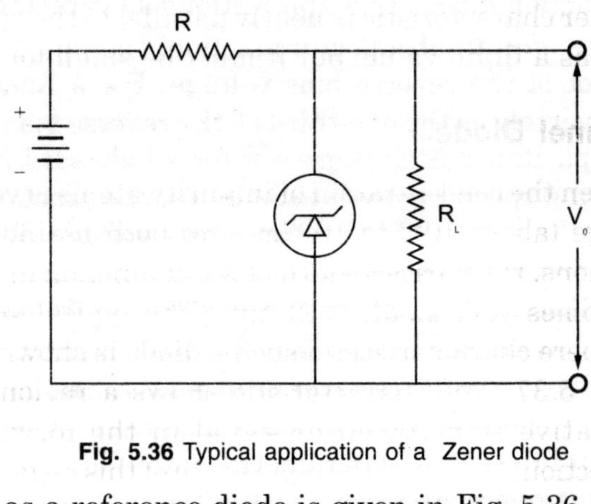

Fig. 5.36 Typical application of a Zener diode

The Zener characteristic is almost parallel to the current axis indicating that the voltage across the diode remains almost constant although the current through it may vary. The voltage across a Zener diode thus serves as a reference and the diode is referred to as a reference diode.

A typical application of a Zener diode as a reference diode is given in Fig. 5.36. The voltage V and the resistance R are determined so that the diode current is within the specified limit and the diode is operated in the breakdown region. The voltage V_0 across the load resistance R_L is a constant here although the supply voltage V and the load resistance R_L may vary. The upper limit of the diode current is determined by the power dissipation rating of the diode.

In Fig. 5.36, if I be the current drawn from the source, and I_z and I_L be the currents through the Zener diode and load resistance respectively, then Kirchhoff's current and voltage laws give

$$I = I_z + I_L \quad ...(5.2.54)$$

and $$V_0 = V - IR \quad ...(5.2.55)$$

Also $$V_0 = I_L R_L \quad ...(5.2.56)$$

Let the supply voltage V remain constant and the load resistance R_L be varied. Since the Zener voltage V_0 tends to remain constant, equation (5.2.55) gives $dI = 0$. We then obtain from equation (5.2.54)

$$dI = dI_z + dI_L = 0$$

i.e., $$dI_z = -\,dI_L \quad ...(5.2.57)$$

Thus, when the load resistance is increased but the supply voltage is fixed, the current I_L decreases and the current I_z increases by an equal amount so that the total I remains constant.

Now suppose that the load resistance R_L is fixed and the supply voltage V is varied. Since V_0 tends to remain constant, we obtain from equation (5.2.55)

$$dV = R\,dI$$

We also know from equation (5.2.56), $dI_L = 0$, and from equation (5.2.54), $dI = dI_z$. That is, when the supply voltage is altered but the load resistance is kept fixed, the total current I and the Zener current I_z change equally to keep the load current I_L constant.

Note that though the Zener voltage V_0 is fixed, the Zener current I_z may vary. Hence the DC resistance of the zener diode, *i.e.*, V_0/I_z is not a constant. The Zener dynamic resistance r_z is given by the inverse of the slope of the characteristic, *i.e.*, by dV_0/dx. As the Zener characteristic is nearly parallel to the current axis, $dV_0/dx = 0$, *i.e.*, $r_z \approx 0$. In practice, r_z has a finite value, but it must be small for a good Zener diode.

Tunnel Diode

When the concentration of impurity atoms is very large (about 10^{24} to $10^{25}/m^3$) in both *p*- and *n*-regions, the barrier-width of a *p*-*n* junction diode becomes very small (≈10 nm). The typical volt-ampere characteristic of such a diode is shown in Fig. 5.37. The characteristic shows a region of negative slope when operated in the forward direction. Note that the occurrence of this negative slope cannot be explained by the mechanism discussed in Section XX. The quantum mechanical process, known as tunneling, gives a satisfactory explanation of the above characteristic. Hence the diode is called a tunnel diode.

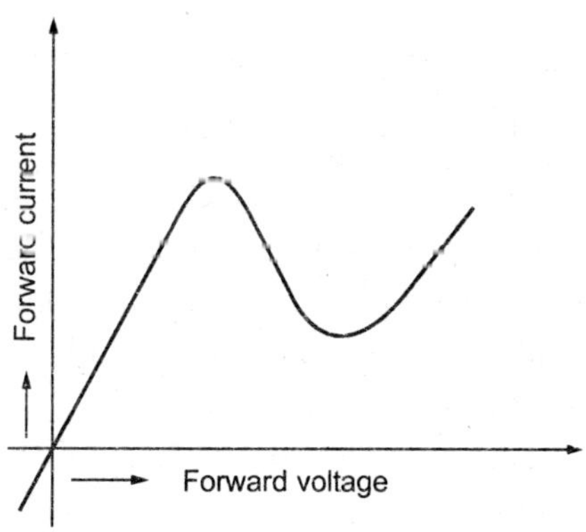

Fig. 5.37 A typical tunnel diode characteristic

The negative slope, *i.e.*, the negative differential resistance (the differential resistance of any point in the volt-ampere characteristic is given by the reciprocal of the slope of the characteristic at that point. If the slope is negative, evidently the differential resistance is negative) exhibited by a tunnel diode can be utilised in the construction of amplifiers, oscillators or switching devices. Since the tunneling process is very fast, tunnel diode devices can work at very high frequencies (about 10 GHz).

Photo Diode

When a light radiation is allowed to fall on a reverse-biased *p-n* junction diode additional-hole pairs are created in both *p*-and *n*-regions. This produces a very small change in majority carrier concentration and a very large change in minority carrier concentration. These additional minority carriers enhance the reverse current since they fall down the barrier potential. It is found that the current through the diode varies almost linearly with the light flux. The diodes designed to operate on this principle are called photo diodes. Such diodes are used in light detection, in light operated switches, reading of computer punched cards and taps.

Light Emitting Diodes (LED)

When a *p-n* junction diode is forward biased, carriers are injected across the junction to establish excess carriers above their thermal equilibrium values. These excess carriers recombine and release electromagnetic energy. In the case of GaAs, the energy appears in the form of infrared radiation. For the alloy semiconductor, $GaAS_{1-x} - P_x$, the energy appears in the form of visible light. The *p-n* junction diode constructed with such semiconductors is called the *light-emitting diode* (LED). Here we get optical energy from the biasing electrical energy. The electrons that are injected in the *p*-side make a downward transition from the conduction band to the valence band and recombine with holes, emitting photons with energy $h\upsilon = E_g$. The corresponding emission wavelength is given by

$$\lambda = \frac{hc}{E_g}$$

where h is Planck's constant, c is the velocity of the electromagnetic wave and E_g is the band gap. Light emission in the n-side of the junction takes place similarly except that here the holes are excess carriers. The downward transition of electrons from the conduction band to the valence band and subsequent emission of photons due to recombination with holes are shown in Fig. 5.38.

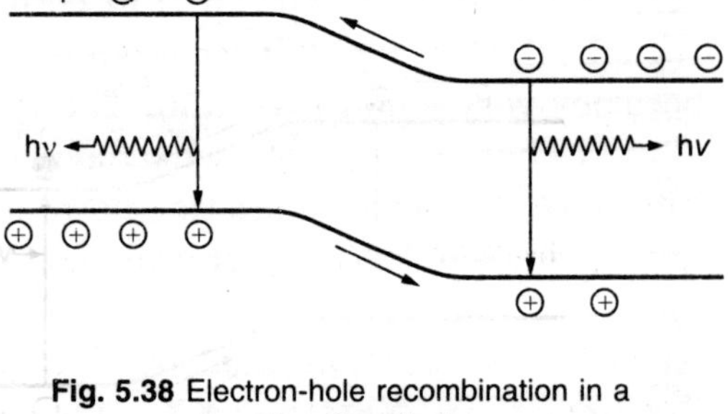

Fig. 5.38 Electron-hole recombination in a *p-n* junction diode

Diodes built with $GaAs_{1-x} - P_x$ are most commercially available because of lower cost and easy fabrication. $GaAs_{1-x} - P_x$ light-emitting diodes are used in the manufacture of signal and displays. The infrared LED is a potential source for optical fibre communication. Under some conditions the emitted light is coherent. Such a diode is referred to as an *injection junction* laser.

Solar Cell

The solar cell is basically a *p-n* junction diode that converts sunlight directly to electricity with large conversion efficiency. The action of the solar cell is explained as follows:

(*i*) When a *p-n* junction diode is exposed to light, photons are absorbed and electron-hole pairs are generated in both the *p*-side and *n*-side of the junction, as shown in Fig. 5.39 (*a*). The electrons and the holes that are produced over a small distance from the junction reach the space-charge region X by diffusion [Fig. 5.39 (*b*)].

(*ii*) The electron-hole pairs are then separated by the strong barrier field that exists across the region X. The electrons in the *p*-side slide down the barrier potential to move to the *n*-side while the holes in the *n*-side move towards the *p*-side [Fig. 5.39 (*c*)].

(*iii*) When the *p-n* junction diode is open-circuited, the accumulation of electrons and holes on the two sides of the junction gives rise to an open-circuit voltage V_0. If a load resistance is connected across the diode, a current will flow in the circuit [Fig. 5.39 (*a*)]. The maximum current, called the short-circuit current is obtained when an electric short is connected across the diode terminals. Note that the current flows as long as the diode is exposed to sunlight and the magnitude of the current is proportional to the light intensity.

Solar cells are used extensively in satellites and space vehicles as most important long-duration power supply. Solar cells are constructed with silicon, gallium arsenide, cadmium sulphide and with many other semiconductors, and in various device configurations.

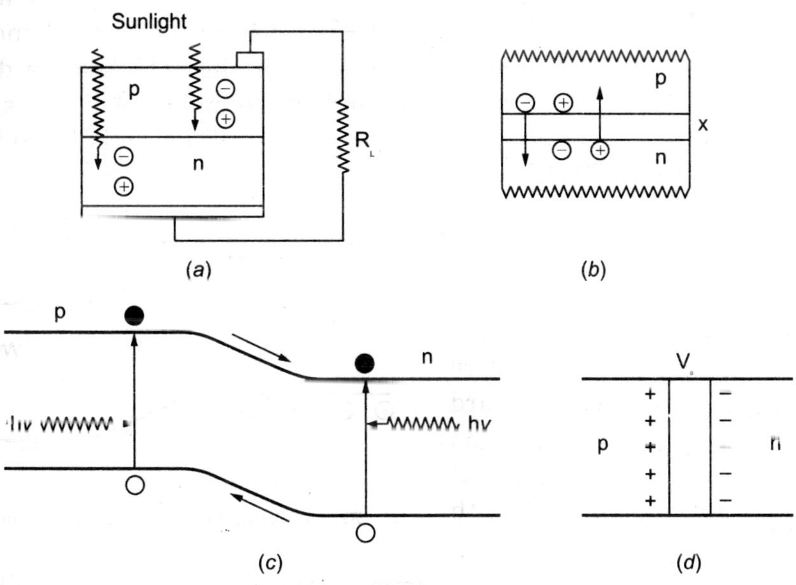

Fig. 5.39 (*a*) *p-n* junction solar cell with load resistance, R_L
(*b*) Diffusion of electrons and holes
(*c*) Energy band diagram corresponding to (*b*)
(*d*) Formation of the open circuit voltage V_0

SHORT QUESTIONS

5.1.1 What are density of energy states in a semiconductor?

(Anna U, BE, November 1992, April 1994)

5.1.2* A pure semiconductor has E_g = 0.7 eV, $N_p = 25 \times 10^{24}$, $m_e^* = m_p^* = m = 9.1\times10^{-31}$ *kg*. T = 27 °C. Get density of holes *(University of Surrey, U.K, B.Tech, Winter 1998).*

5.1.3 How the energy gap in a semiconductor is estimated?

5.1.4 How an intrinsic semiconductor is converted into extrinsic semiconductor?

(MU, BE, April 1999, Madras U, BE, November 1992)

5.1.5 Distinguish between intrinsic and extrinsic semiconductors.

(JNTU, BE, November 2000)

5.1.6 A crystal of a doped semiconductor is electrically neutral though it has an excess of electrons or holes as charge carriers. How do you explain this?

(VTU, BE, May 2002)

5.1.7 Semiconductors are opaque to visible light but transparent to infrared light. How do you explain this?

(IIT Bombay, B. Tech, April 1997, M.K.U, Part-time BE, November 1998)

5.1.8* If silicon is transparent to radiation up to a wavelength 1100 nm, find the energy gap.

5.1.9 Explain why silicon with an energy gap of 1.1eV is opaque to visible light.

5.1.10* Give the salient features of semiconductors with that of metals for reference.

5.1.11* Locate the Fermi level in a germanium crystal containing 10^{22} donor atoms/m^3. Assume E_g =0.72eV and the ionization energy of donor atom is 0.012 eV.

(JNTU BE, May 2001, VTU, B.Tech, November 2002)

5.1.12 What will happen for the graph showing the variation of Fermi level with temperature in a *p*-type semiconductor if donor concentration goes up from 10^{21} to 10^{23}?

(Anna U, M.Sc, April 1995)

5.1.13 Derive the variation of concentration of charge carriers in an *n*-type semiconductor with temperature.

5.1.14 Explain the two kinds of scattering in semiconductors and how they influence the mobility of the carriers.

5.1.15 The resistance of semiconductors is very sensitive to temperature. What are the main uses of this behaviour of semiconductors? *(JNTU, BE, May 2003)*

5.1.16* Discuss some of the uses of Thermistors.

5.1.17* Discuss the advantages and disadvantages of semiconductor devices.

5.1.18 If the Hall coefficient is positive for an *n*-type semiconductor, how you will explain it?

5.1.19 What is meant by life time of minority charge carriers? *(Anna U., B.Tech, April 1995)*

5.1.20 When you will say that a p-n junction is having (*i*) high resistance? (*ii*) low resistance? Explain.

5.1.21 Explain the working of a zener diode. *(M.K.U, Part-time BE, April 1992)*

5.1.22 Discuss the variation of Fermi level with temperature in an *n*-type semiconductor.

(Madras U, BE, November 1994, Anna U, BE, April 1995)

5.1.23* Discuss the salient features of semiconductors.

5.1.24 Explain photoconductivity in semiconductors. *(Bangalore University, BE, April 1990)*

5.1.25 Explain the terms:

(*i*) Zener break down, (*ii*) Avalanche break down

5.1.26 Explain the difference between metals and semiconductors from the consideration of temperature coefficient of resistance. *(JNTU, BE, April 2002)*

5.1.27* A zener diode is a *p-n* junction. Yet it is different from an ordinary *p-n* junction. Explain.

Answers to the starred questions

5.1.2*

$$n = p = 24.9 \times 10^{24} \exp(-0.65/2 \times 0.025)$$

i.e.,

$$\boxed{p = 9.3 \times 10^{19}/\text{m}^3} \quad \textbf{Answer}$$

5.1.8*

$$E_g = h\nu = \frac{ch}{\lambda}$$

$$= \frac{3\times10^{8}\times6.62\times10^{-34}}{1100\times10^{-9}} = \frac{3\times6.62\times10^{-26}}{1100\times10^{-9}}$$

$$E_g = \frac{3\times6.62\times10^{-17}}{1100\times1.6\times10^{-19}} \text{ eV}$$

$$E_g = 1.13 \text{ eV} \quad \textbf{Answer}$$

5.1.10*

1. Semiconductors have resistivity and their temperature coefficient of resistance is negative while it is positive in metals.
2. They are generally hard and brittle having metallic appearance.
3. The conducting properties of a semiconductor change appreciably when some suitable impurity is added to it.
4. Semiconductors control the flow of electricity and convert electrical energy into other types of energy while metallic conductors only transport electric energy.
5. When temperature is increased, semiconductivity increases, while electric conductivity decreases in metals.
6. Semiconductors become dielectric at low temperatures.
7. Semiconducting devices require only small space unlike valves.
8. No voltage is required to heat any filament unlike valves.
9. The operation in semiconducting devices starts instantaneously.

5.1.11*

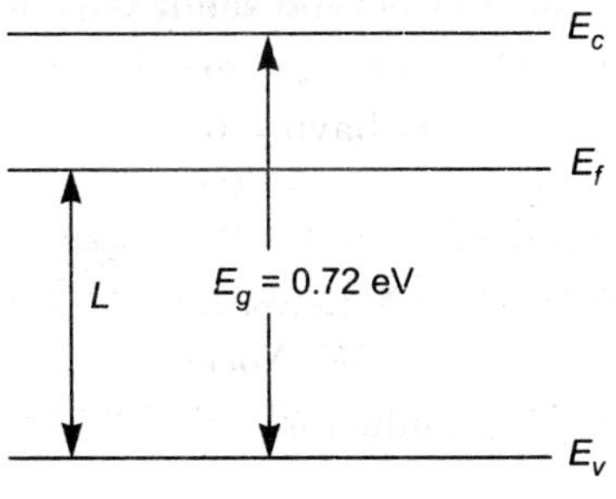

$$(E_c - E_f) = k_B T \, ln\left(\frac{N_e}{N_d}\right)$$

$$= 0.025 \times ln\left(\frac{4.83 \times 300^{3/2} \times 10^{21}}{10^{22}}\right)$$

$$= 0.203 \text{ eV}$$

$$(E_f - E_v) = E_g - (E_c - E_f) = 0.72 - 0.203$$

$$L = 0.52 \text{ eV} \quad \textit{Answer}$$

5.1.16* The important features of a thermistor are its high thermal sensitiveness and compactness. Thermistors find a number of uses listed below:

(*a*) to measure very low temperatures of the order of 10 K or less

(*b*) to protect electronic circuits against surges of current

(*c*) to measure varying temperatures

(*d*) in voltage stabilizers

(*e*) in temperature control units

(*f*) to protect the windings in transformers, motors and generators from surges of heavy currents and

(*g*) to measure wind velocity and altitude

5.1.17*

(*a*) Advantages of semiconductor devices over vacuum tubes:

1. Semiconductor devices are very compact and portable. Therefore circuits using semiconductor devices are very compact and also portable unlike devices with valves.
2. These devices are operated relatively at low voltages, therefore the power consumption is less and hence hand batteries can be used.
3. In semiconductor devices there is no warm up time. Therefore these devices are set into operation as soon as the circuit is switched on.
4. These devices have almost unlimited life unless they are subjected to wrong voltage.
5. Transistors can withstand mechanical shocks as they are solid crystals.
6. Semiconductor devices are low priced as compared to vacuum tubes.
7. In transistors there is no vacuum deterioration trouble as no vacuum has to be created. Vacuum tubes however suffer from such defects.

(*b*) Disadvantages

1. These devices are temperature sensitive. Their characteristics change considerably above the permissible ambient temperature (above 50°C). This is due to the breaking of more covalent bonds which increases the percentage of minority carriers.
2. These devices give excessive reverse current and consequently noise level is high.

3. The power output is relatively low.
4. There is less uniformly in characteristics among the individuals of the same type.

5.1.23* (1) Semiconducting materials have high resistivity and their temperature coefficient of resistance is always negative.
(2) They are generally hard and brittle having metallic appearance
(3) The conducting properties of semiconductors change appreciabley when some suitable metallic impurities are added.
(4) Semiconductors control the flow of electricity and convert electrical energy into other types of energy. (Metallic conductors only transport electric energy)
(5) When temperature is increased, semiconductivity increases.
(6) Semiconductors become dielectrics at low temperatures.

5.1.27* A *p-n* junction allows current to flow freely in one direction only. In the opposite direction it offers a very high resistance. This is the most characteristic feature of a *p-n* junction.

But a zener diode is designed to operate in the breakdown region without damage. By adjusting the doping level it is possible to produce zener diodes with a breakdown voltage from 2 V to 800 V. If the applied voltage exceeds the breakdown voltage the diode acts as a constant voltage source.

REVIEW QUESTIONS

5.1.1 Obtain the expression for the carrier concentration in an intrinsic semiconductor. What are donor and acceptor atoms?

(JNTU, BE April 2000, Anna U, BE, November 1997)

5.1.2 Describe how the carrier concentration can be determined by observing the Hall Voltage for a *p*-type semiconductor.

Describe a method for the estimation of mobilities of charge carriers in semiconductors. *(M.K.U, BE, April 1996, Mysore U, BE, April 1997)*

5.1.3 What is meant by 'intrinsic carrier density in a pure specimen of silicon? Explain how the energy gap of this specimen can be determined by conducting a small experiment.

(IIT, Delhi, B. Tech, April 1996, Anna U, M.Sc. April 1999)

5.1.4 What are *n*-type and *p*-type semiconductors? Obtain the expression for the electron density in the conduction band of an *n*-type semiconductor at low temperatures.

(Newcastle Institute of Technology, London, B.Tech, April 1979)

5.1.5 Mark the Fermi level for

(*i*) an intrinsic semiconductor, (*ii*) *p*-type semiconductor and (*iii*) *n*-type semiconductor.

Discuss the variation of the Fermi level with temperature in these cases.

Describe a method for the evaluation of energy gap in an intrinsic semiconductor.

5.1.6 Sketch the variation of conductivity with temperature in the case of intrinsic and extrinsic semiconductors. What do you infer from the slope of the graphs at different regions? Describe a method of finding the conductivity of semiconductors.

(REC Durgapur, B.Tech, April 1999)

5.1.7 Show that the electron current from *p*-side to *n*-side is equal to that from *n*-side to *p*-side when the *p*-*n* junction is in equilibrium.
When will you say that *p*-*n* junction is having (*i*) high resistance and (*ii*) low resistance? Explain. *(Anna U, M.Sc, April 1992)*

5.1.8 Explain the mechanism of intrinsic conduction in semiconductors. Obtain the equation for the conductivity of an intrinsic semiconductor in terms of their carrier concentration and carrier mobilities.
Explain how intrinsic conductivity various with temperature.
(University of Maryland, US, B. Tech, April 1977, JNTU, BE, November 1996)

5.1.9 Give the theory of direct recombination of carriers in semiconductors and hence obtain the expression for the decay of minority carriers injected into a semiconductor *(VTU, BE, April 2002)*

5.1.10 Obtain the quadratic equation for the Fermi level in an *n*-type semiconductor. Solve it for low temperature range and hence show that the density of electrons in the conduction band is directly proportional to the square root of donor concentration.

5.1.11 Obtain the expression for the decay of minority carriers injected into a semiconductor. Deduce Einstein relation connecting the diffusion constant and mobility of charge carriers in a semiconductor.

5.1.12 What are *n*-type and *p*-type semiconductors? How are they produced?
Prove that in the case of extrinsic semiconductors, the product of the majority and minority carrier densities is equal to the square of its intrinsic concentration.
(REC, Calicut, M. Tech, April 2004, Anna U, M.Sc, November 1998)

5.1.13 (*a*) Explain the effect of temperature and dopants on the Fermi level in a semiconductor.
(*b*) (*i*) Find the conductivity of intrinsic silicon at 300 K. It is given that n_i at 300 K in silicon is $1.5 \times 10^{16}/m^3$ and the mobilities of electrons and holes in silicon are 0.13 m^2/V-s and 0.05 m^2/V-s respectively.
(*ii*) If donor type impurity is added in the extent of one impurity in 10^8 Si atoms, find the conductivity.
(*iii*) If acceptor impurity is added to the extent of one impurity in 10^8 Si atoms, find the conductivity. *(JNTU, BE, April 2002)*

5.1.14 Explain the difference between metals and semiconductors from the consideration of temperature coefficient. The electron and hole mobilities in a Si sample are 0.135 and 0.048 m^2/V-s respectively. Determine the conductivity of intrinsic Si at 300 K if the intrinsic carrier concentration is 1.5×10^{16} atoms/m^3. The sample is then doped with 10^{23} phosphorus atoms/m^3. Determine the equilibrium hole concentration, conductivity and position of the Fermi level to the intrinsic level.

5.1.15 Derive the expressions for the densities of electrons and holes in the conduction band and valence band respectively in an intrinsic semiconductor. Show that the Fermi level in an intrinsic semiconductor lies approximately half way between the top of the valence band and the bottom of the conduction band.
(VTU, Karnataka BE (NR) April 2001, Anna U, BE, November 1998)

5.1.16 Discuss the physical mechanism of conduction in *p*-type semiconductors. Derive the expressions for conductivity at very low and high temperatures.
(NIT Calicut, B. Tech, April 2004)

5.1.17 Derive an expression for the carrier density in an n-type semiconductor and give a qualitative discussion of its variation with temperature.

(VTU, BE, April 2002, NIT, Calicut BE, April 2003)

5.1.18 Describe how a p-n junction behaves under forward bias and reverse bias conduction. Give the equation for the current flow in each case.

5.1.19 Derive an expression relating to the shift in the Fermi level and the carrier density in an extrinsic semiconductor. What you understand by drift and diffusion currents in the case of a semiconductor?

(Anna U, Part-time BE, (Electrical) April 1995)

PROBLEMS AND SOLUTIONS

5.1.1 The electrical resistivity of pure Si is 2300 Ω m at 27°C, what will be its resistivity at 200°C ? Given: E_g = 1.1eV, k_B = 8.62×10^{-5} eV/K.

(Bangalore University, BE, April 1990)

Solution:

$$\sigma_{300} = \frac{1}{\rho} = \frac{1}{2300} = 4.35\times10^{-4}\,\Omega^{-1}\,m^{-1}$$

$$\frac{E_g}{2k_BT_1} = \frac{1.1\times10^5}{2\times8.62\times300} = 21.27$$

and

$$\frac{E_g}{2k_BT_2} = \frac{1.1\times10^5}{2\times8.62\times473} = 13.49$$

Now

$$\sigma_{300} = A\exp(-21.27) \quad\text{and}\quad \sigma_{473} = A\exp(-13.49)$$

$$\frac{\sigma_{473}}{\sigma_{300}} = \frac{e^{-13.49}}{e^{-21.27}} = 0.238\times10^4$$

$$\sigma_{473} = \sigma_{300}\times0.238\times10^4 = \frac{0.238\times10^4}{2300} = 1.03$$

$$\boxed{\sigma_{473} = 1.03\,\Omega^{-1}\,\mathrm{m}^{-1}}$$ ***Answer***

5.1.2 Distinguish the temperature dependence of a metal with that of a semiconductor. Show that the ratio of maximum resistivity to intrinsic resistivity is expressed as

$$\frac{\rho_{max}}{\rho_i} = \frac{(\mu_e+\mu_p)}{2\sqrt{\mu_e\mu_p}}$$

Solution:

Intrinsic conductivity

$$\sigma = n_ie\,(\mu_e+\mu_p) = e\,(n_i\,\mu_e + n_i\,\mu_p)$$

But

$$np = n_i^2; \text{ or } p = \frac{n_i^2}{n}$$

Thus

$$\sigma_i = e\left(n\,\mu_e + \frac{n_i^2}{n}\mu_p\right)$$

For resistivity to maximum, conductivity must be minimum

Hence,

$$\frac{d\sigma}{dn} = 0 = e\left(\mu_e - \frac{n_i^2\mu_p}{n^2}\right)$$

$$\mu_e - \frac{n_i^2\mu_p}{n^2} = 0$$

$$n^2 = \frac{n_i^2\mu_p}{\mu_e}; \; n = n_i\sqrt{\frac{\mu_p}{\mu_e}}$$

Also

$$p = \frac{n_i^2}{n} = \frac{n_i^2}{n_i\sqrt{\mu_p/\mu_e}}$$

$$p = n_i\sqrt{\frac{\mu_e}{\mu_p}}$$

Thus

$$\sigma_{\min} = e\left\{n_i\left[\sqrt{\frac{\mu_p}{\mu_e}}\right]\mu_e + n_i\left[\sqrt{\frac{\mu_e}{\mu_p}}\right]\mu_p\right\}$$

$$\sigma_{\min} = e\left\{n_i\sqrt{\mu_e\mu_p} + n_i\sqrt{\mu_e\mu_p}\right\}$$

$$\sigma_{\min} = (2\,e\,n_i)\sqrt{\mu_e\mu_p}$$

Now

$$\rho_i = \frac{1}{n_i e(\mu_e + \mu_p)} \text{ and } \rho_{\max} = \frac{1}{2\,e\,n_i\sqrt{\mu_e\,\mu_p}}$$

Thus

$$\frac{\rho_{\max}}{\rho_i} = \frac{n_i\,e(\mu_e + \mu_p)}{2\,e\,n_i\sqrt{\mu_e\mu_p}}$$

$$\boxed{\frac{\rho_{\max}}{\rho_i} = \frac{(\mu_e + \mu_p)}{2\sqrt{\mu_e\,\mu_p}}}$$ ***Answer***

5.1.3 The band gap of some materials are given below. Which substance would be photoconductors over the entire range of visible wavelengths?

Substance:	Si	Ge	CdS
Band gap:	1.9×10^{-19} J	1.3×10^{-19} J	3.8×10^{-19} J

Solution:

For minimum energy, we have to use maximum wavelength:

$$E = h\upsilon = \frac{ch}{\lambda} = \frac{3\times10^{8}\times6.62\times10^{-34}}{600\times10^{-9}}$$

$$\boxed{E = 3.3\times10^{-19} J} \quad \textbf{\textit{Answer}}$$

This is higher than E_g of Si and Ge and hence they can be used in photoconductors.

5.1.4 The alternating current-gain of a transistor in common base arrangement is 0.98. Find the change in collector-current corresponding to a change of 5.0 mA in emitter-current. What would be the change in the base-current?

(MKU., Part-time BE, April 1997)

Solution:

Current gain,

$$\alpha = \frac{\Delta i_c}{\Delta i_e}$$

$$\Delta i_c = a\ \Delta i_e = 5 \times 0.98 = 4.9 \text{ mA}$$

$$\boxed{\Delta i_c = 4.9\,\text{mA}} \quad \textbf{\textit{Answer}}$$

Further the base current is emitter current–the collector current

$$I_b = i_e - i_c = \Delta\ i_e - \Delta\ i_c = 5 - 4.9$$

$$\boxed{I_b = 0.1 \text{ mA}} \quad \textbf{\textit{Answer}}$$

5.1.5 In a silicon *p-n* junction the current at 300 K for a reverse bias voltage V such that eV >> k_BT is 10μA, calculate the current for a forward bias of 0.2 volt.

(JNTU, BE, November 2000, AU, BE, April 1996)

Solution:

$$I_F = \exp[-\Delta E_1 / k_BT]\exp[(\text{eV} / kL_BT) - 1]$$

and

$$I_R = \exp[-\Delta E_1 / k_BT]\exp[(-\text{eV} / kL_BT) - 1]$$

When eV >> k_BT

$$\frac{I_F}{I_R} = \exp\ (eV / k_BT)$$

$$I_F = I_R \exp[(1.6\times10^{-19}\times0.2)/1.38\times10^{23}\times300)]$$

$$I_F = 10 \times 10^{-6} \times 22.7 \times 10^2$$

$$I_F = 22.7 \text{ mA} \quad \textbf{\textit{Answer}}$$

5.1.6 A sample of intrinsic germanium at room temperature has a carrier concentration of $2.4 \times 10^{19}/\text{m}^3$. It is doped with antimony at a rate of one antimony atom per million atoms of germanium. If the concentration of germanium atom is $4 \times 10^{28}/\text{m}^3$, find the hole concentration and the conductivity of the semiconductor. Assuming the complete ionization of the antimony atoms. Mobility of electrons is $0.35 \text{ m}^2 \text{ V}^{-1} \text{ s}^{-1}$

(NIT Calicut, BE, May 2002)

Solution:

For 10^6 atoms of Ge one antimony atom is added. For 4×10^{28} atoms of Ge, antimony atoms added is 4×10^{22}. Antimony is donor impurity.

$$n\,p = n_i^2$$

$$p = \frac{n_i^2}{n} = \frac{2.4^2 \times 10^{38}}{4 \times 10^{22}}$$

$$p = 1.4 \times 10^{16}$$

$$\sigma = n e \mu_e = 4 \times 10^{22} \times 1.6 \times 10^{-19} \times 0.35$$

$$\boxed{\sigma = 1.4 \times 10^3\ \Omega^{-1}\text{m}^{-1}} \quad \textbf{\textit{Answer}}$$

5.1.7 Compute the carrier concentration and conductivity of intrinsic germanium at room temperature (300K). Given: $m_e^* = m_p^* = m = 9.1 \times 10^{-31}$kg and $E_g = 0.68$ eV, $\mu_e = 0.38$ m^2/V-sec and $\mu_p = 0.18$ m^2/V-sec. If the germanium sample is doped with 10^{20} donor atoms/m^3, compute the electrons and hole concentration and compare the conductivities $\left(\frac{\sigma_i}{\sigma_d}\right)$.

Solution:

$$n_i = C\,T^{3/2} \exp(-E_g / 2 k_B T)$$

$$2k_B T = \frac{2 \times 1.38 \times 10^{-23} \times 300}{1.6 \times 10^{-19}} = 0.0517$$

$$n_i = 24.9 \times 10^{24} \times \exp(-\,0.68 / 0.0517)$$

$$n_i = 4.83 \times 10^{19}/\text{m}^3$$

$$\sigma_i = 4.83 \times 10^{19} \times 1.6 \times 10^{-19} (0.38 + 0.18)$$

$$\boxed{\sigma_i = 4.32\ \Omega^{-1}\,\text{m}^{-1}} \quad \textbf{\textit{Answer}}$$

Now

$$n_i^2 = np = p \times 10^{20}$$

$$p = \frac{n_i^2}{10^{20}} = \frac{4.83^2 \times 10^{38}}{10^{20}}$$

$$p = 2.3 \times 10^{19}$$

$$\sigma_d = ne\mu_e + pe\mu_p = e\{10^{20} \times 0.38 + 2.3 \times 10^{19} \times 0.18\}$$

$$\sigma_d = e \times 10^{20}\left[0.38 + \frac{0.414}{10}\right]$$

$$\sigma_d = e \times 10^{20}[0.42] = 1.6 \times 10^{20} \times 10^{-19} \times 0.42$$

$$\boxed{\sigma_d = 6.7\ \Omega^{-1}\text{m}^{-1}}$$ ***Answer***

$$\boxed{\frac{\sigma_d}{\sigma_i} = \frac{6.7}{4.32} = 1.42}$$ ***Answer***

5.1.8 Get the equation for Hall coefficient when both holes and electrons are present. If $b = 10$, show that Hall coefficient has its maximum value when $p = 18.1\ n_i$ and its minimum value when $p = 1.75\ n_i$, where n_i is the intrinsic carrier concentration, $b = \frac{\mu_n}{\mu_p}$. Show that $R_H(\text{max}) = \frac{0.036}{n_i e}$ and $R(\text{min}) = \frac{0.966}{n_i e}$

(NIT, Nagpur BE, April 2001, JNTU BE, March 2002)

Solution:

$$R'_H = \frac{\mu_n}{\sigma}$$

$$R''_H = \frac{\mu_p}{\sigma}$$

$$R_H = R'_H + R''_H\ ;\ R_H = \frac{(\mu_p - \mu_n)}{\sigma} = \frac{(\mu_p - \mu_n)}{e(n\mu_n + p\mu_p)}$$

$$R_H = \frac{(\mu_p - b\mu_p)}{e(n\mu_n + p\mu_p)} = \frac{(\mu_p - b\mu_p)}{e(bn\mu_p + p\mu_p)}$$

$$R_H = \frac{(1-b)}{e(nb+p)} = \frac{(1-b)(nb+p)}{e(nb+p)^2}$$

$$R_H = \frac{nb + p - nb^2 - pb}{e(p+nb)^2} = \frac{(p - nb^2)}{e(p+nb)^2}$$

assuming in a compensated material $nb = bp$ and $n = \dfrac{n_i^2}{p}$. In this case $p = 18.1\ n_i$ and

hence $n = \dfrac{n_i^2}{p} = \dfrac{n_i^2}{18.1\,n_i} = 0.0552\ n_i$

Thus

$$R_H(\text{max}) = \frac{1}{e}\,\frac{(18.1 \times n_i - 100 \times 0.0522\ n_i)}{(18.1 \times n_i + 10 \times 0.0522\ n_i)^2}$$

$$= \left\{\frac{1}{n_i\,e}\right\}\left[\frac{12.58}{347.9}\right] = 0.036$$

Similarly R_H (min) can be calculated in a similar way

$$R_H(\text{min}) = \frac{0.966}{n_i\,e}$$

$$\boxed{R_H(\text{max}) = \frac{0.036}{n_i\,e}} \quad \textbf{\textit{Answer}}$$

EXERCISE

5.1.1 Calculate the diffusion current density in a piece of germanium having concentration gradient of 1.5×10^{22} electrons/m^3, the diffusion coefficient $D_n = 0.0012$ m^2/s.

(AMIE, winter 1999)

(***Ans:*** 2.88 A/m^2)

5.1.2 A germanium sample is oriented normal to a 0.5 tesla magnetic field when a current of 1 mA is passed through it by applying a p.d of 400 mV. The length of the sample is 2 cm and its width and breadth are 1 cm each. The Hall voltage measured normal to the current and the magnetic field is 6 mV. Determine the concentration of majority carriers. *(JNTU, BE, November 2000, Anna U., BE, April 1998)*

(***Ans***: 5.2×10^{19} m^3)

5.1.3 The life time for electron is 78 μs. Assume an electron mobility of 0.36 m^2/V-sec. Find the diffusion length T = 300 K. *(MKU, part-time BE, April 1996)*

(***Ans***: 85×10^{-5} m)

5.1.4 A *p-n* junction in series with a resistance of 5000 Ω is connected across a 50 volt d.c source. The forward and reverse bias resistances are 50 Ω and 1 mega ohm respectively. Calculate the forward and reverse bias current in the circuit.

(***Ans:*** 9.9 mA and 4.98×10^{-2} mA)

5.1.5 A zener diode is specified to have a breakdown voltage of 9.2 volt. The maximum power dissipation is 350 mW. What is the maximum current that the diode can handle? (***Ans***: 38 mA)

5.1.6 Calculate the current produced in a small germanium plate of area 1 cm^2 and of thickness 0.3 mm when a potential difference of 2 volt is applied across the faces. Given: concentration of free electron in germanium is $2 \times 10^{19}/m^3$ and mobilities of electrons and holes are 0.36 m^2/V-s and 0.17 m^2/V-sec respectively.

(Anna U., BE, November, 2000, VTU, BE, April 2002).

(***Ans***: 1.13A)

5.1.7 The resistivity of a semiconductor material was known to be 0.00912 Ω m at room temperature. The flux density in the Hall model was 0.48 Wb/m^3. Calculate the Hall angle for a Hall coefficient of 3.55×10^{-4} m^3/C. *(NIT Calicut, BE, April 2002).*

(***Ans***: 1° 4′)

5.1.8 The slope of the curve in the impurity region connecting $ln\ \rho$ and $\frac{1}{T}$ is 0.0246×10^4, show that the ionisation energy of the donor atom is 0.042 eV.

5.1.9 If n_1 is the charge carriers in diamond at 300 K and n_2 that at 3000 K. Find the relation between n_1 and n_2. Given : E_g = 6 eV.

(Anna U., BE, April 1994, JNTU, BE, November 2001)

(***Ans*** $n_1 = n_2 \times 0.2 \times 10^{-46}\ n_2$)

Chapter **6**

UNIT 6

Dielectric and Magnetic Properties of Materials

6.1 DIELECTRIC PROPERTIES

6.1.1 INTRODUCTION

Insulating materials are also termed as insulators or dielectrics in the wider sense. They are almost non-metallic materials with high specific resistance. They are used to prevent leakage of current from one conductor to another or to earth. Almost all insulating materials have a *negative temperature coefficient of resistance. i.e.*, their resistivity is reduced as the temperature increases. The function of any insulator is to prevent the flow of electricity through it when a difference of potential is applied across its ends. Insulating materials obstruct the flow of current without any appreciable power loss. *Dielectric materials* are those which are used in condensers to store electrical energy. e.g. for power factor correction in single phase motors, in tube lights, etc. Dielectric materials are essentially insulating materials. The function of an insulating material is to obstruct the flow of electric current while that of a dielectric material is to store electrical energy. Thus, insulating materials and dielectric materials differ in their functions. The theory of dielectric was begun by Faraday and subsequently developed by Maxwell. From Faraday's experiment we see that if in a charged condenser the air space between the plates is lowered and thus the capacitance of the condenser is increased by a factor $\in_r$ which is greater than one. This factor is known as *dielectric constant* or *specific inductivitive capacitance* and is independent of the shape or size of the capacitor, but varies widely for different materials. The value of $\in_r$ for some selected materials is given in the following table.

Material	ϵ_r	*Material*	ϵ_r
Vacuum	1	Germanium at 20°C	16
Air (1 atm) at 20°C	1.006		
Air (100 atm) at 20°C	1.0548	Glycerin at 20°C	42.5
Benzene at 20°C	2.284	Mica at 25°C	3 – 6
Polystyrene at 20°C	2.6	Backelite at 27°C and at 88°C	5 – 10
Paraffin at 20°C	2.1		
Petroleum oil at 20°C	2.2	Liquid ammonia (–78°C)	25
Ice at –5°C	2.9		
Rubber at 27°C	2.94	Water at 25°C	78.5
Paper at 25°C	3.5	Metal	∞

The comparatively high value of $\in_r$ of water suggests that water is a poor insulator; it is in fact a semiconductor and infinity value for copper yields that it is a metal.

In insulators charge cannot flow freely and if the insulators are subjected to high electric fields, then a superimposition of the applied electric field on the internal electric field of the dielectric gives rise to the lattice deformation and internal distortion which can be studied in detail if one has the knowledge of the dielectric constant. For that, the basic starting point is the Maxwell equations which in its usual form are

$$\begin{aligned} \nabla \times \vec{H} &= \vec{J} + \frac{\partial \vec{D}}{\partial t} \\ \nabla \times E &= -\frac{\partial \vec{B}}{\partial t} \\ \nabla \bullet \vec{D} &= \rho \\ \nabla \bullet \vec{B} &= 0 \end{aligned} \quad \text{...(6.1.1)}$$

$\vec{B}$ and $\vec{H}$ are related by the expression $\vec{B} = \mu \vec{H}$ where μ is the permeability of the medium where as capacity or the dielectric constant $\in$ of the medium relates $\vec{D}$ and $\vec{E}$ by the relation $\vec{D} = \in \vec{E}$. These relations can be more specially written as

$$\begin{aligned} \vec{B} &= \mu_o(\vec{H} + \vec{M}) = \mu_o(1 + \chi_m)\vec{H} = \mu_o\mu_r\vec{H} = \mu \vec{H} \\ \vec{D} &= \in_o \vec{E} + \vec{P} = \in_o (1 + \chi_e)\vec{E} = \in_o\in_r \vec{E} = \in \vec{E} \end{aligned} \quad \text{...(6.1.2)}$$

Here μ_0 and ϵ_0 are the permeability and the dielectric constant of the free space respectively, μ_r and ϵ_r the *relative permeability* and the *relative dielectric constant* with respect to free space respectively and χ_m and χ_e represent the *magnetic susceptibility* and the *dielectric susceptibility* respectively. $\vec{M}$ is the *magnetic moment per unit volume* or the *magnetization* where as $\vec{P}$ is the *electric dipole moment per unit volume or the polarization.*

6.1.2 DIELECTRIC—A SEMI ATOMIC VIEW

An ideal dielectric material has no free electrons, electrons are all tightly bound to the nucleus of the atom. The molecules of a dielectric are classified as *polar* and *non-polar*. In any molecule there will be distribution of protons and a distribution of electrons. The protons can be considered to act as the equivalent positive electric charge at some specific point in the molecule, *the centre of gravity* of the protons. Similarly there is a point at which a negative charge is equivalent to the distributed electrons, the centre of gravity of electrons. If these two points coincide then the molecule is *non polar* and if they are separated by a short distance then the molecule is *polar molecule*. Symmetrical molecules (e.g., H_2, N_2 and O_2) are non-polar. On the other hand molecules H_2O and N_2O, etc. are polar as both hydrogen atoms or both nitrogen atoms lie on the same side of the oxygen atom. Thus the polar molecules work as *dipoles* and have *dipole moments*. If the dielectric is placed in an electric field, the charges of a non-polar molecule become displaced. The molecules are then said to become *polarized* by the field and are called *induced dipoles*. The displacement is however limited by strong restoring force produced by the charge configuration in the molecule. When a dielectric consists of polar molecules (*permanent dipoles*), these dipoles are oriented at random in the absence of an external electric field. When an electric field is applied, the forces on a dipole give rise to a couple, whose effect is to orient the dipole along the direction of the electric field. The stronger the field, the greater is the aligning effect. The orientation of the induced dipoles or of the permanent dipoles in an external electrostatic field is such as to set the axis of the dipole along the field. This phenomenon is known as *electric polarization*.

Dielectric Constant

In a large number of dielectrics, the polarization P is proportional to the applied electric field E,

$$P \propto E; \text{ or } P = \epsilon_0 \chi E$$

where χ is dimensionless constant called electric susceptibility of the dielectric and ϵ_0 the permeability of free space, has been introduced to make χ dimensionless. Also, we know that the capacity of a parallel plate capacitor increases if the gap between the plates is filled with dielectric material of dielectric constant ϵ. This increase in capacity of a capacitor is measured in terms of a quantity known as relative permittivity ϵ_r.

i.e.,

$$\epsilon_r = \frac{\text{Capacity with dielectric material}}{\text{Capacity with vacuum}}$$

$$\in_r = \frac{\in A/d}{\in_o A/d} = \frac{\in}{\in_o}$$

Thus $\in = \in_o \in_r$ is called the *absolute permittivity* of the medium.

Electric Displacement Density

As the electric field intensity depends upon the medium $\left(E = \frac{q}{4\pi \in r^2}\right)$, we can define another important term *electric displacement vector D.* This new quantity $\vec{D}$ gives added insight into problems involving fields in the presence of dielectric. It depends only on the free charges (free charge density $\frac{Q}{A}$), in contrast to E, which depends on both free and polarization charges. Even in the presence of dielectric, polarization charges are not involved in D and it is very useful of solving electrostatic problems.

Thus,

$$D = \frac{q}{4\pi r^2}; \quad \text{or} \quad \vec{D} = \in\left[\frac{1}{4\pi \in}\frac{q}{r^3}\hat{r}\right] = \in \vec{E} \qquad \text{...(6.1.3)}$$

Here vector D is analogous to E and is measured in coulomb/m^2. The term $D.\ dA$ gives the flux of normal displacement through surface dA and remains unaltered in any medium.

Extract of Gauss Law

The field of a single positive point charge surrounded by a closed surface of arbitrary shape will have the electric intensity $\vec{E}$ at every point of the surface directed radially outward from the surface. Let us consider any sufficiently small area dA of the surface for which $\vec{E}$ can be considered to have same magnitude and direction. If θ be the angle between $\vec{E}$ and the outward normal to the surface at any point of this area. Now the product of E_n the component of $\vec{E}$ normal to the surface and the area dA is

$$E_n\, dA = E \cos\theta\, dA = \frac{q}{4\pi \in_o}\frac{\cos\theta\, dA}{r^2} = \frac{q}{4\pi \in_o} d\omega$$

$$\int E_n dA = \frac{q}{4\pi \in_o}\int_o^{4\pi} d\omega$$

or

$$\int E_n dA = \frac{q}{\in_o}$$

The product $E_n\, dA = E \cos\theta\, dA = E\, .\, dA$

Thus,

$$\int E_{\bullet}dA = \frac{q}{\epsilon_o}$$

or

$$\boxed{\begin{aligned}\int \epsilon_0 E_{\bullet}dA &= q \\ \int D_{\bullet}dA &= q\end{aligned}} \quad \text{...(6.1.4)}$$

That is the total electric flux through a closed surface is equal to the charge in coulomb enclosed by the surface (Gauss theorem)

Dielectrics and Gauss Theorem

Let us now apply *Gauss theorem* to a parallel plate condenser without a dielectric and then with a dielectric.

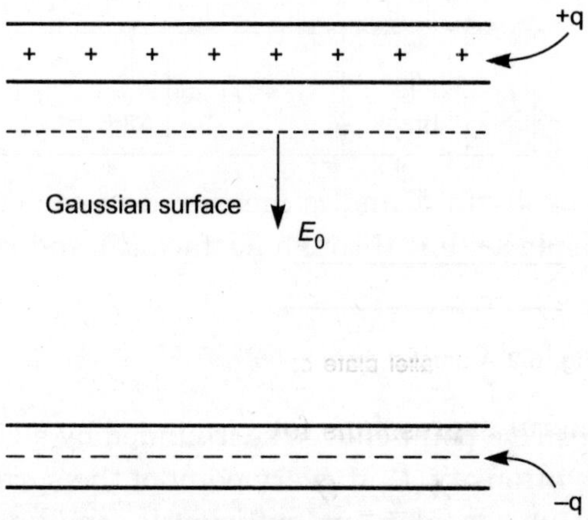

Fig. 6.1 A Parallel plate condenser with no dielectric

In the first case, when no dielectric is present, the electric field E_0 at any point on the *Gaussian surface* is given by

$$\int \vec{E}_0 \bullet d\vec{A} = \frac{q}{\epsilon_o}$$

$$\boxed{E_0 = \frac{q}{A \epsilon_o}} \quad \text{... (6.1.5)}$$

In the second case, when dielectric is present, the net charge within the *Gaussian surface* is $q - q'$, where q' is the induced surface charge, and the field $\vec{E}$ is given by

$$\int \vec{E} \bullet d\vec{A} = \frac{(q - q')}{\epsilon_o}$$

$$E\,A = \frac{q}{\epsilon_o} - \frac{q'}{\epsilon_o} \qquad ...(6.1.6)$$

$$E = \frac{q}{A\,\epsilon_o} - \frac{q'}{A\,\epsilon_o}$$

Equations 6.1.6 tells us that the induced charge q' tends to weaken the original field (*i.e.*, $E < E_o$) which reveals itself a reduction in the potential difference when dielectric is present and not present respectively, then

$$\frac{E_o}{E} = \frac{V_o}{V_d} \qquad ...(6.1.7)$$

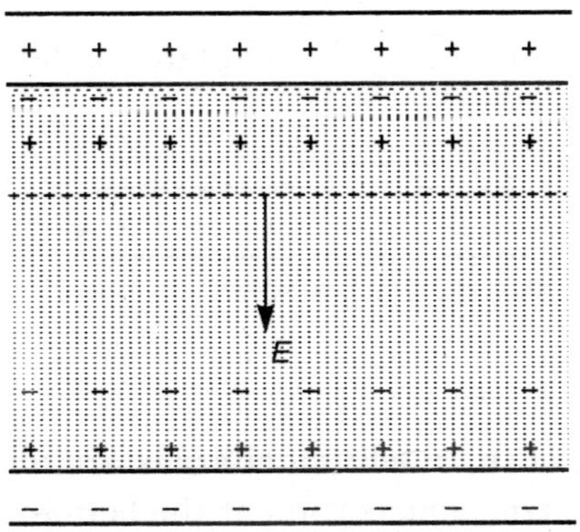

Fig. 6.2 A parallel plate condenser with dielectric

The corresponding general expressions for capacitances are:

$$C_o = \frac{A\,\epsilon_o}{d} = \frac{Q}{V_o}$$

$$C = \frac{A\,\epsilon_o \epsilon_r}{d} = \frac{Q}{V_d}$$

$$\frac{C}{C_o} = \epsilon_r = \frac{V_o}{V_d}$$

Thus Eqn. (6.1.7) bocomes

$$\frac{E_o}{E} = \epsilon_r$$

Substituting for E_o from Eqn. (6.1.5)

$$E = \frac{E_o}{\epsilon_r} = \frac{q}{A\,\epsilon_o \epsilon_r} \qquad ...(6.1.8)$$

Comparing equation (6.1.8) with (6.1.6), one gets

$$\frac{q}{A \in_o \in_r} = \frac{q}{A \in_o} - \frac{q'}{A \in_o}$$

or

$$\boxed{q - q' = \frac{q}{\in_r}} \quad \text{...(6.1.9)}$$

This shows that the induced charge q' is less than the free charge q and is zero when $\in_r = 1$ or the dielectric is absent.

Refer Eqns. (6.1.6) and (6.1.9)

$$\int \vec{E} \cdot d\vec{A} = \frac{q - q'}{\in_o} = \frac{q}{\in_o \in_r}$$

i.e.,

$$\boxed{\in_o \in_r \int \vec{E} \cdot d\vec{A} = q} \quad \text{...(6.1.10)}$$

This is nothing but *Gauss law* in the presence of the dielectric

Electric Flux Density

Owing to the polarization of the medium, charges are induced at the surfaces of the dielectric, the charge due to electric polarization at the boundary near the positive plate of the capacitor being negative and vice-verse. If q is the charge on the plate and q' is the induced charge on the boundary of the dielectric, then we can write Eqn. 6.1.9 as

$$\frac{q}{A} = \frac{q}{A \in_r} + \frac{q'}{A}$$

$$\boxed{\frac{q}{A} = \in_o \left[\frac{q}{A \in_o \in_r}\right] + \frac{q'}{A}} \quad \text{...(6.1.11)}$$

The last term of this equation gives induced surface charge per unit area called electric *polarization*, *P*. It is nothing but the induced dipolemoment for unit volume. Substituting $E = \frac{q}{A \in_o \in_r}$ from Eqn. (6.1.8) in Eqn. (6.1.11), one gets

$$\frac{q}{A} = \in_o \vec{E} + \vec{P}$$

i.e.,

$$\boxed{\vec{D} = \vec{E} \in_o + \vec{P}} \quad \text{...(6.1.12)}$$

In free space without any dielectric this equation becomes,

$$\vec{D} = \vec{E} \in_o \quad \text{as } \vec{P} = 0 \text{ and } \frac{q}{A} = D$$

Thus

$$D = \frac{q}{A} = \frac{q \in_o \in_r}{A \in_o \in_r}$$

Using the relations 6.1.2, 6.1.3 and 6.1.8,

$$D = \left[\frac{q}{A \in_o \in_r}\right](\in_o \in_r) = E \in_o \in_r = E \in$$

Thus Eqn. 6.1.12 gets the form

$$E \in = E \in_o + P$$

or

$$P = E \in_o \in_r - E \in_o = E \in_o (\in_r - 1)$$

$$(\in_r - 1) = \frac{\vec{P}}{\vec{E} \in_o} = \chi \qquad ...(6.1.13)$$

where χ is the *electric susceptibility* of the dielectric medium.

6.1.3 ELECTRONIC POLARIZATION

The displacement of charged particles under the action of the electric field is called *dielectric polarization*. This displacement of electric charges results in the formation of electric dipolement in atoms, ions or molecules of the material. The three important types of polarization are (*i*) *electronic polarization*, (*ii*) *ionic polarization* and (*iii*) *orientation polarization*.

Electronic Polarizability

Consider an atom placed in a space between the two plates of a capacitor. In the absence of external electric field the centres of gravity of the positive and negative charges coincide and the atom has no dipole or dipole moment. When the atom is placed in an electric field, the positive charge being firmly in the nucleus, remains in its original position but the *electron cloud* surrounding the positive charge gets shifted with respect to positive charge.

Let *Ze* be the value of the positive charge (nuclear charge) and a smeared out electron cloud of charge – *Ze*. Let *R* be the radius of the electron cloud. An electric field is applied in a direction as shown in Fig. 6.3. Let *e* be the amount of each type of charge. This cloud shifts through a distance *x* under the field *E*. However the actual magnitude of this shift is very small. Now the force on the positive charge is

$$F = Ze\,E \qquad (6.1.13a)$$

The –ve sign indicates the force is acting opposite to the field.

$$\boxed{\vec{F} = Ze\vec{E}} \qquad ...(6.1.14)$$

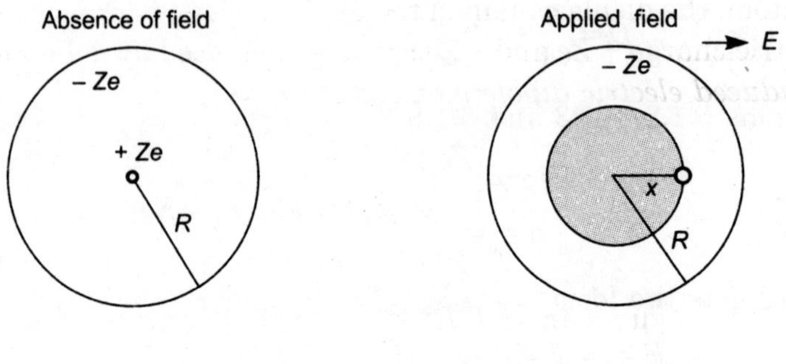

Fig. 6.3 Displacement of electron cloud under a field E

This force is balanced by the force on the *effective electron cloud due to Coulombic attraction*

i.e.,

$$\text{Coulomb force} = Ze \times \frac{\text{charge enclosed in the sphere of radius } x}{4\pi \in_o x^2}$$

If $\rho = -\dfrac{Ze}{(4/3)\pi R^3}$ is the surface density, then charge enclosed in the sphere is

$$\frac{4}{3}\pi x^3 \rho = \frac{4}{3}\pi x^3 \left[-\frac{Ze}{(4\pi/3)R^3}\right] = -\frac{Zex^3}{R^3}$$

Thus coulomb force is

$$\frac{Ze}{4\pi \in_o x^2}\left[-\frac{Ze\,x^3}{R^3}\right] = -\frac{Z^2e^2x}{4\pi \in_o R^3}$$

or

Coulomb force,

$$\boxed{C_e = -\frac{Z^2e^2x}{4\pi \in_o R^3}} \qquad ...(6.1.15)$$

Equating 6.1.14 and 6.1.15, one gets

$$ZeE = \frac{Z^2e^2x}{4\pi \in_o R^3}$$

or

$$x = \frac{4\pi \in_o E R^3}{Ze} \quad ...(6.1.16)$$

i.e., for a given atom, the displacement of the electron cloud is proportional to the field E. Thus the two electric charge $+ Ze$ and $- Ze$ are now displaced by a distance x under the field constitute an *induced electric dipole moment* given by

$$\mu_e = Ze\,x = \frac{Ze\, 4\pi \in_o E R^3}{Ze}$$

or

$$\mu_e = 4\pi \in_o E R^3$$

$$\mu_e = \alpha_e E \quad ...(6.1.17)$$

or

$$\mu_e \propto E$$

where $\alpha_e = 4\pi\in_o R^3$ is called *electronic polarizability* and the *electronic polarizability* is proportional to the applied field. The dipole moment for unit volume is called *electronic polarization.*

i.e.,

$$P_e = N\mu_e = N\alpha_e E \text{ where } N \text{ is the number of atoms/m}^3$$

Also

$$P_e = E\in_o (\in_r - 1) = N\alpha_e E \text{ (Refer Eqn. 6.1.13)}$$

Thus

$$E \in_0 (\in_r - 1) = N\alpha_e E$$

$$(\in_r - 1) = \frac{N\alpha_e}{\in_0} \quad ...(6.1.18)$$

> **N.B:** Ionic polarization can be discussed similar to electronic polarization. However ionic polarizability is about ten times smaller than electronic polarizability and hence not discussed here.

6.1.4 ORIENTATION POLARIZATION AND LANGEVIN'S THEORY

Liquids and solids which entertain asymmetric molecules whose permanent dipole moment can be aligned by an external electric field is shown in Fig. 6.4.

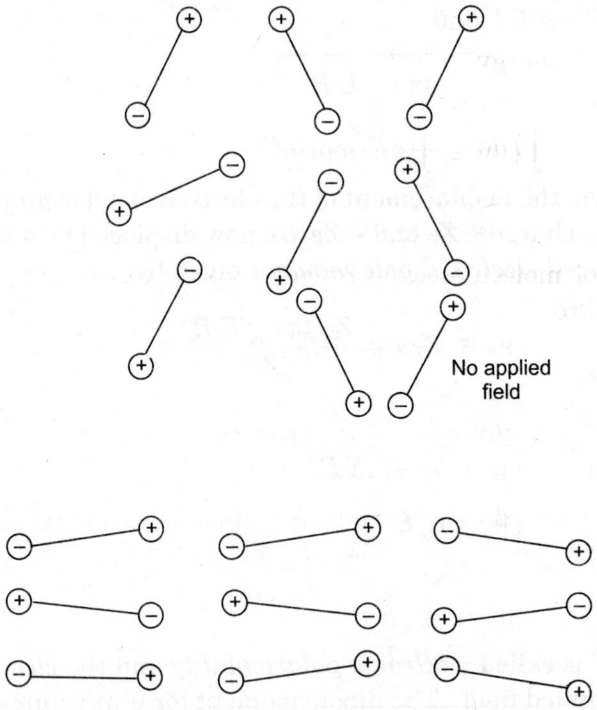

Fig. 6.4 Orientation polarization, the field orients the permanent dipoles

Debye in 1927 conceived the idea that molecules such as HCl and H_2O may possess permanent dipole moment due to some kind of symmetry in their structure. Statistical mechanics gives the probability of finding a molecule with particular molecular energy W at a temperature T is proportional to $e^{-\frac{W}{k_B T}}$, where W is the P.E. associated with the molecule.

The potential energy associated with each dipole is obtained as follows:

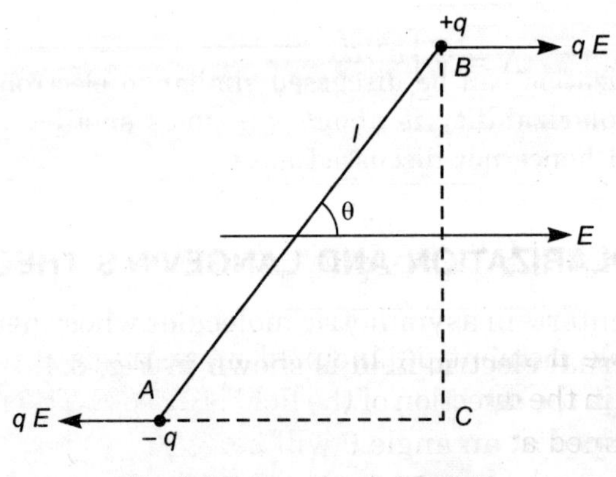

Fig. 6.5 Polar molecule in a field

Torque = $(q\ E)\ BC = q\ E\ l \sin\theta$

Therefore potential energy

$$\int_{90}^{\theta} T d\theta = \int_{90}^{\theta} (qEl) \sin\theta\, d\theta$$

$$W = -\mu_m E \cos\theta$$

But the number of molecules (dN) having inclinations θ and $(\theta + d\theta)$ and energy value W is proportional to

$$\exp(-W/k_B T) \sin\theta\, d\theta$$

or

$$dN = Ce^{-W/k_B T} (\sin\theta d\theta)$$

Substituting for W

$$dN = C \exp(\mu_m E \cos\theta / k_B T)[\sin\theta d\theta]$$

Let

$$\alpha = \frac{\mu_m E}{k_B T}$$

$$\boxed{dN = C \exp(a \cos\theta) \sin\theta d\theta} \quad \text{...(6.1.19)}$$

Thus

$$N = C \int_{o}^{\pi} e^{a \cos\theta} \sin\theta d\theta$$

Let $x = \cos\theta$; $dx = -\sin\theta\, d\theta$

Thus

$$N = C \int_{1}^{-1} e^{ax} (-dx)$$

$$N = C \int_{-1}^{1} e^{ax} dx$$

or

$$\boxed{C = \frac{aN}{(e^{a} - e^{-a})}} \quad \text{...(6.1.20)}$$

The molecular dipole moment μ_m inclined at an angle θ to the field, the resolved component of its moment in the direction of the field is $\mu_m \cos\theta$. Thus the resolved components of all the dN dipoles inclined at an angle θ will be

$$dM = dN \mu_m \cos\theta$$

$$\int dM = P_o = \int_0^{\pi} \mu_m \cos\theta dN$$

Substituting for dN from Eqn. 6.1.19, we get

$$P_o = \int_0^{\pi} \mu_m \cos\theta\, C \exp[a\cos\theta] \sin\theta d\theta$$

$$= C\mu_m \int_0^{\pi} \exp[a\cos\theta] \cos\theta \sin\theta d\theta$$

Let $x = \cos\theta$

$dx = -\sin\theta\, d\theta$

$$P_o = C\mu_m \int_1^{-1} x\, e^{ax} (-dx)$$

$$P_o = C\mu_m \int_{-1}^{1} x\, e^{ax}\, dx$$

Substituting C from Eqn. (6.1.20)

$$P_o = \frac{\mu_m a N}{(e^a - e^{-a})} \int_{-1}^{1} \frac{x}{a} d(e^{ax}) = \frac{\mu_m N}{(e^a - e^{-a})} \int_{-1}^{1} x\, d(e^{ax})$$

$$P_o = \frac{\mu_m N}{(e^a - e^{-a})} \left\{ \left[(x e^{ax}) \right]_{-1}^{1} - \int_{-1}^{1} e^{ax}\, dx \right\}$$

$$P_o = \frac{\mu_m N}{(e^a - e^{-a})} \left\{ (e^a + e^{-a}) - \frac{(e^a - e^{-a})}{a} \right\}$$

$$P_o = \mu_m N \left\{ \frac{(e^a + e^{-a})}{(e^a - e^{-a})} - \frac{1}{a} \right\}$$

The orientational polarization

$$P_o = P_o = \mu_m N \left[\mathrm{Coth}\, a - \frac{1}{a} \right] = \mu_0 N L(a) \qquad \ldots(6.1.21)$$

where $L(a)$ is known as *Langevin function*. $\mu_m N$ is the dipole moment per m^3 when all the molecules are turned in the direction of the applied electric field and hence it gives the *saturation value* of P_o say P_s (which depends on the nature of the gas). A plot of $L(a)$ versus a is shown in Fig. 6.6.

Thus,

$$\frac{P_o}{P_s} = L(a) = \left(\coth a - \frac{1}{a}\right) \qquad ...(6.1.22)$$

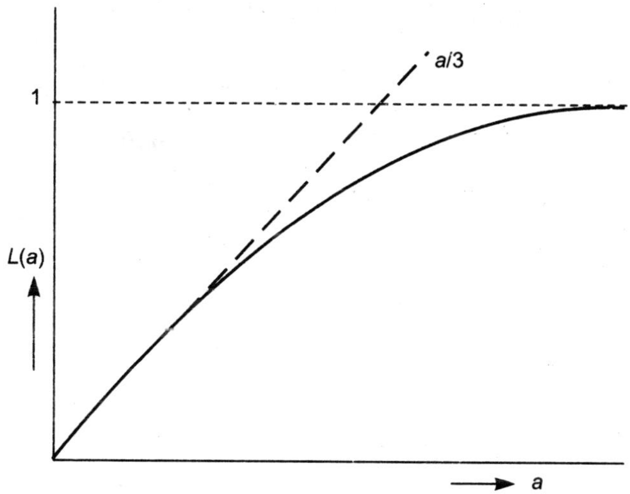

Fig. 6.6 Langevin curve

It is obvious that $L(a)$ has a limit unity. (For greater field and smaller temperature). However when a is small (smaller field and at higher temperature), the value of $L(a)$ becomes $a/3$. Thus the orientational polarization.

$$P_o = \frac{P_s\, a}{3} = \frac{\mu_m N \mu_m E}{3 k_B T}$$

$$P_o = \frac{N \mu_m^2 E}{3 k_B T}$$

$$P_o = \alpha_o E N \qquad ...(6.1.23)$$

$$\alpha_o = \frac{\mu_m^2}{3 k_B T}$$

where, α_0 is the orientational polarizability

Now *total polarization* of a polyatomic gas is given by

$$P = P_e + P_i + P_o$$

$$P = N E \alpha_e + N E \alpha_i + N E \alpha_o = E \in_o (\in_r - 1)$$

$$\in_o (\in_r - 1) = N(\alpha_e + \alpha_i) + \frac{N \mu_m^2}{3 k_B T}$$

Determine $\in_r$ at various temperatures and draw a graph between $\in_0(\in_r - 1)$ and $\frac{1}{T}$.

The slope of the graph gives $\frac{N\mu_m^2}{3k_B}$ from which μ_m can be determined.

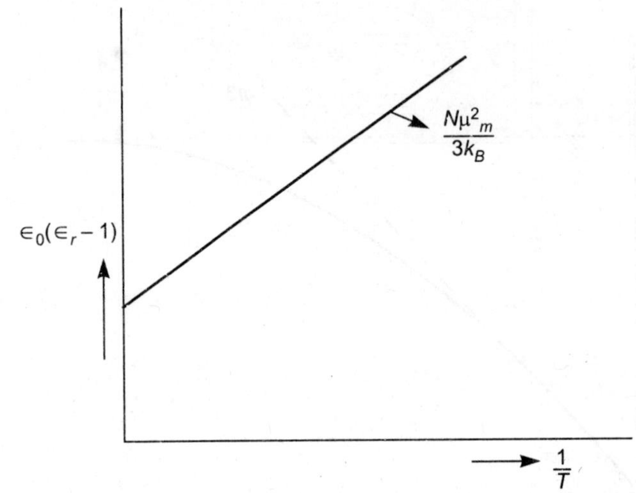

Fig. 6.7 Curve between $\in_0\,(\in_r - 1)$ and $\frac{1}{T}$ for a polar gas

The unit of dipole moment is Debye and it is $3.33 \times 10^{-30}\ C-m$.

The important equation is

$$(\in_r - 1) = \chi = \frac{P}{E\in_o} = \frac{N}{\in_o}\left(\alpha_e + \alpha_i + \frac{\mu_m^2}{3k_B T}\right) \qquad ...(6.1.24)$$

6.1.5 SOLID DIELECTRICS AND LOCAL FIELD

The behaviour of solid dielectrics is quite different from that of gases and liquids. In a solid the molecule or atom experiences a force on account of the fields produced by the neighbouring dipoles in addition to the external field. The resultant field is called the *Local field* (or internal field), and is responsible for polarizing the individual molecules or atoms of the crystal.

Evaluation of Local Field

Lorentz suggested a method to calculate the local field. According to this method, we select a small spherical region from the dielectric with the atom at the centre for which the local field must be calculated. The radius of the sphere is chosen large enough to consider the region outside the sphere as a continuum while the region inside the sphere as the actual structure of the substance. We suppose that the given dielectric has been uniformly polarized (on a macroscopic scale) by placing it in a uniform field between two oppositely charged parallel plates.

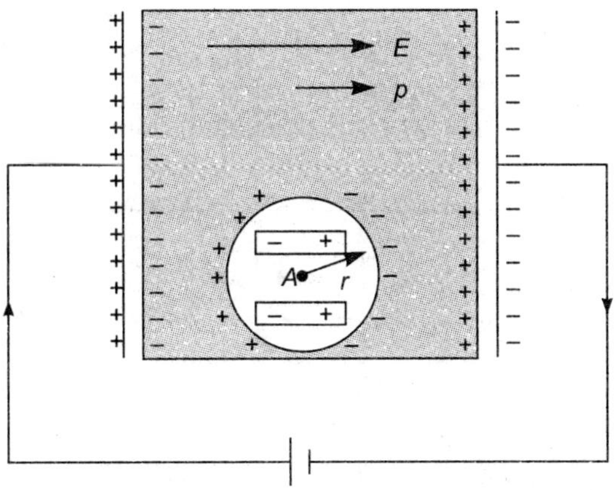

Fig. 6.8 Calculation of the local field

Now the field experienced by a molecule of the dielectric is

$$\vec{E}_L = \vec{E}_1 + \vec{E}_2 + \vec{E}_3 + \vec{E}_4 \qquad ...(6.1.25)$$

where,

(i) E_1 is the field due to charge density on the plates of the capacitor (*i.e.* with no dielectric)

(*ii*) E_2 is the field at the atom due to the polarized charges (induced charge) on the plane surfaces of the dielectric (*i.e.* depolarizing field)

(*iii*) E_4 is the field due to polarized charges on the surface of the spherical cavity

(*iv*) E_3 is the field due to all dipoles inside the spherical cavity

Now

$$D = \vec{E}_1 \in_0 = \vec{E} \in_0 + \vec{P}$$

$$\boxed{\vec{E}_1 = \frac{\vec{D}}{\in_0} = \vec{E} + \frac{\vec{P}}{\in_0}} \qquad ...(6.1.26)$$

and

$$\boxed{\vec{E}_2 = -\frac{\vec{D}}{\in_0} = -\frac{\vec{P}}{\in_0}} \text{ (letting } E = 0) \qquad ...(6.1.27)$$

E_3 is the field at A due to other atoms contained in the cavity. For cubic structure $E_3 = 0$ because of symmetry

Computation of E_4

The enlarged view of the cavity is shown in Fig. 6.9. If dA is the surface area of the disc lying between θ and $\theta + d\theta$, where θ is the direction with reference to the direction of the applied field,

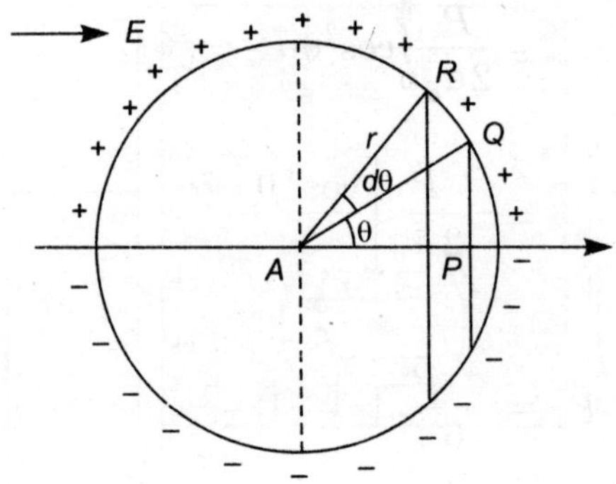

Fig. 6.9 Cavity enlarged and inverted

then,

$$dA = 2\pi(PQ)(QR)$$

and

$$\sin\theta = \frac{PQ}{r};\ PQ = r\sin\theta$$

Also

$$d\theta = \frac{QR}{r};\quad QR = r\,d\theta$$

Thus

$$dA = 2\pi r^2 \sin\theta d\theta$$

The charge dq on the surface dA is equal to the normal component of the polarization multiplied by the surface area. Therefore,

$$dq = P\cos\theta\, dA = P(2\pi r^2 \sin\theta\cos\theta d\theta)$$

Now the field due to this charge at 'O' denoted by dE_4 in the direction $\theta = 0$ is

$$dE_4 = \frac{dq \times 1 \times \cos\theta}{4\pi\epsilon_0 r^2} = \frac{P\cos\theta\, 2\pi r^2 \sin\theta d\theta \cos\theta}{4\pi\epsilon_0 r^2}$$

$$d\vec{E}_4 = \frac{\vec{P}\cos^2\theta\sin\theta d\theta}{2\epsilon_0}$$

or

$$\vec{E}_4 = \frac{\vec{P}}{2\epsilon_0}\int_0^{\pi}\cos^2\theta\sin\theta d\theta$$

$$= \frac{\vec{P}}{2\in_o}\int_0^{\pi}\cos^2\theta d(-\cos\theta)$$

$$= -\frac{\vec{P}}{2\in_o}\left[\frac{\cos^3\theta}{3}\right]_0^{\pi}$$

Thus

$$\vec{E}_4 = -\frac{\vec{P}}{6\in_o}[-1-1] = \frac{P}{3\in_o}$$

Thus

$$\vec{E}_L = \vec{E} + \frac{\vec{P}}{\in_o} - \frac{\vec{P}}{\in_o} + 0 + \frac{\vec{P}}{3\in_o}$$

$$\boxed{E_L = E + \frac{P}{3\in_o}} \qquad ...(6.1.28)$$

Equation 6.1.28 states that the Lorentz field is larger than the applied field by an amount that is directly proportional to the polarization density.

6.1.6 CLAUSIUS-MOSSOTTI RELATION

The original equation for polarization is now modified as

$$\vec{P} = N\,\alpha\,\vec{E}_L$$

But

$$\vec{P} = \vec{E}\in_o(\in_r - 1)$$

or

$$\vec{E}\in_o(\in_r - 1) = N\alpha\vec{E}_L$$

$$\boxed{\in_r = 1 + \frac{N\alpha\vec{E}_L}{\vec{E}\in_o}} \qquad ...(6.1.29)$$

Eqn. 6.1.28 speaks that

$$\vec{E}_L = \vec{E} + \frac{\vec{P}}{3\in_o} = \vec{E} + \frac{N\alpha\vec{E}_L}{3\in_o}$$

or

$$\vec{E} = \vec{E}_L - \frac{N\alpha\vec{E}_L}{3\in_o} = \vec{E}_L\left[1 - \frac{N\alpha}{3\in_o}\right]$$

$$\frac{\vec{E}_L}{\vec{E}} = \frac{1}{\{1-(N\alpha/3\in_o)\}}$$

Substituting this value in Eqn. 6.1.29,

$$\in_r = 1+\frac{N\alpha}{\in_o}\left[\frac{1}{\left(1-\dfrac{N\alpha}{3\in_o}\right)}\right] = \frac{\in_o\left\{1-\dfrac{N\alpha}{3}\right\}+\dfrac{N\alpha\in_o}{\in_o}}{\left[1-\dfrac{N\alpha}{3\in_o}\right]} \qquad \text{...(6.1.29a)}$$

$$\in_r = \left[\frac{1+(2/3)\left(\dfrac{N\alpha}{\in_o}\right)}{1-(1/3)\left(\dfrac{N\alpha}{\in_o}\right)}\right] \qquad \text{...(6.1.30)}$$

Now one can find by substitution and simplification, the final simplified equation as

$$\left[\frac{\in_r -1}{\in_r +2}\right] = \frac{N\alpha}{3\in_o} \qquad \text{...(6.1.31)}$$

This is the well known Clausius-Mossotti relation which relates the macroscopic dielectric constant with the microscopic polarizabilities. The above equation was modified by Debye and the resulting equation is given below:

Multiply both sides of the above equation by the molar volume of the dielectric:

$$\left[\frac{(\in_r -1)}{\in_r +2}\right]\frac{M_A}{\rho} = \frac{N\,M_A}{3\in_o\rho}(\alpha_e+\alpha_i+\alpha_o) = \frac{N_A}{3\in_o}(\alpha'+\alpha_o)$$

Since the orientational polarizability, $\alpha_o = \dfrac{\mu_m^2}{3k_BT}$ where μ_m is the permanent dipole moment, we thus write

$$\left[\frac{(\in_r -1)}{\in_r +2}\right]\frac{M_A}{\rho} = \frac{N_A}{3\in_o}\left(\alpha'+\frac{\mu_m^2}{3k_BT}\right)$$

$$\left[\frac{(\in_r -1)}{\in_r +2}\right]\frac{M_A}{\rho} = \frac{N_A\alpha'}{3\in_o}+\frac{N_A\mu_m^2}{9\in_o k_BT} = a+\frac{m}{T} \qquad \text{...(6.1.32)}$$

This is Debye's equation. By conducting some simple experiments (bridge method with capacity determination), it is possible to determine $\in_r$ at different temperatures and a simple tabulation as given below is possible.

Table 6.1 Dielectric constant at different temperatures

T	$\frac{1}{T}$	$\in_r$	$\left(\frac{\in_r - 1}{\in_r + 2}\right)\frac{M_A}{\rho}$	*Average Slope*
T_1	$\frac{1}{T_1}$	$(\in_r)_1$	K_1	
T_2	$\frac{1}{T_2}$	$(\in_r)_2$	K_2	m_1
T_3	$\frac{1}{T_3}$	$(\in_r)_3$	K_3	m_2
T_4	$\frac{1}{T_4}$	$(\in_r)_4$	K_4	m_3
T_5	$\frac{1}{T_5}$	$(\in_r)_5$	K_5	Average (m)

Computation of Permanent Dipole Moments

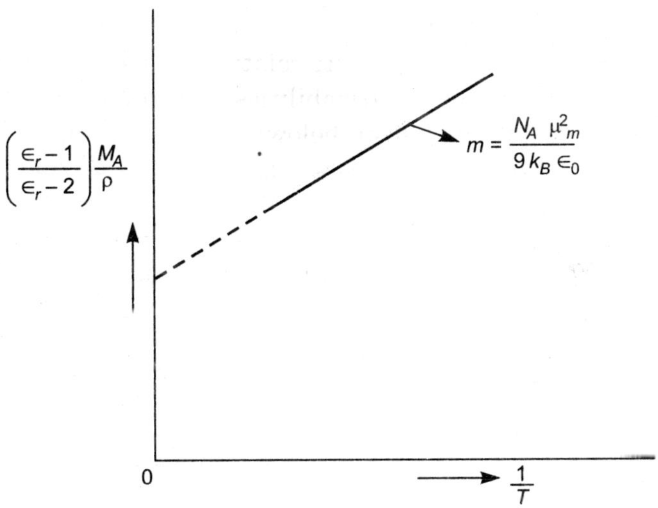

Fig. 6.10 Temperature versus dielectric constant

A graph is drawn between $\left(\frac{\in_r - 1}{\in_r + 2}\right)\frac{M_A}{\rho}$ and $\frac{1}{T}$

Thus

$$m = \frac{N_A \mu_m^2}{9 k_B \in_o}.$$

or

$$\mu_m = \sqrt{\frac{9\, k_B\, \epsilon_o\, m}{N_A}} \qquad ...(6.1.33)$$

Measurement of Dielectric Constant

Dielectric constant of a given dielectric is generally determined by measuring and comparing the capacity C_d of a condenser filled with the substance and the capacity C_0 of the condenser with air as the dielectric. The ratio $\frac{C_d}{C_0}$ gives ϵ the dielectric constant of the medium. The two capacitors C_d and C_0 may be measured conveniently by a resonance method as shown in Fig. 6.11 where C_s is a calibrated variable condenser and C is the condenser in which the given substance with the same length and breadth of the plates may be placed. By varing C_s so as to keep the resonance frequency, $\omega_0 = \frac{1}{\sqrt{L(C_s + C)}}$ constant when C is empty and then filled, we may determine C_0 and C_d and hence ϵ. The voltmeter V measures the response of the resonant circuit. This method of measuring dielectric constant is generally applicable for frequencies up to about 100 MHz. At the microwave region (~ 10^3 to ~ 10^5 MHz) the frequencies are so high that the dimensions of the apparatus are comparable with or greater than, the wavelength, and the specimen then can no longer be treated as if it were in quasi-static fields. Rather, it has to be thought of as a medium for the propagation of *e-m* waves. Here one may measure the dielectric constant of the specimen by measuring essentially the wavelength of the microwave radiation in the specimen and using the relation

$$\frac{\lambda(\text{vacuum})}{\lambda(\text{specimen})} = (\epsilon\, \mu)^{1/2}$$

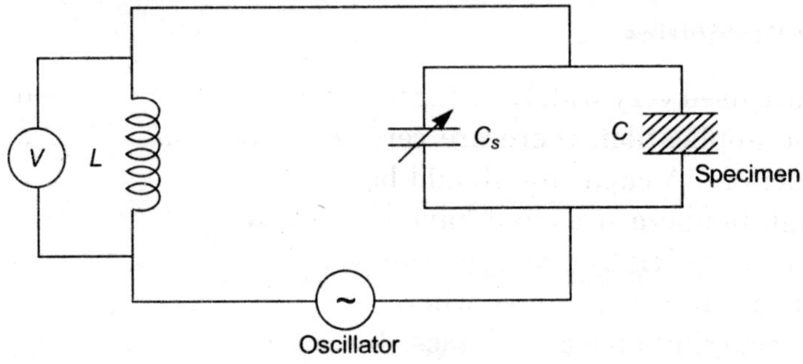

Fig. 6.11 Resonance method of measuring capacities

For many solids the permeability $\mu = 1$. For optical and infrared frequencies, ϵ can be measured by measuring the refractive index, n as

$$n^2 = \mu\epsilon \approx \epsilon$$

Relation between Dielectric Constant and Refractive Index

The idea that there exists a definite relation between dielectric constant and refractive index comes from the propagation of electromagnetic waves through a medium. The electromagnetic waves as developed by Maxwell by the idea that electromagnetic induction consists of varying electric and magnetic fields with time. The electric and magnetic vectors in such waves are perpendicular to each other and also perpendicular to the direction of propagation. Maxwell has shown that the velocity of propagation of such waves for an unbounded medium is given by

$$v = \frac{1}{\sqrt{\mu \in}}$$

where μ is the magnetic permeability of the medium and $\in$ is the absolute permittivity. The permeability for non magnetic media is μ_o. Hence the velocity in such a medium is $\frac{1}{\sqrt{\mu_o \in}}$. The refractive index of the medium is given by

$$n = \frac{\text{Velocity of waves in vacuum}}{\text{Velocity of waves in medium}}$$

$$n = \frac{1/(\mu_o \in_o)^{1/2}}{1/(\mu_o \in)^{1/2}} = \sqrt{\frac{\in}{\in_o}} = \sqrt{\in_r}$$

Thus $n = \sqrt{\in_r}$ where $\in_r$ is the relative permittivity. Using this value of $\in_r$ in Eqn. 6.1.31, one gets

$$\boxed{\frac{n^2 - 1}{n^2 + 2} = \frac{N\alpha}{3 \in_o}} \qquad \text{...(6.1.34)}$$

This relation is known as Lorentz-Lorentz equation.

Applications of Dielectrics

Dielectrics are used very widely as capacitors. Although the actual requirements vary depending on the application, there are certain characteristics which are desirable for their use in capacitors. A capacitor should be small, have high resistance, be capable of being used at high temperatures and have long life. From a commercial point of view it should also be cheap. Specially prepared thin kraft paper free from holes and conducting particles is used in power capacitors where withstanding high voltage stresses is more important than incurring dielectric losses. In addition, the kraft paper is impregnated with a suitable liquid such as chlorinated diphenyl. This increases the dielectric constant and thus reduces the size of the capacitor. In addition the breakdown strength is increased.

In addition to paper capacitors for general purpose, other types of capacitors are also used. In the film capacitors, thin films of teflon, mylar or polythene, etc. are used. These not only reduce the size of the capacitor but also have high resistivity. Teflon is used at high frequencies as it has low loss. In electrolytic condensers, an electrolyte is deposited on the impregnating paper. The size of such a capacitor is small as the film is very thin. Polarity and the maximum operating voltage are important specifications for these capacitors.

Ceramic materials occupy an important position since their properties such as dielectric constant and its temperature coefficient of resistance can be varied through materials processing. Some ceramics can be used as temperature compensators in electronic circuits. High dielectric constant materials, where small variations in dielectric constant with temperature can be tolerated, help miniature capacitors. Barium titanate and its modifications are the best examples of such materials.

Piezoelectric materials have a variety of applications. These can be used as oscillators, the most important material for such a use being quartz. It is frequently used to control frequency in circuits. It has an extremely high dielectric strength and has good mechanical properties such as hardness. When suitably cut, the temperature coefficients and some of the electric constants are zero with the result that very high stability is attained. Quartz crystals are also used in selective band-pass filters in telephone industry. Piezoelectric materials when used as transducers measure pressures very accurately. They are also used to produce ultrasonic waves and as phonograph pick-ups and microphones.

Effects of Dielectrics

Dielectric Relaxation

Dielectrics take a finite time for the polarization to reach its maximum value. This is due to the forces between adjacent molecules which tend to prevent the alignment along the external field. The average time for relaxation varies from substance to substance, values from 10^{-10} sec to several thousand years. Due to this *relaxation*, capacitor connected to a p.d of V acquires charge at a finite rate. Discharging process is also slow. If a capacitor is charged and discharged quickly and then left with terminals open, it will be found to have acquired a potential difference between its plates. Phenomenon of hysterisis with alternating fields is also the consequence of the relaxation time.

Dielectric Absorption

On account of the finite conductivity of the dielectric materials noticeable energy loss is observed in the form of heat. This loss and the hysterisis loss are referred jointly as *dielectric losses*. Similar to the relaxation phenomenon absorption effect is observed in capacitors. If an initially charged capacitor is discharged by short circuiting it, it is found that the capacitor still retains a charge when the short circuit is removed. This is due to *electric absorption*. It is when the dielectric is not perfectly homogeneous. For a perfectly homogeneous dielectric, this residual charge should not occur. Thus a gas or a perfectly homogenous crystal like calcite shows no residual charge. If the capacitor is repeatedly charged and discharged, dielectric absorption causes heat to be produced. If this process is rapid, as with high frequency A.C., energy losses in the dielectric are low, *i.e.*, the dielectric absorption is small, as the molecules are unable to rotate to any extent in this very short time (*i.e.*, the time of one cycle).

Dielectric Breakdown

It is the most important property of real dielectrics. For any material there is a maximum field intensity beyond which damage occurs that results in conduction, sparking and other break down phenomena. At relatively high fields, the electrons in the dielectric gain enough energy to knock other charged particles and make them available for conduction. This multiplying process occurs along certain paths in the material. Along these paths, local heating also occurs and a permanent damage results.

Electrostatic apparatus at high potentials, requires a good insulation with respect to break down. Air is the most common insulating material. Its breakdown occurs, when dry and at N.T.P, at fields of about 3×10^6 V/m. Fields somewhat less than this may produce some steady discharging currents known as *corona*. When the field is not very large over most of the region, it may come to or above the break down value near the points. The lines of force near the point are like those for a sphere, hence at the surface of the point, $E = \frac{V}{a}$. Thus break down can occur even at 300V at the point of 0.1mm radius. The operation of lightning rods and brush discharge points in electrostatic generator is based on this principle. After relative permittivity and loss factor the third important property of the dielectric is the *dielectric strength*. The dielectric strength of an insulator is the maximum voltage gradient which the dielectric can withstand without failure, or the intensity of the electric field at which breakdown occurs is called *dielectric strength.*

i.e.,

$$\text{Dielectric strength} = \frac{\text{breakdown voltage}}{\text{thickness of the dielectric}}$$

The geometry of the specimen, the type of the electrodes and procedure of the testing may affect the dielectric strength resulting into a *large scatter* in the test result. Dielectric strength is most commonly measured in kilovolt per mm.

Power-loss

From the definition of the capacity given by

$$V = \frac{Q}{C}$$

and

$$I = \frac{dQ}{dt} \text{ and } Q = CV$$

we get

$$\boxed{I = C\frac{dV}{dt}} \quad \ldots(6.1.35)$$

If the voltage is sinusoidal then

$$\boxed{V = V_o \sin \omega t} \quad \text{and} \quad \frac{dV}{dt} = V_o\ \omega \cos \omega t \quad \ldots(6.1.36)$$

Here V_0 is the maximum value of V and $\omega = 2\pi\upsilon$ where υ is the frequency of the sinusoidal wave, and t is the time, then from Eqn 6.1.35 and 6.1.36:

$$I = CV_0\omega\cos\omega t \quad \text{...(6.1.37)}$$

From Eqns. 6.1.36 and 6.1.37, it is seen that a phase difference of 90° exists between the current and the voltage. However, this applies only to the ideal case that is when no loss takes place in the dielectric of the capacitors. In a real dielectric capacitor current leads the voltage by (90° – δ), where δ is a measure of dielectric power loss (see Fig. 6.12). It can be shown that power loss is given as

$$P = \pi\upsilon V_0^2 \in_r \tan\delta$$

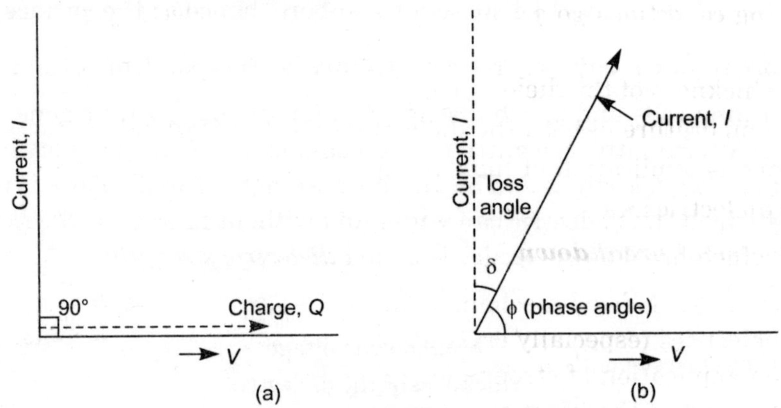

Fig. 6.12 (a) Vector diagram of perfect capacitor
(b) Vector diagram of real capacitor

The product $\in_r$ tanδ is called *loss factor* and tanδ is known as *dissipation factor*. Angle δ is called *dielectric loss angle*. The loss factor is characteristic of dielectric material or an insulator. For both, dielectric and insulator, the dissipation factor should be small. The dissipation factor is also sometimes known as *loss tangent*. Loss factor is a measure of the electric energy lost as heat energy by a capacitor in an a.c circuit. The sin of the loss angle is called *power factor*. It may also be cosine of phase angle ϕ.

Breakdown in Solid Dielectrics

The phenomenon of breakdown in solids is the most complicated one and is least understood. The following three kinds of breakdown are commonly possible in solid dielectrics

(*i*) Electrothermal breakdown
(*ii*) Purely electric breakdown
(*iii*) Electrochemical breakdown

(i) Electrothermal breakdown

This type of breakdown is caused due to heat produced by the dielectric loss which is proportional to the intensity of the electrostatic field and the frequency. It is seen that the heat generated due to dielectric loss varies exponentially with temperature whereas the heat dissipated to the cooler surrounding medium will be directly proportional to (*i*) temperature difference, (*ii*) the thermal conductivity of the dielectric and (*iii*) the co-efficient of heat transmission of the surface of the dielectric to the ambient medium.

Because of this low conductivity the solid dielectrics are poor conductors of heat. At a given applied voltage if the heat generated due to the losses is greater than the heat dissipated and if the voltage is applied for a long enough period then the dielectric will come unable to reach a state of internal thermal equilibrium resulting in an electro thermal breakdown of the dielectric.

The following conditions go a long way in causing the electrothermal breakdown of the dielectric:

(*i*) Large thickness of the dielectric

(*ii*) High temperature of both the dielectric and the surrounding medium

(*iii*) Continuous application of high voltage

(*iv*) Large dielectric loss

(ii) Purely electrical breakdown

Purely electric break down mechanism (also called intrinsic break down mechanism) in many solid dielectrics (especially crystalline types) is caused by *collision ionization* by electrons. On the application of strong electric field the free electrons in the crystals are accelerated and a part of the kinetic energy so obtained is lost through interaction with the lattice points. Experimentally it has been observed that direct proportionality relationship exists between the lattice energy and dielectric strength. So *collision ionization* may take place, it is necessary that the accelerated free electrons should gain an amount of kinetic energy equal to a certain portion of the lattice energy.

The purely electrical breakdown may occur if the following requirements are met with:

(*i*) Moderate temperature

(*ii*) Low frequency

(*iii*) Small thickness of dielectric

However, electric breakdown (as a rule only) will occur when *impulse voltages* are applied.

(iii) Electro chemical breakdown

This type of breakdown normally occurs when the temperature is very high and the surrounding air has *high humidity*. The insulation resistance of a material is reduced due to *electrolytic processes* developed in the dielectric material. This happens when a d.c. voltage is continuously applied to a dielectric whose leakage current due to ionic conductivity is adequately large.

6.1.7 FERROELECTRIC MATERIALS

There are a few materials which have a special crystal structure which permits *spontaneous polarization*. Such materials are known as *ferroelectrics.* Their spontaneous polarization is like spontaneous magnetization of *ferromagnetic materials.* However, there is no relationship between ferroelectric and ferrous materials.

- If the centre of gravity of the positive and negative charges do not coincidence in the absence of an applied electric field, the substance has an electric dipole moment and is said to be spontaneously polarized. Such a substance is a ferroelectric one. It contains small regions which are polarized in different directions even in the absence of an electric field. When the temperature exceeds a certain value called the *Curie point*, the substance loses its ferroelectric properties. Above the curie point the formation of small regions - domains - is disrupted by thermal agitation and permittivity is reduced to usual small value.
- Ferro electrics used to have the following main characteristics
  - (*i*) They have a high dielectric constant which is non-linear. *i.e.*, it depends to a considerable extent on the intensity of the electric field
  - (*ii*) They exhibit hysterisis loops, *i.e.*, the polarization is not linear function of applied electric field.

When an electric field is applied to a specimen of a ferroelectric crystal, the polarization first rises rapidly with applied field to a value above which the dependence is linear. Linear extrapolation to zero field gives P_s, the *saturation* or *spontaneous polarization*. On subsequently reducing the field to zero, *remanent* (*residual*) polarization P_r remains. The negative field to reduce the polarization to zero is called the *coercive field* and represented by E_c. The existence of a dielectric hysteresis loop in a dielectric material implies that the substance possesses a spontaneous polarization and the value of P_s (depending upon the shape of the hysteresis loop) depends upon a number of factors such as the dimensions of the specimen, the temperature, the texture of the crystal, the thermal and electrical properties of the crystal. The hysteresis loop of a ferroelectric material changes its shape as the temperature is increased. The height and width decrease with increase of temperature. At a certain temperature known as ferroelectric Curie temperature, the loop merges in to a straight line and the ferroelectric behaviour of the material disappears.

Spontaneous polarization in a ferroelectric material disappears above the Curie temperature. If the temperature is below the Curie temperature, the polarization or dielectric constant is not a linear function of the field and $\in_r$ is not a constant and so the equation $P = E\in_o (\in_r - 1)$ cannot be applied directly. In such cases $\in_r$ for a ferroelectric material may be defined for a vergin cure as a differential quantity,

$$\frac{dP}{dE} = \in_o (\in_e - 1) \qquad \text{...(6.1.38)}$$

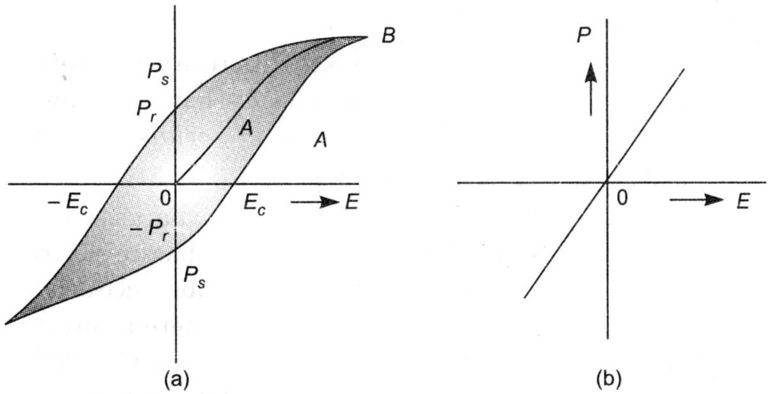

Fig. 6.13 (a) Hysteresis loop for electric materials
(b) $P-E$ relation above Curie temperature

We know that the internal field

$$E_i = E + \frac{\gamma P}{\epsilon_o}$$

$$P = N\alpha E_i = NE_i(\alpha_e + \alpha_i + \alpha_0)$$

$$P = (N\alpha)\left[E + \frac{\gamma P}{\epsilon_o}\right]$$

$$P\left[1 - \frac{N\alpha\gamma}{\epsilon_o}\right] = N\alpha E$$

$$\boxed{P = \frac{N\alpha E}{\left[1 - \frac{N\alpha\gamma}{\epsilon_o}\right]}} \qquad \text{...(6.1.39)}$$

We may thus conclude that when $\frac{N\alpha\gamma}{\epsilon_o} = 1$, one gets a non-vanishing solution for P, spontaneous magnetization exists and the dielectric constant of the substance becomes infinite. $\gamma = \frac{1}{3}$ for cubic structure (Internal field constant).

Dipole Theory of Ferroelectricity

The interaction between adjacent dipoles in a ferroelectric is large and a dipole moment has a tendency to align itself in a direction parallel to that of its neighbour. Assuming that the permanent diploes are responsible for spontaneous polarization, we can write

$$P = P_o = N\alpha_o E_i = N\mu_m L(a)$$

For higher temperatures,

$$P = P_o = N\alpha_o E_i = N\mu_m L(a)$$

and at higher temperatures,

$$a = \frac{\mu_m E_i}{k_B T} \quad \text{is small}$$

and hence

$$L(a) = \frac{a}{3}$$

Thus

$$P = P_o = \frac{N\mu_m^2 E_i}{3k_B T} = \frac{N\mu_m^2}{3k_B T}\left(E + \frac{\gamma P}{\epsilon_o}\right)$$

$$P\left[1 - \frac{N\mu_m^2 \gamma}{3\epsilon_o k_B T}\right] = \frac{N\mu_m^2 E}{3k_B T}$$

i.e.,

$$\boxed{\frac{P}{E} = \frac{N\mu_m^2/3k_B}{\left[T - \frac{N\mu_m^2\gamma}{3k_B\epsilon_o}\right]} = \frac{N\mu_m^2/3k_B}{[T - \theta_c]}} \qquad \text{...(6.1.40)}$$

where

$\theta_c = \dfrac{N\mu_m^2\gamma}{3k_B\epsilon_o}$ is known as *ferroelectric Curie temperature.*

or

$$\frac{P}{E} = \frac{C\epsilon_o}{(T - \theta_c)}$$

with

$$C\epsilon_o = \frac{N\mu_m^2}{3k_B}$$

and so

$$C = \frac{N\mu_m^2}{3k_B\epsilon_o} = \frac{\theta_c}{\gamma}$$

Now

$$\boxed{\frac{P}{E\epsilon_o} = (\epsilon_r - 1) = \frac{C}{T - \theta_c}} \qquad \text{...(6.1.41)}$$

This equation is known as Curie-Weiss law and we can show with the help of simple theory that a ferroelectric material obeys this equation at sufficiently large temperatures. The Curie point for Rochelle salt is about 24°C and this substance has a very narrow temperature range of about 40°C within which it is ferroelectric. We shall now show that spontaneous polarization is possible in a ferroelectric material below the Curie temperature.

We know that

$$\boxed{P = P_s = N\mu_m L(a)} \quad ...(6.1.42)$$

For very low temperatures, a is large (or for high electric fields) and hence $L(a) \rightarrow 1$. The polarization is called the *saturation polarization* P_s.

i.e., when a is large $L(a) = 1$ and $N\mu_m = P_s$ is known as saturation polarization. Thus Eqn. 6.1.42 becomes

$$P = P_s L(a)$$

$$\frac{P}{P_s} = L(a)$$

$$L(a) = L\left[\frac{\mu_m E_i}{k_B T}\right]$$

$$\boxed{\frac{P}{P_S} = L\left\{\frac{\mu_m}{k_B T}\left[E + \frac{\gamma P}{\epsilon_o}\right]\right\}} \quad ...(6.1.43)$$

Spontaneous polarization exists if there is a non-vanishing solution for P in Eqn. 6.1.43 with the applied field, $E = 0$

Hence

$$\frac{P}{P_S} = L\left[\frac{\mu_m \gamma P}{k_B T \epsilon_o}\right]$$

$$\boxed{\frac{P}{P_S} = L(x) = \left[\coth x - \frac{1}{x}\right]} \quad ...(6.1.44)$$

with

$$x = \frac{\mu_m \gamma P}{k_B T \epsilon_o}$$

$$\boxed{P = \frac{k_B T \epsilon_o x}{\gamma \mu_m}} \quad ...(6.1.45)$$

$$\frac{P}{P_S} = \frac{P}{\mu_m N} = \frac{k_B T \epsilon_o x}{\mu_m^2 N \gamma}$$

or

$$\frac{P}{P_S} = \left[\frac{k_B T \in_o}{\mu_m^2 \; N\gamma}\right] x = \left[\frac{T}{3\,\theta}\right] x \qquad ...(6.1.46)$$

Eqn. (6.1.44) gives the *Langevin curve*, while the second one (Eqn. 6.1.46) a straight line passing through the origin whose slope is $\left(\frac{T}{3\theta}\right)$ and is equal to $\left(\frac{T}{3}\right)\left(\frac{3\,k_B \in_o}{N\,\mu_m^2 \gamma}\right) = \left[\frac{k_B T \in_o}{\mu_m^2 \, N\gamma}\right]$ and the slope increases with temperature. The Langevin curve intersects with the first line at the point A. *i.e.*, they intersect each other at the origin 0 and at another point A, so that the solutions are $\frac{P}{P_S} = 0$ or AN. $\frac{P}{P_S} = 0$ cannot be a true solution, because there is spontaneous polarization. Hence the only possible solution is $\frac{P}{P_S} = AN$, that is to say, a point of intersection of the two curves other than origin. As the temperature increases the degree of spontaneous polarization decreases and at very high temperature $T > \theta_c$, there is no spontaneous polarization, because the straight line does not intersect the $L(x)$ curve. Measured values of γ are: Rochelle salt 2.1; $BaTiO_3$ 0.064; $KH_2\,PO_4$ 0.37.

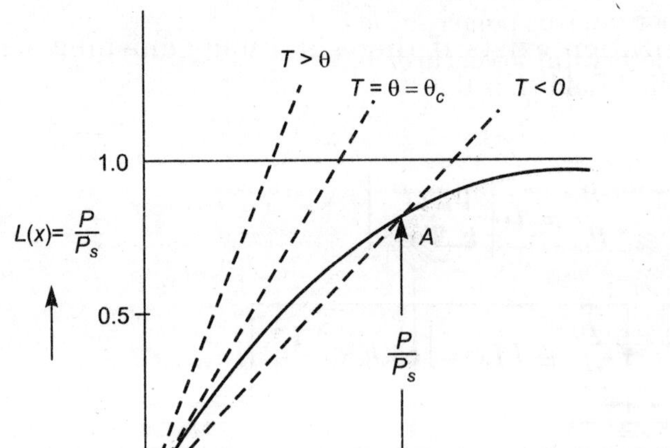

Fig. 6.14 Langevin curve

The spontaneous polarization, which is the most important characteristic property of a ferroelectric material, usually vanishes above a certain temperature θ_c called the *ferroelectric Curie temperature.* In the ferro-electric region *i.e.*, below θ_c, the dielectric constant is evidently a function of the field strength and is no longer a constant on the basis of equation (6.1.38):

$$\in_o (\in_r - 1) = \frac{dP}{dE} \quad ...(6.1.47)$$

Generally the derivative $\frac{dP}{dE}$ is taken for the *virgin curve* at the origin. The electric constant $\in_r$, as defined by the above equation may become extremely high as the temperature approaches the ferroelectric curie temperature. It may also be noted that in the ferroelectric region, the dielectric constant $\in_r$ is a function of the field, $\in_r$ being higher for higher values of field. However, for temperatures exceeding the ferroelectric Curie temperature θ_c, $\in_r$ does not vary with field. For temperature exceeding the ferroelectric curie temperature, the variation of absolute dielectric constant $\in = \in_r \in_o$ with temperature is given by the Curie law:

$$\in = \frac{C}{(T - \theta)} \quad ...(6.1.48)$$

where C is a constant and θ is the characteristic temperature for the material, only a few degrees smaller than the ferroelectric Curie temperature θ_c.

An Example of Ferroelectric Crystal

Ferroelectrics show spontaneous polarization due to slight irregularity which is induced by thermal agitation in the crystal structure. Barium titanate ($BaTiO_3$) is the classical example of ferroelectric.

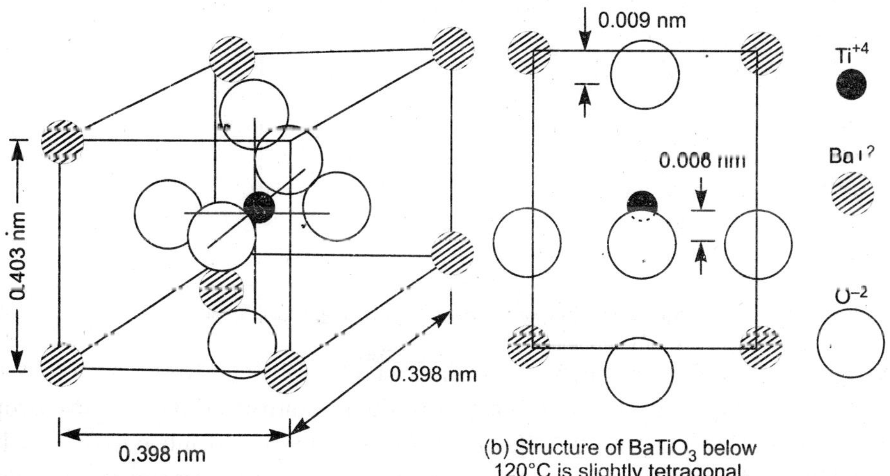

(a) The structure of $BaTiO_3$ above 120°C is cubic

(b) Structure of $BaTiO_3$ below 120°C is slightly tetragonal. Central atom Ti^{+4} ion shifts with respect to O^{-2} ions

Fig. 6.15 Structure of $BaTiO_3$

Its crystal structure is shown in Fig. 6.16(a). The permanent polarization is because the central titanium ion is displaced towards one of the oxygen ions. The oxygen ions are also slightly displaced with respect to barium ions as shown in Fig. 6.16(b). Parallel displacement occurs throughout the crystal lattice. In a polycrystalline structure the individual displacements are aligned by the application of a strong external field and thus the sample develops an overall electric dipole moment. Sufficient thermal energy may, however, destroy such an arrangement and the permanent polarization may be lost. The temperature at which this occurs is termed *ferroelectric Curie temperature*. At temperatures below the Curie point the individual moments are easily aligned, thereby increasing the permittivity. The ferroelectric Curie temperature of $BaTiO_3$ is 120°C which is one of the highest. When the temperature is lowered up to 120°C the polarization begins at scattered points throughout the material. If no field is present the orientation at these points is random. As polarization spreads from these nuclei, domains are formed in which the directions of polarization are same as those of the nuclei. Since there is a possibility of more than one direction, the domains form in such a way that the net polarization is zero.

The spontaneous polarization-temperature curve is like a ferromagnetic one and the spontaneous polarization vanishes at the curie temperature, above which the material is said to be *paraelectric*. See Fig. 6.16(c). There are two classes of ferroelectric: the order-disorder type involving the movement of hydrogen atoms, for example KH_2PO_4, and the displacive type such as the perovskites, typified by $BaTiO_3$. Only the latter will be discussed as they are by far the most important. (These materials or derivatives from them have recently been found to be superconductive at extraordinarily high temperatures).

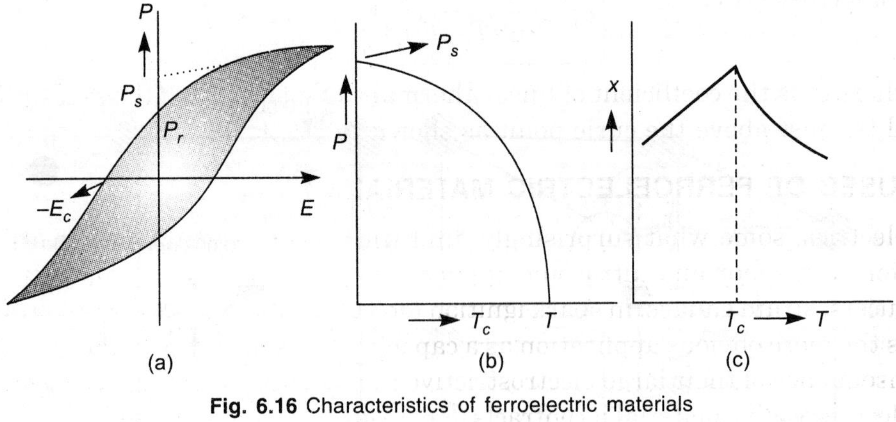

Fig. 6.16 Characteristics of ferroelectric materials

Perovskite is a mineral with the formula $CaTiO_3$. But the first important perovskite ferroelectric is $BaTiO_3$, but its relatively low curie temperature ($\approx$ 400 K) led to the development of superior materials, especially *Pb* (Zr, Ti)O_3, which has $T_c \approx$ 650 K. The magnitude of the ionic displacements in the unit cell can be calculated from the saturation polarization, P_s (which is the dipole moment/m^3). The unit cell of $BaTiO_3$ is cubic above the Curie temperature and the lattice parameter is about 0.4 nm. At 300 K the saturation polarization is 0.26 C/m^2, so the dipole moment per unit cell, μ, is $0.26 \times (4 \times 10^{-10})^3$, which is 1.7×10^{-29} Cm. The charge on the Ti^{4+} and Ba^{2+} ions is 6*e* (= 9.6×10^{-19} *C*) per unit cell, so their displacement with respect to the O^{2-} ions, $l\left(=\frac{\mu}{q}\right)$, must be $1.7 \times 10^{-29}/9.6 \times 10^{-19}$, or 1.8×10^{-11} m (0.018 nm). This displacement causes the structure to become tetragonal below T_c though the actual ratio is close to unity.

The Catastrophe Theory

Displacive ferroelectrics can be interpreted in terms of a *polarization catastrophe* in which the local field produced by the displacement of the ions is larger than the restoring forces, so that the displacement, and hence the polarization, tends to infinity. The non-linearity of the restoring forces results in a finite, though very large polarization. While deriving the Clausius-Mossotti equation, we find that

$$\epsilon_r = \left[\frac{1+2\beta}{1-\beta}\right] \qquad ...(6.1.49)$$

where $\beta = \sum \alpha N / 3\epsilon_0$, the sum of the contributions of all the ions in the structure to the polarization. When $\beta = 1$, the relative permittivity becomes infinite. Since the derivation of Clausius-Mossotti equation relies on an isotropic, or cubic, ionic environment, the structural distortion as β approaches unity restricts the value of ϵ_r. Nevertheless, the relative permittivity reaches extraordinarily high values (over 10000) is some pervoskites. It may be shown that in the paraelectric state ($T > T_c$), the relative permittivity is given by

$$\epsilon_r = \frac{1}{\alpha'(T - T_c)} \qquad ...(6.1.50)$$

where α' is the coefficient of linear thermal expansion. Thus the susceptibility suffers a rapid fall just above the curie point as shown in Fig. 6.16(c).

6.1.8 USES OF FERROELECTRIC MATERIALS

Ferroelectrics, some what surprisingly, find wide use as sound transducers (1 kHz to10 MHz, for both sonor and ultrasonic applications), accelerometers, delay lines, positional transducers, strain gauags, in spark ignition (domestic gas appliances) and pressure gauges, besides the more obvious application as a capacitor material. Most of these applications are the consequence of their large electrostrictive coefficients. When an electric field is applied to an electrostrictive material it contracts or expands, depending on the sign of the coefficient, along the polarization direction. The polarity of the field is immaterial, unlike a piezoelectric material such as quartz (once used in sound transducers, until superseded by magnetostrictive nickel transducers and then by ferroelectrics). The advantage of ferroelectric materials lies in the possibility of their having a built-in biasing field, which can be produced by a process called *poling*. The ceramic is heated to little below T_c and a large field (perhaps 1 – 5 MV/m) is applied for a few minutes. The ferroelectric then has a remanent polarization which can only be destroyed by heating to near T_c or employing fields in excess of the *coercive field* (about 1 MV/m). After poling, the ferroelectric behaves just like a piezoelectric material, though strictly it is a polarized, electrostrictive material.

Ferroelectric materials are also good *pyroelectric materials*. The pyroelectric coefficient is defined by

$$\lambda = \frac{dP}{dT}$$

A change in temperature of the material causes a change in polarization and hence a change in surface charge. In a 'good' material, such as barium niobate, $\lambda \approx 3 \times 10^{-3}$ C/m^2/ K, and as a charge of 10^{-16}C is detectable, it can be seen that pyroelectric devices can register temperature changes of the order of 1μK. Extensive use is made of this in military hardware. responding to infra-red radiation (heat). Thermal-imaging devices are also made from pyroelectric materials.

Conclusion

Ferroelectric materials are the dielectrics analogous to ferromagnetic materials. Their uses are parallel to those of ferromagnetic materials in such applications as high permeability materials. The other advantage of ferroelectric material is its large dielectric constant, which permits the use of physically small capacitors. In ferroelectrics a high electric field applied to a device can cause voltage breakdown. The general rule is that, the higher dielectric constant, the lower the dielectric strength. A figure of 2.5 kV/mm at room temperature for high-dielectric constant materials (4000 to 6000) is normally accepted value, and this can be increased to 15 kV/mm for low dielectric constant, (≈ 30) dielectrics. These values would have to be derated for elevated temperatures. Capacitors with very high dielectric constants, in the order of 100000 and higher, have been developed by treating $BaTiO_3$ in a reducing atmosphere. These capacitors are semiconductor in nature and rather loosy and are therefore, limited to very-low voltage applications such as those in transistor circuits.

6.1.9 ANTI-FERROELECTRIC MATERIALS

Anti-ferroelectric materials are the materials of ionic crystals having lines of ions spontaneously polarized; with neighbouring lines polarized in antiparallel direction. They are very similar to the antiferromagnetic materials. The similar law is followed by the anti-ferroelectric materials. Tungsten oxide (WO_3) was discovered first to show such anti-ferroelectric properties. A thorough study of $PbZrO_3$ was done to show the antiferroelectric properties. The antiferroelectric properties of lead zirconate were shown on room temperature by many workers.

6.1.10 DEPENDENCE OF PERMITTIVITY ON DIFFERENT FACTORS

1. ***Dependence of permittivity on frequency***

As we know from the theory of polarization that time of setting of electronic polarization and ionic polarization is very little in comparison with the time of variation of the sign of voltage that is one half period of alternating voltage even for the highest frequency which are used practically in electrical and radio engineering. Therefore the polarization of dielectrics having only deformational mechanism of polarization is completely set for the time, which is very little in comparison with half period of voltage and permittivity will be independent of frequency. We may say that neutral dielectrics have the permittivity independent of frequency within a very wide range of frequency as in Fig. 6.17.

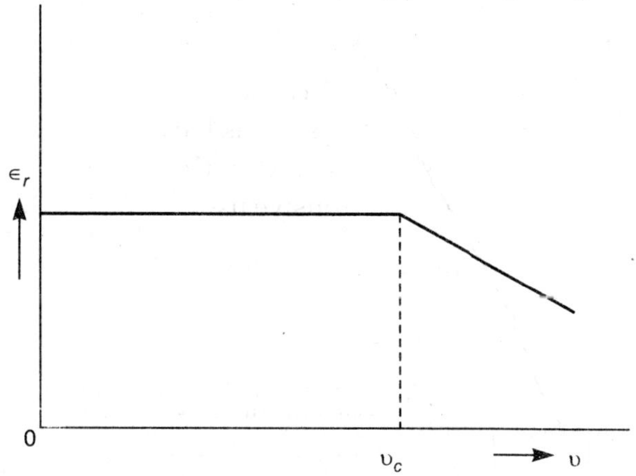

Fig. 6.17 Frequency dependence of permittivity

The case is somewhat different with dipole polarization. If we increase the frequency of ac voltage the permittivity of a dipole material at first is constant but decreases with the increase of frequency starting from critical frequency υ_c, when the polarization already does not have time completely set for one half period. The critical frequency υ_c is given by the relation

$$\upsilon_c \approx \frac{k_B T}{8\pi f R^3}$$

If τ_0 is the relaxation time, then in τ_0 sec half oscillation will be completed and hence in 1 sec $\frac{1}{\tau_0}(1/2) - \upsilon_c$ will be completed

or

$$\frac{1}{2\tau_0} \approx \frac{k_B T}{8\pi f R^3}$$

i.e.,

$$\tau_o = \frac{4\pi f R^3}{k_B T} = \frac{3 f V}{k_B T}$$

as

$$V = \frac{4}{3}\pi R^3$$

where f is the *restoring force constant* and V = molecule volume. The phenomenon of dependence of permittivity on frequency is called some times *dispersion of permittivity* which is analogous to the dispersion of light. R is the radius of the molecule.

2. *Dependence of permittivity on temperature*

Temperature does not affect the electronic polarization. Hence electronic polarization in separate molecules is independent on temperature as we have known already. But thanks to thermal expansion of the material, the quantity of molecules which can be polarized per unit of dielectric volume decreases if temperature increases and therefore the value of $\in$ decreases. If liquids are heated they will form gas and $\in$ will be about to those of gases quite near to unity. Solid ionic dielectrics have different dependence of permittivity on temperature. Generally ionic polarization increases the permittivity, like inorganic glasses and in ceramic materials. Permittivity decreases with increase of temperature for rutile TiO_3 and perovskite ($CaTiO_3$).

In polar dielectrics in the range of low temperatures, molecular orientation is impossible in majority of cases as we know, in polarization. With the increase of temperatures at high value, dipole orientation is possible that rises permittivity, but further rise of temperatures reduces their orientation and after a peak value of $\in$, it will be down.

Let us say $T_{c\in}$ be the temperature coefficient of permittivity

i.e.,

$$T_{c\in} = \frac{1}{\in}\frac{d\in}{dT} \quad ...(6.1.51)$$

If T_{ce} is temperature coefficient of capacity, then

$$T_{ce} = \frac{1}{C}\frac{dC}{dT} \quad ...(6.1.52)$$

and $T_{c\rho}$ the temperature coefficient of resistivity, is given by

$$T_{c\rho} = \frac{1}{\rho}\frac{d\rho}{dT} \quad ...(6.1.53)$$

Similarly the temperature coefficient of linear expansion is

$$T_{cL} = \frac{1}{L}\frac{dL}{dT}$$

We can write

$$T_{ce} = T_{c\in} + T_{cL} = T_{c\in} + \alpha_L$$

But

$$T_{c\in} = \frac{1}{\in}\frac{d\in}{dT} = \frac{1}{\in}\frac{(\in_2 - \in_1)}{(T_2 - T_1)}$$

i.e.,

$$\boxed{\frac{d\in}{dT} = \frac{(\in_2 - \in_1)}{(T_2 - T_1)}} \quad ...(6.1.54)$$

For highly polar liquids the temperature coefficient of permittivity is given by

$$T_{c\in} = \frac{1}{T}$$

3. *Dependence of permittivity on pressure*

In certain dielectrics the permittivity increases with increase of pressure. For the case of gases we have

$$\in = 1 + AP$$

where A is a constant and P is pressure

Thus

$$\frac{d\in}{dP} = A$$

Hence

$$\boxed{\in = 1 + P\frac{d\in}{dP}} \quad ...(6.1.55)$$

or

$$\boxed{\frac{d\in}{dP} = \frac{(\in - 1)}{P}} \quad ...(6.1.56)$$

In non-polar gases the permittivity increases linearly with the increase of pressure

$$P_{c\in} = \frac{1}{\in}\frac{d\in}{dP}$$

From Eqn. (6.1.55),

$$P_{c\in} = \frac{(\in - 1)}{P\in}$$

For the normal conditions of pressure, temperature and humidity, permittivity of air is:

$$\in_{air} = 1.00053$$

$$P_{(c\in)} = 0.00058/\text{atmosphere}$$

For pressures the variation of permittivity is given by A. R. Volpert's formula

$\in = 1 + (688 - 93H) \times 10^{-6}$ (summer seasons)

$\in = 1 + (648 - 81H) \times 10^{-6}$ (winter seasons)

H = sea level of H km

4. *Dependence of permittivity on humidity*

Hygroscopic dielectrics have higher permittivity. The permittivity grows considerably with humidification, and

1. Resistivity decreases
2. Losses increase
3. Electric strength is lowered

However in a great majority of important cases, it is difficult to calculate the value of permittivity of humidified dielectrics. Water has permittivity of about, $\epsilon_r = 80$ greater than other materials.

5. *Dependence of permittivity on voltage*

Clearly expressed dependence of permittivity upon voltage applied to dielectric is the character of ferroelectrics. For linear dielectrics the permittivity is independent of voltage.

For polar dielectrics especially liquids and gases one may expect effect of saturation. For example, water, the ratio is

$$\frac{\Delta \epsilon}{\epsilon} = 0.0011$$

$$\Delta \epsilon = 0.0011 \times \epsilon$$

Due to saturation the dielectric constant falls down. In some cases effect of saturation, may increase permittivity called the positive effect by B. Piekara, and permittivity increases in strong fields.

6.1.11 FREQUENCY DEPENDENCE OF POLARIZATION

At low frequencies, in the radio-frequency region of the electromagnetic spectrum, the dipolar molecules in some media may be capable of rotation in response to the oscillations of an alternating electric field. This rotation will be opposed by damping force, such as viscous drag, which results in a characteristic relaxation time for the dipoles, which is given by

$$\tau = \frac{1}{\omega_o}$$

where ω_o is the frequency of maximum power absorption, the resonant frequency. When $\omega << \omega_o$, the dipoles are in phase with the field and no energy is absorbed and when $\omega >> \omega_o$, the dipoles cannot respond to the field at all and again no energy is absorbed. At low frequencies the dipoles make maximum contribution to the permittivity and at high frequencies they make no contribution.

The Complex Dielectric Constant

We were considered with the dielectric constant when the dielectric was subjected to a static electric field. Let us now consider the dielectric under an alternating electric field. How will a dielectric material behave when subjected to an alternating electric field? Will the polarisation, electric displacement and the dielectric constant be different for different frequency ranges? The response of each type of polarization to the impressed frequency (of the field) is different. For example, while electronic polarization can be quickly built up in $10^{-14} - 10^{-15}$ sec, ionic polarization may require a time period as long as $10^{-11} - 10^{-13}$ sec. As a consequence, while both ionic and electronic polarizabilities respond in the regions from microwave to infrared, only the electronic polarizability contributes to the polarization in the optical to ultraviolet regions.

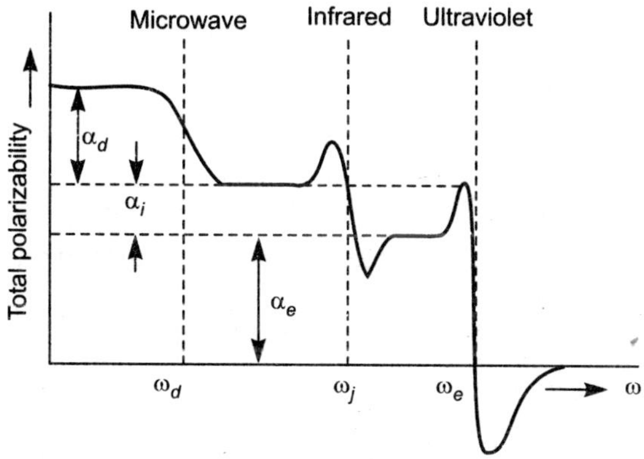

Fig. 6.18 Frequency dependence of different polarizabilities

Figure 6.18 depicts the frequency dependence of the different types of polarizabilities. It may be noted that the total polarization and, hence the dielectric constant decreases in steps as the frequency is steadily increased. In the first instance, one may ask why does this step-wise decrease occur? Does it mean that a particular type of polarizability simply "switched off" when the impressed frequency becomes too high for that type to respond? Clearly, that should be the case. But more importantly, we can understand, the detailed shape of the polarizability curve from the Fig. 6.18. When we consider the dielectric constant under an alternating electric field, we come across two main situations. They are

(*i*) when there is no measurable phase difference between $\vec{D}$ and $\vec{E}$, then polarization is in phase with the alternating electric field and $\vec{D} - \vec{E} \in$ is a valid relation and

(*ii*) when there is a phase difference between $\vec{D}$ and $\vec{E}$, then polarization $\vec{P}$ is not in phase with the alternating field and $\vec{D} = \in \vec{E}$ is not a valid relation. The basic difference between these two situations is that in the first possibility no energy is absorbed by the dielectric from the electric field, whereas in the second possibility energy is absorbed by the dielectric, which is known as dielectric loss. In order to see this, let us apply an alternating voltage to dielectric between the plates of plane capacitor

$$\boxed{\vec{E} = \vec{E}_o \cos\omega t} \qquad ...(6.1.57)$$

The true surface density σ on the capacitor plates, which is equal to $\vec{D}$, gives the current density as

$$\vec{J} = \frac{\partial \vec{D}}{\partial t} \qquad \text{...(6.1.58)}$$

In the first case, when the electric displacement $\vec{D}$ is in phase with $\vec{E}$, then

$$\vec{D} = \vec{D}_o \cos \omega t \qquad \text{...(6.1.59)}$$

giving

$$\vec{J} = -\omega \vec{D}_o \sin \omega t \qquad \text{...(6.1.60)}$$

Thus, the electric current density is out of phase by $\frac{\pi}{2}$ from $\vec{E}$. The dissipated energy per unit volume per second of the dielectric is

$$W = \frac{\omega}{2\pi} \int_0^{2\pi/\omega} \vec{J} \cdot \vec{E}\, dt \qquad \text{...(6.1.61)}$$

with $\frac{\omega}{2\pi} = \frac{1}{T}$ and $T = \frac{2\pi}{\omega}$ and $\omega = 2\pi\nu$.

Substituting $\vec{j}$ and $\vec{E}$ from (6.1.60) and (6.1.57) respectively, we get

$$W = \frac{\omega}{2\pi} \int_0^{2\pi/\omega} (-\omega \vec{D}_o \sin \omega t)(\vec{E}_o \cos \omega t)\, dt = 0 \qquad \text{...(6.1.62)}$$

From this we find that there is no dissipation of energy when $\vec{D}$ and $\vec{E}$ are in phase.

Let us now discuss the results when they are out phase by δ. The expression for electric displacement takes the form

$$\vec{D} = \vec{D}_o \cos(\omega t - \delta)$$

or

$$\vec{D} = \vec{D}_o \cos \delta \cos \omega t + D_o \sin \delta \sin \omega t \qquad \text{...(6.1.63)}$$

Putting the value of $\frac{\partial \vec{D}}{\partial t}$ in Eqn. (6.1.58), we get

$$\vec{J} = \omega\vec{D}_o \cos\omega t \sin\delta - \vec{D}_o \omega \sin\omega t \cos\delta \qquad \text{...(6.1.64)}$$

This helps to get the equation for the energy dissipated per unit volume per second as

$$W = \frac{\omega}{2\pi}\int_0^{2\pi/\omega} \vec{E}\,\vec{J}\,dt$$

$$W = \frac{\omega}{2\pi}\int_0^{2\pi/\omega} \vec{E}_o \cos\omega t\,\{\omega\vec{D}_o \cos\omega t \sin\delta - D_o\omega \sin\omega t \cos\delta\}\,dt$$

$$W = \frac{\omega}{2}\vec{D}_o\,\vec{E}_o \sin\delta \qquad \text{...(6.1.65)}$$

This is nothing but the dielectric loss and therefore the term sinδ is usually termed as the *loss factor* (or *power factor*) and δ is known as the *loss angle* (or phase angle). In case of phase difference between $\vec{D}$ and $\vec{E}$, it is meaningful to use complex notations; then D is the real part of $D_o e^{i(\omega t-\delta)}$ and the field E is the real part of $E_o e^{i\omega t}$. When the phase difference δ is independent of E, then the ratio between the complex quantities $D_o e^{i(\omega t-\delta)}$ and $E_o e^{i\omega t}$ will yield a complex dielectric constant say ϵ^*. It is given by

$$\epsilon^* = \frac{D_o\, e^{i(\omega t-\delta)}}{E_o\, e^{i\omega t}}$$

or

$$\epsilon^* = \frac{D_o\, e^{-i\delta}}{E_o} \qquad \text{...(6.1.66)}$$

If the real and imaginary parts of ϵ^* are ϵ' and $-\epsilon''$ respectively, then

$$\epsilon' = \frac{D_o}{E_o}\cos\delta \text{ and } \epsilon'' = \frac{D_o}{E_o}\sin\delta; \text{ or } \sin\delta = \frac{E_o\,\epsilon''}{D_o} \qquad \text{...(6.1.67)}$$

remembering $e^{-i\delta} = (\cos\delta - i\sin\delta)$

Thus

$$\frac{\sin\delta}{\cos\delta} = \tan\delta = \frac{\epsilon''}{\epsilon''}$$

and

$$\epsilon'^2 + \epsilon''^2 = \frac{D_o^2 \cos^2\delta}{E_o^2} + \frac{D_o^2}{E_o^2} \sin^2\delta$$

or

$$\frac{D_o^2}{E_o^2}(\sin^2\delta + \cos^2\delta) = \epsilon'^2 + \epsilon''^2$$

i.e.,

$$\boxed{\sqrt{\epsilon'^2 + \epsilon''^2} = \frac{D_o}{E_o}} \quad ...(6.1.68)$$

Take the value of $\sin\delta = E_o\epsilon'' D_o$ from Eqn. (6.1.67) to Eqn. (6.1.65) and write the expression for energy dissipation per m^3 per sec as

$$\boxed{W = \frac{\omega}{2} E_o^2 \epsilon''} \quad ...(6.1.69)$$

6.1.12 RELAXATION TIME IN DIELECTRICS

We are well aware that the net or total polarization say P in a dielectric is due to three contributions: electronic, ionic (or generally atomic) and orientational polarization. One should know that when a dielectric material is subjected to an external electric field (static), sufficient time is required for the electric polarization to get its final value. It is observed that electronic and ionic polarization is attained the maximum value instantaneously, if we consider high frequencies ($10^7 - 10^{11}$ Hz) and not the optical frequencies. Hence the dielectric loss must be mainly due to the relaxation effect of the permanent dipoles. Hence, first we will consider the transient effects, in which this relaxation effect of the permanent dipoles is characterized by a relaxation time τ and then switch over the discussion to the situation of applying an alternating field. Refer Eqn. 6.1.2. Since it is under static field the symbol s is used wherever required. Thus Eqn. 6.1.2. is re-written as

$$D_s = \epsilon_{rs}\epsilon_o E \text{ and } P_s = D_s - \epsilon_o E = E\epsilon_o(\epsilon_{rs} - 1) \quad ...(6.1.70)$$

The total polarization may now be written as the sum of only two terms, one in which the polarization is attained instantaneously, denoted by P_∞ and the other in which relaxation effects are important, denoted by P_{os}.

$$\boxed{P_s = P_\infty + P_{os}} \quad ...(6.1.70a)$$

Now

$$\boxed{P_\infty = E\epsilon_o(\epsilon_\infty - 1)} \quad ...(6.1.71)$$

Instantaneously on the application of static field, the polarization is P_∞, and then let us consider that at the time instant t, P_s part out of P_{os} is build up. Hence at the time instant t, we have

$$P_t = P_\infty + P_s$$

In general, in relaxation processes, one assumes that the increase $\frac{d(P_s)}{dt}$ is directly depending on the difference between the final value P_{os} and the actual value P_s.

i.e.,

$$\frac{dP_s}{dt} = \frac{(P_{os} - P_s)}{\tau} \quad \text{...(6.1.72)}$$

where τ is a constant, known as the *relaxation time*, a measure of the time lag. Integration of Eqn. (6.1.72) with the boundary condition $t = 0$, $P_s = 0$, one gets

$$P_s = P_{os}(1 - e^{-t/\tau}) \quad \text{...(6.1.73)}$$

In the case of alternating field also, it is assumed that Eqn. 6.1.72 is valid. * is used to denote complex quanties. Thus we have

$$\frac{dP_s^*}{dt} = \frac{1}{\tau}\left[P_{os}^* - P_s^*\right] \quad \text{...(6.1.74)}$$

With the help of equations 6.1.70, 6.1.70*a* and 6.1.71, we write,

$$P_{os}^* = P_s^* - P_\infty^* = \epsilon_o E_o e^{i\omega t}(\epsilon_{rs} - 1) - E_o \epsilon_o e^{i\omega t}(\epsilon_\infty - 1)$$

$$P_{os}^* = P_s^* - P_\infty^* = \epsilon_o (\epsilon_{rs} - \epsilon_\infty) E_o e^{i\omega t} \quad \text{...(6.1.75)}$$

Substituting Eqn. 6.1.75 into 6.1.74 and integrating, we obtain

$$P_s^* = C e^{-t/\tau} + \frac{\epsilon_o (\epsilon_{rs} - \epsilon_\infty)}{(1 + i\omega\tau)} E_o e^{i\omega t} \quad \text{...(6.1.76)}$$

After sometime, the first term on the right hand side is small that it can be neglected. Now the total polarization in an altering field is

$$P^* = [P_\infty + P_s^*]$$

Referring equations 6.1.71 and 6.1.76

$$P^* = P_\infty + P_s^* = E^* \epsilon_o (\epsilon_\infty - 1) + \frac{E^* \epsilon_o (\epsilon_{rs} - \epsilon_\infty)}{(1 + i\omega\tau)} \quad \text{...(6.1.76a)}$$

But the electric displacement (with $E^* = E e^{i\omega t}$) is given by

$$D^* = \epsilon_o E^* + P^* = \epsilon_o E^* + E^* \epsilon_o \epsilon_\infty - E^* \epsilon_o + \frac{E^* \epsilon_o (\epsilon_{rs} - \epsilon_\infty)}{(1 + i\omega\tau)} \quad \text{...(6.1.76b)}$$

But

$$D^* = \in^* E^* = E^* \in_o \in_\infty + \frac{E^* \in_o \in_{rs} - E^* \in_o \in_\infty}{(1+i\omega\tau)}$$

or

$$\boxed{\in^* = \in_o \in_\infty + \frac{(\in_o \in_{rs} - \in_o \in_\infty)}{(1+i\omega\tau)}} \quad ...(6.1.77)$$

where $\in^*$ complex dielectric constant in the alternating field

$$\in^* = \in_o \in_\infty + \frac{(\in_o \in_{rs} - \in_o \in_\infty)(1-i\omega\tau)}{(1+\omega^2\tau^2)}$$

$$= \in_o \in_\infty + \frac{(\in_o \in_{rs} - \in_o \in_\infty) - i\omega\tau(\in_o \in_{rs} - \in_o \in_\infty)}{(1+\omega^2\tau^2)}$$

The real part of the dielectric constant

$$\boxed{\in'(\omega) = \in_o \in_\infty + \frac{(\in_o \in_{rs} - \in_o \in_\infty)}{(1+\omega^2\tau^2)}} \quad ...(6.1.78)$$

The imaginary part

$$\boxed{\in''(\omega) = \frac{(\in_o \in_{rs} - \in_o \in_\infty)\omega\tau}{(1+\omega^2\tau^2)}} \quad ...(6.1.79)$$

These two equations are called *Debye's equations* and they can be written as

$$\frac{[\in'(\omega) - \in_o \in_\infty]}{[\in_o \in_{rs} - \in_o \in_\infty]} = \frac{1}{(1+\omega^2\tau^2)}$$

and

$$\frac{\in''(\omega)}{(\in_o \in_{rs} - \in_o \in_\infty)} = \frac{\omega\tau}{(1+\omega^2\tau^2)}$$

The left hand side of these equations is plotted as a function of ωt shown in Fig. 6.19. All these relations are in satisfactorily agreeing with experimental observations.

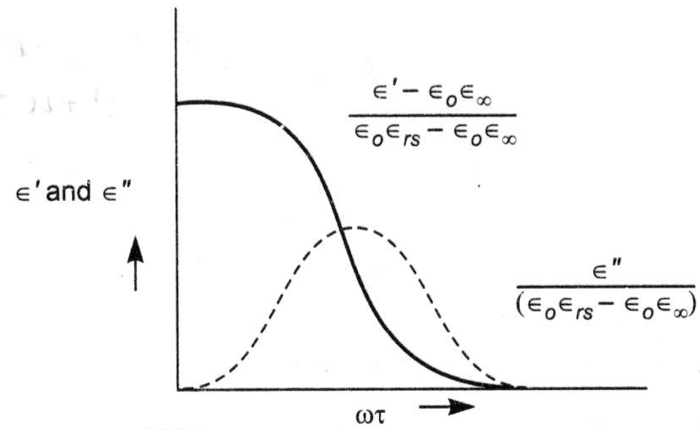

Fig. 6.19 Frequency dependence of the real and the imaginary part of the dielectric constant

6.1.13 FREQUENCY DEPENDENCE OF ELECTRONIC POLARIZATION

Let us consider a simple model of the atom in which a heavy nucleus of charge Ze is symmetrically surrounded by a homogeneous spherical electron cloud of charge $-Ze$ and radius a. When the centres of the negative charges and positive charges are displaced by x, a restoring force acts on the electron cloud similar to the Hook's law force. The electron cloud become a harmonic oscillator whose equation of motion is given by

$$m\frac{d^2x}{dt^2} = -\frac{Z^2e^2x}{4\pi\epsilon_o a^3} = -m\,\omega_o^2\,x \quad ...(6.1.80)$$

The natural frequency ω_0 of the system is given by

$$\omega_o = \sqrt{\frac{Z^2e^2}{4\pi\epsilon_o\,m\,a^3}}$$

where a is the radius of the atom. Substituting the appropriate values for the different quantities we get the natural frequency of the electron cloud as 10^{11} radian/sec, and this frequency lies in the ultra-violet region. From the above, it appears the electron cloud would go on oscillating indefinitely. But in actual practice radiation and other losses are present which damp the oscillations. Consider such a damped atomic oscillator placed in an alternating electric field E_o exp $(i\omega\, t)$. The electron cloud is now acted on by a force $Ze[\exp(i\omega\, t)]$. The damping due to radiations and collisions can be represented by some sort of frictional force proportional to the velocity $\left(\frac{dx}{dt}\right)$. So the equation of motion of the electron cloud under these conditions becomes:

$$m\frac{d^2x}{dt^2} + 2b\frac{dx}{dt} + m\,\omega_o^2\,x = -ZeE_o\exp(i\omega t)$$

The constant $2b$ (called dissipation constant of the material) is related the natural frequency ω_o as

$$2b = \frac{\mu_o e^2 \omega_o^2}{6\pi m c}.$$

If we choose $Z = 1$, then

$$m \frac{d^2 x}{dt^2} + 2b \frac{dx}{dt} + m\,\omega_o^2\, x = -e E_o \exp(i\omega t) \quad ...(6.1.81)$$

The solution is of the form

$$x(t) = \text{Real}\,[A^* e^{i\omega t}] \quad ...(6.1.82)$$

Thus

$$\frac{dx}{dt} = \text{Real}\,\left[A^* e^{i\omega t}(i\omega)\right]$$

and

$$\frac{d^2x}{dt^2} = \text{Real}\,\left[A^* e^{i\omega t}(i\omega)^2\right].$$

Substituting these values in Eqn. (6.1.81) and simplifying, we get

$$A^* = \frac{-e E_o / m}{[(\omega_o^2 - \omega^2) + i(2b\omega/m)} \quad ...(6.1.83)$$

Now the induced dipole moment as a function of time is

$$\mu_m(t) = -e\,x(t) \quad ...(6.1.84)$$

Substituting the values from Equations 6.1.83 and 6.1.82, we get

$$\mu_m(t) = \text{Real}\left[\frac{(e^2/m) E_o e^{i\omega t}}{(\omega_o^2 - \omega^2) + i(2b\omega/m)}\right]$$

This is the relation between the *induced dipole moment* and applied alternating field.

Thus

$$\mu_m(t) = \text{Real}\,[\alpha_e^* E_o e^{i\omega t}]$$

where α_e^* is the complex electronic polarizability and its value is given by

$$\alpha_e^* = \frac{e^2/m}{(\omega_o^2 - \omega^2)\times i(2b\omega/m)}$$

i.e.,

$$\alpha_e^* = (e^2/m)\left[\frac{(\omega_o^2 - \omega^2) - i(2b\omega/m)}{(\omega_o^2 - \omega^2)^2 + (4b^2\omega^2/m^2)}\right]$$

or

$$\alpha_e^* = \frac{e^2}{m}\left[\frac{(\omega_o^2 - \omega^2)}{(\omega_o^2 - \omega^2)^2 + (4b^2\omega^2/m^2)} - \frac{i\,(2b\omega/m)}{(\omega_o^2 - \omega^2)^2 + (4b^2\omega^2/m^2)}\right]$$

$$\boxed{\alpha_e^* = \alpha_e' - i\alpha_e''} \qquad \text{...(6.1.85)}$$

where α_e' and α_e'' are respectively the real and imaginary parts of polarizability.

Since $\in_o(\in_r - 1) = N\alpha_e$, we write the real and imaginary part of the dielectric constants using $\gamma = 2b$

$$\boxed{\in_r'(\omega) = 1 + \frac{Ne^2}{m\in_o}\left\{\frac{(\omega_o^2 - \omega^2)}{(\omega_o^2 - \omega^2)^2 + (4b^2\omega^2/m^2)}\right\}} \qquad \text{...(6.1.85a)}$$

and

$$\boxed{\in_r''(\omega) = \frac{Ne^2}{m\in_o}\left[\frac{2b\omega/m}{(\omega_o^2 - \omega^2)^2 + (4b^2\omega^2/m^2)}\right]} \qquad \text{...(6.1.86)}$$

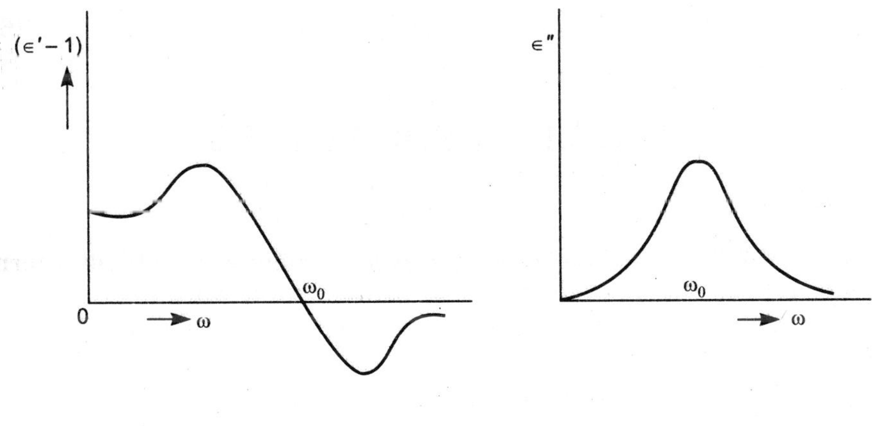

Fig. 6.20 Variation of ∈' and ∈" as function of frequency

Conclusion

The energy absorbed by the dielectric is proportional $\in''(\omega)$, as is obvious from equation (6.1.69). Equation (6.1.86) indicates the $\in''$ contains $2b$, the damping constant. Hence, if there is no damping there would be no absorption. $2b$ is some times represented by the letter γ.

One illustration

Discuss the dependence of dielectric loss angle tan δ and the permittivity, $\in_r$ on the temperature. The ebonite material has a permittivity 6 and independent of temperature approximately. The dielectric loss angle at 20°C is 0.012 and 0.02 at 60°. Find the value of dielectric loss angle at a temperature of 50°C. Calculate alpha characteristics of the material and give its units.

Answer

The dependence of tanδ and permittivity on temperature is given by

$$\in_1 \tan\delta_1 \propto \exp(\alpha T_1) \text{ and } \in_2 \tan\delta_2 \propto \exp(\alpha T_2)$$

Since permittivity does not vary much with temperature $\in_1 = \in_2$

$$\frac{\tan\delta_2}{\tan\delta_1} = \frac{\exp(\alpha T_2)}{\exp(\alpha T_1)} = \exp[\alpha(T_2 - T_1)]$$

$$\frac{0.02}{0.012} = \exp[\alpha(60-20)];\ \exp[40\alpha] = 1.7$$

$$\boxed{\alpha = 0.0128 \text{ per degree}} \quad \textbf{\textit{Answer}}$$

Also at 50°C

$$\tan\delta_3 = \tan\delta_1 \exp[\alpha(50-20)] = 0.012 \times \exp[0.0128(50-20)]$$

$$\boxed{\tan\delta_3 \text{ at } 50°\text{C is } 0.0176} \quad \textbf{\textit{Answer}}$$

6.2 MAGNETIC PROPERTIES

6.2.1 INTRODUCTION

Magnetism is one of the topics of this subject in which considerable advances have been made, both in theoretical understanding and in experimental work over the last two-three decades. The *magnetic induction B* is directly proportional to applied magnetic field *H*. Thus

$$B = \mu_0 H \qquad ...(6.2.1)$$

Here μ_0 is *permeability of free space*. If a solid is inserted in the magnetic field, the flux density or the magnetic induction is again proportional to the applied field but here the constant of proportionality will be different from μ_0. The relationship between B and H will now be

$$\boxed{B = \mu H} \quad \text{...(6.2.1}a\text{)}$$

where μ is the permeability of the solid material through which the lines of magnetic force pass. The general equation in this case is

$$\boxed{B = \mu_0(H + M)} \quad \text{...(6.2.2)}$$

M in the above equation is termed the magnetization of solid. The magnetization is directly proportional to the applied field and hence

$$M = \chi H$$

The constant of proportionality χ (called *magnetic susceptibility*) is a dimensionless quantity. In diamagnetic materials some lines of forces are repelled by the medium and they have negative susceptibility or magnetization. Paramagnetic materials have small positive value of susceptibility and these materials weekly attract the lines of forces. In ferromagnetic solids the susceptibility is large and positive value of the order of $10^2 - 10^4$. These materials strongly attract the magnetic lines of forces as shown in Fig. 6.21.

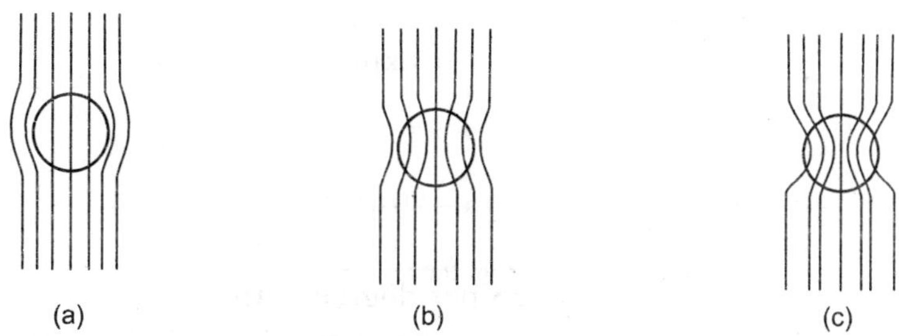

Fig. 6.21 (*a*) A diamagnetic solid repels slightly the magnetic lines of forces
(*b*) A paramagnetic solid weakly attracts the lines of force
(*c*) A ferromagnetic solid attracts lines of force very strongly

Table 6.2. Magnetic susceptibilities of some common elements

Diamagnetic elements	χ *in* 10^{-4}	*Para magnetic elements*	χ *in* 10^{-4}
Cadmium	– 0.18	Aluminium	0.65
Copper	– 0.086	Calcium	1.10
Silver	– 0.20	Oxygen	106.20
Tin	– 0.25	Platinum	1.10
Zinc	– 0.157	Titanium	1.25

6.2.2 CLASSIFICATION OF MAGNETIC MATERIALS

The magnetic properties of the magnetic mterials depend upon the way of orientation of the dipoles with respect to each other and hence the magnetic materials are classified as:

(*i*) Diamagnetic, (*ii*) Paramagnetic, (*iii*) Ferromagnetic, (*iv*) Antiferromagnetic, and (*v*) Ferrimagnetic.

1. Diamagnetic Substances

The substances which acquire feeble induced magnetism in a direction opposite to the external field, when placed in it are called diamagnetic substances. Example of diamagnetic substances are bismuth, antimony, copper, gold, mercury, water, alcohol, air and hydrogen. The behviour of diamagnetic substances when subjected to an external field is shown in Fig. 6.22.

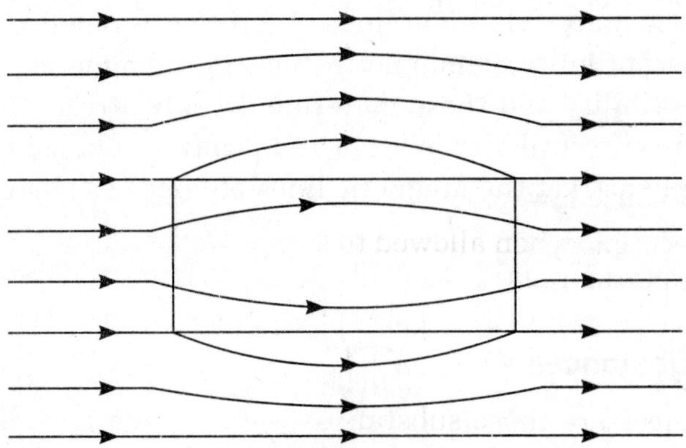

Fig. 6.22

Some of the important characteristics of diamagnetic substances are as follows:

(*i*) When subjected to a strong uniform magnetic field, the magnetism induced is in a direction opposite to that of the external field.

(*ii*) In a non-uniform field, they have tendency to move from sronger to weaker part of the field.

(*iii*) The diamagnetic substances are characterised by negative susceptibility.

(*iv*) The bar of a diamagnetic substance stays at right angle when suspended between the pole pieces of a magnet.

(*v*) A diamagnetic liquid shows a dipression in U-tube placed between the pole-pieces of a strong magnet.

(*vi*) The diamagnetic susceptibility is independent of temperature.

2. Paramagnetic Substances

Paramagnetic substances are those substances, which when plced in a stong magnetic field, get weakly magnetised in the same direction as the applied field. The atomic dipoles get aligned with the applied magnetic field and so the resultant magnetic field in the magnetised material becomes greater than the applied field.

Examples of paramagnetic substances are platinum, aluminium, chromium, oxygen, manganese and solutions of salts of iron and nickel.

Some of the characteristics of paramagnetic material are as follows:

(*i*) The relative permeability is slightly greater than unity and suscetibility is positive, of the order of 10^{-4}.

(*ii*) When a bar of paramagnetic substance is suspended between pole pieces of a magnet, it aligns parallel to the lines of force.

(*iii*) A paramagnetic liquid shows a rise in U-tube when placed between pole-pieces of a strong magnet.

(*iv*) In a non-uniform magnetic field, the paramagnetic substance tends to move from weaker part of stronger part of the field.

(*v*) The susceptibility of a paramagnetic substance is inversely proportional to absolute temperature (T), *i.e.*,

$$\chi \propto \frac{1}{T}.$$

This is called Curie law.

(*vi*) A paramagnetic gas when allowed to ascend between the pole pieces of a magnet, it spreads along the field.

3. Ferromagnetic Substances

Ferromagnetic substances are those substances, which when placed in a relatively weak magnetic field, get magnetised and are capable of developing north and south poles. The ferromagnetic materials show magnetism even in the absence of any external field. This arises due to the alignment of permanent dipoles in atoms that result from unpaired electrons in the outer orbit.

The relative permeability of ferromagnetic substances is very high of the order of hundreds and thousands. The susceptibility is positive and very high. The ferromagnetic substances has a chracteristic temperature, called Curie temperature, above which they behave as paramagnetic and below which they are ferromagnetic. Examples of ferromagnetic substances are iron, steel nickel and cobalt.

Some of the other characteristics of ferromagnetic subsances are as follows:

(*i*) Ferromagnetic substances show almost all properties of paramagnetic substances.

(*ii*) The susceptibility of ferromagnetic substances remains constant for smaller values of H, increases in the intermediate region and then decreases for very large values of H as shown in Fig. 6.23.

(*iii*) The intensity of mgnetisation I is proportional to the small value of magnetising field H and becomes constant for large values of H as also shown in Fig. 6.23.

4. Anti-ferromagnetic Substances

Anti-ferromagnetic substances are those in which the spin magnetic moments of neighbouring atoms are aligned in an anti-parallel order as shown in Fig. 6.24. It is obvious from the figure that the spin moments of alternate atoms are parallel to each other but the adjacent atoms have anti-parallel spin moments.

Anti-ferromagnetic substances are feebly magnetised when subjected to a strong magnetic field. Examples of anti-ferromagnetic substances are MnO, FeO, CoO, NiO, MnS, $FeCl_3$, MnO_4, MnF_2, Cr and Mn.

Anti-ferromagnetism was discovered by Neel and Bitter. The magnetisation of anti-ferromagnetic substances depends on temperature. It increases with temperature and becomes maximum at a particular temperature called Neel temperature. Above the Neel temperature the magnetisation decreases as shown in Fig. 6.25 and the material becomes a paramagnetic substance.

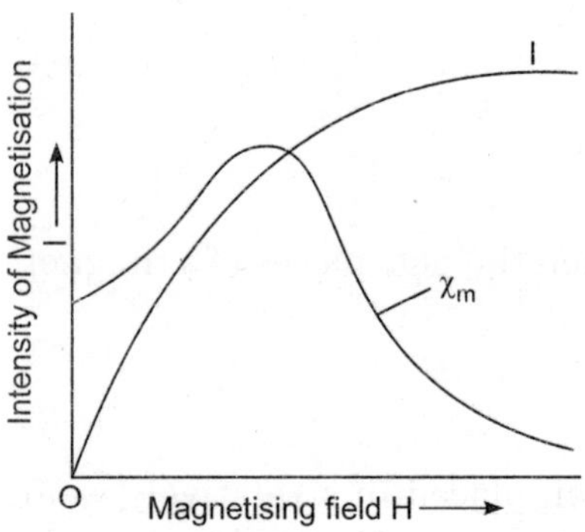

Fig. 6.23

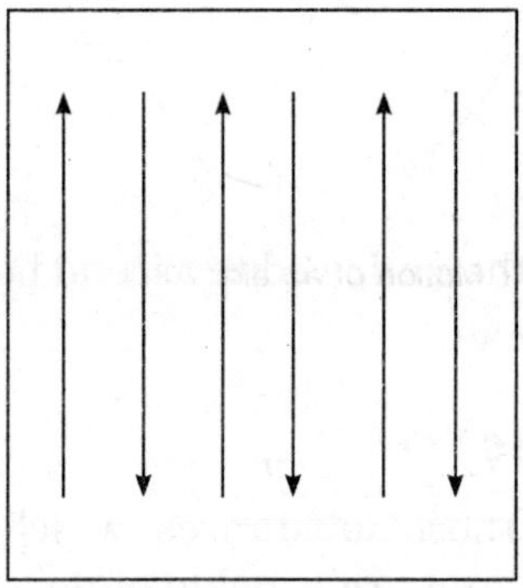

Fig. 6.24

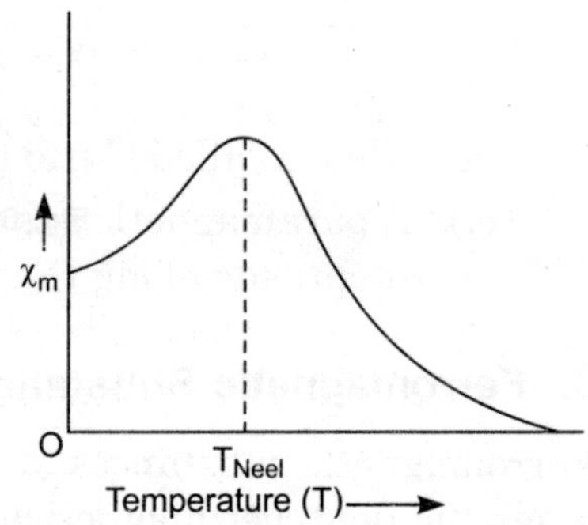

Fig. 6.25

5. Ferrimagnetic Substances

Ferrimagnetic substances are those in which the atomic or ionic dipoles in one direction are of unequal magnitude from those lined up in opposite direction as shown in Fig. 6.26. This alignment of dipoles gives a net magnetisation and is the property of those magnetic substances which have two or more different kind of atoms.

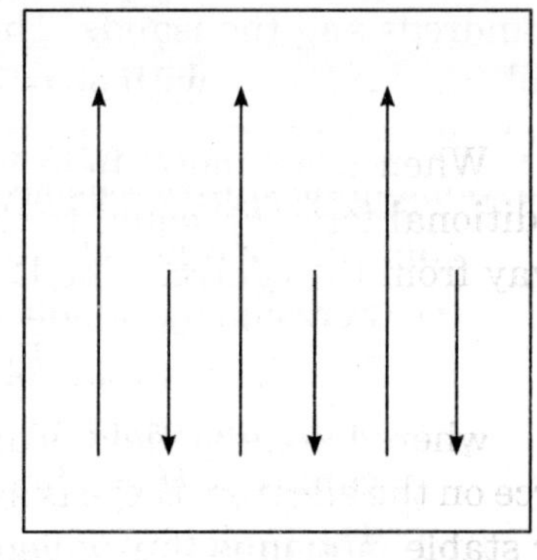

Fig. 6.26

In ferrimagnetic materials, there may be large net magnetisation as compared to anti-ferromagnetic materials, due to the resultant of anti-parallel alignment of neighbouring dipoles of unequal magnitude.

Ferrimagnetic materials, generally known as ferrites are represented by the formula $XOFe_3O_4$, where X stands for Mn, Co, Ni, Mg, Zn, Cu, Cd, etc.

6.2.3 LANGEVIN'S CLASSICAL THEORY OF DIAMAGNETISM

We know that the diamagnetism is the occurrence of negative susceptibility and magnetization is antiparallel to the field. The negative susceptibility arises from *Lenz's law* which states that the flux in an electric circuit is changed, an induced current is set up in a direction opposite to the flux change.

Consider an electron of mass m and charge e revolving about the nucleus of charge Ze in a circular orbit of radius r. Let ω_0 be the angular velocity of the electron. Now in the absence of an external field the *centripetal force* acting on the electron due to nucleus (F_0) is

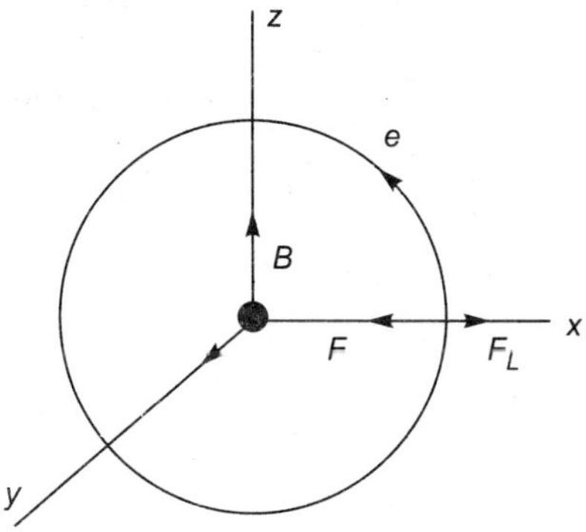

Fig. 6.27 Motion of an electron in a circular orbit

$$F_0 = m\omega_0^2 r = \frac{mv^2}{r} = \frac{Ze^2}{4\pi\epsilon_0 r^2} \quad \text{...(6.2.3)}$$

$$\omega_0 = \sqrt{\frac{Ze^2}{4\pi\epsilon_0 mr^3}} \quad \text{...(6.2.4)}$$

The magnetic moment

$$\vec{\mu} = \text{current } x \text{ area} = \frac{e\omega_0}{2\pi} \times \pi r^2 = \frac{e}{2}\omega_0^2 r$$

$$\text{with } \omega_0 = 2\pi v_0 = \frac{2\pi}{T} \text{ and } \frac{e}{T} = \text{current}$$

When a magnetic field is applied perpendicular to the xy plane (along z axis), an additional force F_L equal to the *Lorentz force* starts acting on the electron in a direction away from the centre. The Lorentz force by acting on the negatively charged electron is

$$F_L = -e(\vec{v} \times \vec{B}) = -eBr\omega \quad \text{...(6.2.5)}$$

where magnetic field is applied perpendicular to r. Because of the change in the net force on the electron, there is a change in the angular frequency from ω_0 to ω. The condition for stable motion is thus given by

$$F = F_0 - F_L = \frac{Ze^2}{4\pi\epsilon_0 r^2} - eB\omega r$$

$$\frac{mv^2}{r} = \frac{Ze^2}{4\pi\epsilon_0 r^2} - eB\omega r$$

$$mr\omega^2 = \frac{Ze^2}{4\pi\epsilon_0 r^2} - eB\omega r$$

$$\omega^2 = \frac{Ze^2}{4\pi m\epsilon_0 r^3} - \frac{eB\omega}{m}$$

$$\omega^2 + \frac{eB\omega}{m} - \frac{Ze^2}{4\pi m\epsilon_0 r^3} = 0 \quad \text{(Refer Eqn. 6.2.4)}$$

$$\omega = \frac{-\frac{eB}{m} \pm \sqrt{\left(\frac{eB}{m}\right)^2 + \frac{4Ze^2}{4\pi m\epsilon_0 r^3}}}{2}$$

$$\boxed{\omega = -\frac{eB}{2m} \pm \sqrt{\left(\frac{eB}{2m}\right)^2 + \omega_0^2}} \quad \text{...(6.2.6)}$$

If $\omega_0 >> \frac{eB}{2m}$, then

$$\boxed{\omega = -\frac{eB}{2m} \pm \omega_0 = \pm\omega_0 - \omega_L} \quad \text{...(6.2.7)}$$

where $\omega_L = \frac{eB}{2m}$ is called the *Larmor frequency.*

The $\pm$ sign on ω_0 implies that those electrons whose orbital moment are parallel to the field are slowed by ω_L and $\omega = +\omega_0 - \omega_L$. Similarly those electrons whose moments are antiparallel to the field are speeded by the same amount or $\omega = -\omega_0 - \omega_L$. The change in the frequency of the electronic motion as a result of the application of magnetic field produces magnetization. In the absence of the field the electron motion being spherically symmetrical, produces no net current or flux. On the other hand, in the field, the electron's motion is no longer spherically symmetric but processes about the field and produces a net current I. Due to the change in frequency ω_L, the current for each electron is

$$I = \text{charge} \times \text{revolution per unit time}$$

$$= -\frac{e\omega_L}{2\pi}$$

$$= -\frac{e^2B}{4\pi m} \quad \text{(Refer Eqn. 6.2.7)}$$

If Z is the atomic number, then

$$\boxed{I = -\frac{Ze^2 B}{4\pi m}} \qquad \text{...(6.2.8)}$$

Now we know that magnetic moment, $\mu = I \times A$

i.e.,

$$\boxed{\mu_m = -\frac{Ze^2 B\pi\bar{\rho}^2}{4\pi m} = \frac{Ze^2\beta\bar{\rho}^2}{4\pi m}} \qquad \text{...(6.2.9)}$$

where $\bar{\rho}^2$ is the mean square radius of the projection of the orbit on a plane perpendicular to the field axis. See Fig. 6.28. The field is acting parallel to the z-axis therefore.

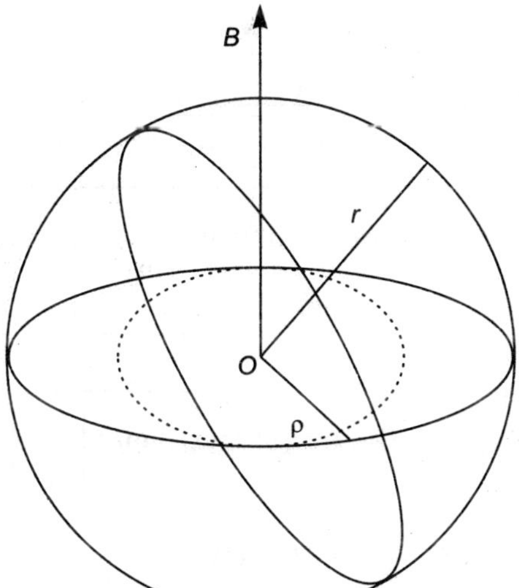

Fig. 6.28 Projection of the orbit on a plane perpendicular to the field

Now $\bar{\rho}^2 - \bar{x}^2 + \bar{y}^2$. The orbits may have different orientations with respect to the external field. Let $\bar{x}$, $\bar{y}$, $\bar{z}$ be the average values of radii for all the electrons along the three axes. Then, the mean square distance of the electron from the nucleus is

$$r^2 = x^2 + y^2 + z^2$$

For spherical symmetric charge distribution

$$x^2 = y^2 = z^2$$

or

$$\bar{r}^2 = 3x^2$$

and

$$\bar{\rho}^2 = 2x^2$$

or

$$\frac{\bar{\rho}^2}{\bar{r}^2} = \frac{2}{3} \quad \text{or} \quad \bar{\rho}^2 = \frac{2}{3}\bar{r}^2$$

Thus $$\mu_m = -\frac{ze^2\beta\pi}{4\pi m}\left[\frac{2}{3}\bar{r}^2\right]$$

$$\boxed{\mu_m = -\frac{ze^2\beta}{6m}\bar{r}^2} \qquad \text{...(6.2.10)}$$

$M = N\,\mu_m$, then magnetic susceptibility per unit volume is

$$\chi = \frac{M}{H} = -\frac{Ze^2\mu_0 H\bar{r}^2 N}{6mH}$$

$$\boxed{\chi = -\frac{\mu_0 Ze^2 N\bar{r}^2}{6m}} \qquad \text{...(6.2.11)}$$

This is the classical Langevian equation for diamagnetism. It is clear that the diamagnetic susceptibility depends upon *N*, the size of the atom. Also $\chi \propto \bar{r}^2$ indicating that outer electron make the largest contribution to the diamagnetic susceptibility.

For normal solids $N = 5\times10^{28}\ m^{-3}$ and $\bar{r}^2 = 10^{-20}\ m^2$; substituting these values in Eqn. 6.2.11, one gets $\chi = Z\times10^{-7}$.

This also shows the temperature invariance of diamagnetic susceptibility. This phenomenon is common to all elements, however, it is too weak to be detected in para and ferromagnetic materials.

6.2.4 LANGEVIN'S THEORY OF PARAMAGNETISM

It is assumed that each atom behaves like a magnet with moment μ or μ_m. There is no interaction between individual magnetic moments, *i.e.*, each moment can interact with the external magnetic field or flux density. In the absence of the field, the magnetic moments of individual atoms or ions are randomly oriented due to *thermal agitation*. Only under the influence of the field will the individual moments tend to align parallel to the field, giving rise to a net magnetic moment. Classical Maxwell-Boltzmann statistics classically describes the potential energy of a magnetic dipole making an angle θ with the magnetic field. The relation is

$$\boxed{W = -\mu_m B\cos\theta} \qquad \text{...(6.2.12)}$$

Now on classical statistics, the number of atoms making an angle between θ and $(\theta + d\theta)$ is

$$dn \propto \exp\,(\mu_m B\cos\theta / k_B T)$$
$$dn = C\exp\,(\mu_m B\cos\theta / k_B T)\sin\theta\, d\theta$$

Let $$\frac{\mu_m B}{k_B T} = a$$

Then $$dn = C \exp(a \cos\theta) \sin\theta \, d\theta$$

Hence the total number of atomic magnets in unit volume of the paramagnetic material is

$$n = \int_0^\pi dn = C\int_0^\pi \exp(a \cos\theta)\sin\theta d\theta$$

Let $$x = \cos\theta$$
$$-dx = \sin\theta \, d\theta$$

or $$n = C\int_1^{-1} \exp(ax)(-dx) = C\int_{-1}^{1} \exp(ax)dx$$

i.e., $$n = C\left[\frac{e^{ax}}{a}\right]_{-1}^{1} ; C = \frac{an}{(e^a - e^{-a})}$$

$$\boxed{C = \frac{an}{(e^a - e^{-a})}} \quad ...(6.2.13)$$

The component of each *dipole moment* parallel to B is $\mu_m \cos\theta$. The total magnetic moment of all the n atoms contained in unit volume of the gas is the magnetization M. It is now given by

$$M = \int_0^\pi \mu_m \cos\theta dn = \int_0^\pi \mu_m \cos\theta C \exp(a\cos\theta)\sin\theta d\theta$$

Let $\cos\theta = x$, then $dx = -\sin\theta d\theta$

Thus $$M = C\mu_m \int_1^{-1} x\exp(ax)(-dx)$$

$$M = C\mu_m \int_{-1}^{1} x\exp(ax)dx - \left[\int_{-1}^{1} \frac{xe^{ax}}{a} d(e^{ax})\right] C\mu_m$$

$$= \frac{C\mu_m}{a}\left\{\left[xe^{ax}\right]_{-1}^{1} - \int_{-1}^{1} e^{ax}dx\right\}$$

$$= \frac{C\mu_m}{a}\left\{(e^a + e^{-a}) - \frac{(e^a - e^{-a})}{a}\right\}$$

Substituting the value of C from Eqn. (6.2.13)

$$M = \frac{an\mu_m}{a(e^a - e^{-a})}\left[(e^a + e^{-a}) - \frac{(e^a - e^{-a})}{a}\right]$$

$$M = n\mu_m\left[\frac{(e^a + e^{-a})}{(e^a - e^{-a})} - \frac{1}{a}\right] = n\mu_m L(a) \quad ...(6.2.14)$$

or

$$M = n\mu_m\left[\coth a - \frac{1}{a}\right] \quad ...(6.2.15)$$

The term $L(a) = \left[\coth a - \frac{1}{a}\right]$ is called the *Langevin function.*

Case 1: At low temperatures or large applied field.

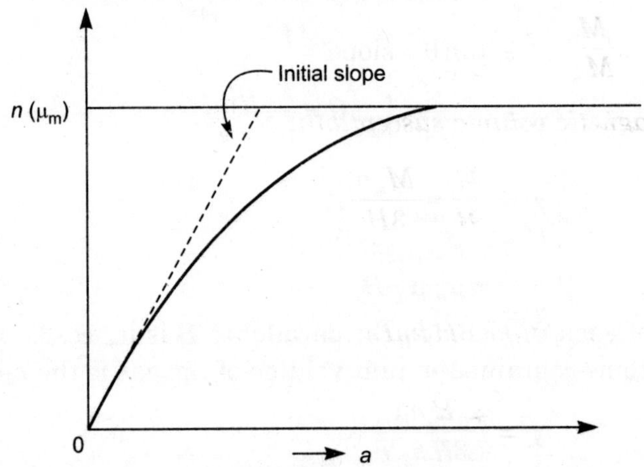

Fig. 6.29 Langevin function

$L(a) \to 1$

Hence, magnetization M in this case will be

$$M = n\,\mu_m = M_s$$

Thus saturation is reached when all the atomic dipoles are parallel to B.

Case 2: Under normal conditions a is small. Then,

$L(a) = \left(\coth a - \frac{1}{a}\right) = \frac{a}{3}$. This is explained below:

Refer Eqn. (6.2.14) we get

$$\left[\frac{(1 + a + \frac{a^2}{!2} + ...) + (1 - a + \frac{a^2}{!2} + ...)}{[1 + a + \frac{a^2}{!2} + \frac{a^3}{6} + ...] - [1 - a + \frac{a^2}{!2} + ...]} - \frac{1}{a}\right]$$

$$= \left[\frac{2 + \frac{2a^2}{2}}{2a + \frac{2a^3}{6}}\right] - \frac{1}{a}$$

$$= \frac{1+\frac{a^2}{2}}{a\left(1+\frac{a^2}{6}\right)} - \frac{1}{a} = \frac{\left(1+\frac{a^2}{2}-1-\frac{a^2}{6}\right)}{a\left(1+\frac{a^2}{6}\right)} \approx \frac{2a^2}{6a}$$

Thus $$L(a) = \frac{a}{3} = \frac{M}{M_S} = \frac{\mu_m B}{M_S} = \frac{\mu_m B}{3k_B T}$$

$$\frac{M}{M_S}\Big/ a = \tan\theta = \text{slope} = \frac{1}{3}$$

Now the *paramagnetic volume susceptibility*

$$\chi = \frac{M}{H} = \frac{M_s a}{3H}$$

$$\chi = \frac{n\mu_m \mu_m B}{3H k_B T}$$

$$\chi = \frac{\mu_0 H n\mu_m{}^2}{3H k_B T}$$

$$\boxed{\chi = \frac{\mu_0 n\mu_m^2}{3k_B T} = \frac{C}{T}} \qquad (6.2.16)$$

where $C = \dfrac{\mu_0 n\mu_m{}^2}{3k_B}$ is called the *Curie constant.*

Failure of Langevin Theory

(*i*) Langevin's theory was unable to explain a more complicated dependence of susceptibility upon temperature exhibited by several paramagnetics such as highly compressed and cooled gases, very concentrated solutions of salts etc.

(*ii*) Langevin's theory could not account for the intimate relation between para and ferromagnetism.

Weiss modification: Langevin's theory applies strictly only to gases, where the molecules are sufficiently far apart for their mutual interactions to be negligible. In liquids and solids such interactions may be large, and many substances obey the modified Curie-Weiss law,

$$\boxed{\chi = \frac{C}{(T-\theta)}} \qquad ...(6.2.17)$$

θ is called the *Curie temperature* and is characteristic of the substance. Eqn. (6.2.17) holds good only at temperatures where $T > \theta$. Eqn. (6.2.17) is of the same as Eqn. (6.2.16), except that the origin of the temperature is shifted from 0 to θ.

6.2.5 FERROMAGNETISM

Ferromagnetism is that characteristic which results in *spontaneous magnetization* or *magnetic alignment* even in the absence of an external magnetic field. This moment is many orders of magnitude larger than that induced in paramagnetic. Further, the induced paramagnetic moment vanishes on removing the external field. Ferromagnetism is due to electron structure within the atoms.

The most important of the ferromagnetic materials are all metallic, and, iron, cobalt, nickel, and gadolinium are the only elements which are ferromagnetic at room temperature. The origin of the force which aligns the moments in metals is beyond our scope, but it is known to arise from the wave mechanical nature of electrons and it is related to the force which aligns the spins of electrons in a partly filled atomic shell.

There is a further peculiarity about metallic bonding. It is found that in the case of iron, cobalt, and nickel the 3d electrons are able to wander through the lattice just like the 4s electrons. One effect of this is to reduce the effective magnetic moment of each electron so that instead of the expected values of the atomic magnetic moment calculated, the measured values are as given below:

Metal	*Number of Bohr magneton per atom (calculated)*	*Number of Bohr magneton per atom (measured)*
Fe	4	2.22
Co	3	1.72
Ni	2	0.61

If one of these elements is bonded ionically or covalently in a compound, however, the moment is not reduced because none of the 3d electrons can move through the lattice.

Ferromagnetism is the existence of magnetization even in the absence of external magnetic field. It is the magnetic analogue of ferroelectricity. Some of the experimentally observed facts are:

1. Magnetization M increases when an external field is applied to the specimen. Moreover, the relationship between magnetization M and applied field H is non-linear and shows *hysteresis effect*.
2. the strong magnetism is found to disappear at a critical temperature T_c, the specimen becomes demagnetized, behaves like a paramagnet and shows magnetic effects only when external field is applied.

In order to explain these experimental facts, Weiss put forward a theory of ferromagnetism based on the following hypothesis.

1. Ferromagnetic materials are made of large number of '*domains*' (each domain contains about 10^9 – 10^{15} atoms), which are spontaneously magnetized. The

magnitude of spontaneous magnetization (refers to a single domain) of the specimen is determined by the vector sum of magnetic moments of the individual domains.

2. Within a given domain, at the dipoles (spins) are aligned in the same direction, so that magnetization is maximum possible value for a given material and temperature.

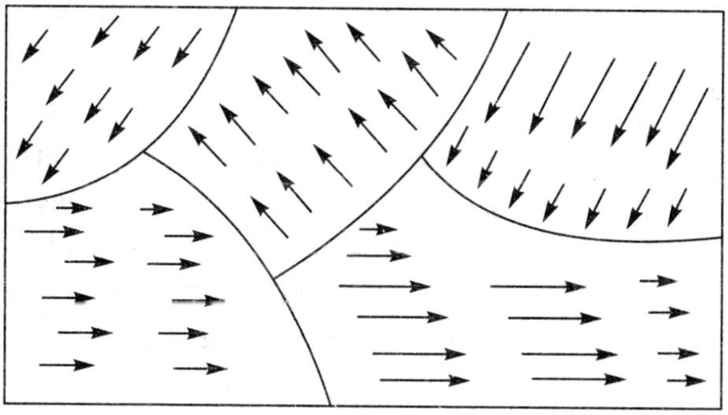

Fig. 6.30 Domains

6.2.6 THE WEISS MOLECULAR FIELD

For explaining the sources of the large magnetization in ferromagnetic materials, *Weiss* in 1907, postulated the existence of an *internal field* proportional to the magnetization in the material. In this classical treatment, hence, the field seen by an atomic dipole is the sum of the applied, B_0 and the internal field B_i. *i.e.*,

$$\boxed{B = B_0 + B_i} \quad ...(6.2.18)$$

with

$$\boxed{B_i = \lambda M} \quad ...(6.2.19)$$

where λ is known as *Weiss constant*. If B is such that $\frac{g\mu_B B}{k_B T}$ is small, then the equation used to compute the susceptibility is

$$\chi_{\text{ferro}} = \frac{\mu_0 M}{B_0} = \left[\frac{\mu_0}{B_0}\right] \frac{Ng^2\mu_B^2 J(J+1)(B_0 + B_i)}{3k_B T}$$

$$\chi_{\text{ferro}} = \frac{\mu_0 N g^2 \mu_B^2 J(J+1) B_0 \left(1 + \dfrac{B_i}{B_0}\right)}{B_0 3k_B T}$$

with $B_i = \lambda M$

$$\chi_{\text{ferro}} = \mu_0 \frac{Ng^2\mu_B^2 J(J+1)\left(1+\dfrac{\lambda M}{B_0}\right)}{3k_B T}$$

$$= \frac{C}{T}\left(1+\frac{\lambda M}{\mu_0 H}\right)$$

or
$$\chi_{\text{ferro}} = \frac{C}{T}\left(1+\frac{\lambda}{\mu_0}\chi_{\text{ferro}}\right)$$

$$\chi_{\text{ferro}} - \frac{C\lambda\chi_{\text{ferro}}}{T\mu_0} = \frac{C}{T}$$

$$\chi_{\text{ferro}}\left(1-\frac{C\lambda}{T\mu_0}\right) = \frac{C}{T}$$

$$\chi_{\text{ferro}} = \frac{C}{\left(T-\dfrac{C\lambda}{\mu_0}\right)} \qquad \text{...(6.2.20)}$$

$$\chi_{\text{ferro}} = \frac{C}{(T-\theta)}$$

with $\theta = \dfrac{C\lambda}{\mu_0}$

Here θ is the *Curie temperature* at which the susceptibility tends to ∞. This means that M has a finite value when B_0 is zero which is an evidence of spontaneous magnetization. In a similar way we write the equation for magnetization as

$$M = Ng\mu_B J B_J\left[\frac{Ng\mu_B J}{Nk_B T}(B_0+\lambda M)\right]$$

$$= M_s(0)B_J\left[\frac{M_s(0)}{Nk_B T}(B_0+\lambda M)\right] \qquad \text{...(6.2.21)}$$

where $M_s(0) = Ng\mu_B J$ represents the maximum value of magnetization at 0 K.

When there is no magnetic field, $B_0 = 0$

$$M = M_s(0) B_J \left[\frac{M_s(0)}{N k_B T} \lambda M \right] = M_s(0) B_J(y) \quad \text{...(6.2.22)}$$

with

$$y = \frac{g\mu_B J B}{k_B T} = \frac{g\mu_B J}{k_B T}(B_0 + \lambda M) \quad \text{...(6.2.23)}$$

In case of no magnetic field, $B_0 = 0$ and hence

$$M = \frac{k_B T y}{g\mu_\beta J\lambda} \quad \text{...(6.2.24)}$$

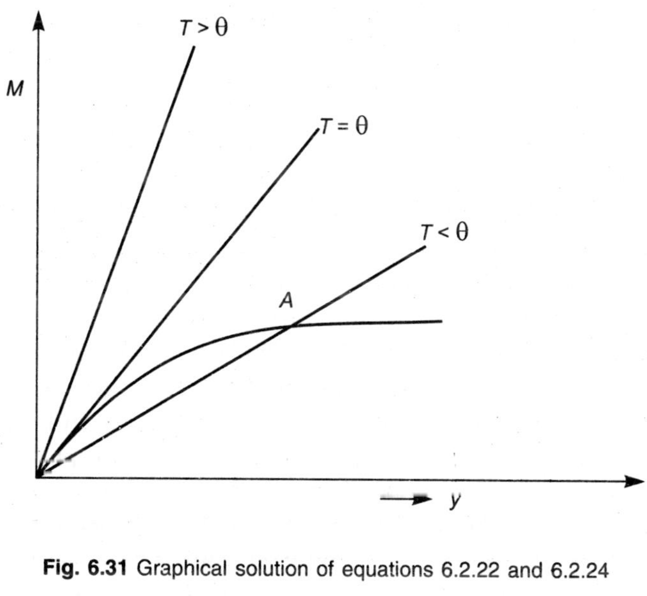

Fig. 6.31 Graphical solution of equations 6.2.22 and 6.2.24

Since M must satisfy Eqns. (6.2.22) and (6.2.24), the two M versus y curves are plotted in Fig. 6.31. Equation (6.2.24) is a straight line between M and y and for various temperatures. *i.e.*, $T < \theta$, $T = \theta$ and $T > \theta$ is shown in Fig. 6.31. The Eqn. (6.2.22) represents a curve which intersects the straight line for $T < \theta$ at the point A. This gives a non-vanishing value of M even if the external field B_0 is zero. This spontaneous magnetization below the *Curie temperature* is shown in Fig. 6.32 where the magnetization decreases from a saturated value M_s at $T = 0$ to zero at $T = \theta$. At $T = \theta$, the straight line given by equation (6.2.24) is the tangent to the curve given by Eqn. (6.2.22) at the origin. Thus there is no spontaneous magnetization for $T \geq \theta$.

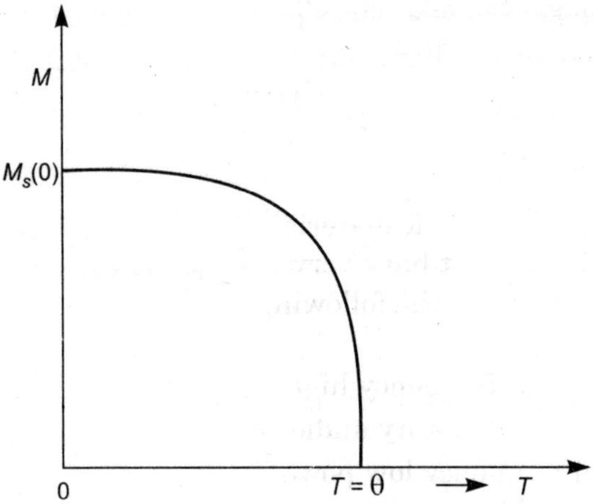

Fig. 6.32 Spontaneous magnetization below Curie temperature

6.2.7 HARD AND SOFT MAGNETIC MATERIALS

Hard Magnetic Materials

The description hard as applied to magnetic materials is an indication that the cohesive force is high generally above 10^4 A/m. If, in addition the *remanent flux density* is high the material is suitable as a permanent magnet. In a *polycrystalline material* the grains are randomly oriented. In a randomly oriented array only few grains will be favourably oriented in the direction of easy magnetization and so the process of magnetization is slowed down. Further *magnetostriction effects* prevent the easy movement of the grains thereby affecting magnetization in a complicated way. When a permanent magnet is required you must look for materials with the highest possible saturation magnetization, remanance and coercive force. This means that the hysteresis loop must have a large area. When the area of the hysteresis loop is large there is little loss of induced magnetism when the applied field is removed. Again you must look for a material of large *magnetostrictive coefficient*, since large magnetostriction gives rise to large *coercive force*. When magnetostriction is large, it is more difficult to change the direction of magnetization once it has reached saturation in a particular direction, since any further rotation involves expenditure of work against elastic restoring forces. Another useful criterion is the energy product given by BH, since the useful energy which a permanent magnet can deliver is determined by the BH product. Larger area for the hysteresis loop takes care of this criterion. The most important commercial hard materials are the Alnico type alloys. A steady improvement in fabrication has been achieved in Alnico alloys and Alnico V introduced in the year 1940 has very many interesting properties. Its composition is 8% Al, 14% Ni, 24% Co, 51% Fe and 3% Cu. It has a coercive field of 50000 ampere turn and a BH product of 45000 J/m^3. The alloy is cast and solidified directionally along the direction of easy magnetization [100]. The alloy is annealed at an elevated temperature of 1300°C for a prolonged period during which a new phase precipitates out of the grains and the material is mechanically hardened. This is called *precipitation hardening* and it increases the *coercive force.*

Soft Magnetic Materials

For transformer cores a ferromagnetic material having a small hysteresis loop is required. This means that the material must have very low magnetostrictive coefficient so that the material is practically stress free. The following materials are found to be suitable for the different frequencies.

1. Fe – Si alloys for low frequency high powr applications
2. Fe – Ni alloys for high quality audio transformers
3. Ferrites for high frequency low power applications

The addition of Si to Fe improves its magnetic quality in two ways (1) by increasing the resistivity and thereby reducing the *eddy current* losses and (2) by decreasing the hysteresis losses. An alloy with 2% Si is used for ordinary quality induction motors. Another alloy with high quality performance contains 4% Si. The rolling and annealing are carefully controlled so that the crystallites get oriented in the preferred direction. In that case the domains are large and regular in shape. The direction of easy magnetization also lies in the direction of rolling. The Ni – Fe alloy which goes under the commercial name *Hypernik* is made of two different compositions. One grade contains 50% Ni and 50% Fe. Another grade called the *permalloy* contains 78% Ni and 32% Fe. Both have high permeabilities and low hysteresis losses. The 78% permalloy has in addition zero *magnetostrictive coefficient*. *Suppermally* is made of 78% Ni, 5% Mo and 17% Fe. This alloy has higher resistivity and hence lower eddy current losses.

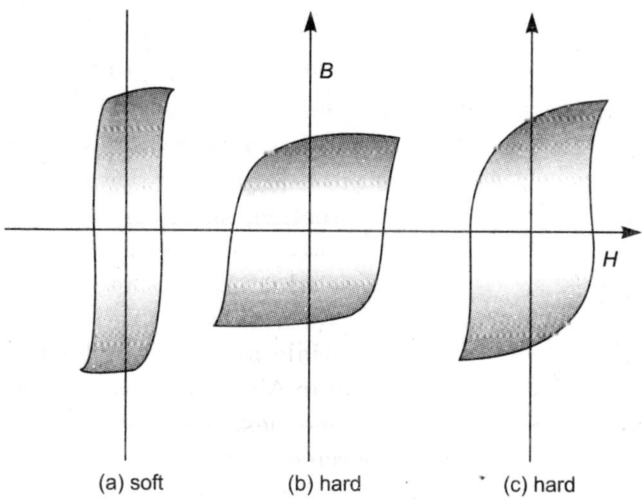

Fig. 6.33 Typical hysteresis (*a*) small coercive force and large B (*b*) large coercive force and small B, and (*c*) large coercive force and large B

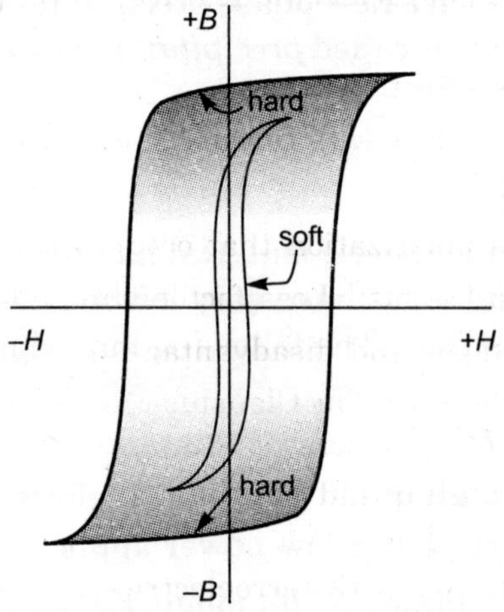

Fig. 6.34. Magnetization curves for hard and soft magnetic materials

SHORT QUESTIONS

6.1.1* Give the salient features of dielectrics.

6.1.2* Write a small note on piezoelectric effect.

6.1.3 Your Master likes to have an extract of Gauss law in electrostatics. Oblige him with a few equations. *(University of Groningen, Netherlands, U.G., 1992)*

6.1.4 Get the relation $(\in_r - 1) = \dfrac{\vec{P}}{\vec{E} \in_o} = \chi$. The symbols denote the usual meanings.

(JNTU, B. Tech., April 2002, Anna U. BE, April 1997)

6.1.5 Can you explain orientation polarization in liquids and solids?

6.1.6 Draw the Langevin curve for polar substance. Discuss its behaviour at low and high temperatures. *(M.K.U., BE, April 1992, AU, BE, November 2000)*

6.1.7 Explain local electric field in solid dielectrics. *(Anna U. BE, May 1992)*

6.1.8 Suggest a method to evaluate the permanent dipole moment of a polar molecule. *(I.I.T. , Delhi, B. Tech, April 2000)*

6.1.9 Sort out some of the important uses of dielectrics.

(Metallurgy division, University of Manchester, UK, B. Tech, April 1983, IIT Bombay, May 1992).

6.1.10* Help me to write a paragraph on ionic polarizability.

6.1.11 Write a small note on dielectric strength.

(MU, BE, December 1996)

6.1.12 Explain power loss and dissipation of energy in dielectrics.

(VTU, BE, July 2004)

6.1.13 Explain some of the characteristic features of ferroelectric materials.
(Anna U., BE, April 1979)

6.1.14* Explain pyroelectric effect.

6.1.15 What is dielectric break down? Discuss the reasons of their breakdown.
(M.K.U., BE, April 1997)

6.1.16 Explain the types of polarization that occurs in dielectrics.

6.1.17 Explain the following terms: Loss factor, dissipation factor and loss tangent.

6.1.18* Discuss the advantages and disadvantages of piezoelectric devices/transducers.

6.1.19 State and explain the terms in Clausius-Mossotti relation.
(VTU, BE (NR), April 2003)

6.1.20 Distinguish between polar and non-polar substances with suitable example.
(JNTU, BE, April 2002)

6.1.21* Compare piezoelectricity with ferroelectricity.

6.1.22* Write a note on dielectric loss angle (tanδ).

6.2.1 What are the basic characteristics of paramagnetic substances?
(University of Surrey, UK, B. Tech., April 1990, J.N.T.U., BE, 2000)

6.2.2 Explain the origin of ferromagnetic domains.
(REC Kurukshetra, BE, November 1999, M.K.U. BE, April 1995)

6.2.3 Distinguish between hard and soft magnetic materials
(Anna U., BE, November 2000 JNTU, BE, April 2001)

6.2.4 How does magnetic susceptibility vary with temperature in the case of anti ferromagnetic materials?
(University of Maryland, Mid summer B. Tech., June 1978, Anna U. BE, April 1995)

6.2.5 What are hard ferrites and soft ferrites? Indicate the specific applications of each.
(JNTU, BE, April 2002)

6.2.6* Explain the significance of internal field and Weiss constant.
(Anna U., BE, November 1998)

6.2.7 Discuss the origin of atomic magnetism.
(Imperial College of Science and Technology UK, B. Tech Winter 1979, JNTU BE, April 2000)

6.2.8 Write a note on ferrires. *(Anna U., BE, November 1999)*

6.2.9 How ferrites are produced?

6.2.10 Describe the composition and properties of ferrites.
(University of Minnesota, U.S, B. Tech., November 1989, A U., BE, April 1997)

6.2.11 Discuss magnetic anisotropy and magnetostriction.
(I.I.T. Bombay, B.Tech., April 1990)

6.2.12 What are the various uses of ferrites?

6.2.13 What is meant by hysteresis in magnetic materials? *(Anna U, BE, November 1997)*

6.2.14 Point out some of the industrial application of ferrites.
(REC, Durgapur, BE, April 1999)

6.2.15 What are hard and soft magnetic materials? *(REC, Calicut BE, April 1997)*

6.2.16 What are the distinguishing features of ferromagnetism?

6.2.17* Explain what is meant by magnetocrystallnic anisotropy.
(REC, Jamshedpur BE, April 2000 JNTU, BE, November 2001)

6.2.18* Explain the variation of magnetic susceptibility with temperature in the main types of magnetic materials. *(JNTU, BE, April 2001)*

6.2.19* In the table below, which of the elements is given the wrong value of magnetic moment in the solid state. *(M.K.U., BE, April 1996)*

	A	*B*	*C*	*D*	*E*	*F*	*G*
Element	Ca	V	Cr	Mn	Fe	Ni	Cu
No. of 3d electrons	1	3	5	5	6	8	10
Atomic magnetic moment in Bohr magneton	1	3	4	6	4	2	1

6.2.20 Discuss the variation of magnetization with applied field in a ferromagnetic material.

6.2.21* Give a graphical representation of the susceptibility with temperature.

6.2.22* Write a note on ferrites in computer memory devices.

Answers to the starred questions

6.1.1*

- Dielectrics as distinct from conductors have no free electrons. Hence when a source of emf is connected across a dielectric no current flows. However, since no dielectric is perfect it contains a small number of free electrons and very small current flows through it when electric field is applied. Capacitors therefore have a small leakage conductance.
- Dielectric materials are essentially insulating materials. However, while the function of an insulating material is to obstruct the flow of current, the function of dielectric material is to store electrical energy. This difference of function demands that the material to be used as dielectric must have properties some of which may not be answered by a material ordinarily used as an insulating material.
- The behaviour of real dielectric is primarily electrostatic; electrodynamic effects, that also take place due to their non-zero conductivity, are quite insignificant in normal operating conditions. In all kinds of electrical devices and circuits, dielectrics serve to insulate one current-carrying part from another whenever they operate with a difference in electrical potential relative to each other. In capacitors dielectrics serve to provide some needed value of capacitance

  The dielectrics used in electrical engineering may be (*i*) solids, (*ii*) liquids, (*iii*) gases.

6.1.2* Some crystalline solids exhibit electric polarization when strained elastically. This is piezoelectric effect in which a local strain gives rise to a local electric field. The converse effect of piezoelectricity effect is that when such crystalline solids are subjected to an external electric field along specific orientations, then these solids get strained. *i.e.*, the material expands or contracts as shown in the figure.

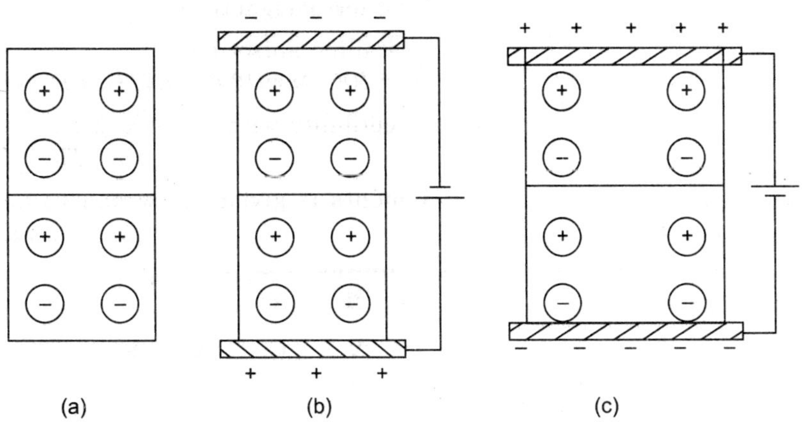

Dimensional changes in piezoelectric material

When the field is applied as in Fig.(c), the positive charged ions are repelled by positive electrode of electric field, and crystal thus contracts. When the field direction is reversed, the crystal expands or elongates as in Fig.(b). Now if alternate electric field is applied, the crystal vibrates (alternately expands and contracts). The frequency of vibration depends upon the voltage difference. This electric energy is changed to mechanical energy and such devices, which changes electrical energy to mechanical energy are called *transducers*. They are used to produce high frequency sound waves. If pressure is applied on ceramic crystal as SiO_2, $PbZrO_3$ or $BaTiO_3$; the equilibrium position of electric dipole changes, consequently the voltage across the crystal changes. Thus if the crystal is subjected to variation or some type of alternate mechanical movements, the vibrational energy is transformed to electrical potential. The devices which change mechanical energy to electrical energy are called piezoelectrics.

6.1.10* Ionic polarization is induced in those materials which are ionically bonded. The negative ions are attracted towards the positive side of the field while the positive ions are attracted towards the negative side. This situation is depicted in the figure. The relative displacement of the negative and positive charges induces the polarization in the material. Ionic polarizability is about ten times smaller than electronic polarizability.

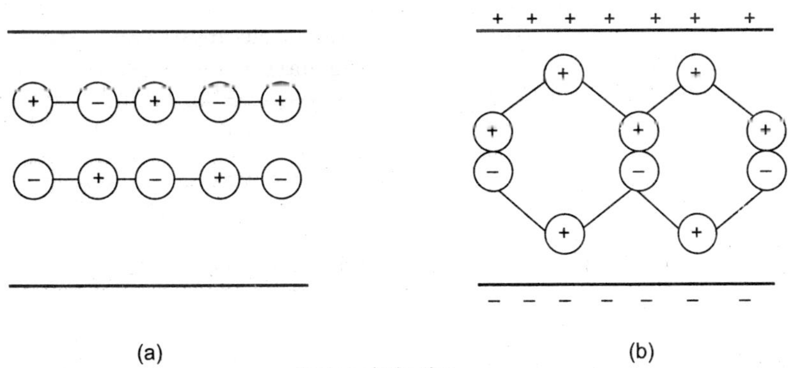

Ionic polarization

6.1.11* The pyroelectric effect, as the name implies, is connected with heat. It was known, as long ago as the 17th century, that natural crystal tourmaline acquired an electric charge when heated, and such crystals are named pyroelectric. It is known that out of the 20 commonly available piezoelectric crystal classes ten are pyroelectric and clearly the effect must be also associated with low symmetry. In fact they are crystals in which there is *spontaneous polarization*. This arises when, because of the low symmetry, the centres of gravity of the positive and negative charges in the unit cell are separated, giving a permanent dipole moment, and, in addition, all the individual dipoles are aligned such that the overall dipole moment is non-zero for the whole, macroscopic crystal.

Tourmaline and polyvinylidene fluoride are pyroelectric materials. They are used:

(*i*) as detectors for faulty system traps in industry

(*ii*) as infrared detector

(*iii*) for building insulation

6.1.18*

Advantages	*Disadvantages*	*Application*
1. High frequency response	1. Output affected by changes in temperature	1. Accelerometer
2. Small size	2. Cannot measure static condition	2. Pressure cells
3. High output		3. Force cells
4. Rugged construction		4. Ceramic microphones
5. Negligible phase shift		5. Phonograph pick-up
		6. Catridges
		7. Industrial cleansing apparatus
		8. Under-water detection system

6.1.19*

Sl No.	*Piezoelectricity*	*Ferro electricity*
1.	The crystal is polarized by the application of an external stress	The source of polarization is the dipole interaction energy itself
2.	The phenomenon occurs in non-centro symmetrical crystals, occurs in almost all the 20 crystals	The phenomenon occurs in non-centrosymmtrical crystals, it occurs only in 10 crystals, namely, those which provide a favourable axis of polarity
3.	All piezoelectric crystals are not surely ferroelectrics (*e.g.* Tourmaline is piezoelectric, but not ferroelectric at all)	All ferroelectrics are piezoelectrics

6.1.20* Under alternating field across dielectrics the electrical energy is absorbed by the material and is dissipated in the form of heat. The energy absorbed per unit of time that is power is called dielectric loss. If this loss were not present voltage and current in a dielectrics will be at 90° in phase difference. Due to the loss in the material the phase angle is not exactly 90° but at an angle ϕ. The complementary angle $(90 - \phi) = \delta$ is called the dielectric loss angle. The loss angle δ or $\tan\delta$ is a very important parameter for insulating materials.

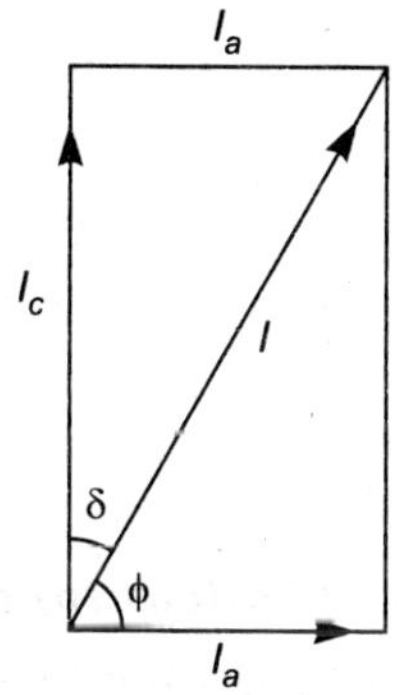

$$\tan\delta = \frac{I_a}{I_e}$$

6.2.6* Weiss suggested that in a ferromagnetic material, an internal, or molecular, field acting on the atomic moments to produce alignment even without an applied field. This internal magnetic field H_i (Weiss field) may be estimated as follows:

Beyond the Curie temperature, a ferromagnet becomes a permanent magnet and follows the Curie-Weiss law. Therefore, the thermal energy just balances the magnetic potential energy at $T = T_c$. Thus

$$\mu_B H_i = k_B T_c$$

For iron

$$T_c = 1000K$$
$$H_i = 10^9\ A/m$$

Weiss assumed that this internal field is directly proportional to the magnetization

i.e., $H_i = \lambda M$

where λ is the so callod molecular field constant. Since the units of H_i and M are difforont, we shall prefer the following expression to the above one

$$H_i = \frac{\lambda M}{\mu_0}$$

so that λ will be a dimensionless constant.

6.2.17* Magnetization vector preferentially lies in one or more specific crystallographic directions. It is true from the figure that much higher fields are required to produce magnetic saturation in the [1 1 1] direction than need to be applied in [1 0 0] direction. In other words, higher fields are required to twist the saturation magnetization into [1 1 1] direction from the [1 0 0] direction. Also it is easy to magnetize the iron along [100] direction than along [111] direction; for this reason [1 0 0] and [1 1 1] directions are referred to as easy and hard directions of magnetization respectively. The excess energy required to magnetize the specimen in a given direction over that required for the easy direction is called the anisotropy energy.

Example: An order 10^5 ergg/cm^2 is required to magnetize iron in its hardest [1 1 1] direction.

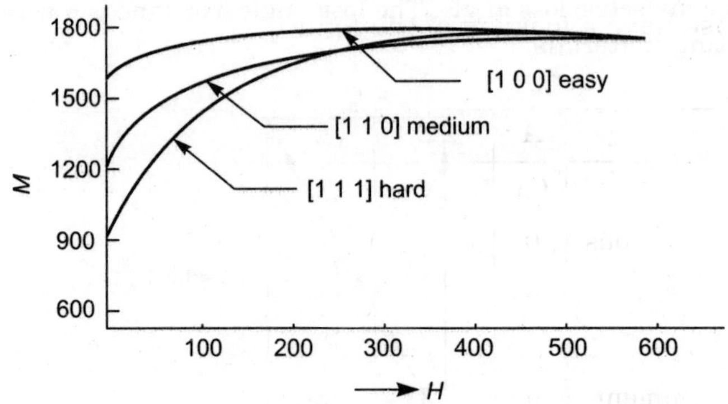

6.2.18* The energy of interaction making adjacent dipoles line up is called the exchange energy. The energy resulting from the dipole moment of the material is called the magnetostatic energy. The variations of susceptibility of the ferromagnetic material is shown in the first figure.

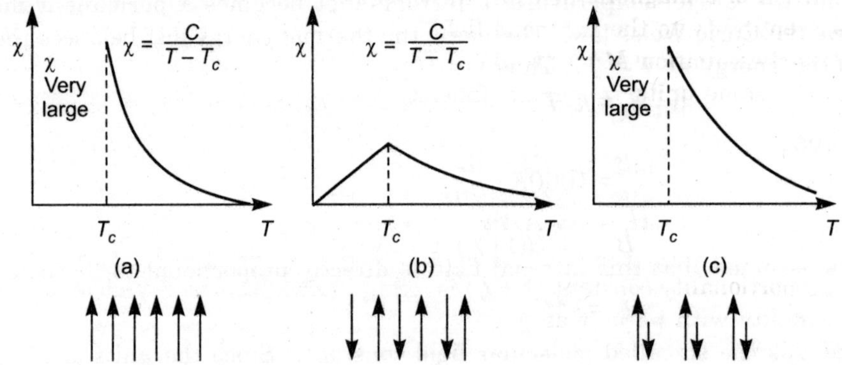

Magnetic susceptibility *versus* temperature for (*a*) ferromagnetic, (*b*) antiferromagnetic and (*c*) ferrimagnetic materials with magnetic moment alignments indicated for each case

The exchange energy some times has the opposite effect to ferromagnetism so that the spins align opposite to each other rather than in the same direction, this phenomenon is called antiferromagnetism. The susceptibility in antiferromagnetism is positive and increases with temperature as against the susceptibility in the ferromagnetism which decreases with temperature. However, the increase in susceptibility of anti ferromagnetic materials reaches a peak at a temperature and then begins to decrease. Thus thermal agitation disrupts the anti parallel moment arrangements and makes the line up with the field. The peak temperature in the susceptibility curve is called the Neel temperature and corresponds to Curie temperature in ferromagnetic materials.

There is yet another class of materials known as spinals which exhibit ferromagnetic interaction and yet have anti parallel spins. However, in these materials the nearest neighbours have opposing spins but their moments are not equal. This phenomenon is known as ferrimagnetism and consequently results in a net moment. The spinals which are ferromagnetic materials are complex oxides having the composition $MOFe_2O_3$ where M is a metal. Please have a lookism of the figure.

6.2.19*

	A	*B*	*C*	*D*	*E*	*F*	*G*
Element	Ca	V	Cr	Mn	Fe	Ni	Cu
No. of 3d electrons	0	3	5	5	6	8	10
Spin directions	0	↑↑↑	↑↑↑↑↑	↑↑↑↑↑	↑↑↑↑↑ ↓	↑↑↑↑↑ ↓↓↓	↑↑↑↑↑ ↓↓↓↓↓
Magnetic moment in Bohr magneton	0	3	5	5	4	2	0

The values for *A*, *C*, *D* and *G* are wrong

6.2.20* In practice a piece of iron is generally unmagnetized in the natural state. On application of a magnetic field, *H*, it, becomes magnetized to an extent determined by its susceptibility to the magnetic field. The intensity of magnetization, *M*, is related to *H* by the equation $M = \chi H$, where χ is the magnetic susceptibility. Since *M* and *H* share the same units, χ has no dimensions-it is just a number. The general relation is,

$$B = \mu_0 (H + M)$$

$$B = \mu_0 (H + \chi H)$$

or

$$B = \mu_0 H(1+\chi)$$

The proportionality constant $(1 + \chi)$ is often lumped into one symbol: μ_r – the relative permeability with no unit as χ.

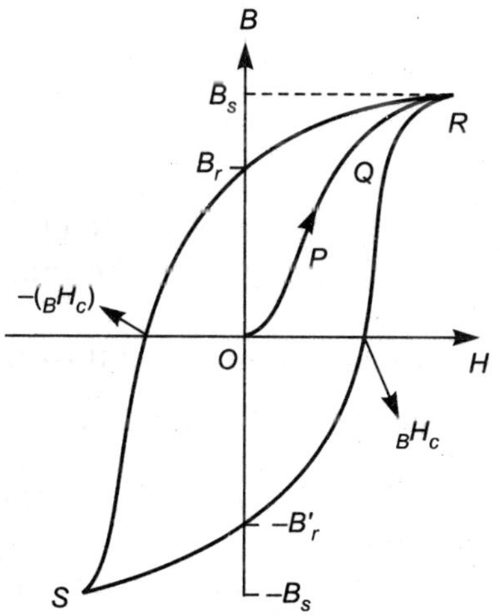

Starting with the iron unmagnetized, we apply an increasing field and measure the flux density produced in the material. The flux B rises along the line O –P–Q –R, reaching a saturation value, B_{sat}, beyond which any increase in the field has a negligible effect on B. If now the field is reduced again, B does not follow the same path but decreases more slowly along the line R–B_r; reversing the field and increasing it in the opposite direction causes the flux to follow the line $B_r - (-H_c) - S$ at which point the iron is again saturated but with the magnetization in the opposite direction.

Reduction and re-reversal of the field carries the flux along the path $S - B_r' - H_c - R$ and finally closes the loop. The closed path so described is called a hysteresis loop and it shows that when the field is removed after reaching the point R, the material retains a remanent flux density, B_r, and behaves as a permanent magnet. In order to destroy the magnetization, it is then necessary to apply a reversed field equal to the coercive force H_c. The value of B_r and H_c may vary widely from material to material.

6.2.21* Diamagnetism is present in all substances but is very weak so that it is not normally observed if other effects are present. It is produced by circulation of electrons in an atom or molecule. Atoms or molecules with closed shells of electrons are diamagnetic. Unpaired electrons, however, give rise to paramagnetism. Simple paramagnetic behaviour is found for substances such as liquid oxygen or transition metal complexes in which the unpaired electrons on different centres are isolated from each other. In a magnetic field the magnetic moments on different centres tend to align with the field and hence with each other, but this is opposed by the randomizing effect of thermal energy and in the absence of a field, the unpaired electrons on different centres are aligned randomly. The interplay of applied field and thermal randomization leads to the temperature dependence described by the Curie law : $\chi = \dfrac{C}{T}$

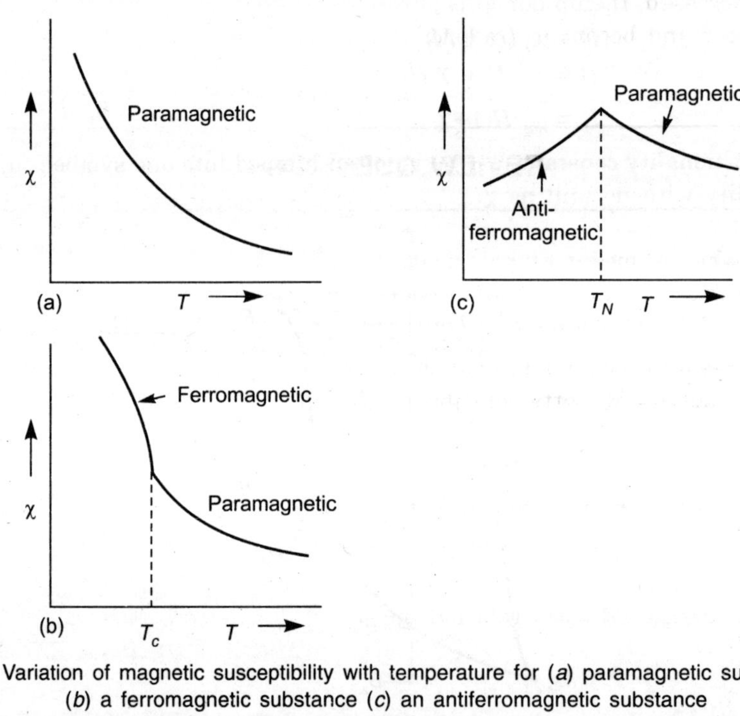

Fig. Variation of magnetic susceptibility with temperature for (*a*) paramagnetic substance (*b*) a ferromagnetic substance (*c*) an antiferromagnetic substance

where χ is the magnetic susceptibility, C is a constant known as the Curie constant and T is the temperature in kelvin.

Different temperature dependence is observed when there is co-operative behaviour. The change over from independent to co-operative behaviour is associated with a characteristic temperature. For ferromagnetism, the Curie law becomes

$$\chi = \frac{C}{(T - T_c)}$$

where T_c is the Curie temperature. For antiferromagnetism, the temperature dependence is of the form

$$\chi = \frac{C}{(T + T_N)}$$

where T_N is the Neel temperature. These two behaviours are illustrated in the figure.

6.2.22* In early computers, ferrites were employed as memory devices. For this purpose, they needed to switch cleanly between two different states which are read as 0 and 1 and to stay in one state until switched to the other. An almost square hysteresis possessed by some ferrites, fulfils these requirements.

More recently, magnetic bubble devices have been used. These consist of very thin films of magnetic material which are manufactured to be magnetically anisotropic. *i.e.*, their magnetization is not the same in different directions. In the absence of an applied field, magnetic domains magnetized up and down with respect to an axis at right angles to the film are formed. When a magnetic field say, is applied in the down direction, the up domains will shrink and the down domains grow. As the applied field is increased, the up domains shrink further until they are only a few micrometre in diameter and become cylinderical.

REVIEW QUESTIONS

6.1.1 Explain what is meant by polarization in dielectric. Arrive at the relation between the dielectric constant and atomic polarizability of an isotropic solid

(Bangalore University, BE, April 1992, M. K. U., BE, June 1995)

6.1.2 What is meant by local field in a dielectric and how is it calculated for a cubic structure? Deduce Clausius Mosotty relation and explain its use in predicting the dielectric constant of solids.

(University of Groningen, Netherlands, UG, Winter 2000)

6.1.3 Develop the theory for obtaining the expression for orientational polarizability. Mention the approximations made.

(JNTU, BE, March 2002)

6.1.4 Deduce Clausius Mosotty relation and point out its importance.

(M.K.U, Part-time BE, September 1995)

6.1.5 Discuss Lorentz method of evaluating the local field for a cubic structure. How the permanent dipole moment of a solid dielectric is determined?

6.1.6 What are ferroelectric materials? Give the dipole theory relating to ferro-electrics.
(Anna U, BE, (Five Year) April 1980)

6.1.7 Explain how the dielectric constant of a ferroelectric crystal varies with temperature. Name two ferro electric materials. Mention the applications of ferroelectric materials.
(Anna U, M.Sc, April 1997)

6.1.8 Discuss in detail the theory of dielectrics in alternating fields. Obtain an expression for the dielectric constant in terms of the frequency and relaxation time.
(VTU, B.Tech, April 2002, IIT Delhi, BE, November 1995)

6.1.9 Define spontaneous polarization and coercive field. What are the distinguishing features of piezoelectric and ferroelectric crystals. Explain with suitable examples.
(NIT Calicut, M.Tech (Mech), April 2002).

6.1.10 What are the different sources of polarizability in dielectrics? Obtain Clausius - Mosotti equation and explain its significance.
(JNTU, BE, December 2002)

6.1.11 What is orientational polarization? Derive an expression for the mean dipole moment when a polar material is subjected to an external field.

The relative dielectric constant of sulphur is 3.75 when measured at 27°C. Assuming the internal field constant $\gamma = \frac{1}{3}$, calculate the electronic polarizability of sulphur if the density at this temperature is 2050 kg/m^3. The atomic weight of sulphur is 32.
(JNTU, BE, April 2002)

6.1.12 Explain clearly orientational polarization under static electric fields. How does it vary with temperature. Derive Clausius-Mosotti equation
(NIT, Calicut, BE, April 2004)

6.1.13 What do you mean by the polarization of a solid? Derive an expression for the local field that is responsible for polarizing molecules or atoms of a solid.
(Annamalai U, M.Sc, April 1990)

6.1.14 Discuss the frequency dependence of the dielectric constant of a dipolar system. Explain dielectric relaxation and dielectric loss in solids
(Anna U, BE, April 1982)

6.1.15 What are ferroelectric substances? Give the theory of ferroelectricity and mention their applications.

6.2.1 State the criteria for iron to be ferromagnetic. What are hard and soft materials? Give the structure of ferrites.
(IIT, Delhi, B. Tech., April 1992, Anna U, BE, November 1993)

6.2.2 What is antiferromagnetism? Describe the structure of ferrites. Mention the uses of ferrites.
(Imperical College of Science and Technology, U.K, B. Tech., Winter 1987, JNTU BE, April 1992)

6.2.3 Explain the properties of ferromagnetic materials. Discuss the atomic interpretation of spontaneous magnetization and of the Currie-Weiss law.
(M.K.U, Part-time BE, November 1978, Anna U, BE, April 1995)

6.2.4 What are ferrites? Describe their composition and properties. State two applications of ferrites and describe one of them in detail.

6.2.5 What are ferrites? Explain their structure. How they can be produced? What are their uses? *(University of Minnesota, U.S.A, BE, April 1988, JNTU BE, March 2001)*

6.2.6 Describe with examples the different types of magnetic substances. What type of magnetism is exhibited by ferrites? Mention a few industrial applications of ferrites *(IIT Bombay, B. Tech., September 1970, Anna U, BE, April 1990)*

6.2.7 What are the basic characteristics of paramagnetic substances? What are magnetic domains?

6.2.8 Derive the expression for paramagnetic susceptibility based on Langevins's theory. *(M.K.U, BE, April 1990, Anna U., BE, September 1996)*

6.2.9 Explain clearly diamagnetism, paramagnetism and ferromagnetism. Explain the significance of Curie-Weiss law. *(REC, Bhopal BE, May 1992, JNTU BE, September 2000)*

6.2.10 Discuss Weiss theory of ferromagnetism based on molecular field theory. Discuss the merit and demerit of this theory. *(The Open University, U.K Winter 1992, Anna U., BE, May 1993)*

6.2.11 Explain the origin of atomic magnetism. Discuss Weiss theory of ferromagnetism. *(IIT Bombay, B. Tech., April 1993)*

6.2.12 What is hard ferrites and soft ferrites? Indicate the specific application of each. Explain the principle of formation of domains in a ferromagnetic material. *(Anna U., BE, November 1993)*

PROBLEMS AND SOLUTIONS

6.1.1 A parallel plate capacitor consists of free space and two dielectrics of relative dielectric constants $\in_r = 2$ and $\in_r = 3$ respectively. If the plates of the capacitor have a charge density $\sigma_s = 5 \times 10^{-6}$ C/m^2, find the electric displacement, electric field and electric polarization in the three regions *(AMIE, November 1990, Anna U. BE, April 1992)*

Solution:

1. *Free space*

(*a*) electric displacement, $D_1 = \sigma_s = 5\times10^{-6}$ C/m^2

(*b*) electric field, $E_1 = \dfrac{\sigma_s}{\in_o} = \dfrac{5\times10^{-6}}{8.85\times10^{-12}} = 0.56\times10^{6}$ V/m

(*c*) electric polarization, P_1 – zero

2. *First dielectric*

(*a*) $D_2 = \sigma_s = 5\times10^{-6}$ C/m^2

(*b*) electric field, $E_2 = \frac{D_2}{\epsilon_o \epsilon_2} = \frac{5\times10^{-6}}{8.85\times10^{-12}\times2} = 0.28\times10^6$ V/m

(*c*) $P_2 = \epsilon_o (\epsilon_r - 1) E_2 = 8.85\times10^{-12}\times(2-1)\times0.28\times10^6 = 2.48\times10^{-6}$ C/m^3

3. *Second dielectric*

(*a*) $D_3 = 5\times10^{-6}$ C/m^2

(*b*) $E_3 = \frac{5\times10^{-6}}{8.85\times10^{-12}\times3} = 0.188\times10^6$ V/m

(*c*) $P_3 = E_3 \epsilon_o (\epsilon_r - 1) = 8.854\times10^{-12}\times(3-1)\times0.188\times10^6$

$P_3 = 3.3\times10^{-6}$ C/m^2

Conclusions:

The electric displacement remains the same while the electric field goes on decreasing and the electric polarization goes on increasing along free space to dielectric with $\epsilon_r = 2$ to that with $\epsilon_r = 3$

6.1.2 An electric field of 10^5 V/m is applied on a sample of neon at NTP. Calculate the dipole moment induced in each atom. Given dielectric constant for neon = 1.000134. Also find the atomic polarizability of neon gas

(University of Maryland, U.S, U.G. Winter 1989, JNTU BE, April 2003)

Solution:

The general equation is

$$P = E\epsilon_o (\epsilon_r - 1)$$

Also

$$P = n\mu$$

i.e.,

$$\mu = \frac{E \epsilon_o (\epsilon_r - 1)}{n}$$

$$n = \frac{6.02\times10^{26}}{22.4} = 2.69\times10^{25}/\text{m}^3$$

Thus

$$\mu = \frac{10^5\times8.85\times10^{-12}\times0.000134}{2.69\times10^{25}}$$

i.e.,

$$\boxed{\mu = 4.4\times10^{-36}\ \text{Cm}}$$ ***Answer***

Atomic polarizability,

$$\alpha = \frac{\mu}{E} = \frac{4.4\times10^{-36}}{10^5}$$

or

$$\boxed{\alpha = 4.4\times10^{-41}\,\text{C m}^2/\text{V}}$$ ***Answer***

6.1.3 The polarizability of NH_3 molecule in gaseous state, from the measurement of dielectric constant, is found to be 2.5×10^{-39} C^2 m/N and 2×10^{-39} C^2 m/N at temperatures 300 K and 400 K respectively. Calculate the contribution to the polarizability because of deformation of molecules and the contribution because of permanent dipole moment at each temperatures.

Solution:

From Langevin-Debye's theory the total polarizability is

$$\alpha = \alpha_1 + \frac{\mu^2}{3k_BT} = \alpha_1 + \frac{\beta}{T}$$

i.e.,

$$2.5\times10^{-39} = \alpha_1 + \frac{\beta}{300}$$

and

$$2\times10^{-39} = \alpha_1 + \frac{\beta}{400}$$

$$\frac{\beta}{300} - \frac{\beta}{400} = 0.5\times10^{-39}$$

$$\frac{100\beta}{300\times400} = 0.5\times10^{-39}$$

$$\beta = 1200\times0.5\times10^{-39} = 6\times10^{-37}$$

$$\alpha_1 = \alpha - \frac{\beta}{300} = 2.5\times10^{-39} - \frac{6\times10^{-37}}{300}$$

$$\boxed{\alpha_1 = 0.5\times10^{-39}\,\text{C}^2\text{m/N}}$$ ***Answer***

Polarizability at 300 K

$$\alpha_o = \frac{\mu_o^2}{3k_BT} = \frac{\beta}{T} = \frac{\beta}{300}$$

$$\alpha_o = \frac{6\times10^{-37}}{300}$$

$$\boxed{\alpha_o = 2\times10^{-39}\,\text{C}^2\text{m/N}}$$ ***Answer***

at 400 K

$$\alpha_o = \frac{\beta}{400} = \frac{6\times10^{-37}}{400} = 1.5\times10^{-39}\,C^2\,m/N$$

$$\boxed{\alpha_o = 1.5\times10^{-39} C^2 m/N} \quad \textbf{\textit{Answer}}$$

6.1.4 Draw a graph for a ferroelectric material from the following reading obtained by J. Halbutzel

Absolute temperature	255	260	270	280	290	298
Spontaneous polarization	0	1200	2500	2600	2400	0

Mark on it the curie temperature. Derive an expression for spontaneous polarization as a function of temperature.

Solution:

$P = \epsilon_0 E (\epsilon_r - 1)$

Above curie temperatures

$$\epsilon_r = \frac{C}{(T - T_c)}$$

$$P_s = \epsilon_o E\left[\frac{C}{(T - T_c)} - 1\right]$$

$$\boxed{P_s = \frac{C\epsilon_o E}{(T - T_c)} - \epsilon_o E} \quad \textbf{\textit{Answer}}$$

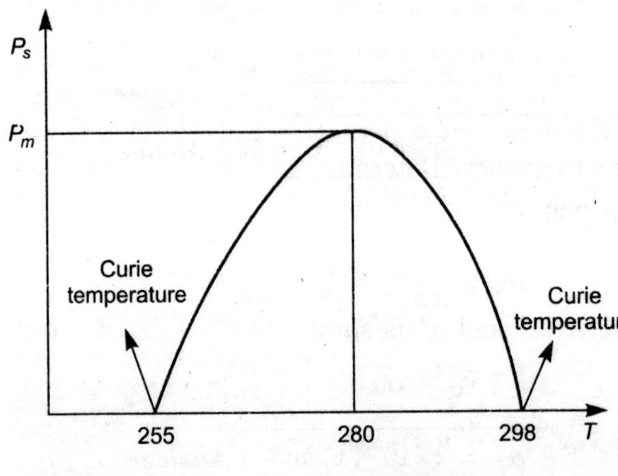

Also

$$\frac{1}{\epsilon_o}\frac{dP_s}{dE} = \frac{C}{(T-T_c)} - 1 = \frac{C-T+T_c}{(T-T_c)}$$

$$\boxed{\frac{1}{\epsilon_o}\frac{dP_s}{dE} = \frac{(C+T_c)-T}{(T-T_c)}}$$ ***Answer***

6.1.5 An atom of oxygen on being polarized produces a dipole moment of 0.5×10^{-22} Cm. If the distance of the centre of negative charge cloud from the nucleus be 4×10^{-17} m, calculate the polarizability of oxygen atom.

(Anna U, Part-time BE, May 1989, JNTU BE, November 2003)

Solution:

In equilibrium position both Lorenz force and the coulomb interaction will be equal. Thus

$$(8e)E = \frac{(8e)(8e)}{4\pi\epsilon_o d^2}$$

$$E = \frac{(8e)}{4\pi\epsilon_o d^2}$$

$$E = \frac{8\times1.6\times10^{-19}}{4\pi\times8.85\times10^{-12}\times16\times10^{-34}} = 2.6\times10^{24} \text{ V/m}$$

Dipole moment,

$$\mu = \alpha E; \quad \alpha = \frac{\mu}{E}$$

$$\alpha = \frac{0.5\times10^{-22}}{2.6\times10^{24}} = 0.19\times10^{-46}$$

$$\boxed{\alpha = 0.19\times10^{-46} \text{ Fm}^2}$$ ***Answer***

6.1.6 The real part of the dielectric constant of bakelite is 6.36 with a loss tangent of 2.8×10^{-2} at 1 MHz frequency. Determine the complex polarizability of the material assuming Lorentz field. $N = 4\times10^{28}/\text{m}^3$

Solution:

The relation between ϵ_r^* and α^* is similar to Clausius-Mossotti relation as

$$\frac{\epsilon_r^* - 1}{\epsilon_r^* + 2} = \frac{N\alpha^*}{3\epsilon_o} \quad ...(1)$$

where,

$$\epsilon_r^* = \epsilon_r - i\epsilon_r'' \text{ and } \tan\delta = \frac{\epsilon_r''}{\epsilon_r'}$$

or

$$\epsilon_r^* = \epsilon_r' - i\epsilon_r' \tan\delta = \epsilon_r' (1 - i \tan\delta)$$

Substituting these values in Eqn. (1)

$$\frac{4.36\,(1 - i \times 0.028) - 1}{4.36\,(1 - i \times 0.028) + 2} = \frac{4 \times 10^{28} \times \alpha^*}{3 \times 8.85 \times 10^{-12}}$$

$$\boxed{\alpha^* = (3.5 - i\,0.06) \times 10^{-40}} \quad \textbf{\textit{Answer}}$$

6.1.7 Consider an electron of charge (– e) moving in a circular orbit of radius a about a charge (+ e) in a field directed at right angles, to the plane of the orbit. Show that the polarizability α_e is approximately $4\pi\epsilon_0 a^3$.

Solution:

Due to polarization, the electric field inside the atom becomes

$$\vec{E} = \frac{\vec{P}}{3\epsilon_o}$$

where P is density of polarization. But the dipole moment,

$$\mu = 4\pi\epsilon_o a^3 \vec{E} = \alpha_e \vec{E}$$

Thus

$$4\pi a^3 \frac{P}{3} = \alpha_e \frac{P}{3\epsilon_o}$$

or

$$\boxed{\alpha_e = 4\pi\epsilon_o a^3} \quad \textbf{\textit{Answer}}$$

6.1.8 Make a capacitor with a capacitance of 1μF working at rms voltage of 50 volt at 1 MHz with a dielectric of thickness 50 μm. The electric field therefore is 10^6 volt/m. Given; $\epsilon = 3.1\,\epsilon_o$ and $\tan\delta = 10^3$. Calculate the power loss and dissipation.

Solution:

$$C = \frac{A\epsilon}{d} \text{ and } \frac{\epsilon}{\epsilon_o} = \epsilon_r = 3.1$$

$$A = \frac{Cd}{\epsilon} = \frac{10^{-6} \times 50 \times 10^{-6}}{3.1 \times 8.85 \times 10^{-12}}$$

$$A = 1.82\,\text{m}^2$$

$$\text{Volume, } v = 1.82 \times 50 \times 10^{-6} = 9.1\times10^{-5}\ \text{m}^3$$

$$\text{Power loss, } P = W = \omega E^2 \in' \tan\delta = \omega E^2 \in_0 \in_r \tan\delta$$

$$= 2\pi \times 10^6 \times 10^{12} \times 8.85 \times 10^{-12} \times 3.1\times 10^3$$

$$\boxed{P = 172\ \text{kW}} \quad \textbf{\textit{Answer}}$$

Power dissipation

$$P_d = P \times v = 172\times10^3 \times 9.1\times10^{-5}$$

$$\boxed{P_d = 15.7 W} \quad \textbf{\textit{Answer}}$$

6.1.9 If a dielectric is introduced between the plates of a parallel plate capacitor, show that the induced charge varies with the dielectric as $\in_r = \left[1 - \frac{\sigma'}{\sigma}\right]^{-1}$. Also show that $\in_r \to \infty$ for a metal dielectric. *(Anna U., BE, April 1998)*

Solution:

The induced charge

$$q' = q - \frac{q}{\in_r} = q\left[1 - \frac{1}{\in_r}\right]$$

$$\frac{q'}{A} = \frac{q}{A}\left[1 - \frac{1}{\in_r}\right]$$

$$\sigma' = \sigma\left[1 - \frac{1}{\in_r}\right]$$

$$\frac{\sigma'}{\sigma} = \left[1 - \frac{1}{\in_r}\right]; \quad \frac{1}{\in_r} = 1 - \frac{\sigma'}{\sigma}$$

$$\boxed{\in_r = \left[1 - \frac{\sigma'}{\sigma}\right]^{-1}} \quad \textbf{\textit{Answer}}$$

In metal dielectric, the induced charge is the same as that of the free charge, but opposite in sign. *i.e.*, $\sigma' = \sigma$ or $\in_r = \infty$.

6.1.10 Determine the percentage of ionic polarizability in the sodium chloride crystal which has the optical index of refraction and the static dielectric constant 1.5 and 5.6 respectively. *(IIT Bombay, B. Tech, April 1995, REC Calicut, BE, May 1996)*

Solution:

According to Clausius-Mossotti equation,

$$\frac{\in_r - 1}{\in_r + 2} = \frac{N(\alpha_e + \alpha_i)}{3 \in_0} \qquad \text{...(1)}$$

At optical frequencies,

$$\frac{n^2 - 1}{n^2 + 2} = \frac{N\alpha_e}{3\epsilon_0} \qquad \text{...(2)}$$

where n is the refractive index

Now

$$\left[\frac{\alpha_i}{\alpha_e + \alpha_i}\right] \times 100 = \left[1 - \frac{\alpha_e}{\alpha_e + \alpha_i}\right] \times 100 \text{ is required}$$

Divide Eqn. 2 by Eqn. 1.

$$\left[\frac{n^2 - 1}{n^2 + 2}\right]\left[\frac{\epsilon_r + 2}{\epsilon_r - 1}\right] = \frac{\alpha_e}{\alpha_e + \alpha_i}$$

Thus

$$1 - \frac{\alpha_e}{(\alpha_e + \alpha_i)} = \frac{\alpha_i}{\alpha_e + \alpha_i} = 1 - \left[\frac{(n^2 - 1)}{(n^2 + 2)}\right]\left[\frac{\epsilon_r + 2}{\epsilon_r - 1}\right]$$

$$\left[\frac{\alpha_i}{\alpha_e + \alpha_i}\right] \times 100 = \left[1 - \frac{1.25}{4.25}\left\{\frac{7.6}{4.6}\right\}\right] \times 100$$

$\boxed{51.4\%}$ ***Answer***

6.2.1 Estimate the paramagnetic contribution to the magnetic susceptibility per m^3 of potassium, for which the Fermi energy is 2.1 eV. Given: $n = 1.33 \times 10^{28}/m^3$

(Meerut U., M. Sc April 1997, Anna U., BE, November 1999)

Solution:

The quantum expression for paramagnetic susceptibility is

$$\chi_p = \chi_c\left[\frac{3T}{2T_F}\right]$$

where χ_c is the classical value of susceptibility which is equal to $\frac{\mu_0\mu_B^2 n}{k_B T}$

Thus $$\chi_p = \left[\frac{\mu_0\mu_B^2 n}{k_B T}\right]\left[\frac{3T}{2T_F}\right] \text{ with } E_F = k_B\, T_F$$

$$= \frac{1.5 \times 4\pi \times 10^{-7} \times 1.33 \times 10^{28} \times (9.27 \times 10^{-24})^2}{2.1 \times 1.6 \times 10^{-19}}$$

$\boxed{\chi_p = 6.4 \times 10^{-6}}$ ***Answer***

6.2.2 A paramagnetic material has 10^{28} atom/m^3. The magnetic moment of each atom is 1.8×10^{-23} A m^2. Calculate the paramagnetic susceptibility at 300 K. What would be the dipole moment of a bar of this material 0.1 m long and 1 sq cm cross-section placed in a field of 8×10^4 A/m?

(Imperial College of Technology, winter 1978, JNTU BE, May 2000)

Solution:

$$\chi = \frac{\mu_0 \mu_B^2 n}{3k_B T}$$

$$= \frac{4\pi\times10^{-7}\times(1.8\times10^{-23})^2\times10^{28}}{3\times1.38\times10^{-23}\times300}$$

$$= \frac{4\pi\times1.8^2\times10^{-2}}{3\times1.38\times300}$$

$$\boxed{\chi = 3.27\times10^{-4}} \quad \textbf{\textit{Answer}}$$

$$\text{Volume} = \pi r^2 h = \frac{1\times0.1}{10^4} = 10^{-5} \text{ m}^3$$

$$M = \mu_m V = \chi H$$

$$\mu_m = \frac{\chi H}{V} = \frac{3.27\times10^{-4}\times8\times10^4}{10^{-5}}$$

$$\boxed{\mu_m = 2.62 \text{ A m}^2} \quad \textbf{\textit{Answer}}$$

6.2.3 The ions in the molecule of magnetite are Fe^{+2}, Fe^{+3} and O_4^{-2}, the subscripts, giving the number of ions per molecule. In the conventional unit cell, which is cubic with a = 0.837 nm, there are eight molecules. The Fe^{+3} magnetic moments cancel and the magnetization that produced by the Fe^{+2} ions alone. If the saturation magnetitie is 5.2×10^5 A/m, calculate the moment per Fe^{+2} ion in Bohr magneton.

(University of Maryland US, UC 1976, Anna U., M.Sc, May 1998)

Solution:

There are eight magnetite molecules in a unit cell with one Fe^{2+} ion per molecule.

Thus
$$N = \frac{8}{a^3} = \frac{8}{(0.837\times10^{-9})^3}$$

$$N = 1.36\times10^{28}/\text{m}^3$$

$$M_s = Nn\mu_B$$

$$n = \frac{M_s}{N\mu_B}$$

$$= \frac{5.2\times10^5}{1.36\times10^{28}\times9.27\times10^{-24}}$$

$n = 4$

Thus the magnetic moment per Fe^{+2} is

$\boxed{4\mu_B}$ ***Answer***

6.2.4 The curie temperature of iron is 1043 K. Assume that iron atoms have moment of two Bohr magneton per atom. Determine the saturation magnetization, Curie constant, Weiss field constant and the magnitude of the internal field. The radius of the atom is 0.123 nm

(IIT Bombay, B. Tech, April 1990, M.K.U., Part-time BE, May 1993)

Solution:

$$a = \frac{4r}{\sqrt{3}} = \frac{4\times0.123}{1.732} = 0.284 \text{ nm}$$

Saturation magnetization,

$$2\mu_B N = 2\mu_B\frac{2}{a^3}$$

$$= \frac{4\times9.27\times10^{-24}}{(0.284\times10^{-9})^3}$$

$\boxed{2\mu_B N = 15.8\times10^5 \text{ A/m}}$ ***Answer***

with $N = \dfrac{2\times10^{27}}{0.284^3} = 8.7\times10^{28}/\text{m}^3$

$$\text{Curie constant} = \frac{N\mu_0\mu_B^2}{k_B}$$

$$= \frac{8.7\times10^{28}\times4\pi\times10^{-7}\times(9.27\times10^{-24})^2}{1.38\times10^{-23}}$$

$\boxed{C = 0.789}$ ***Answer***

$$\text{Weiss field constant} = \frac{\text{Curie temperature}}{\text{Curie constant}}$$

$$= \frac{1043}{0.789}$$

$\boxed{1321}$ ***Answer***

Magnitude of the internal field = μ_o × Weiss field constant × saturation magnetization

$$= 4\pi \times 10^{-7} \times 1321 \times 15.8 \times 10^5$$

$$\boxed{B = 2.6 \times 10^3 \text{weber}/m^2}$$ ***Answer***

6.2.5 The saturation magnetic induction of iron is 0.65 Wb/m^2. If the density of nickel is 8906 kg/m^3 and its atomic weight 58.7, calculate the magnetic moment of nickel in Bohr magneton.

(Massachusetts Institute of Technology, U. S, B. Tech., winter 1982, Anna U., BE, April 1993)

Solution:

The number of Nickel atoms/m^3 is

$$N = \frac{\rho N_A}{M_A}$$

$$= \frac{8906 \times 6.023 \times 10^{26}}{58.7}$$

$$= 9.14 \times 10^{28}/\text{m}^3$$

We know that

$$\chi = \frac{M_s}{H_s} \text{ with } B_s = \mu H_s$$

$$\chi = \frac{\mu M_s}{B_s} = \frac{\mu_0 \mu_r M_s}{B_s} \qquad ...(1)$$

$$\chi = (\mu_r - 1)$$

or, $$\mu_r = 1 + \chi$$

Substituting the value of μ_r in equation (1), one gets

$$\chi = \frac{M_s \mu_0 (1+\chi)}{B_s} = \frac{\mu_m N \mu_0 (1+\chi)}{B_s}$$

$$\mu_m = \frac{B_s \chi}{N \mu_0 (1+\chi)}$$

Since χ is very large for Ni, $\chi >> 1$

$$\mu_m = \frac{B_s}{N \mu_0}$$

$$= \frac{0.65}{9.14 \times 10^{28} \times 4\pi \times 10^{-7}}$$

$$= 5.66 \times 10^{-24} A/m^2$$

$$\frac{5.66 \times 10^{-24}}{9.27 \times 10^{-24}} \text{ Bohr magneton}$$

$$\boxed{\mu_m = 0.61 \mu_B}$$ ***Answer***

6.2.6 Calculate the frequency of the radiation which must be incident on a substance placed in a magnetic field of strength $[5\times10^5/\pi]$ amp/m, so that the electron can absorb energy. *(JNTU, BE, November 2002, VUT, BE, April 2003)*

Solution:

Electron can absorb energy if the frequency of radiation is such that

$$h\upsilon = 2\mu_B B$$

$$\upsilon = \frac{2B\mu_B}{h} = \frac{2\mu_0 H\mu_B}{h}$$

$$= \frac{2\times4\pi\times10^{-7}\times5\times10^5\times9.27\times10^{-24}}{\pi\times6.626\times10^{-34}}$$

$$= \frac{2\times4\times5\times9.27\times10^{-26}}{6.626\times10^{-34}}$$

$$\boxed{\upsilon = 5.6\times10^9/s}$$ ***Answer***

6.2.7 A paramagnetic system of electronic spin magnetic dipole moment is placed in an external field of 10^5 amp/metre. Calculate the average magnetic moment per dipole at 300K. Also calculate the fractional number of spins which are parallel to the field *(REC, Warangal, BE, April 2001, M.U, BE, September 1995)*

Solution:

$$\bar{\mu}_m = \frac{M}{N} = \frac{\mu_0 H\mu_B^2}{k_B T}$$

$$\bar{\mu}_m = \frac{4\pi\times10^{-7}\times10^5\times(9.27\times10^{-24})^2}{1.38\times10^{-23}\times300}$$

$$\bar{\mu}_m = 2.6\times10^{-27}\,Am^2$$

$$\frac{N_p}{N} = \frac{\exp(\mu_0\mu_B H/k_B T)}{[\exp(\mu_0\mu_B H/k_B T)+\exp(-\mu_0\mu_B H/k_B T)]}$$

$$\frac{N_p}{N} = \frac{\exp(2.8\times10^{-4})}{[\exp(2.8\times10^{-4})+\exp(-2.8\times10^{-4})]} = 0.5$$

or $$\boxed{\frac{N_p}{N} = 50\%}$$ ***Answer***

EXERCISE

6.1.1 NaCl is a cube with lattice parameter 0.564 nm and four formula units per unit cell. The polarizabilities Na and Cl are 0.16×10^{-40} and 3.3×10^{-40} F/m^2. Calculate the refractive index of NaCl.

(University of Maryland, U.S., B.Tech winter 1992, M.K.U., BE, April 1996)
(**Ans:** 1.49)

6.1.2 The polarizability of ammonia molecule is found approximately by the measurement of dielectric constant as 2.42×10^{-39} C^2 m/N and 1.74×10^{-39} C^2 m/N at 339 and 448 K respectively. Calculate for each temperature the orientation polarizability.

(Anna U., BE, April 1998)
(**Ans:** 2.192×10^{-39} C^2 m/N and 1.512×10^{-39} C^2 m/N)

6.1.3 The bakelite is found to have the real part of its complex dielectric constant of 2.8×10^{-2} of frequency 1MHz. Calculate the complex polarizability of the material assuming Lorentz field. Given: $N = 4 \times 10^{28}$/m^3.

(Anna U., Part-time BE, April 1998)
(**Ans:** $(3.5 - i \times 0.06) \times 10^{-40}$)

6.1.4 What is temperature coefficient of permittivity? How will you determine the temperature coefficient of permittivity for highly polar liquid insulating materials? Find the temperature coefficient of permittivity of highly polar liquid at 25°C.

(GATE, April 1995, Anna U., December 1997)

(**Ans:** $\dfrac{(\epsilon_r - 1)(\epsilon_r + 2)\alpha_l}{\epsilon_r}$ and 0.0034)

6.1.5 The energy stored in a capacitor per unit volume is given by $U = \frac{1}{2}\epsilon_0\epsilon_r E^2$. If the maximum possible field is 5 MV/m, what volume of capacitor with $\epsilon_r = 1000$ is required to store 1 kW.

(REC Durgapur B. Tech, December 1998, JNTU BE, April 2001)
(**Ans:** 33m^3)

6.1.6 The electronic polarizability of helium is 0.18×10^{-40} F m^2. Calculate the value of the radius of the electron orbit and hence the permittivity for 2.7×10^{25} atoms in one cubic metre. (**Ans:** 5.46×10^{-11} m, 1.000057776)

6.1.7 The following data relate to liquid $CHCl_3$ (Chloroform). Calculate the dipole moment.

(Loughborough University of Technology)
(**Ans:** 2.6 D)

ϵ_r	6.76	6.12	5.61	4.81	3.71	3.33	2.93
T (°C)	– 60	– 40	– 20	20	100	140	180

6.2.1 A ferromagnet becomes paramagnet and follows Curie-Weiss law. Therefore, the thermal energy just balances the magnetic potential energy. Get the value of the molecular field and express it in oersted. Given: T_c = 1000 K.

(Anna U., BE, April 1995)

(***Ans***: 0.149×10^4 A/m, 18.7 oersted)

6.2.2 Anhydrous nickel ($NiCl_2$) has a paramagnetric susceptibility of 78.5×10^{-8} kg at 20°C. The atomic weight of nickel and chlorine are 58.7 and 35.5 respectively. Using simplified Langevin formula, calculate the magnetic moment of the Ni^{2+} ion in Bohr magneton. Given: molecular weight of $NiCl_2$ = 129.7.

(Royal Military College, UK, B. Tech., March 1988, JNTU BE, March 1996)

(***Ans.***4.4 μ_B)

6.2.3 A magnetic material has a magnetizatin of 3200 A/m and flux density 0.005 weber/m^2. Calculate the magnetizing force and the relative permeability of the material.

(M.K.U, Part-time BE, April 1989)

(***Ans***: 5.5)

6.2.4 Consider an atom, placed in a magnetic field of 1 weber/m^2, with l = 2. Calculate (*a*) rate of precision (*b*) the torque of the atom, given that the magnetic moment makes an angle 45°.

(***Ans:*** 8.8×10^{10} revolutions/s, 1.3×10^{-23} J)

6.2.5 A typical magnetic field achievable with an electromagnet with iron core is about 1 tesla. Compare the magnetic interaction energy, $\mu_B B$ of an electron spin magnetic dipole moment with $k_B T$ at room temperature and show that at ordinary temperatures the approximation $\frac{k_B T}{\mu_B B} >> 1$ valid.

(REC Nagpur, BE, April 1999, JNTU, BE, April 2001)

Chapter 7

UNIT 7

Lasers and Fiber Optics

7.1 LASERS

7.1.1 INTRODUCTION

A laser is an oscillator that operates at very high frequencies. These optical frequencies range to values several orders of magnitude higher than can be achieved by the *conventional approaches* of solid-state electronics or electron tube technology. In common with electronic circuit oscillators, a laser is constructed from an amplifier with an appropriate amount of positive feed back. The acronym LASER stands for *light amplification by stimulated emission of radiation*.

The wide spread practical use of lasers and optical devices in applications such as communications and increasingly in areas such as signal processing and image analysis has lead to the use of the term *photonics*. Whereas, electronics uses electrons in various devices to perform analog and digital functions, photonics aims to replace the electron with photons. Because photons have zero mass, do not interact with each other to any significant extent, and travel at the speed of light, photonic devices promise small size and high speed.

Historically the laser is the outgrowth of maser which means *microwave amplification by stimulated emission of radiation*. If the stimulated radiation lies in optical region, then the device is called optical maser or laser. In 1960 laser was introduced on an appleting method for the production of intense, monochromatic, collimated and highly coherent beam of light.

7.1.2 CHARACTERISTIC OF LASERS

Following are the main characteristics of laser radiation (*a*) coherence (*b*) very narrow bandwidth (*c*) high directionality (or narrow angular speed) (*d*) extreme brightness. Some of the important characteristics of lasers are discussed below:

(*i*) **Coherence :** A conventional light source such as in candescent lamp or a natural source such as the sun produces incoherent light since they emit random wavelength light waves with no common phase relationships. On the other hand, the waves emitted by a laser source will be in phase and are of the same frequency.

(*ii*) **Directionality:** The conventional sources emit light in all directions. Lasers emit light only in one direction as the photons travelling along the optical axis of the system is selected and augmented with the help of the optical resonator.

(*iii*) **Divergence:** Light from conventional sources spreads out in the form of spherical wave fronts and hence it is highly divergent. The divergence or angular spread of the laser beam is extremely small. The little divergence that exists in it arises out of the wave properties of light. When the light issues out from the front mirror, it undergoes diffraction because the semitransparent mirror acts as a circular aperture. Accordingly, it spreads out and the angular spread is given by

$$\Delta\theta = \frac{1.22\lambda}{d}$$

where d is the diameter of the front mirror. In the case gas lasers $\Delta\theta$ is small as 10^{-5} to 10^{-6} radian.

(*iv*) **Intensity:** The intensity light from a conventional source decreases rapidly with distance, as it spreads in the form of spherical waves. One can look at the source without any harm to his eyes. In contrast, a laser beam emits light in the form of a narrow beam which propagates in the form of plane waves. As the energy is concentrated in a very narrow region, its intensity would be tremendously high. It is estimated that light from a typical 1 mW laser is 10000 times brighter than the light from the sun at the earth's surface. The intensity of the laser beam stays nearly constant with distance as the light travels in the form of plane waves.

(*v*) **Monochromacity:** The light from normal monochromatic sources spreads over a wavelength range of the order of 100 Å to 1000 Å. The laser light is highly monochromatic. The spread is of the order of a few angstroms (< 10 Å) only. Such a vast difference arises because conventional sources emit wave trains of very short duration and length, whereas lasers emit continuous waves of very long duration.

7.1.3 PRINCIPLE AND EVENTS

Let us consider two energy states 1 and 2 of an atomic system. The black dots represent the number of atoms in each state and three different events (*a*), (*b*), and (*c*) are shown in the Fig. 7.1.

These three important events are described in a nut-shell below:

1. *Absorption*: Ordinarily atoms of a substance are in their ground quantized state. When energy is given to the substance, one of its atoms passes from state 1 to state 2 by absorbing a photon of appropriate energy (Fig. 7.1(*a*)).
2. *Spontaneous emission*: One of the atoms passes spontaneously from state 2 to state 1, emitting a radiation of frequency υ given by $\upsilon_{12} = \dfrac{E_2 - E_1}{h}$, E_1 and E_2 being the energy of an atom in state 1 and state 2 respectively. See Fig. 7.1(*b*).

3. *Stimulated or induced emission*: A photon of correct energy of the surrounding can stimulate that one of the atoms of the substance passes from state 2 to state 1; so that the two photons of same energy $h\upsilon_{12}$ are emitted.

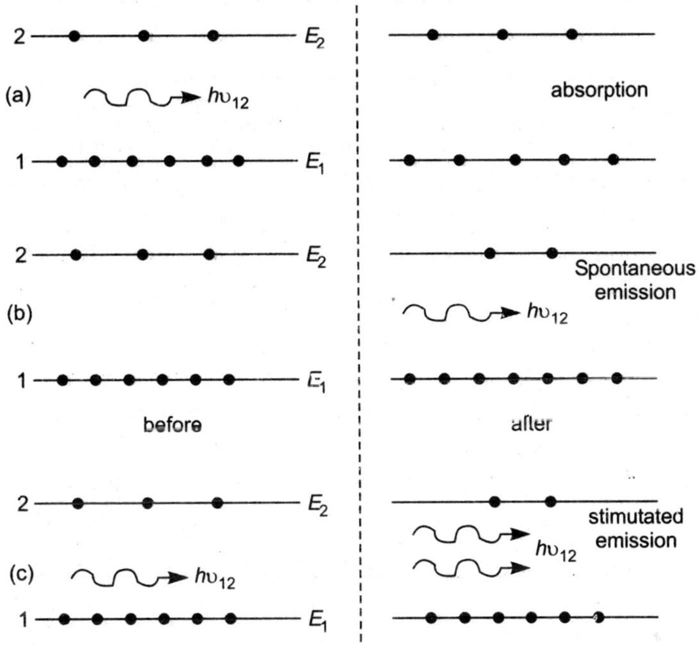

Fig. 7.1 Spontaneous and stimulated emission

The events of spontaneous emission and absorption were already known. In 1917 Einstein suggested that the stimulated transition 2 → 1 can happen in the same way as 1 → 2. He gave the relative probabilities of the three events (*a*), (*b*) and (*c*) in terms of the population N_1 and N_2 of the number of atoms in the states 1 and 2 respectively and the radiation density ρ_{12} of radiation of frequency υ_{12} in the region of interest. The rates at which the three processes, occur are given by

$$r_b \text{ (spontaneous emission)} = A_{12}\, N_2$$

$$r_a \text{ (absorption)} = B_{12}\, N_1\, \rho_{12}$$

$$r_c \text{ (stimulated emission)} = B_{12}\, N_2\, \rho_{12}$$

with ρ_{12} as the density of electromagnetic radiation.

The coefficients A and B are called *Einstein's coefficients*. The main point to be noted is that the same coefficient B_{12} determines the rates for stimulated emission and absorption. That is why stimulated emission is also called *inverted absorption*. The remarkable feature of stimulated emission is that the *emitted photon is coherent* with the stimulating photon. But in ordinary sources of light the spontaneous emission, which is incoherent, dominates. The reason is the following. In any substance the equilibrium of atoms among energy levels is given by Boltzmann's formula.

$$\boxed{\frac{N_2}{N_1} = \exp[-(E_2 - E_1)/k_B T]} \qquad \text{...(7.1.1)}$$

For any positive value of T, if $E_2 > E_1$; $N_2 < N_1$. In other words the population of atoms in higher energy levels is less than that in lower energy levels. The stimulated emission dominates over spontaneous emission and absorption if (*i*) $N_2 > N_1$, (*ii*) ρ_{12} is large. The first condition makes $r_c > r_a$ *i.e.*, stimulation results in net emission and second condition makes $r_c > r_b$; so that stimulated emission exceeds the spontaneous emission.

The establishment of a situation in which number of atoms in higher energy state is greater than that in lower energy state is called *population inversion*. The process of creating a population inversion is called *pumping*. When the condition of population inversion is achieved, if the incident photon stimulates only one atom to emit a photon, the number of identical coherent photons grows from one to two. If each of these two stimulates the emission of other photon, the result is four identical coherent photons. The process thus builds up and finally we get an intense beam of directed radiation. The process is called *light amplification by stimulated emission of radiation*.

Optical Pumping

In optical pumping the process involves three energy levels. Atoms are excited to highest energy state E_3 from lowest energy state E_1 by radiation of energy $h\upsilon_{13}$. Some excited

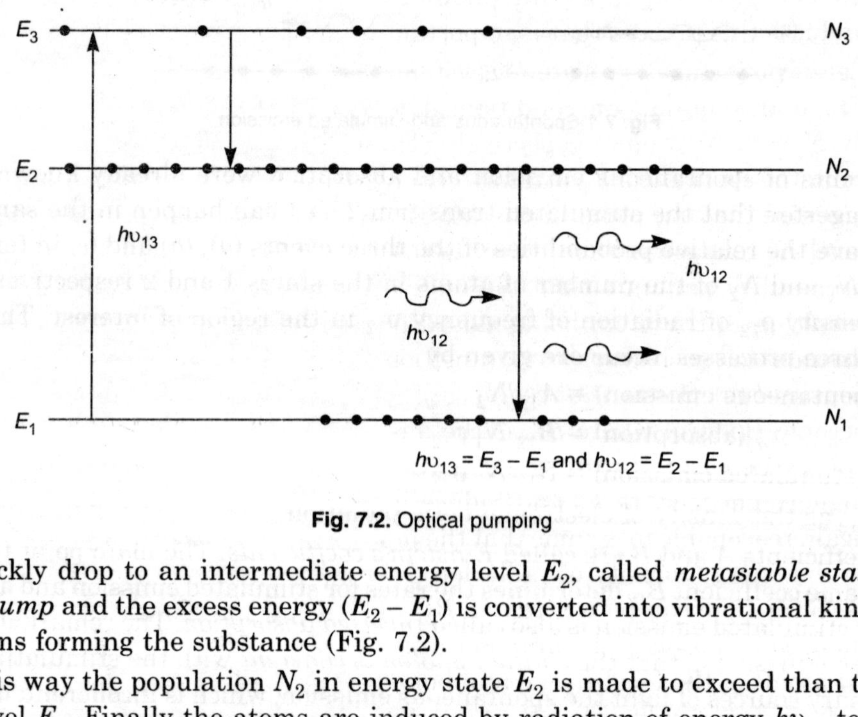

Fig. 7.2. Optical pumping

atoms quickly drop to an intermediate energy level E_2, called *metastable state* in a *non radiative jump* and the excess energy $(E_2 - E_1)$ is converted into vibrational kinetic energy of the atoms forming the substance (Fig. 7.2).

In this way the population N_2 in energy state E_2 is made to exceed than that (N_1) in ground level E_1. Finally the atoms are induced by radiation of energy $h\upsilon_{12}$ to drop from level E_2 to E_1 emitting *laser* of frequency.

$$\upsilon_{12} < \upsilon_{13}$$

Laser operations have been obtained with different type of solid state materials with mixture of gases. Accordingly various types of lasers are in use. *e.g.*, Ruby laser, gas laser, semiconductor laser etc.

7.1.4 SPONTANEOUS AND STIMULATED EMISSION PROBABILITIES (STUDY OF EINSTEIN'S COEFFICIENTS)

In general radiation interacts with matter in the following ways:

1. When the radiation strikes an atom in the lower energy state the energy of the radiation may be absorbed, thereby raising the atom to the higher energy state. If nothing further happens to the atom; it may fall again spontaneously to the lower energy state re-emitting the photon which it has absorbed earlier. This process is called *spontaneous emission*. The spontaneous emission is a statical process; *i.e.*, we cannot say when exactly a particular transition is going to take place. We can only talk about the probability of spontaneous transition. In this, it resembles the phenomenon of radioactivity.
2. If the photon strikes an excited atom, that is, an atom in the higher energy state, it may stimulate the atom to emit its stored energy by falling to the lower energy state, provided the energy difference between the two states is exactly equal to the energy of the impinging photon. This process is called stimulated emission and it is this process that is important in maser/laser action.

   Let us now consider an atomic system placed in a radiation field of energy density $E(\upsilon)$ under equilibrium conditions. Let E_1 and E_2 be the two energy levels occupied by N_1 and N_2 atoms respectively. The photons corresponding to the transition from the level E_2 to E_1 will have energy $(E_2 - E_1) = h\upsilon$. There are three possible processes taking place in the atomic system. Spontaneous transition from the level E_2 to E_1 may take place. Let A_{21} be the spontaneous emission transition probability per unit time. Secondly, absorption of energy $h\upsilon$ can take place resulting in the atom being raised from the level E_1 to E_2. It is reasonable to assume that such absorption transitions per unit time is proportional to the energy density $E(\upsilon)$. So the *induced absorption transition probability* per unit time will become $B_{12}\,E(\upsilon)$ where B_{12} is the absorption probability per unit time per unit energy density. Thirdly the radiation field may interact with an atom in the higher energy state E_2 and induce it to fall to the lower energy state E_1. It is again reasonable to assume that the probability of induced emission is proportional to the energy density of radiation field. So the probability of induced emission per unit time can be written as $B_{21}\,E(\upsilon)$ where the constant B_{21} represents the induced emission probability per unit time per *unit energy intensity*. If there are N_2 atoms in the level E_2 the number of atoms that fall from the level E_2 to E_1 per unit time is equal to $[A_{21}+B_{21}\,E(\upsilon)]\,N_2$ and the number of atoms that rise up from the level E_1 to E_2 per unit time is $B_{12}\,E(\upsilon)\,N_1$. So the net rate of change of atoms

   $\dfrac{dN_2}{dt}$ in the level E_2 per unit time is given by

$$\frac{dN_2}{dt} = B_{12}\,E(\upsilon)\,N_1 - [A_{21} + B_{21}\,E(\upsilon)]\,N_2 \qquad ...(7.1.2)$$

Under equilibrium conditions the net rate of change of atoms in any level must be zero. So

$$\frac{dN_2}{dt} = 0$$

or

$$B_{12}\,E(\upsilon)\,N_1 = [A_{21} + B_{21}\,E(\upsilon)]\,N_2 \qquad ...(7.1.3)$$

$$\frac{N_2}{N_1} = \frac{B_{12}\,E(\upsilon)}{[A_{21} + B_{21}E(\upsilon)]} \qquad ...(7.1.4)$$

Boltzmann has shown that the atomic population at different energy levels at a given temperature T is given by the relation.

$$N_2 = N_1 \exp[-(E_2 - E_1)/k_B T] = N_1 \exp[(-h\upsilon/k_B T)] \qquad ...(7.1.5)$$

Under normal conditions the population in the lower energy level is greater than the population in the higher energy level. Substituting the value of $\frac{N_2}{N_1}$ from Eqn. (7.1.5) in Eqn. (7.1.4)

$$\exp[-h\upsilon/k_B T] = \frac{B_{12}\,E(\upsilon)}{[A_{21} + B_{21}\,E(\upsilon)]}$$

$$B_{12}\,E(\upsilon)\times\exp[h\upsilon/k_B T] = A_{21} + B_{21}\,E(\upsilon)$$

Divide by B_{12}

$$E(\upsilon)\exp[h\upsilon/k_B T] = \frac{A_{21}}{B_{12}} + \frac{B_{21}}{B_{12}}\,E(\upsilon)$$

$$E(\upsilon)\left[\exp(h\upsilon/k_B T) - \frac{B_{21}}{B_{12}}\right] = A_{21}/B_{12}$$

$$E(\upsilon) = \frac{A_{21}/B_{12}}{\left[\exp(h\upsilon/k_B T) - \frac{B_{21}}{B_{12}}\right]} \qquad ...(7.1.6)$$

The Planck's radiation formula gives the energy density of radiation at a given temperature as

$$E(\upsilon) = \frac{8\pi h\upsilon^3}{c^3}\left[\frac{1}{\exp(h\upsilon/k_B T)-1}\right] \quad ...(7.1.7)$$

Comparing equation (7.1.7) and (7.1.6), we get

$$\frac{A_{21}}{B_{12}} = \frac{8\pi h\upsilon^3}{c^3} \text{ with } \frac{B_{21}}{B_{12}} = 1 \quad ...(7.1.8)$$

These relations were obtained first by Einstein in 1917. The constants A_{21}, B_{12} and B_{21} are called *Einstein coefficients*. Substituting $B_{21} = B_{12}$ in Eqn. (7.1.6), we get the relation between the *spontaneous emission probability* and the *induced emission probability*, when matter is in equilibrium with radiation.

i.e.,

$$\frac{\text{Spontaneous emission probability}}{\text{Induced emission probability}} = \frac{A_{21}}{B_{21}E(\upsilon)} = \exp[(h\upsilon/k_B T)-1] \quad ...(7.1.9)$$

remembering $B_{12} = B_{21}$.

If $h\upsilon > k_B T$ spontaneous emission is much more probable than induced emission. This holds true in the case of electron transitions in atoms and molecules and in the case of *radiative transitions* in nuclei. If $h\upsilon < k_B T$ *induced emission* or *stimulated emission* may become important and this is possible in the case of atomic transitions in the microwave as well as in the visible region.

Let us now find the conditions under which the radiative emission can be induced or stimulated continuously. We have seen that when radiation and matter are in thermal equilibrium there is no net absorption or emission since the number of absorptions and emission transitions per unit time is the same. Though the induced transmission probability for absorption is considerably less than the spontaneous transition probability for emission, absorption equals the emission because of the large number of atoms in the lower energy level which can absorb the energy. Such equilibrium conditions are not suitable for maser/laser action. In the case when matter is being continuously bombarded by the photons and *non-equilibrium conditions* prevail, we can write the ratio of the *emission rate* to the *absorption rate* as

$$\frac{\text{Emission rate}}{\text{Absorption rate}} = \frac{[A_{21} + B_{21}E(\upsilon)]N_2}{B_{12}E(\upsilon)]N_1}$$

$$= \frac{N_2}{N_1}\left[1 + \frac{A_{21}}{B_{21}E(\upsilon)}\right] \quad \text{as } B_{21} = B_{12}$$

$$= \left[1 + \frac{A_{21}}{B_{21}\,E(\upsilon)}\right] \frac{N_2}{N_1} \qquad ...(7.1.10)$$

If the energy difference between the two levels is small such that the ratio $h\upsilon/k_BT$ is small, the quantity $A_{21}/[B_{21}E(\upsilon)]$ becomes very small as compared to unity as seen from the equation (7.1.9) and so the Eqn. (7.1.10) becomes

$$\frac{\text{Emission rate}}{\text{Absorption rate}} = \frac{N_2}{N_1} \qquad ...(7.1.11)$$

At any fixed temperature the atomic population N_2 in the higher energy level E_2 is always less than the atomic population N_1 in the lower energy level E_1. If by some means we are able to effect population inversion so that $N_2 > N_1$, we can maintain the emission rate higher than the absorption rate. This means that, when a radiation of energy density $E(\upsilon)$ passes through a system in which $N_2 > N_1$, the radiation comes out with more photons of frequency υ than that are incident on the system. This situation is referred to as the wave application by stimulated emission of radiation. If the radiation is in the microwave region, the phenomenon is the *maser*, and if the radiation is in the visible region, the phenomenon is the *laser*. Since more atoms jump to the lower energy state than the atoms which reach the higher energy state, the upper level gets depleted and the amplification decreases. In order to maintain the amplification level constant, the population ratio must be maintained at the same level by pumping the atoms at the lower energy level to the higher energy level. Though the phenomenon of stimulated emission was known as early as 1917, yet a practical device could be constructed only in 1954 and the credit for this discovery goes to Prof. C. H. Towns of the Columbia University for which he was awarded the Nobel prize.

In summary, we can state the necessary conditions for sustained maser action as follows:

1. There must be at least pair of energy levels ($E_2 > E_1$) separated by the desired radiation which is to be stimulated.
2. There must be means of inverting the population of the two energy levels so that the population N_2 at higher energy level E_2 is always kept greater than the population N_1 at the lower energy level E_1.
3. There must be an electromagnetic system which can retain the photon of energy $(E_2 - E_1)$ with low losses so that they can be used for stimulation.

7.1.5 POPULATION INVERSION

The life time of an atom in an excited state is of the order of 10^{-8} sec. So before an excited atom can be stimulated to emit a photon it is most likely to make a spontaneous emission. The photons emitted by spontaneous emission are not coherent. The ratio of the number n^* of excited atoms to that n in the ground state is given by Boltzmann's equation

$$\frac{n^*}{n} = \exp(-W/k_BT) \qquad ...(7.1.12)$$

where W is the energy difference between the ground state and the excited state. Usually the population of higher energy level (say N_2) is less than the population N_1 of lower energy level. Making $N_2 > N_1$, *i.e.*, the number of particles N_2 more in higher energy level than the number of particles N_1 in lower energy level is called *population inversion* or *inverted population*. When the population in the higher energy state is more in comparison with the lower energy state, the states of system are called *negative temperature states*. Here the negative term temperature is not a physical quantity, but a convenient mathematical expression, signifying the equilibrium state of the system. A system in which population inversion is achieved is called an *active system*. The method of raising a particle from lower energy state to higher energy state is called *pumping*. This can be done by a number of ways. We now consider the pumping of ammonia gas. It has two states with a separation corresponding to $\lambda = 1.27$ cm (microwave region). Ammonia is first heated and then allowed to pass through a non-uniform electric field. The inhomogeneous electric field acting on the ammonia molecules aligns the higher energy molecules along the axis and pushes the lower energy molecules away from the axis. These lower energy molecules are readily removed by a vacuum pump. The higher energy molecules enter a resonant microwave cavity which is tuned to the same frequency as the energy separation of the two energy levels of the ammonia molecule. This causes $N_2 > N_1$ in one region.

Suppose radiation of appropriate energy is incident on a mixture of excited and unexcited atoms. The probability of induced absorption by an unexcited atom is exactly the same as that of stimulated emission by an excited atom. The population of unexcited atoms is very much larger than that of excited atoms. The life time of the excited atom is very short. For these two reasons most of the photons of the incident radiation cause *induced absorption*. There will be very few stimulated emissions.

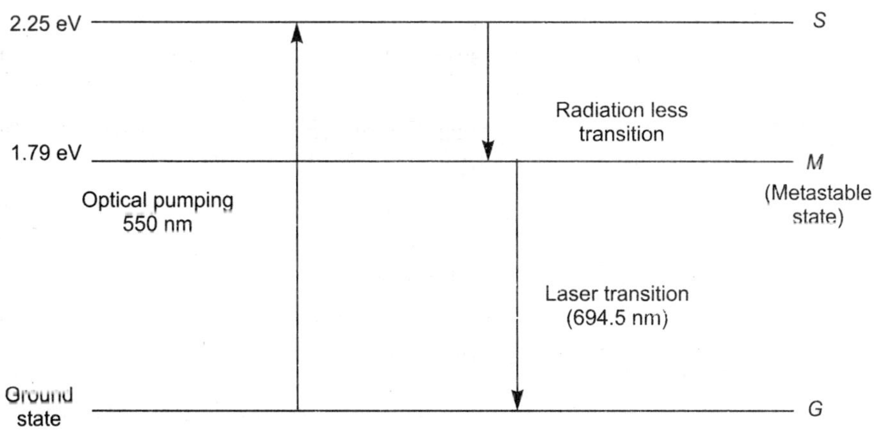

Fig. 7.3. Excitation of atoms

A solution to the above problem can be found in an atom which has three energy levels as shown in Fig. 7.3. The atoms that are originally in the ground state G are raised to the higher excited state S by an external agency. This excited state is short lived. The atoms cannot remain in the state S for long. They rapidly decay into an intermediate state M called a *metastable state*. The above process of transferring atoms from the ground state to the metastable state is called the *optical pumping*. The life time of an atom in the metastable state is of the order of 10^{-3} sec which is much larger than the life time in the excited state S which is of the order of 10^{-8}sec. A passing photon produces lasing transition from M to the ground state. The above process in which the ratio of the population in a higher state (M) to that in the ground state is increased considerably is called population inversion as seen earlier. It is therefore important to note that the following are some of the *important requirements* for laser action.

(*i*) Creation of photons of certain *resonant energy* in a cavity containing the laser medium

(*ii*) Interaction of photons with atoms of laser medium

(*iii*) Population inversion and stimulated emission of photons

7.1.6 COMPONENTS OF THE LASER DEVICES

The three main components of many laser devices are listed below:

1. *The Pump*: The pump is an external source that supplies energy and helps in obtaining the population inversion. The excitation of atoms may occur directly or through atom–atom collisions. The pumps can be optical, electrical or thermal. In the ruby laser we use optical pumping. In the He–Ne laser, *electric discharge pumping* is used. The energy supplied by the pump excites the atoms to higher energy levels and through spontaneous emission or through non-radiative processes, the population inversion occurs. The life time of the metastable energy state, in which population inversion occurs, must be very large as compared to the normal life time of the excited atom in any other energy state.
2. *Laser Medium*: It is the material in which the laser action is made to take place. It may be solid, liquid or gas. Many lasers are named after the material used: ruby laser, He–Ne laser, CO_2 laser etc. Laser wavelength may extend from ultraviolet region to the infrared region. Population inversion in the laser medium is the essential requirement. If N_1 and N_2 are the number of atoms in the energy states E_1 and E_2, then

$$\frac{N_2}{N_1} = exp\,(-h\upsilon / k_B T)$$

with $h\upsilon = (E_2 - E_1)$. Hence, N_2 is in general less than N_1. Because of this reason rigorous pumping may be required for sustaining the population inversion and so only certain pairs of energy levels with appropriate life times can be inverted.
3. *The Resonator*: It consists of a pair of plane or spherical mirrors having common principal axis. The reflection coefficient of one of the mirrors is very near to 1 and that of the other kept some what less than 1. It enables a part of the internally reflected beam to escape out as a laser beam. The resonator is basically a feedback device, that directs the photons back and forth through the laser medium and in the process, the number of photons is multiplied due to stimulated emission.

Thus creation of photons of certain resonant energy in a cavity containing the laser medium, interaction of photons with atoms of laser medium and population inversion and stimulated emission of photons are the features of laser action.

7.1.7 DIFFERENT KINDS OF LASERS

Ruby Laser: It is one of the first operating lasers using three energy level scheme of population inversion and a solid state laser element. It consists of three parts:

(*i*) The working substance in the form of a rod of ruby crystal.

(*ii*) The resonant cavity made of a fully reflecting plate at the left and a partially reflecting plate at the right, both optically plane and accurately parallel.

(*iii*) The optical pumping system consisting of a helical xenon discharge tube.

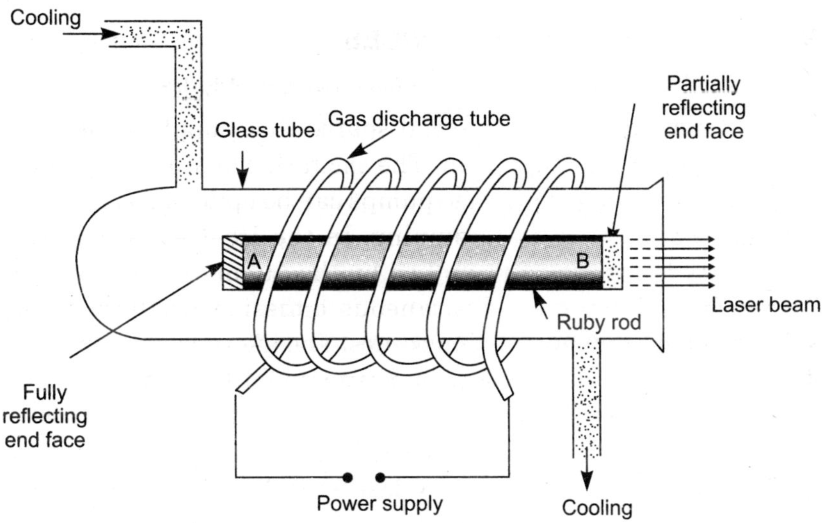

Fig. 7.4 Ruby laser

The ruby rod is a crystal of aluminium oxide (Al_2O_3) doped with 0.05% chromium oxide (Cr_2O_3), so that some of the Al^{+++} ions are replaced by Cr^{+++} ions. These impurity chromium ions give pink colour to the ruby and give rise to the laser action. The ruby crystals are grown in special furnaces and then shaped into rods of nearly 10 cm in length and 0.8 cm in diameter. Flat end faces *A* and *B* of the rod are made strictly parallel, plane and silvered so that the end face *A* becomes fully reflecting and the end face *B* partially reflecting. The rod is surrounded by a cylindrical glass tube through which liquid nitrogen or water circulates to keep the rod cool.

The ruby rod is arranged along the axis of a helical xenon flash tube so that the coils of the tube surround the rod. The flash tube is provided with a suitable power supply. The tube is made to flash for several millisecond. Only a small part of energy is used in pumping. *i.e.*, in exciting the Cr^{+++} ions; the rest energy goes to heal up the apparatus, thereby explaining the need of cooling.

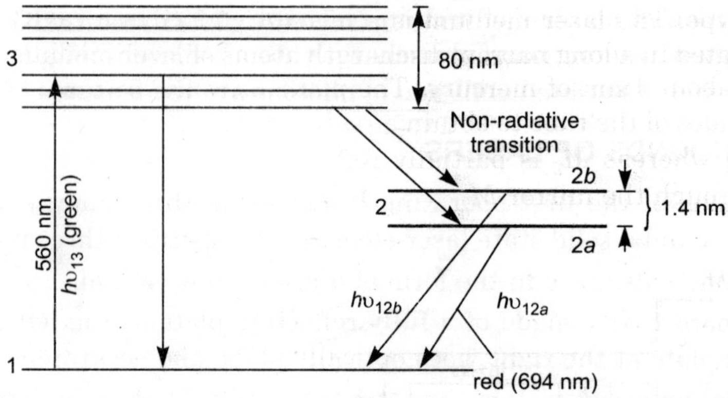

Fig. 7.5 Energy level diagram of a Ruby laser

The energy diagram corresponding to operating principle is shown in Fig. 7.5. The lines 1, 2, 3 corresponds to energy level of chromium atom. When light from the flash tube is made to fall upon the ruby rod, the chromium atom absorbs the green component 560 nm and get excited from level 1 to energy level 3 which has a bandwidth nearly 80 nm. The chromium atom has a double energy level 2 with a difference of 1.4 nm. From energy level 3 some excited atoms return to the ground state 1; whereas the other moves to the level 2. The probability of transition from 3 to 2 is much higher than that from 3 to 1 whereas the probability of transition from 3 to 1 is much smaller than that from 2 to 1. As a result the energy level 2 becomes more populated than 1. Population inversion thus takes place and therefore provide proper conditions for stimulated emission.

Such a system with population inversion is not a stable one; because one spontaneous transition may cause an induced transition resulting in two photons, which may again cause induced transitions resulting in four photons and so on. The process builds up, helped by multiple reflections, in the cavity. The coherent radiation (laser beam) thus grows. A stage is reached where the population inversion caused by one-flash of Xe tube is used up. The coherent laser beam then ceases, till the next flash of Xe tube repeats the process. Thus the ruby laser is a pulsed laser. It means that the active system (Cr^{+++} ions) is excited by xenon helical discharge in pulses and the system emits laser light in pulses. The duration of xenon pulse is several millisecond and that of laser pulse is less than a millisecond. The instantaneous power output of the emitted pulse reaches the order of mega watt.

The level 2 in ruby is actually a doublet of separation 1.4 nm. The spontaneous transition would give two weak lines at 694.3 nm and 692.9 nm, each of width nearly 0.6 nm. But under the lasting conditions the line 694.3 nm dominates over the other line.

Principle of Gaseous Laser: Usually in a gaseous laser a mixture of neon and helium gases are used as a working substance. Unlike ruby laser its operation involves four energy level – three in neon and one in helium. The atoms of the working substance are excited to higher levels by means of high frequency electromagnetic field. Light from this laser is emitted continuously rather than in pulses.

He–Ne Laser: The main drawback of ruby laser is that the output beam is not continuous though very intense. For continuous laser beam, gas lasers are used. He–Ne laser is one such type. The laser medium is a mixture of Helium and Neon in the ratio 10:1. It is encapsulated in a long narrow discharge tube as shown in Fig. 7.6. The pressure inside the tube is about 1mm of mercury. The plane or concave mirrors M_1 and M_2 are placed on the two sides of the tube to obtain a *resonator system*. One of the mirrors say M_1 is highly reflecting whereas M_2 is partially reflecting and partially transparent. Laser beam comes out through the mirror M_2.

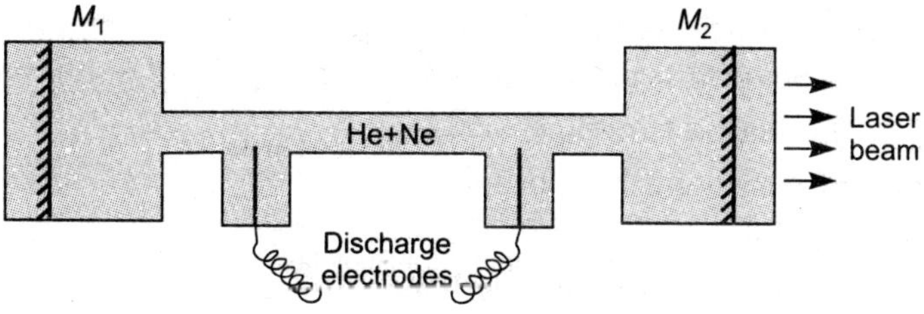

Fig. 7.6 Schematic diagram of He–Ne laser

The first few energy levels of He and Ne are shown in Fig. 7.7. The levels E_2 and E_3 of helium have nearly the same energy as the levels E'_4 and E'_6 of neon. However, the levels E_2 and E_3 are metastable. When discharge is passed through the mixture of gas, the electrons collide with helium atoms and excite them to the metastable states E_2 and E_3. This may be termed as *pumping*. The He atoms continue in these states for long time.

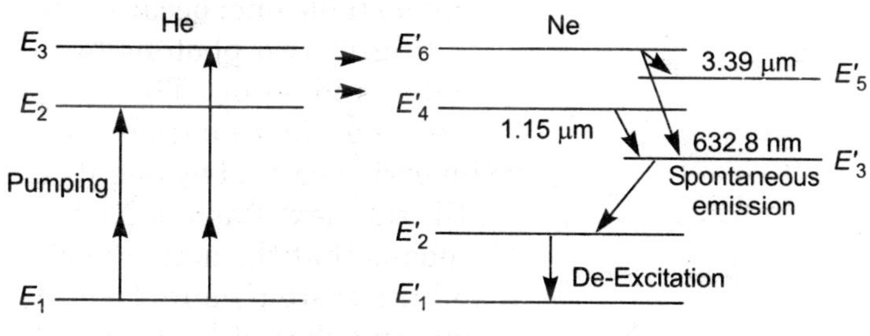

Fig. 7.7 Energy level diagram for He–Ne laser

These helium atoms collide with unexcited neon atoms raising them (Ne atoms) to the levels E'_4 and E'_6. This results in a large number of neon atoms in the energy states E'_6 and E'_4 and this population inversion occurs with respect to the levels E'_3 and E'_5.

Therefore, the following three transitions occur.

1. $E'_6 \rightarrow E'_5$ producing laser of 3.39 μm
2. $E'_6 \rightarrow E'_3$ producing laser of 632.8 nm
3. $E'_4 \rightarrow E'_3$ producing laser of 1.15 μm.

To eliminate the photons of wavelengths 1.15μm and 3.39μm, mirrors are designed to have low reflectivity for these wavelengths. From the level E'_3 the Ne atoms drop to level

E'_2 through spontaneous emission and then de-excited to ground level through collisions. Thus, they are again ready to go through the cycle. This is achieved by making the tube narrow (≈ 0.6 cm) so that the neon atoms are de-excited by collision with the walls of the tube.

The power output from He–Ne laser is rather small (≈ 100 mW maximum), however, the radiation is extremely useful in a wide range of applications because, it is highly collimated, coherent and has extremely narrow bandwidth.

Gas Lasers: Gas lasers produce coherent light beams by an electrical discharge in the laser cavity or gas chamber. Several types of gas lasers have gained popularity because of

(*a*) the simplicity of the lasing function,

(*b*) the wide variety of wavelengths available,

(*c*) they are less expensive,

(*d*) some, for example the CO_2, are very efficient and

(*e*) very high power outputs are available from relatively small units.

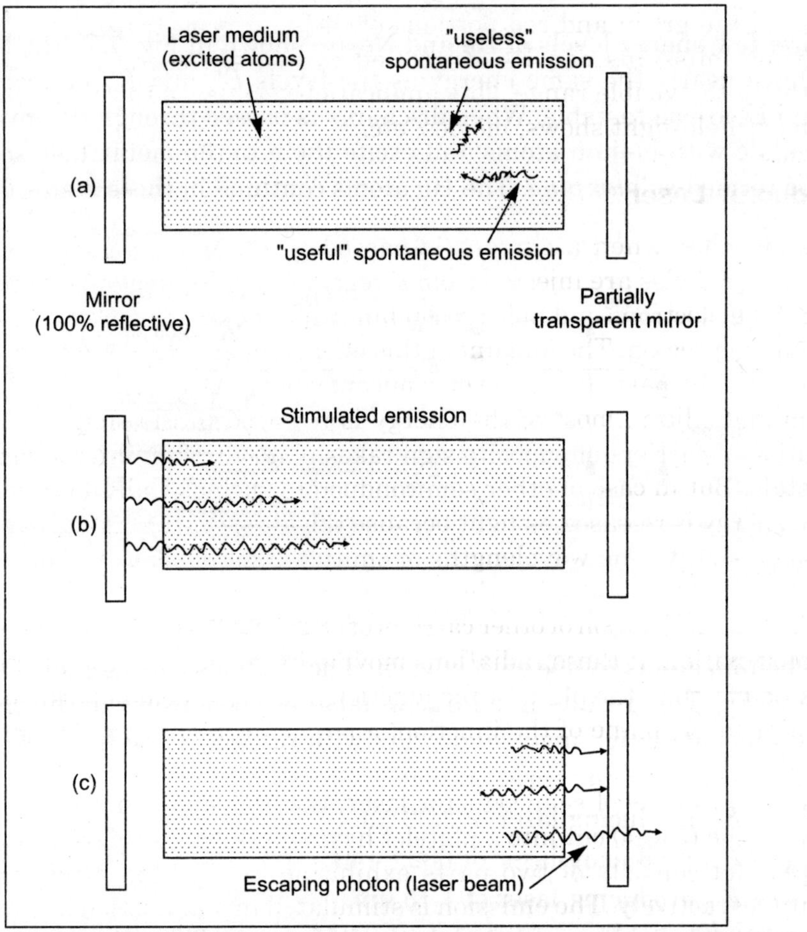

Fig. 7.8 Diagrammatic sketch showing (*a*) excitation of gaseous laser medium (*i.e.*, electrical discharge), (*b*) reflection of photons during stimulated emission, and (*c*) exciting photons in a laser beam

Gas lasers can be explained by the following parts: the gas cavity, excitation of the gas by electrical charge, and stimulated emission of the laser light on discharge of the electrical pulse. A variety of "*pumping methods*" are used in gas laser technology whereby the atoms and molecules of the lasing gas are excited. In some gas mixtures, such as the most popular of all, helium-neon, the electrical discharge excites the helium atoms which transfer their energy to the neon atoms which emit red light. Other gas mixtures, such as carbon dioxide, nitrogen, and argon, are energized by the CO_2 molecules absorbing the electrical energy and vibrating at a high energy level.

The helium-neon laser cannot produce much power, typically only a few milli-watt, but this versatile laser can emit laser light continuously for many thousands of hours and has a plethora of uses.

The CO_2 laser can be used to produce tera watt (10^{12} W) in pulses of ultra short pulse width (10^{-12} sec). Amplification is possible by making multiple passes of the beam through the huge gas cavities developed to produce clustures of beams of several feet in diameter. Continuous beams of CO_2 lasers of high intensity are also available for such applications as welding or cutting metals. Other popular gas lasers are the argon and krypton lasers which emit in the green and red portions of the spectrum, respectively. Mixed together, the resulting emissions can be controlled to provide four individual and separate wavelengths in the visible range. One application popular with this combination is in the entertainment field-light shows, movies, etc.

Semiconductor Laser

We know that when a current is passed through a *p-n* junction, *p*-region being positively biased, holes are injected from *p*-region into *n*-region and electron from *n*-region to *p*-region. The electrons and holes recombine and release of energy takes place in or very near the junction region. The amount of this energy, called the activation energy or energy gap, depends on the particular type of semiconductor. In case of some semiconductors like germanium and silicon, most of the energy is released in the shape of heat because the recombination of carriers of opposite sign takes place through interaction with the atoms of the crystals. But in case of other semiconductors such as Gallium Arsenide (GaAs) and others, the energy is released as light because the atoms of the crystals are not involved in the release of energy. The wavelengths of emitted light depends on the activation energy of the crystal. Photons emitted at the moment of recombination of an electron with a hole will stimulate recombination of other carriers of electric charges. The result will be stimulated emission of radiation. If these radiations moving in the plane of the junctions are made to move back and forth in the plane of the junction by reflection at opposite parallel sides and perpendicular to the plane of the junction, a powerful laser beam of stimulated radiation can be produced.

In the first models of semicoductor laser, the active medium was a single crystal of Gallium Arsenide (GaAs) cut into a platelet having a thickness of 0.5 mm as shown in Fig. 7.9. The platelet consists of two parts exhibiting an electron conductivity and a hole conductivity respectively. The emission is stimulated in *p-n* junction layer which, of course, is very thin. Electric field is applied to the crystal platelet through a strip electrode fixed to its upper surface.

When exciting currents are small, only a small part of carriers undergo recombination. The process is spontaneous. The laser radiation is random and incoherent. But when the current density is increased the emission becomes more and more coherent and the radiation intensity markedly increases. The efficiency of GaAs lasers reaches 40 percent. When cooled to 20 K, semiconductor lasers have delivered an output of more than two watt of continuous power, which is the most continuous power produced by a laser. It is believed that semiconductor lasers may reach 100% efficiency. They are capable of ensuring a high stability of the output frequency, which is the characteristic only of gas lasers.

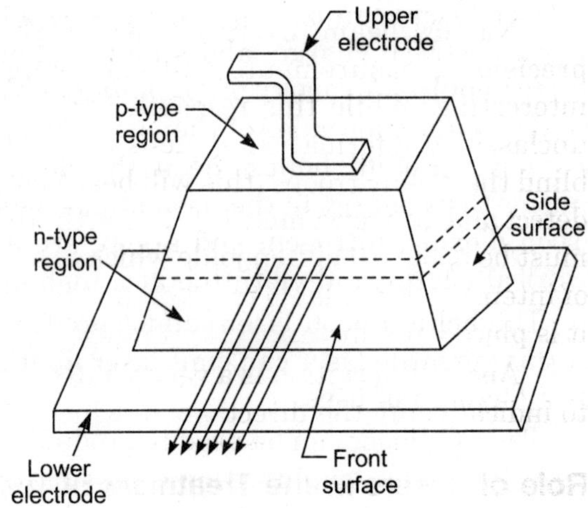

Fig. 7.9 Design of a semiconductor laser emitting element

Laser Applications

Laser beams have a variety of remarkably important applications in different fields of science and technology. We shall here describe only few applications.

Because they are coherent, they can be used to carry signals by amplitude modulation. Hence development for communication purposes is active. It can however be said even at this stage that use of lasers have extended the upper limit of communication frequencies from 10^{11} c/s to 10^{14} c/s. As regards the voice channels, it can be said that 80 million channels can be contained in the spectral region corresponding to that of visible light. The light intensity and *monochromaticity* of laser light makes it useful in photographic and spectroscopic research. Among the conventional applications of this type are in the high speed photography. (bullets traveling at 200 m/s can be well photographed), and in obtaining the spectral information of inert gases and rare earths as well as in establishing and confirming transition rates between different allowable electronic energy levels in atoms.

Since the laser beam can be easily focused to a small area, it has been used for certain surgical and metallurgical purposes where highly localized heating is mandatory. As regards surgical applications, the laser is now used somewhat routinely for photocoagulation, such as one may wish in operations on the human retina where detachment has occurred. If the retina becomes loose, vision is gradually lost in that eye. The laser may thus be used for re-attachment. Human and animal early stage cancers disappear after two to four weeks exposure. It has also been used for cauterization in humans and for localized treatment of skin growth and blemishes. Reference of metallurgical applications are many. Technological applications include machining, welding, drilling, and soldering. A prime interest in drilling with lasers is in the forming of holes and slots in exotic metals and refractory materials, such as tungsten and metal carbides. There is also some interest in the drilling of diamonds.

Narrow beam divergence and high intensity also make the laser beam useful in precision measurements and surveying, and weaponry. Its use in weaponry is very interesting. While this research is in its infancy and is being done in secrecy, certain unclassified information alone is available. For example, the laser beam may be used to blind the enemy troops; this will bear heavily on the troops. Further, they may be aimed to detect and destroy a missile from earth. However, the intensity of the beam for this purpose must be about 10 billion joule which is presently not available. But a sufficient magnitude of intensity is available to set on fire easily the inflammable material of the enemy, even if it is placed at a distance of two miles.

Above description is a very small collection of the applications of laser beam, designed to indicate only the directions in which its use is presently advancing.

Role of Lasers in the Treatment of Diseases

(Journal of Biomedicine, June 1995, 2, 13 Dr. S.O. Pillai and Dr. O.S. Pillai, Institute of Paediatrics, Chennai-8)

Abstract of the Review Paper

This review work gives some important details of lasers that are used in the field of medicine. The main advantages of laser therapy are:

1. No danger of infection. The skin is neither hurt nor in any way damaged as the laser beam has no pricking effect.
2. Laser beam can be applied with utmost precision to all the parts of the body.
3. Aimed irradiation. Individual organs or part of the organs can be treated through points.
4. Short periods of therapy. The irradiation time in many cases is from 5 sec to 50 sec.
5. Precise point of positioning.
6. Controlled therapy.
7. Complete lack of pain and hence children always feel at home while irradiation.

Biology

One of the areas of biology to benefit the most from lasers is the discipline of bio-chemistry and in particular spectroscopic study of fast chemical reactions and also in short-lived species. Much progress has been made in clinical immunology through the use of laser cytofluorometry. Plasma study and cellular kinetics in leukaemia are some potential areas for study. Use of CO_2 laser for non-invasive analysis of blood sugar, urea and lipids has been demonstrated.

Ophthalmology

Ophthalmologists use lasers to treat a variety of eye problems, including retinal bleeding and the excessive growth of blood vessels in the eye caused by diabetes. Argon lasers are used for *spot-welding* - reattaching retinas that have become partly detached from the back surface of the eye. Treatment of glaucoma, cataracts and a host of other disorders are there.

Acupuncture System

Laser systems are used to stimulate the acupuncture points by intermittent irradiation of pulsed laser beam through flexible optical fibre. Laser stimulation of multiple acupuncture points simultaneously resulted in remarkable alleviation of headache and stiff neck due to muscle contraction. This therapy proved to be effective for the treatment of frozen shoulders and to promote rehabilitation of paralytic limbs.

Urology and Neurosurgery

Significant results by laser beams in kidney bladder and prostate using CO_2 and Neodymium-YAG laser use obtained by M. Missiggaung even in the year 1970. The endoscopic applications of lasers is a open book. The first successful tumour extirpation on human using CO_2 laser was by P.N. Ascher and F. Heppner. Lasers are also widely used in two more areas namely Gynaecology and Revascularization of the heart.

The above cited paper may be referred for further details.

7.2 FIBER OPTICS

7.2.1 INTRODUCTION

Optical fibers are best known for their use in signal transmission for communications. However, they were originally developed to serve as optical elements. Single fibers and bundles of randomly arranged fibers can serve as light pipes, conveying light from one point to another along sometimes convoluted paths. Bundles of fibers in which the fibers maintain their relative alignments at the input and output faces can serve other functions. For example, a simple image inverter can be made by twisting the bundle 180°. A lens can do the same sort of image inversion, but not in such a short linear distance. Aligned or 'coherent' fiber bundles can be made rigid or flexible, and both types have been used in many applications, such as in medical endoscopes to view inside regions of the human body not otherwise accessible without surgery. Although most recent writings on fiber optics concentrate on communications, there are some good references on the design and applications of fiber bundles as optical elements.

7.2.2 ACCEPTANCE ANGLE AND ACCEPTANCE CONE

A monochromatic light ray *SA* from a laser source enters the optical fiber as shown in Fig. 7.1. Let the angle of incidence at air-core interface be i and r be the angle of refraction at air-core interface. Let θ be the angle of incidence at core-cladding interface. Let n_0, n_1 and n_2 be the refractive indices of air, core and cladding materials respectively. Thus

$$\frac{\sin i}{\sin r} = \frac{n_1}{n_0} \quad \text{or} \quad \sin i = \frac{n_1}{n_0} \sin r$$

i.e.,
$$\sin i = \frac{n_1}{n_0}\sin(90° - \theta) = \frac{n_1}{n_0}\cos\theta$$

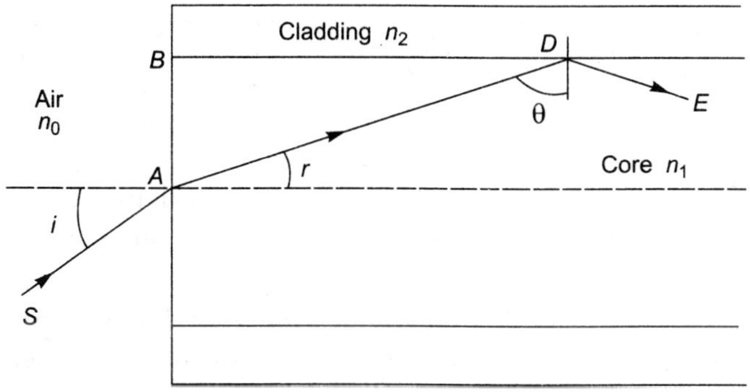

Fig. 7.10 Propagation of light ray through optical fiber

We know that as i increases, r will increase and hence θ will decrease. However, when the angle of incidence (i) at air-core interface is maximum *i.e.,* equal to the acceptance angle, i_{max} , the angle of incidence (θ) at core-cladding interface becomes equal to the critical angle C of core (denser medium) with respect to the cladding (rarer medium) the refracted ray in cladding will just grace the core-clad interface and travel along the surface of optical fiber thus when $i = i_{max}$, $\theta = C$ and hence the above equation becomes

$$\sin i_{max} = \frac{n_1}{n_0}\cos C$$

or
$$\sin i_{max} = \frac{n_1}{n_0}\sqrt{1-\sin^2 C} \qquad ...(7.2.1)$$

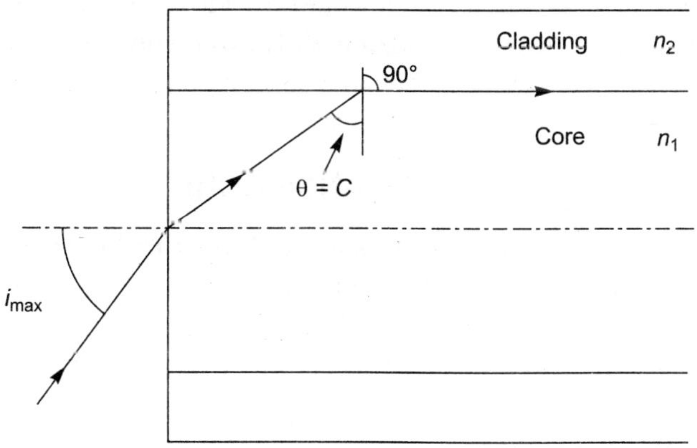

Fig. 7.11 Light ray incident at critical angle at core-cladding interface

From Fig. 7.11 we have

$$\frac{\sin C}{\sin 90^\circ} = \frac{n_2}{n_1} \quad \text{or} \quad \sin C = \frac{n_2}{n_1}\sin 90^\circ$$

$\sin C = \frac{n_2}{n_1}$. Substituting this value in Equation (7.2.1), we get

$$\sin i_{max} = \frac{n_1}{n_0}\sqrt{1 - \frac{n_2^2}{n_1^2}}$$

$$= \frac{n_1}{n_0}\sqrt{\frac{n_1^2 - n_2^2}{n_1^2}}$$

$$= \frac{\sqrt{n_1^2 - n_2^2}}{n_0}$$

$$i_{max} = \sin^{-1}\left(\frac{\sqrt{n_1^2 - n_2^2}}{n_0}\right)$$

The maximum angle of incidence, i_{max} is called the acceptance angle (or) half-angle of acceptance core. For air $n_0 = 1$ and so

$$i_{max} = \sin^{-1}\left[(n_1^2 - n_2^2)\right]^{1/2} \qquad \text{...(7.2.1}a\text{)}$$

7.2.3 TOTAL INTERNAL REFLECTION

The growing use of fiber optics in optical instrumentation, communication, and laser beam delivery for surgical and other important applications is a open book. Optical fibers may be either glass or plastic. The operation of simple cptical fibers is based on the phenomenon of *total internal reflection.*

Since total internal reflection is probably the important phenomenon used in fiber optical communication, we shall revise it in few lines with small diagrams. If light of given wavelength from a medium of refractive index n_1 strikes a second medium with $n = n_2$, with θ as the angle of refraction, then by *Snell's law*:

$$\boxed{n_1 \sin i = n_2 \sin\theta} \qquad \text{...(7.2.2)}$$

where i is the angle of incidence.

(*a*) The ray is refracted as $i < \theta_c$.

(*b*) The ray traverses along the interface as $i = \theta_c$.

(*c*) The ray is totally internally reflected as $i > \theta_c$.

When a ray of light is travelling from a optically dense medium (high index of refraction) to a less dense medium (low index of refraction), it will not be refracted if it strikes the surface at an angle equal to or greater than a particular angle called '*critical angle*'. Instead it will be totally reflected at the surface, between the two media (Fig. 7.12). Such a phenomenon occurs because when a ray of light travels from a dense medium to a less dense medium, the refracted ray bends away from the normal. Consider the case when $\theta = 90^\circ$ for a particular value of $i = \theta_c$ then we have

$$n_1 \sin\theta_c = n_2 \sin 90^\circ$$

$$\boxed{\sin\theta_c = \frac{n_2}{n_1}} \qquad \text{...(7.2.3)}$$

with $(n_1 > n_2)$

i.e., for any ray whose angle of incidence is greater than the critical angle, *total internal reflection occurs*. Of course, it must be emphasised here that the ray is travelling from a medium of higher refractive index to a medium of lower refractive index.

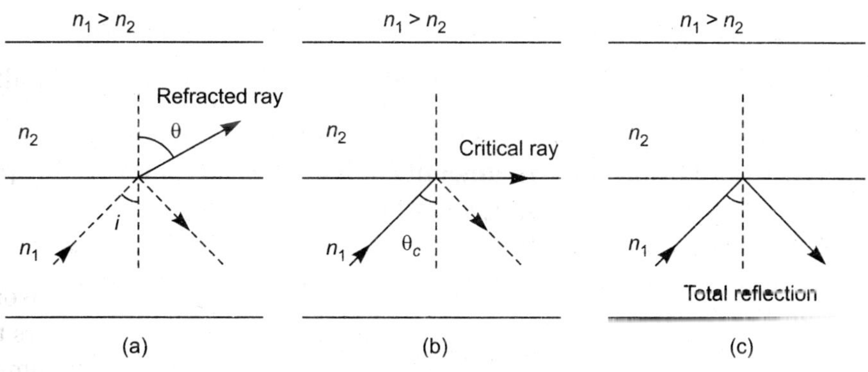

Fig. 7.12 Total internal reflection

7.2.4 STRUCTURE OF AN OPTICAL FIBER AND NUMERICAL APERTURE

The most widely used structure is the single solid dielectric cylinder of radius 'a' and index of refraction n_1 shown in Fig. 7.13. This cylinder is known as the *core of the fiber*. The core is surrounded by a similar material but of lower refractive index n_2 (*i.e.*, $n_2 < n_1$), known as *cladding*. Although in principle, a cladding is not necessary for light to propagate along the core of the fiber, it serves several purposes.

Fig. 7.13 Representation of fiber structure. A circular solid core of refractive index n_1, is surrounded by a cladding having refractive index $n_2 < n_1$

The cladding reduces scattering loss resulting from dielectric discontinuities at the core surface, it adds mechanical strength to the fiber and it protects the core from absorbing surface contaminants with which it could come in contact. The whole assembly is encapsulating in a plastic buffer known as buffer coating. This adds further strength to the fiber and mechanically isolates the fibers from small geometrical irregularities, distortions or roughness of adjacent surfaces. These perturbations could otherwise cause scattering losses induced by random microscopic bends that can arise when the fibers are incorporated into cables or supported by other structures. Optical fibers are made in lengths of 1 km. Fibers can be joined using suitable connectors. The thickness of an optical fiber is much less than that of human hair. Its outer diameter ranges from 0.1 to 0.15 mm. Core diameters range from 5 to 600 μm, whereas cladding diameters vary from 125 to 750 μm. To keep the light within the core, the cladding must have a minimum thickness of one or two wavelength of light. The protective jacket may add as much as 100 μm in diameter to the fiber's total diameter. Optical fiber is made of silica (SiO_2) glass. But during drawing of the fiber, cladding is formed by the addition of small amounts of boron, germanium or phosphorous in the outer layer to reduce the refractive index. Cladding is necessary to provide proper light guidance *i.e.*, to retain the light wave within the core as well as to provide high mechanical strength and safety to the core from scratches. Let us now consider the path of the rays as shown in Fig. 7.14.

Assuming the outside medium to have a refractive index n_0 (for most cases it is unity), we can write

$$\boxed{\frac{n_1}{n_0} = \frac{\sin i}{\sin \theta}} \qquad \text{...(7.2.4)}$$

Here

$\sin\phi < \sin 90°$. It is possible only when

$$\sin\phi > \frac{n_2}{n_1}$$

i.e.,

$$\sin\phi\ (= \cos\theta) > \frac{n_2}{n_1}$$

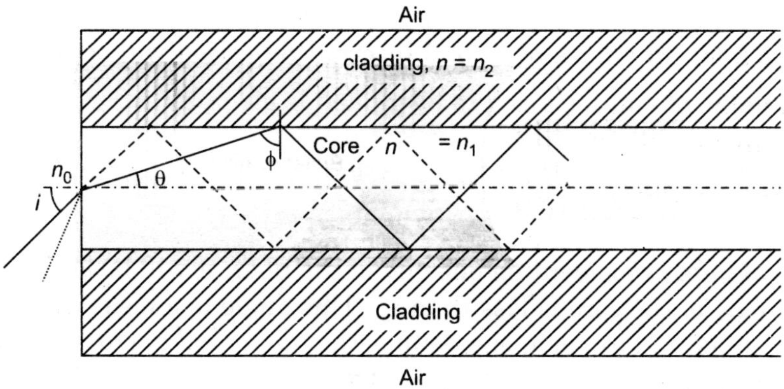

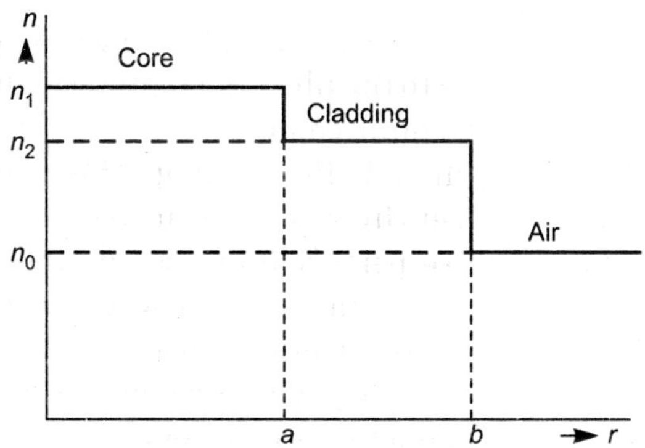

Fig. 7.14 Light propagation in fibers

Thus

$$\sin\theta < \left[1-\left(\frac{n_2}{n_1}\right)^2\right]^{1/2} \quad \text{...(7.2.5)}$$

From Eqn. (7.2.4)

$$\sin i = \frac{n_1 \sin\theta}{n_0}$$

$$\sin i < \frac{n_1}{n_0}\left[1-\left(\frac{n_2}{n_1}\right)^2\right]^{1/2} \quad \text{...(7.2.6)}$$

or

$$\sin i < \frac{n_1}{n_0}\left[\frac{n_1^2 - n_2^2}{n_1^2}\right]^{1/2} < \left\{\frac{n_1^2 - n_2^2}{n_0^2}\right\}^{1/2} \quad \text{...(7.2.7)}$$

If $n_1^2 - n_2^2 \geq n_0^2$, then for all values of i, total internal reflection will occur. Assuming $n_o = 1$, the maximum value of sin$i$ for a ray to be guided is given by

$$\sin i_m = \begin{cases} \left(n_1^2 - n_2^2\right)^{1/2} & \text{when } n_1^2 - n_2^2 < 1 \\ 1 & \text{when } n_1^2 - n_2^2 > 1 \end{cases} \quad ...(7.2.8)$$

Thus, if a course of light is incident on one end of the fiber, it will be guided through it provided the semi-angle of the core is less than i_m. This angle is a measure of the light gathering power of the fiber and as such, one defines the *numerical aperature* (*NA*) of the fiber by the following equation

$$NA = \left(n_1^2 - n_2^2\right)^{1/2} = n_1(2\Delta)^{1/2} \quad ...(7.2.9)$$

where

$$\Delta = \frac{(n_1^2 - n_2^2)}{2n_1^2} = \frac{(n_1 - n_2)}{n_1} \quad [i.e., \text{ when } n_1 = n_2]$$

and in the last step we have assumed $n_1 = n_2$ which is indeed the case for all practical optical fibers.

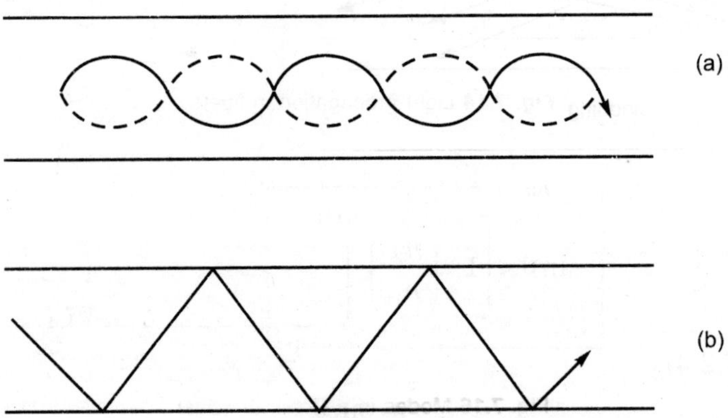

Fig. 7.15 Methods of optical confinement (*a*) Refractive confinement, (*b*) Reflective confinement

For a typical optical fiber $n_2 = 1.458$, $\Delta = 0.01$ and the corresponding NA is ≈ 0.2. Thus the fiber would accept light incident over a cone with a semi-angle $\sin^{-1}(0.2) \approx 11.5^o$ about the axis. Optical fibers guide light by either reflection or refraction, depending on the type of the fiber used. In the reflective types light travels in a zig-zag fashion as shown in Fig.7.15(*b*). In the refractive types light rays travel in a continuous curve (Fig. 7.15(*a*)). In either case light rays are confined to the core.

Optical fibers have great advantages over copper wire for communications purposes:

- Much greater bandwidth

- Immunity from electromagnetic interference
- Smaller and lighter
- Stable or declining price

There are disadvantages in getting signals into and in joining fibers, but these are relative minor problems. But the more important consideration with optical fibers is bandwidth, which is largely determined by dispersion. Dispersion is the term given to phenomena which tend to a spread out an optical signal over time and frequency. In a *binary pulse-code modulation (PCM) system*, the energy in a light pulse which is 'on' is spread out so that if 'on' pulses are too close together they cannot be separated.

7.2.5 SOURCES OF DISPERSION IN OPTICAL FIBERS TYPES OF OPTICAL FIBERS

There are two chief sources of dispersion in optical fibers: modal (or intermodal) dispersion and intramodal dispersion (made up of material and wave guide dispersion). We shall deal first with modal dispersion in two types of fiber in turn.

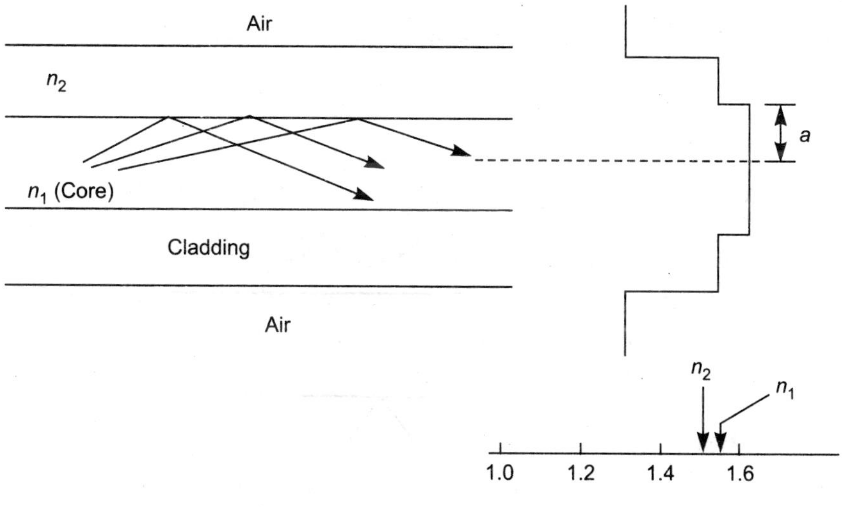

Fig. 7.16 Modes in a step index fibor

1. Step-index Fibers

A step-index fiber consists of a cylinder of material, $RI = n_1$, which is clad with a material of slightly lower RI, n_2. Modal dispersion is much the most important source of dispersion in these fibers and arises because there arc many different paths down the fiber for light rays. Figure 7.7 shows several modes in a fiber. There will be an angle of incidence at which the ray will leave the fiber, that is when $\theta_i < \theta_c$, and such modes will not propagate. A thick fiber will have more modes than a thin one.

The energy of the light ray, and hence its signal content, moves with the group velocity of the ray, v_g, so the group delay in a fiber of length, L, for this ray will be $\tau = \frac{L}{v_g}$.

The dispersion in arrival times because of multimode propagation will be the difference between the longest and shortest group delays. It turns out that the longest group delay is for rays which travel entirely in the central region for which $v_{g1} \approx \frac{c}{n_1}$, and the shortest group delay is for *evanescent rays* which travel in the cladding with group velocity, $v_{g2} \approx \frac{c}{n_2}$. The dispersion is then

$$\Delta\tau \approx \frac{L}{v_{g1}} - \frac{L}{v_{g2}} = \frac{L(n_1 - n_2)}{c} \qquad ...(7.2.10)$$

For a fiber 1 km long, with $n_1 = 1.52$ and $n_2 = 1.50$, the modal dispersion works out to 67 ns. In fibers rather longer than 1 km, the dispersion given by equation (7.2.10) is too great, because defects and kinks allow slower modes to catch up so that

$$\Delta\tau \approx \Delta\tau_1 \sqrt{L}$$

where $\Delta\tau_1$ is the dispersion for 1 km according to Eqn. (7.2.10). The maximum *bit-rate, R,* for PCM signal is given by

$$R \approx (4\Delta\tau)^{-1}$$

while the bandwidth for amplitude modulated signals is roughly $\frac{R}{2}$. Thus over a 50 km length of the fiber in this example, $\Delta\tau = 67 \times \sqrt{50} = 474$ nsec, and $R = 530$ kbits/sec, a modest bite rate. Monomodal step index fibers can be made to eliminate modal dispersion, but the rather small core diameter makes joining and signal insertion much more difficult. A better approach is the use of *RI* which varies in the core material.

2. Graded-index Fibers

The modal dispersion of step-index fibers can be reduced by using several cladding layers and making *multi-step-index* fibers, but enormous reductions in modal dispersion may be achieved by having a continuously changing *RI* from the centre of the fiber to the cladding. These are called *graded-index fibers.* Because the *RI* changes gradually as a ray passes through the core to the cladding, it does not change direction abruptly, but gradually, so that rays of longer pathlength spend a greater proportion of their time in a medium of lower *RI* and hence travel faster. This is the secret of the *low modal dispersion* in graded index fibers.

The *RI* in the central zone of the fiber can be written as

$$n(r) = \begin{cases} n_1\left[\sqrt{1-2(r/a)^{\alpha}}\right] & \textit{for } r < a \\ n_2 & \textit{for } r > a \end{cases} \qquad ...(7.2.11)$$

where r is the distance from the centre of the fiber, a is the radius of the core, $\Delta = \frac{(n_1 - n_2)}{n_1}$. n_1 is the *RI* at the centre, n_2 is the *RI* of the cladding and α is the *grading parameter* (α = 1 for linear grading, $\alpha = 2$ for parabolic and $\alpha = \infty$ for a step-index fiber). Figure 7.17 shows n (r) for various values of α. The optimal value for α is given by

$$\alpha_{opt} = 2\ (1 - \Delta)$$

that is, a little less than two. For optimal α the dispersion is given by

$$\Delta\tau_{min} = \frac{L n_1 \Delta^2}{8c} \qquad ...(7.2.12)$$

when α is little different from its optimal value, the dispersion is

$$\Delta\tau = \frac{L(n_1 - n_2)\,[\alpha - \alpha_{opt}]}{c\ (\alpha + 2)}$$

Suppose $n_1 = 1.5$ and $\Delta = 0.01$, then $\alpha_{opt} = 1.98$, $\frac{\Delta\tau_{min}}{L} = 63\ \text{ps/km}$, and if $\alpha = 2$, $\frac{\Delta\tau}{L} = 0.25\ \text{ns/km}$. The dispersion has a very sharp minimum at α_{opt}, which is easily missed in manufacturing the fiber.

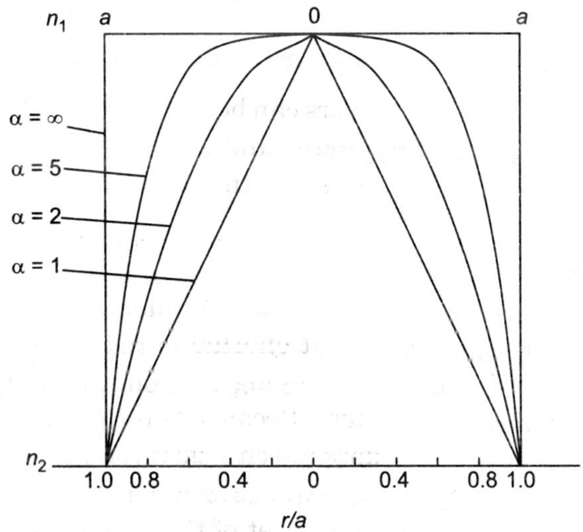

Fig. 7.17 *RI* profiles for some values of the grading parameter, α

The number of modes supported by the graded-index is approximately given by

$$N \approx \frac{(k_1 a)^2 \alpha \Delta}{(\alpha + 2)} \qquad ...(7.2.13)$$

where $k_1 = \dfrac{2\pi n_1}{\lambda}$. When $\alpha = 2$, N is half that for a step-index fiber, which leads to half the energy input to it compared with the step-index fiber, as the energy in each propagating mode is the same.

3. Intramodal Dispersion

Intramodal dispersion is made up of *material dispersion* and *waveguide dispersion.* The former is caused by the RI being a function of wavelength, and would occur regardless of the fiber dimensions. Waveguide dispersion is caused by variation in the propagation constant of the wave guide with frequency. Since waveguide dispersion is proportional to $\dfrac{1}{a^2}$, where a is the fiber radius, it is most important for monomodal fibers. Wave guide dispersion is negative while material dispersion is positive, so in these fibers it is possible to eliminate intramodel dispersion at a particular wavelength.

In silica fibers, the material dispersion is negligible at λ = 1.3 μm, and is roughly $\Delta\tau \leq 30\ (\Delta\lambda)L$ μs elsewhere in the wavelength region from 1 to 2 μm, where $(\Delta\lambda)$ is the spectral width of the light source. Intermodal and intramodal dispersions are not directly additive, but must be combined as follows:

$$\Delta\tau = \sqrt{\Delta\tau_a^2 + \Delta\tau_e^2}$$

where $\Delta\tau_\alpha$ and $\Delta\tau_e$ are the respective intramodal and inter modal dispersions for the entire length of fiber. Intramodal dispersions can be reduced by using a source of narrow spectral range, such as a laser.

7.2.6 ATTENUATION IN OPTICAL FIBERS

At first, attenuation was the most important obstacle to the use of optical fibers, since about 1000 dB/km was the best obtainable one in the 1960s. Subsequent improvements have resulted in an attenuation coefficient for silica fibers as shown in Fig. 7.18, which is very close to the theoretical minimum. The long-wavelength in the infra-red is caused by electronic lattice transitions, especially the strong absorption around 10 μm, which has a large 'tail' down to 1.5 μm. This limit is an intrinsic property of silica, and can only be overcome by changing to other materials, such as alkali halides or fluoride glasses, which have no infrared absorption up to longer wavelength than 10 μm. The absorption peak at 1.4 μm is due to hydroxyl ions in the silica, which cannot be eliminated. At short wavelength the limit is set by Rayleigh scattering. Silica is a glass and as such consists of small volumes in which the atoms (about a dozen in number) are correctly positioned, joined to other such groups of atoms in haphazard way. These groups scatter light because of micro-variations in the RI. The scattering losses go as λ^{-4}, amounting to 4.5 dB/km at 0.6 μm and 0.3 dB/km at 1.2 μm. With reasonable sources and receivers, losses of 0.3 dB/km means that repeaters are not necessary in fibers up to 100 km long. Coupling and bending losses are more significant in most cases.

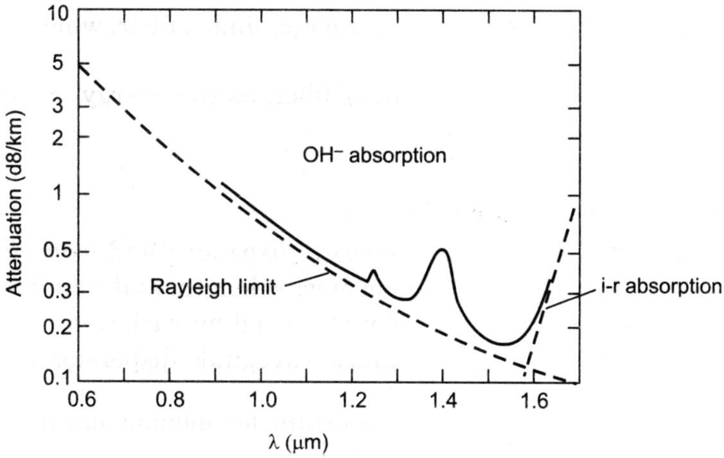

Fig. 7.18 Attenuation against wavelength for silica fibers

7.2.7 A NOTE ON THE MANUFACTURE OF OPTICAL FIBERS

The basic process is simple in principle. The core material is melted in a container with a hole in the bottom which is centrally placed in another vessel containing the molten cladding material, as shown in Fig. 7.19. Graded index fibers may be obtained by adding dopants as GeO_2 to the molten SiO_2 which diffuses into the central core by an amount which depends on the temperature and time taken to cool. The *RI* profile is not easily controlled.

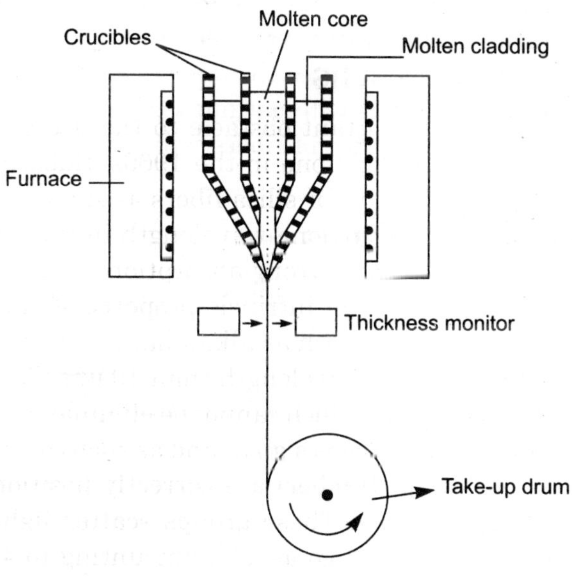

Fig. 7.19 An apparatus for producing cladded fibers

A second process offering potentially greater control of *RI* profile is a CVD (Chemical Vapour Deposition process) in which doped SiO_2 is deposited on the walls of a tube of pure silica under the influence of heat which can be applied to a narrow moving zone of the tube. Volatile reactants such as $SiCl_4$ and $GeCl_4$ are used to transport germanium and silicon, which are oxidized by oxygen at 1200°C to (Ge, Si)O_2 of the desired *RI*. Repeated passes with differing gas compositions can be made to give the desired profile. The finished tube is then heated to beyond its softening point (about 1800°C) so that it collapses into a solid rod, which can then be drawn down to the desired diameter. This is largely determined by the temperature of the undrawn fiber (through its very pronounced influence on viscosity), though the draw-rate is also important and can be controlled by feeding back thickness information to the drum motor. Deposition can also be carried out on the outside of a rod or tube in much the same way to produce the required index gradient.

7.2.8 SALIENT FEATURES OF FIBER OPTICAL SYSTEM OF COMMUNICATIONS

Very large information transmission capacity and large repeater spacing are the main features of fiber optic communication systems. They also offer many advantages compared with the usual metallic transmission process. Since the fibers are composed of dielectric materials they are totally immune to extraneous interfering electromagnetic signals. There is virtually no signal leakage from the fibers and therefore cross-talks between neighbouring fibers are almost absent while this is very common in the conventional systems. Since signals do not leak away from the fibers they are resistant to intrusion and are accordingly highly suited to secure good results in defence communication networks. Since these fibers are immune to electromagnetic interference and do not pick up line currents, they can be safely used in a power station and can be laid along side metallic power cables.

They are also much smaller in size than an electrical line of equivalent band width and thus would occupy much less duct space. Also the raw material used in the fabrication of low loss fibers is silica, which is abundantly available in nature (of course in an impure form), whereas copper or aluminium constitutes the basic raw material in coaxial cables. Cost surveys indicate that coaxial cable price will continue to rise due to an almost continuous increase in the price of copper but the price of fibers will come down as the production volume increases.

7.2.9 APPLICATIONS OF OPTICAL FIBERS

Optical fibers and transmission of light (via) an optical fiber have a wide variety of applications. We discuss here some of the applications.

(*a*) They are used for illumination and short distance transmission of images.

(*b*) They are used as wave-guides in telecommunications.

(*c*) They are used in fabricating a new family of sensors.

Traditionally, electronic communications were carried out by sending electrical signal through copper cables, coaxial cables or wave guides. In recent years optical fibers are being used where light signals replace electrical signals. Large number of signals can be passed through optical fibers. They are cheap and light weighted.

Medical applications: Fiber optic endoscope is used to inspect internal organs for diagnostic purposes. In ophthalmology, a laser beam guided by optical fibers is used to reattach detached retina and to correct defective vision. In cardiology, optical energy transmitted through an optical fiber is used to evaporate buildup plaque that is blocking an artery.

Military applications: Fiber guided missiles are used in recent wars. Sensors mounted on the missile transmit video information through the optical fiber to a ground control van and receives commands from the van again. The control van continuously monitors the course of the missile and if necessary correct its course to ensure the missile precisely hits the target.

7.2.10 HOLOGRAPHY

Holography is lensless three-dimensional photography. A conventional photograph is only a flat record of real image projected on to a photographic film. Information about the three-dimensional character of the object is almost entirely lost during the photographic recording process. Only the blurring of the images of objects that are not within the depth of field of the camera preserves any record of the location of the object relative to the camera. A *hologram* on the other hand is a special 'photograph' of an object that remains information about the phase of waves coming from the object. Holography has a much longer history than that of the laser, although it was the invention of the laser that made the production of high quality hologram a reality.

In 1947, Denis Gabor was looking for a way to improve the quality of images that could be obtained with an electron microscope. Although the electron microscopes of that day should have been capable of resolving down to the atomic scale, *aberration* of the *electron optics* prevented them from doing so. Gabor conceived the idea of recording an electron hologram, which would contain amplitude and phase information, and then reconstructing a corrected image of the object by optical means. He did succeed in producing holograms and performing image reconstruction, but the optical sources available to him were not sufficiently temporally coherent to allow high quality images to be achieved. For his pioneering work he was awarded the 1971 Nobel Prize for physics.

The high degree of coherence obtainable with a laser has made the production of holograms and applications of holography into a large subject in its own right. The principles underlying holography share much in common with the phenomena of interference and diffraction.

In conventional photography the photographic emulsion becomes dark according to the local intensity of the real image falling on it. The emulsion is a mixture of silver halide crystals of different sizes, predominantly silver bromide, in a gelatin matrix. Incident light frees electrons from the halide, but the electrons become trapped. Addition of a developer reduces the silver ions in an exposed crystal to silver atoms, leading to a strongly absorbing region where exposure levels were high. This is how a photographic negative is produced. We make a *hologram*, and preserve phase information in the photographic record, by allowing temporally and spatially coherent light to illuminate the object. Light reflected and scattered from the object falls on a photographic film together with a reference wave supplied directly to the film with the same laser as is used to illuminate the object. The schematic way in which this is done is shown in Fig. 7.20.

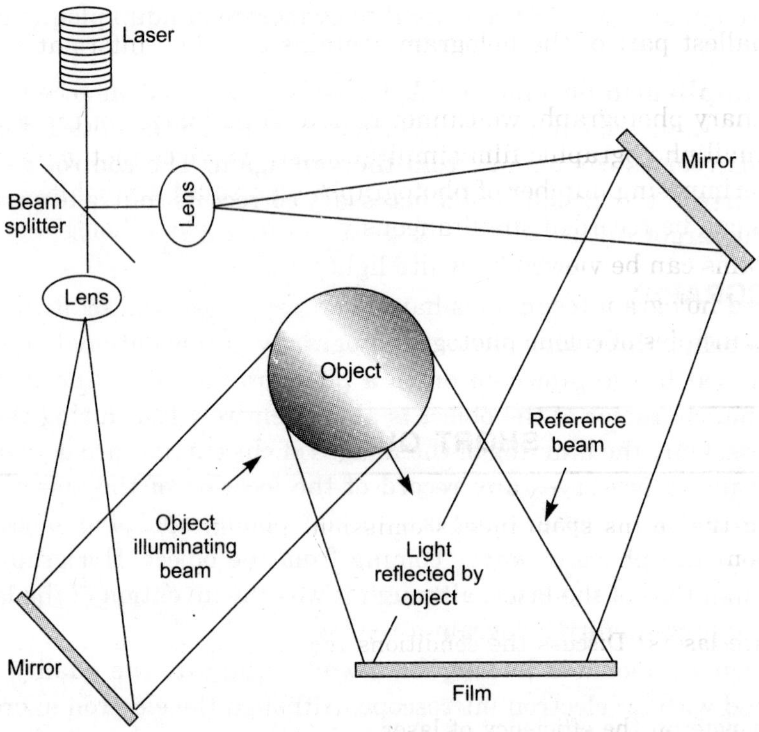

Fig. 7.20 Diagramatic sketch illustrating technique with a laser to produce a hologram on film

Light emitted by a laser splits into two paths, one to create an object beam and the other to serve as a reference beam. It is intersecting, or interference, of the two beams and the resultant beam projected on to a photographic film that forms the holographic image. Because every location on the object is illuminated by the object beam, the reflection of these waves intersects the reference pattern and the resulting image appears suspended in space so that as the viewer moves around the image a three dimensional object is visualizd. The developed hologram does not contain a recognizable image of the object as a conventional photograph does. The hologram generally appears as a collection of bright and dark lines with bright and dark patterns of concentric circles. The stored image can be regarded as a representation of the diffraction pattern of the object. The hologram formed by exposure of a film and the production of a complex pattern of transmission is called *amplitude hologram*. However, it is possible to process the film chemically by bleaching so that the dark silver atoms are converted into silver salts that are transparent but have a different refractive index than the surrounding film. This converts the recorded hologram into a pattern of phase information and is called a *phase hologram*.

7.2.11 APPLICATIONS OF HOLOGRAPHY

1. In common photography, each part of the work has information of the original object separately. Hence if some part of the object is lost or destroyed, then the history of that part of the object is lost. However in the case of a hologram, even

the smallest part of the hologram contains complete information of the whole object.

2. In ordinary photograph, we cannot record many images of the different objects on a small photographic film simultaneously as we cannot get any information by superimposing number of photographs on a small film while large number of images can be recorded simultaneously in the case of a hologram.
3. Holograms can be viewed by white light as well.
4. Coloured hologram techniques have been developed well by Russians using the basic principles of colour photography.

SHORT QUESTIONS

7.1.1 Explain the terms spontaneous emission, stimulated emission and population inversion.

(Anna. U, BE, April 1992)

7.1.2 What are lasers? Discuss the conditions required for laser action.

(JNTU, BE, April 2003)

7.1.3* Write a note on the efficiency of laser.

7.1.4 Explain the process of absorption, spontaneous emission and stimulated emission.

(MU, BE, December 1999)

7.1.5 How is that laser light is described as the purest and most intense form of radiation?

7.1.6* Q-switching. Explain.

(Madras U, BE, November 2001)

7.1.7 What is meant by maser? Explain the process of stimulated emission with energy level diagram.

7.1.8* Discuss the salient uses of ruby lasers. *(JNTU, BE, May 2003)*

7.1.9 What are the main uses of lasers?

7.1.10* Discuss briefly laser Angioplasty.

(University of Maryland, U.K, UG, Winter 1996)

7.1.11* Write a note on the general properties of laser light.

7.1.12* Consider a repetitively pulsed neodyium: YAG laser of average power P_{av} = 5 kW and beam diameter d = 7 mm. Find the beam power density.

7.2.1 Explain the principles involved in fiber-optic communication.

7.2.2* Discuss fiberscope with a suitable diagram.

7.2.3 Discuss the different types of optical fibers.

7.2.4 What is holography?

7.2.5* What are the advantages in using optical fibers in communication and in other areas?

7.2.6* The core of silica optical fibers contains some B_2O_3, GeO_2 and P_2O_5. How do these impurities increase the refractive index?

Answers to starred questions

7.1.3* The output of a small helium-neon laser is of the order of a few milliwatt. But the power required to operate the laser is of the order of 100 watt. Thus efficiency of the laser is very small. But the usefulness of the laser lies in the coherence and directionality of the laser beam. The energy density of the beam is very high. A laser can deliver an intensity of 100 to 1000 W/m^2. Further a laser beam is perfectly parallel. There is very little divergence over long distances. For example the Apollo 2 expedition left a plane mirror on the surface of the moon. A laser beam sent from the earth was narrow enough to be detected after reflection at the mirror. This distance covered by the beam was more than three quarter of a million kilometre.

7.1.6* Q-switching is a technique for obtaining short (microsecond pulses at few kilohertz frequencies), intense bursts of oscillation from lasers. It is a transient phenomenon, *i.e.*, we can obtain a very large peak pulsed power from a laser even if it can only deliver much lower continuous wave power. Single high power pulses can be obtained by introducing time or irradiance dependent losses into cavity. The effects of such losses can be interpreted in terms of the spiking oscillations. If there is initially a very high loss in the laser cavity, the gain due to population inversion can reach a very high value without laser oscillation occurring. The high losses prevents laser action while energy is being pumped into the excited state of the medium. If, when a large population inversion has been reached, the cavity loss is suddenly reduced (*i.e*, cavity Q is switched to a high value), laser oscillations will suddenly commence. On Q-switching the threshold gain decreases immediately while the actual gain remains high because of the large population inversion. Owing to the large difference between the actual and the threshold gain, laser oscillations within the cavity build up very rapidly and all of the available energy is emitted in a single large pulse. This quickly depopulates the upper lasing level to such an extent that the gain is reduced below the threshold and lasing action stops. *Q*-switching is carried out by placing a closed shutter within the cavity, thereby isolating the cavity from the laser medium. After the laser has been pumped, the shutter is opened, so restoring the *Q* of the cavity. A little thought reveals that there are two important requirements for effective *Q*-switching. These are

(*i*) the rate of pumping must be faster than the spontaneous decay rate of the upper lasing level, otherwise the upper level will be empty more quickly than it can be filled so that a sufficiently large population inversion will not be achieved.

(*ii*) the Q-switch must switch rapidly in comparison to the build up of laser oscillations, otherwise the later will build up gradually and a longer pulse will be obtained so reducing the peak power. In practice, the *Q*-switch should operate in a time less than 1ns.

7.1.8* The largest uses of ruby lasers has been in holography. A *Q*-switch can produce one, two, three or more highly coherent pulses during a single flash lamp pump pulse. These 10 to 30 ns pulses can record holograms to freeze the motion of moving objects. Double pulses can record a pair of pulses on the same plate, making it possible to observe deformation in techniques widely used in non destructive testing.

Thomson scattering of laser pulses can record the density and temperature of electrons in high-energy plasmas. Ruby lasers also can perform atmospheric ranging, scattering and lidar measurements where the high peak power and red wavelength are important. Other solid-state lasers have largely supplanted ruby in some old applications such as hole drilling, military range finding, and military target designators. However, ruby's wavelength and its spiking behaviour in long-pulse mode may make it attractive for some specialized applications. Ruby lasers are being investigated for tattoo removal, because the ruby wavelength is strongly absorbed by dyes used in some tattoos. Ruby lasers are also used in general research, including plasma production, studies of liquid flow and fluorescence spectroscopy.

7.1.10* Arferiosclerosis is the constriction-or-blocking, of blood flow through arteries. The material that leads to this problems is called plaque, a composite material containing layers of fat, fibrous tissue, muscle tissue and varying amounts of calcium [as $Ca_3(PO_4)_2$]. There has been great interest, and some success, in using both CW and pulsed lasers to ablate plaque away. A laser beam is directed at the blockage in an artery along an optical fibre inserted through a catheter into the artery. The catheter is threaded over a previously placed guide wire. The use of lasers in this way is called laser angioplasty, in contrast with the earlier (and still more widely used and successful) technique of balloon angioplasty. In this technique a small balloon is threaded into the occluded region of the artery and then filled with fluid, thereby compressing the plaque against the wall of the artery. The problems that have been faced in the use of lasers to remove plaque have involved:

(*i*) Avoidance of damage to the wall of the artery, which is only about 1mm thick. This problem has been alleviated to a large extent by placing a ball lens on the end of the fibre.

(*ii*) The opened region can reocclude within a few months.

(*iii*) Thermal damage to healthy tissue can occur. The use of pulsed lasers seems to avoid thermal damage problem while short intense pulses are effective in photoablating the plaque. Sub-micro second ultraviolet pulses from XeCl lasers at 308 nm are being investigated in this application as is the use of holmium lasers operating at 2.1 μm. These lasers are used in a long pulse mode delivering in a few joules per pulse in 1/4 second long pulses at 1–10 pulses per second. Their wavelength of 2.1 μm is strongly absorbed by water. Both these ultraviolet and near-infrared wavelengths can be delivered to the occluded site along silica fibres. The treatment of the occluded artery is frequently followed up by an application of angioplasty.

7.1.11* It should be remembered that not all of the different types of laser exhibits these properties to the same degree and, this may generally restrict the choice of laser for a given application.

1. *Directionality*:

   Apart from semiconductor junction lasers, lasers emit radiation in a highly directional, collimated beam with low angle of divergence. This is important because it means that the energy carried by the laser beam can be collected easily and focused on to a small area.

2. *Line width*:

   Laser beam is potentially extremely monochromatic but the spectral content of the laser radiation may extent over almost as wide a range as the fluorescent line width of the laser medium. In otherwords, although the line width of an individual cavity mode may be extremely small there may be many modes present in the laser output.

3. *Beam coherence*:
One of the characteristics of stimulated emission is that the stimulated wave is in phase with the stimulating wave, *i.e.*, the spatial and temporal variation of the electric field of the two waves are the same. Thus for a perfect laser we would expect the electric field to vary with time in an identical fashion for every point on the beam cross-section.

4. *Brightness*:
The primary characteristic of laser radiation is that lasers have a higher brighness than any other light source. We define brightness as the power emitted per unit area per unit solid angle.

5. *Focusing*:
The minimum spot size to which a laser beam can be focused is determined by diffraction. A single mode beam can be focused into a spot which has dimension of the order of the wavelength of light, though imperfections in the optical system may mean that we cannot achieve this in practice.

6. *Tuning*:
Some lasers can be tuned to emit radiation over a range of wavelengths. Laser tunability leads to applications in photochemistry, and Raman spectroscopy.

7.1.12* $E = \dfrac{P_{av}}{\pi(d/2)^2} = \dfrac{5}{(\pi \times 3.5 \times 10^{-3})^2}$

i.e. $E = 1.30 \times 10^8$ W/m^2

7.2.2* The initial efforts in fiber optics were concerned with the fabrication of fiberscope in endoscopy for the visualization of internal portions of the human body. The lasers in combination with fiber optic probes not only permit the visualization of internal portions of human body but also the selective cauterization of tissues, if necessary. A typical fiberscope is shown in the figure.

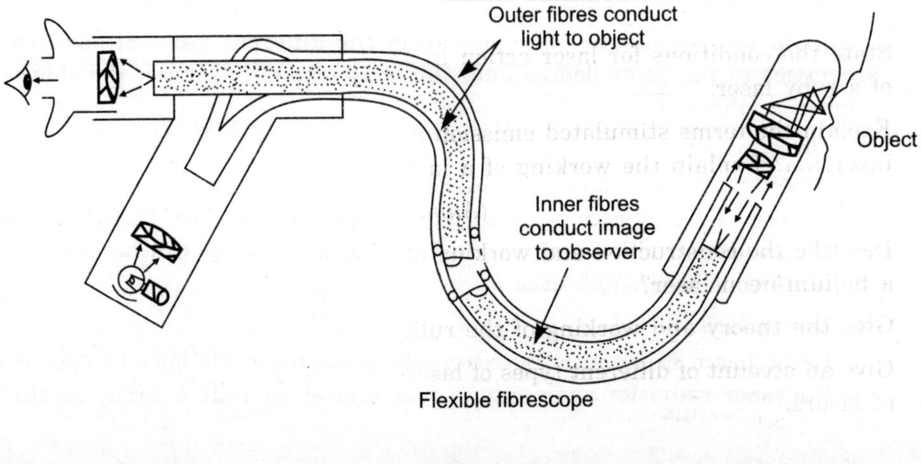

Flexible fibrescope

7.2.5*

1. Extremely wide bandwidth: This absolutely indicates that a greater volume of information or messages can be carried over in a fiber-optic system. This is because the rate at which information can be transmitted is directly related to signal frequency. Light has a frequency in the range of 10^{14}–10^{15} Hz compared to radio frequencies $\approx 10^6$ Hz and microwave frequencies 10^8–10^{10} Hz. Therefore a transmission system that operates at the frequency of light can theoretically transmit information at a higher rate than systems that operate at radio frequencies or microwave frequencies.
2. Smaller and light weight cables: Optical fibers because of their light weight and flexibility can be handled more easily than copper cables.
3. Lack of cross talk between parallel fibers: In conventional communication circuit signals often stray from one circuit to another, resulting in other calls being heard in the background. This cross talk is negligible with fiber optics even when numerous fibers are cabled together.
4. Immunity to inductive interference: Since optical fibers are not metallic, they do not pick up electromagnetic waves. The result is noise free transmission *i.e.*, fiber optic cables are immune to interference caused by lighting or other electromagnetic equipments.
5. Potential of delivering signal at low cost.
6. Much safer than copper cables: This is because only light and not electricity being conducted
7. Longer life span: Their life span is expected to be 20–30 years or more in contrast to copper cables which have a life of 12–15 years.
8. Temperature resistance: In contrast to copper cables, they have high tolerance to temperature extremes as well as to corrosive liquids and gases.
9. Easy maintenance: Optical fibers are more reliable and easy to maintain than copper cables.

7.2.6* The oxides contain small highly charged ions. These have low polarizability and hence impart high refractive index.

REVIEW QUESTIONS

7.1.1 State the conditions for laser action in crystals. Explain the construction and action of a ruby laser. *(MKU., BE, April 1984)*

7.1.2 Explain the terms stimulated emission and spontaneous emission. What is population inversion? Explain the working of a laser. *(Anna U., M.Sc., May 1990, JNTU BE, July 2004)*

7.1.3 Describe the construction and working of ruby laser. In what way is it different from a helium-neon laser? *(Madras U., BE, April 1979)*

7.1.4 Give the theory and working of the ruby laser.

7.1.5 Give an account of different types of lasers and their working. Discuss the applications of lasers. *(Madras U., BE, November 1985)*

7.1.6 Explain with principle, the construction and working of a laser. Write a note on high power laser.

7.1.7 What is meant by "Population inversion"? Explain the principle and action of a He-Ne laser. Mention some of the applications. *(MKU., Part-time BE, May 1977)*

7.1.8 What are lasers? Discuss the conditions required for laser action. Explain the construction and working of a ruby laser with energy level diagram. Discuss holography.

7.1.9 Explain the process of absorption, spontaneous emission, stimulated emission. Obtain an expression for energy density of radiation under equilibrium condition in terms of Einstein's coefficients. Explain the construction of hologram.

(VTU, BE, July 2003)

7.1.10 Deduce the expression for radiant energy density in terms of Einstein's coefficients. Explain the construction and working of He–Ne gas laser.

(Madras U., BE, April 1998)

7.1.11 Outline the methods by which there can be an interaction between energy levels in an atom and electromagnetic radiation.

7.1.12 You are provided with an atomic system in a radiation field of energy density $E(\upsilon)$ under equilibrium conditions. Get the relation between spontaneous emission probability and also induced emission probability. Show that the $\frac{\text{emission rate}}{\text{absorption rate}} = \frac{N_2}{N_1}$ where N_1 and N_2 are the atoms occupied in the states E_1 and E_2 respectively.

7.2.1 Explain briefly the basic principle of an optic fiber. Describe graded index optical fiber and explain the transmission of signal through it.

(Anna U., BE, May 2003)

7.2.2 Explain numerical aperature of a fiber. Describe the step-index fiber and explain the transmission of a signal through it.

(California Institute of Technology, UG, winter 1970)

7.2.3 How step-index fiber differs from graded-index fiber. Explain the propagation of light through step-index fiber.

7.2.4 Describe the structures of different types of optical fibers tracing the path of the rays.

7.2.5 Explain the terms numerical aperature and acceptance angle of a fiber. Derive expression for them.

PROBLEMS AND SOLUTIONS

7.1.1 For a wave of intensity 1 watt/m^2 and wavelength 1 μm in vacuum, find the number of photons. What is the energy of a photon of wavelength 1 μm? Compare this with the characteristic thermal energy at 300 K.

Solution:

$$N = \frac{I}{E} = \frac{I}{h\upsilon} = \frac{I\lambda}{ch}$$

$$N = \frac{1\times10^{-6}}{6.62\times10^{-34}\times3\times10^{8}} = 5.05\times10^{8}/\text{m}^2\,\text{sec}$$

$$\boxed{N = 5.05 \times 10^8 / \text{m}^2 \text{ sec}}$$ ***Answer***

$$E = h\upsilon = \frac{hc}{\lambda} = \frac{6.62 \times 10^{-34} \times 3 \times 10^8}{1 \times 10^{-6}} \text{ joule}$$

$$E = \frac{19.86 \times 10^{-26}}{1.6 \times 10^{-19} \times 10^{-6}} = 1.24 \text{ eV}$$

$$\boxed{E = 1.24 \text{ eV}}$$ ***Answer***

Thermal energy $k_B T = 0.026$ eV (at 300 K)

$$\boxed{\frac{E}{k_B T} = \frac{1.24}{0.026} = 47.7}$$ ***Answer***

7.1.2 A laser source of wavelength 6×10^{-7} m, coherence width 8×10^{-3} metre and power 10 mW shines on a surface 100 metre away. Deduce the illumination. Compare it with that due to a collimated beam from a torch of filament diameter 0.1 cm, lens with focal length 10 cm and power 10 watt.

Solution:

The semi-angle of cone of laser beams, $\theta = \frac{\lambda}{a} = \frac{6 \times 10^{-7}}{8 \times 10^{-3}} = 7.5 \times 10^{-5}$ radian

$$\text{Solid angle } \Delta\omega = \frac{\Delta s}{r^2} = \frac{\pi (r\theta)^2}{r^2} = \pi\theta^2 = 3.14(7.5 \times 10^{-5})^2 \text{ staradian}$$

$$\text{A real spread } \Delta A = r^2\,\Delta\omega = 100^2 \times 3.14 \times (7.5 \times 10^{-5})^2$$
$$= 1.76 \times 10^{-4} \text{ m}^2$$

$$\text{Illumination } E = \frac{P}{\Delta A} = \frac{10 \times 10^{-3}}{1.76 \times 10^{-4}}$$

$$\boxed{E = 57 \text{ watt/m}^2}$$ ***Answer***

For a torch, the angle subtended by the filament size at the lens,

$$\theta = \frac{0.1 \text{cm}}{10 \text{cm}} = 10^{-2} \text{ radian}$$

A real spread, $\Delta A' = \pi(r\theta')^2 = 3.14(100 \times 10^{-2})^2 = 3.14 \text{ m}^2$

$$\text{Illumination, } E' = \frac{P'}{\Delta A'} = \frac{100}{3.14} = 3.2 \text{ W/m}^2$$

$$\boxed{I = 3.2 \text{ W/m}^2}$$ ***Answer***

7.1.3 In a material at 300 K two energy levels have a wavelength separation of 1 μm. Determine

(a) the ratio of upper to lower level occupation densities when the material is in thermal equilibrium.

(b) The effective temperature when the levels are equally populated.

(c) The effective temperature when the upper level is twice as densely populated as the lower (assume unit degeneracy for each level).

Indicate briefly the relevance of population inversion in the operation of lasers and describe one way by which inversion is achieved.

***Solution*:**

At normal temperature, the lower energy level N_1 in a material is more densely populated than a higher energy level N_2 according to the Boltzmann's equation (assuming unit degeneracy) as

$$N_2 = N_1 \exp[-(E_2 - E_1)/k_B T]$$

$$\frac{N_1}{N_2} = \exp[(E_2 - E_1)/k_B T]$$

$$ln\left[\frac{N_1}{N_2}\right] = \frac{(E_2 - E_1)}{k_B T}$$

$$T = \frac{(E_2 - E_1)}{k_B\, ln\left[\frac{N_1}{N_2}\right]}$$

Since $E_2 > E_1$ and $N_1 > N_2$, the temperature is positive as illustrated in the figure. However, it is possible under certain conditions to make $N_2 > N_1$, with $E_2 > E_1$. This is known as population inversion and from the expression for T given above, it will be seen that T has a negative value and the effect is therefore equivalent to a negative temperature. This is also illustrated in the figure.

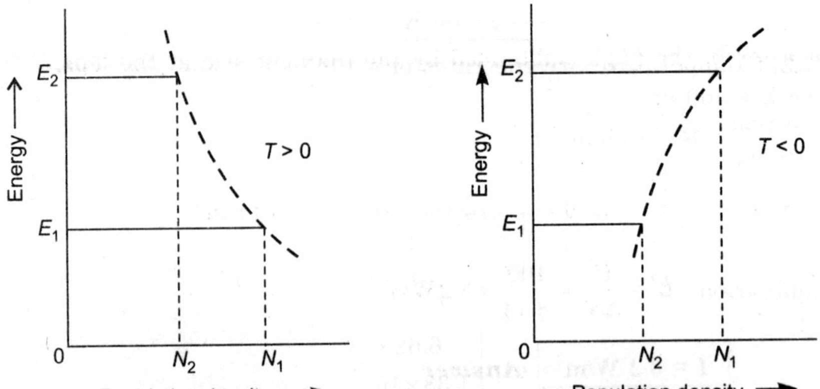

(a) (b)

(a) $\frac{N_2}{N_1} = \exp[-(E_2 - E_1)/k_BT]$ and for the transitions between the energy levels

$$E_2 - E_1 = h\upsilon = \frac{ch}{\lambda}$$

$$\frac{E_2 - E_1}{k_BT} = \frac{ch}{\lambda k_BT} = \frac{3\times10^8 \times 6.626\times10^{-34}}{10^{-6}\times1.38\times10^{-23}\times300}$$

$$= 48.0$$

Thus $$\frac{N_2}{N_1} = \exp(-48.01) = 10^{-21}$$

$$\boxed{\frac{N_2}{N_1} = 10^{-21}}$$ ***Answer***

(b) $$\frac{N_2}{N_1} = \exp[-(E_2 - E_1)/k_BT] = 1$$

$$\frac{-(E_2 - E_1)}{k_BT} = ln\,1 = 0 \text{ or } T \to \pm\infty$$

$$\boxed{T \to \pm\infty}$$ ***Answer***

(c) $$\frac{-(E_2 - E_1)}{k_BT} = ln\,2$$

$$\frac{ch}{\lambda} = -k_BT\ ln\,2$$

$$T = -\frac{ch}{\lambda k_B\,ln\,2} = \frac{3\times10^8\times6.626\times10^{-34}}{10^{-6}\times1.38\times10^{-23}\times0.693}$$

$$\boxed{T = -\ 21000\text{ K}}$$ ***Answer***

7.1.4 At what temperature are the rates of spontaneous and stimulated emission equal. Given λ = 400 nm

Solution:

$$R = \left(e^{\frac{h\upsilon}{k_BT}} - 1\right)$$

$$1 = \exp\left[\frac{6.62\times10^{-34}\times3\times10^8}{1.38\times10^{-23}\times4000\times10^{-9}}\right] - 1$$

$$\exp\left[\frac{0.0359\times10^6}{T}\right] = 2$$

$$T = \frac{0.0359\times10^6}{ln\,2}$$

$$\boxed{T = 51792\ \text{K}}\ \textbf{\textit{Answer}}$$

7.1.5 **A relative population of 1/*e* is often considered in two energy states at 20°C. Determine the wavelength of the radiation emitted at that temperature.**

Solution:

$$\frac{N_2}{N_1} = \exp[-(E_2 - E_1)/k_BT] = \frac{1}{e}$$

i.e.,

$$\exp[(E_2 - E_1)/k_BT] = e$$

or

$$\frac{(E_2 - E_1)}{k_BT} = ln\,e = 1$$

$$E_2 - E_1 = k_BT = h\upsilon = \frac{hc}{\lambda}$$

or

$$\lambda = \frac{ch}{k_BT} = \frac{3\times10^8\times6.62\times10^{-34}}{1.38\times10^{-23}\times293}$$

$$\boxed{\lambda = 0.05\times10^{-3}\text{m}}\ \textbf{\textit{Answer}}$$

7.1.6 **The figure shows the energy levels of Nd^{3+} in yttrium aluminium garnet ($Y_3Al_5O_{12}$) which are involved in the laser action of this crystal (known as Nd YAG laser). Describe the processes that occur when the laser is working.**

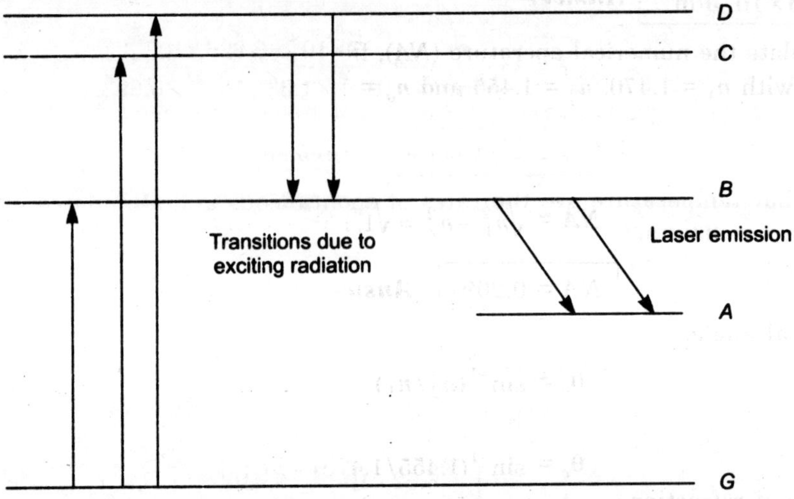

Energy levels of Nd^{3+} in yttrium garnet

***Answer*:**

Radiation from a lamp excites electrons from the ground state, G, to states B, C, D. Electrons in states C and D undergo non-radiative transitions to state B. Radiative transitions from B to lower states are forbidden and so a large population of electrons in state B builds up. Eventually a photon is emitted and an electron goes to state A. This photon induces further emission from B to A, eventually producing a beam of laser light. Transitions from B to both A and G must be forbidden for a large population to build up in B, but the transmission B → A must be less forbidden than the transition B → G as it is B → A that gives rise to laser action, and hence must be most likely to occur.

7.1.7 If the laser is pulsed for 1 nm, 10^{-9} sec, and contains 40 kJ energy, what are the inferences one can add?

***Solution*:**

1. The power $= \dfrac{40{,}000}{10^{-9}}$

40 tara watt

Answer

2. Length of the light bullet for a 10^{-9} pulse is

$$\frac{c}{1/t} = ct = 3\times10^{8}\times10^{-9}$$

0.3 m = 1 ft

Answer

3. If this energy can be (40 kJ) focused on a 10 mm spot on a target, then the spot will

receive $\dfrac{40kJ}{\pi(d/2)^2} = \dfrac{4\times40\times10^{3}}{\pi\times(100\times10^{-6})}$

5×10^{8} J/m^2

Answer

7.2.1 Calculate the numerical aperature (NA), fiber acceptance angle (α_m) for a step index fiber with n_1 = 1.470, n_2 = 1.455 and n_u = 1.

(NIT Calicut, BE, April 2003)

***Solution*:**

$$NA = \sqrt{n_1^2 - n_2^2} = \sqrt{1.470^2 - 1.455^2}$$

NA = 0.209

Answer

Critical angle,

$$\theta_c = \sin^{-1}(n_2/n_1)$$

i.e.,

$$\theta_c = \sin^{-1}(1.455/1.470) = 81.8°$$

Angle of refraction

$$\alpha'_m = 90° - \theta_c = 90° - 81.8° = 8.2°$$

We know

$$n_a \sin\alpha_m = n_1 \sin\alpha'_m$$

$$\sin\alpha_m = \frac{n_1 \sin\alpha'_m}{n_a}.$$

$$\alpha_m = \sin^{-1}\left[\frac{n_1 \sin\alpha'_m}{1}\right]$$

$$\alpha_m = \sin^{-1}[1.470\times\sin 8.2°]$$

$$\boxed{\alpha_m = 12.1°} \quad \textit{Answer}$$

7.2.2 **A 6 km optical link consists of multimode step index fiber with a core of refractive index of 1.5 and a relative refractive index difference of 1%. Estimate the delay difference between the slowest and fastest modes at the first output.**

(Annamalai U., M.Sc, April 1996)

Solution:

In the case of step index fiber, the refractive index of cladding n_2 is less than the core refractive index n_1.

$$\text{Given: } n_1 = 1.5,\ n_1 - n_2 = \frac{1}{100}n_1 = \frac{1.5}{100} = 0.015$$

$$n_2 = n_1 - 0.015 = 1.5 - 0.015 = 1.485$$

Now the pulse dispersion for step-index fiber is

$$\Delta T = \frac{n_1 L}{c}\left(\frac{n_1}{n_2} - 1\right)$$

$$= 1.5\times 6\left(\frac{1.5}{1.485} - 1\right)$$

$$\boxed{\Delta T = 30.3\times 10^{-3} \text{ sec}} \quad \textit{Answer}$$

Thus the delay difference between the slowest and fastest modes at the fiber output is 30.3 m sec.

7.2.3 **In an optical fiber $n_1 = 1.53$ and $n_2 = 1.50$ and the vacuum wavelength $\lambda_o = 1$ μm; calculate the normalized frequency (cut off parameter) and the number of modes N possible if the radius of the core is 50 μm.**

Solution:

$$V = \frac{\pi d}{\lambda_0}(n_1^2 - n_1^2)^{1/2}$$

$$V = \frac{\pi\times 2\times 50\times 10^{-6}\times(1.53^2 - 1.50^2)^{1/2}}{10^{-6}}$$

$$V = 94.72 \quad \textbf{Answer}$$

$$N = \frac{V^2}{2} = \frac{(94.7)^2}{2} = 4486 \quad \textbf{Answer}$$

EXERCISE

7.1.1 For a source radiating at mean $\lambda = 600$ nm, the coherence time τ is 2×10^{-10} sec. Get the order of magnitude of the following: (*i*) coherence length, (*ii*) the spectral width of the line, (*iii*) purity factor.

(California Institute of Technology, UG, April 1979)

(***Ans:*** 6×10^{-2} m, 0.003 nm, 2×10^{5})

7.1.2 Show that there is only one stimulated emission for 1.76 million of spontaneous emission. Given: $\upsilon = 6 \times 10^{14}$ Hz. *(MKU, BE, April 1992)*

7.1.3 A laser beam of wavelength 740 nm has coherence time 4×10^{-5} sec. Deduce the order of magnitude of its coherence length and spectral half-width and purity factor.

(Anna U., M.Sc., April 1995)

(***Ans***: 12×10^{3} m, 0.45×10^{-16} m, 1.6×10^{10})

7.1.4 If a laser is pulsed for one nanosecond, and contains 40 kJ energy, do the power calculation. Also compute the intensity of the beam.

(JNTU, BE, November 2002)

(***Ans***: 40 tara watt, 51 joule/m^2)

7.1.5 What is the diffraction-limited beam divergence of a Q-switched Nd: glass laser (λ = 1.06 μm) having an output aperature $d = 1''$? (***Ans***: 1.02×10^{-4} radian).

7.2.1 Calculate the maximum value of the angle of incidence that a ray can make with the axis of a fiber such that it is guided through the fiber for the following parameters (*i*) $n_1 = 1.6$, $n_2 = 1.5$; (*ii*) $n_1 = 2.1$, $n_2 = 1.5$. Assume outside medium is air *i.e.*, $n_o = 1$. Discuss light gathering power of the fibers.

(***Ans***: 33.8°, A)

7.2.2 A 10 km long fiber cable has loss of 1.5 dB/km and connector loss of 0.8 dB/km. Find the total loss.

(***Ans***: 22.2 dB)

7.2.3 Calculate the numerical aperature and hence the acceptance angle for an optical fiber given that refractive indices of the core and the cladding are 1.45 and 1.46 respectively.

(***Ans***: 0.3775, 22.18°)

Chapter 8

UNIT 8

Acoustics of Building & Acoustic Quieting and Nanotechnology

8.1 ACOUSTICS OF BUILDING & ACOUSTIC QUIETING

8.1.1 INTRODUCTION

During your earlier student days, your grandpa used to be a regular radio listener and so, a couple of carnatic numbers used to reach your ear if you were around home when he tuned into a carnatic concert.

You have taken the morning flight and the two hour flight gives you ample time to read every line of the newspaper. You start reading an analytical article on carnatic music having exhausted everything else in the newspaper. Somehow, it impresses you and arouses your curiosity. The week after that, you land up in a cassette shop to pick up an album.

But one apparent information is that a reasonable percentage of the audience watching the music performance did have lesser knowledge of music; but the pleasant voice in their own language make them comfortable and the neatly clothed female or male singer bring them again and slowly develop in them a taste for music without knowing the complicated theory behind this fascinating area. This is true when many go and enjoy the powerful speeches of great orators in lucid language though the subject matter is totally new to them. However, in both the cases the quality of the hall to clear the transmission of sound across it is the most important prerequisite to make the listeners feel at home.

In short "pleasant noise is music and unpleasant music is noise". Similarly even subjectless talk of an orator in a fully packed auditorium with all the acoustically required parameters will nail the audience till the programme is over.

The study of this branch of science and the analysis of the techniques involved is known as Acoustics of Buildings.

8.1.2 MUSICAL SOUND AND THE ESSENTIAL REQUIREMENTS FOR GOOD ACOUSTIC IN A LARGE AUDITORIUM

A sound of regular frequency is called tone or a musical note. Hence music is a combination of such sounds. A musical sound produces a smooth and pleasing effect to the ear.

A noise is a sound or combination of sound of irregular frequencies. Their wave functions have no definite regularities. It produces a jarring effect on the ear. The sounds produced by a violin or a flute are musical, whereas those like the horn of a motor car or the rattling of the train are noises.

The characteristics of a musical sound are (1) Pitch (2) Loudness and (3) Quality or timbre.

Pitch: Pitch is a characteristics that depends on the frequency of sound. It is the shrillness of sound felt by the ear. Pitch depends on the frequency of the source and the relative motion between source and listener.

Loudness: It is the characteristics of the musical sound which depends on the intensity of sound and also the sensitiveness of the ear.

The amount of sound energy crossing unit area around a point in unit time is known as intensity of sound. Loundness and intensity are related as $L = k \log_e I$, where k is the constant of proportionality. It has been observed that a normal human ear can just hear a note of 1000 Hz if it has intensity equal to 10^{-16} W/cm^2. This energy is said to represent zero level intensity I_0 or threshold hearing. The intensity level L of a sound of intensity I is represented by

$$L = \log_{10}\left(\frac{I}{I_0}\right)$$

when $I = 10\ I_0$, $L = 1$.

The intensity level then is one bel. Hence one bel is the loudness of sound whose intensity is 10 times the zero level intensity. For ordinary sound intensities decibel unit is used.

Quality: Quality is the characteristic of a musical sound which enables us to distinguish between the sounds produced by two different persons or two different musical instruments even if their loudness and pitch are the same. It is because of the quality of musical sound that we can distinguish the sound of a piano from that of a violin.

No person or musical instrument can produce a pure tone of single frequency. In addition to the fundamental frequency (ν) there will be overtones (2ν, 3ν, 4ν, ... etc.) also. The number and the intensity of the overtones present in the sound depend on the shape and size of the musical instrument as well as its mode of vibration. The resultant effect of the fundamental tone and overtones differ from instrument to instrument. It is this resultant effect that determines the quality of the sound.

Reverberation: When sound is produced in a room or an auditorium it spreads out and strikes the surfaces of the various objects in the hall such as ceiling, walls, floor and other material present in the hall. Some of the sound energy is absorbed by these objects but some energy is reflected back. The sound waves received by the listener are (1) direct wave and (2) reflected wave due to multiples reflection at the surfaces of the various objects. Before the direct wave received by the listener dies out, he may receive the reflected wave. This creates the impression of prolongation or persistence of sound for some more time to the listener after the original sound has ceased. This persistence of audible sound even after the source has ceased to emit sound is called reverberation.

Reverberation Period

Reverberation period is defined as the time in which the intensity of a continuous note falls to one millionth of its original intensity.

Echo and Reverberation: Although echo and reverberation are caused due to the reflection of sound, they are different from one another. Echo is the repetition of a note after the original note is dead; whereas reverberation is continuation of the note in the ear of the listener. Echo is possible in large halls only, but reverberation is possible everywhere, be it a small room or a large hall.

Preamble

The main requirements for good acoustic in a large auditorium are listed below:

(*i*) Each syllable of the speech or song should be heard sufficiently loud and intelligible at every point in the auditorium.

In order that sufficiently loud sound reaches every point in the auditorium, a good direct path as well as proper reflection of sound should be ensured. A hard plane wall behind the speaker is good reflector. It is parabolic, the reflected sound reaches every point.

(*ii*) The decay of each syllable should be quick to avoid any overlapping so that the successive syllable may be heard distinctly.

This means that there should not be excessive reverberation. There is certain range of optimum reverberation period for good acoustics. In general, for music, the optimum reverberation time is 0.5 to 1 sec and for speech 1 to 2 sec. A smaller reverberation time than this makes the auditorium 'dead' and a larger one creates confusion. The reverberation time of an auditorium can be calculated using Sabine's formula:

$$T = \frac{0.165V}{\Sigma as}$$

where V is the total volume of the auditorium and s is the surface area having an absorption coefficient 'a'.

The time of reverberation can be adjusted by having:

(*a*) a few open windows (*b*) wall covered with sound absorbing materials like card-board, felt, asbestos, heavy curtain, etc. (*c*) upholstered seats which will absorb sound and will avoid appreciably the effect of variation in number of audience.

(*iii*) **Echoes must be eliminated**

Echoes can be eliminated by using sound absorbing materials and few open windows.

(*iv*) Undesirable focusing of sound due to reflection from walls and ceiling at any place and any zone of silence or region of poor audibility must be avoided. For this concave surfaces such as domes, curved arches or spherical surfaces of walls or the ceiling should be avoided. Moreover, the seats should be gradually elevated so that the sound reaches the listener directly from the source.

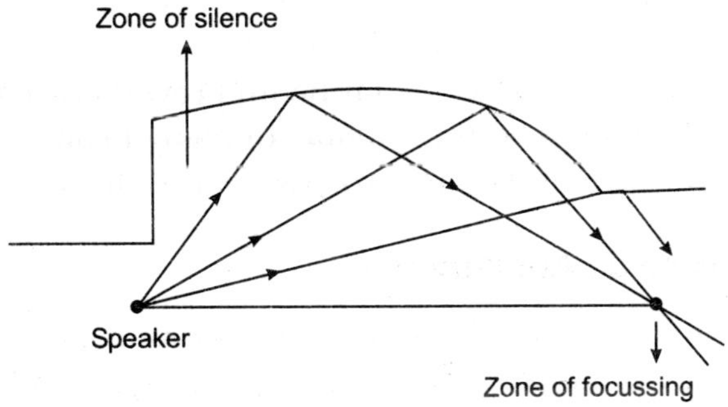

Fig. 8.1 Method to avoid focussing of sound due to reflections.

8.1.3 REVERBERATION AND TIME OF REVERBERATION — A DISCUSSION

When sound waves are produced in a big hall, they spread out, scatter and strike the surfaces of the various objects in the hall such as ceiling, walls, floor, etc. Some of the sound energy is reflected and some absorbed and thus the distribution of sound energy in the hall continues to be affected. In other words the sound from the source does not cease immediately after it is directly received but is lengthened out owing to the energy received

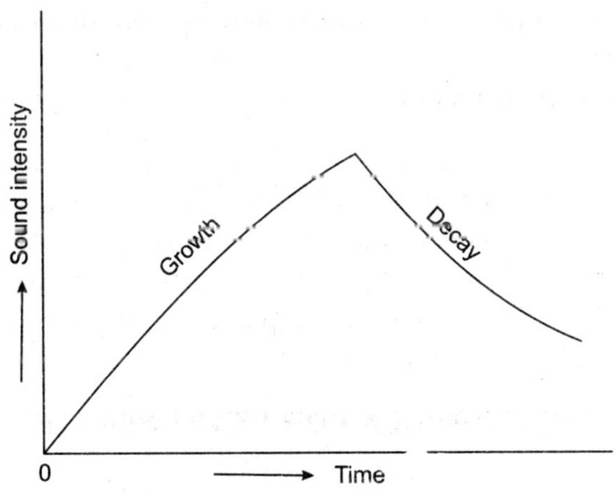

Fig. 8.2 Growth and decay of sound intensity.

by continuous reflections from successive portions of the walls, etc. Thus the original wave (pulse) on breaking up, wanders about for some time in the air of the hall, and even when the direct pulse has ceased or dead, some of the energy is still on its way from the source to the walls and then to the ear, and hence the prolongation of sound. This persistence of audible sound after the source has ceased to emit sound is called reverberation. The time for which the sound thus persists *i.e.*, until it dies out or falls below the threshold of audibility, measured from the instant of its generation (in the case of short sounds), or the stopping of source (in the case of continuous note) is known as reverberation time. Proper adjustment of this time interval forms the chief or central problem of building acoustics. A certain amount of this reverberation is, of course, desirable as it adds a pleasing characteristic to the acoustical qualities of a room, especially for giving richness to music, but too much reverberation is most undesirable.

By using an organ pipe of frequency 512 Hz, Prof. Sabine found that its sound became inaudiable when the intensity fell to one millionth of its intensity just before stopping the organ pipe. Hence, Sabine defined the standard reverberation time as the time taken by sound to fall to one millionth of its intensity just before the source is cut off.

Acoustical Demands on an Auditorium

The following conditions are demanded for obtaining good acoustical properties:

(*a*) Each separate syllable should produce sufficient energy in every part of the hall. *i.e.,* sound should be sufficiently loud and intelligible at every point.

(*b*) Sound of each syllable should soon decay so that the succeeding syllable may be heard distinctly. This means that the auditorium must be free from excessive reverberation.

(*c*) There should not be present echos more than necessary for maintaining continuity.

(*d*) There should be no undesirable focussing of sound due to walls, ceiling nor there should be present zones fence or regions of poor audiability anywhere in the hall.

(*e*) There should not be any unpleasant reinforcement of any of the overtones of a complex sound, so that total quality may not be affected.

Importance of Reverberation Time

If a building is to be acoustically the best then it is very essential that the reverberation time must be of the correct order. It should be neither too long nor too short. A very short reverberation time makes a room 'Sound dead' and the sensation of speaking in such a room is similar to that experienced when speaking from the top of an isolated tower or inside a hall. The reverberation time also depends on the frequency of the note. It also depends directly upon the physical dimensions of the room or hall *i.e.*, to volume.

8.1.4 SABINE'S LAW AND FORMULA FOR REVERBERATION

Sabine and others established from theoretical consideration that the standard 'period of reverberation' viz., the time that the sound takes to fall in intensity to one millionth

(10^{-6}) of its original intensity after the source is stopped, is given by

$$T = \frac{0.165\,V}{\Sigma as} \quad \text{in sec} \qquad \text{...(8.1.1)}$$

i.e.,

$$T = \frac{0.165\,V}{a_1 s_1 + a_2 s_2 + a_3 s_3 + \ldots} \qquad \text{...(8.1.2)}$$

Here V is the total volume of the hall used.

Σas is the sum of the products of absorption coefficient and surface areas of different units. The above equation is called Sabine's formula.

Let us now consider the reception of sound energy by a small area ds at the plane boundary wall of the enclosure. Let us draw two concentric circles of radii r and $r + dr$ with ds as centre in a plane containing the normal to ds. Now consider an element of area enclosed between circles by the radii making angles θ and $(\theta + d\theta)$. The arc of this area is $rd\theta$ and the radial length is dr. Hence its area is $(rd\theta)\, dr$. If the whole figure is rotated about the normal through an angle $d\phi$ (radius of the rotating figure is $r \sin \theta$), the distance travelled by this area will be the circumferential distance x

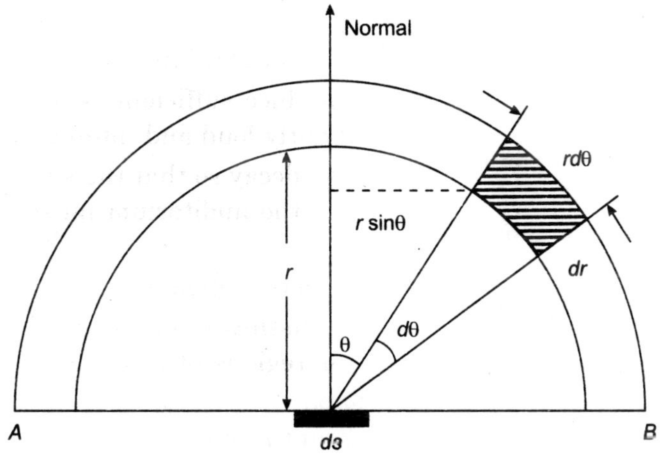

Fig. 8.3 Distribution of sound

i.e.,

$$d\phi = \frac{x}{r \sin \theta}$$

i.e.,

$$x = r \sin \theta \, d\phi$$

Volume traced out by this area element is

$$(rd\theta)\, dr \times x$$

$$= (rd\theta)\, dr\, (r \sin \theta\, d\phi)$$

$$= r^2 \sin \theta\, d\theta\, dr\, d\phi$$

If E be the value of sound energy density *i.e.*, energy per unit volume, then the energy contained in the above volume is

$$Er^2 \sin \theta\, d\theta\, dr\, d\phi$$

This sound energy is travelling through the element equally in all directions. The energy travelling per unit solid angle along any direction is

$$\frac{Er^2 \sin\theta \, d\theta \, dr \, d\phi}{4\pi} \qquad ...(8.1.3)$$

The solid angle subtended by the element of area ds is $\frac{ds\cos\theta}{r^2}$ at the element. Therefore the energy in the element volume that is travelling towards ds is given by

$$\frac{Er^2 \sin\theta \, d\theta \, dr \, d\phi}{4\pi} \times \frac{ds\cos\theta}{r^2}$$

$$= \boxed{\frac{Eds}{4\pi} dr \sin\theta \cos\theta \, d\theta \, d\phi} \qquad ...(8.1.4)$$

Though the final equation required is to give the value of the sound energy reaching ds in unit time, we do the following rough work to achieve the said goal.

1. Integrate the above Equation 8.1.4 with respect to ϕ

$$i.e., \left(\frac{Eds}{4\pi} dr\right)(\sin\theta \cdot \cos\theta d\theta) \int_0^{2\pi} d\phi$$

$$= \left(\frac{Eds}{4\pi} dr\right) \sin\theta \cdot \cos\theta d\theta [\phi]_0^{2\pi} = \frac{Eds}{2} dr \sin\theta \cdot \cos\theta d\theta$$

2. Integrating the above equation with respect to θ, we get

$$\frac{Eds\, dr}{4} \int_0^{\pi/2} 2\sin\theta\cos\theta \, d\theta$$

Since $\int_0^{\pi/2} 2\sin\theta\cos\theta \, d\theta = \int_0^{\pi/2} \sin 2\theta d\theta = 1$, we get

$$\boxed{\frac{Eds\, dr}{4}} \qquad ...(8.1.5)$$

3. Again integrating with respect to r which varies between 0 and C (C = velocity of sound)

$$\frac{Eds}{4} \int_0^C dr = \frac{Eds}{4} C$$

If a is the absorption coefficient of the wall AB whose value ds is a part of it, their energy absorbed for unit time

$$\frac{1}{4} EC(ads)$$

Hence, total absorption at all the surfaces of the wall where the sound is falling is

$$\boxed{\frac{1}{4}EC\sum ads = \frac{ECA}{4}} \quad ...(8.1.6)$$

where $\sum ads = A$, the total absorption on all the surfaces on which sound falls.

The second part of the study is to have an observation on the growth and decay of sound energy in the Hall. Let P be power output *i.e.,* the rate of emission of energy from the source and V the total volume of the room. Then the total energy in the room at the instant when the energy density E will be EV. Thus rate of growth of energy = $\frac{d}{dt}(EV) = V\frac{dE}{dt}$.

But at any instant the rate of growth of energy in space – (rate of supply of energy from the surface – rate of absorption by the surface)

$$= \boxed{V\frac{dE}{dt} = P - \frac{ECA}{4}} \quad ...(8.1.7)$$

When the steady state is attained, $\frac{dE}{dt} = 0$. Hence

$$0 = P - \frac{E_m CA}{4}$$

$$\boxed{E_m = \frac{4P}{CA}} \quad ...(8.1.8)$$

Again from Equation 8.1.7

$$\boxed{\frac{dE}{dt} = \frac{P}{V} - \frac{ECA}{4V}}$$

If we use $\frac{1}{V} = \frac{4\alpha}{CA}$ and $\frac{CA}{4V} - \alpha$, we get

$$\frac{dE}{dt} + \alpha E = \frac{4P}{CA}\alpha$$

or

$$\boxed{\frac{d}{dt}(Ee^{\alpha t}) = \frac{4P}{CA}\alpha e^{\alpha t}} \quad ...(8.1.9)$$

Integrating the above equation, we get

$$Ee^{\alpha t} = \frac{4P}{CA}e^{\alpha t} + K \quad ...(8.1.10)$$

Let us now use the boundary condition to get K when $t = 0$, $E = 0$. Now

$$K = -\frac{4P}{CA}$$

Substituting the value of K in Equation (8.1.10)

$$Ee^{\alpha t} = \frac{4P}{CA} e^{\alpha t} - \frac{4P}{CA}$$

Refer Equation 8.1.8, we get

$$Ee^{\alpha t} = E_m e^{\alpha t} - E_m$$

$$E = E_m - E_m e^{-\alpha t}$$

or $$\boxed{E = E_m(1 - e^{-\alpha t})} \qquad ...(8.1.11)$$

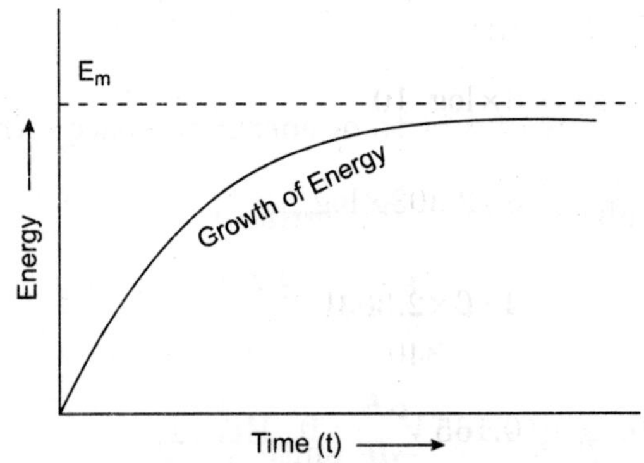

Fig. 8.4 Growth of energy

The second one is decay of energy density. Here the source of sound is cut-off when E has reached the maximum value say E_m, so that $P = 0$ and $t = 0$ and hence

$$E = E_{max} = K$$

Substituting this in Equation 8.1.10, we get

$$E e^{\alpha t} = E_m$$

or $$\boxed{E = E_m e^{-\alpha t}} \qquad ...(8.1.9)$$

Equation (8.1.9) shows the decay of energy with time after the source is cut-off. This decay is shown by the experimental curve as given in Fig. 8.5.

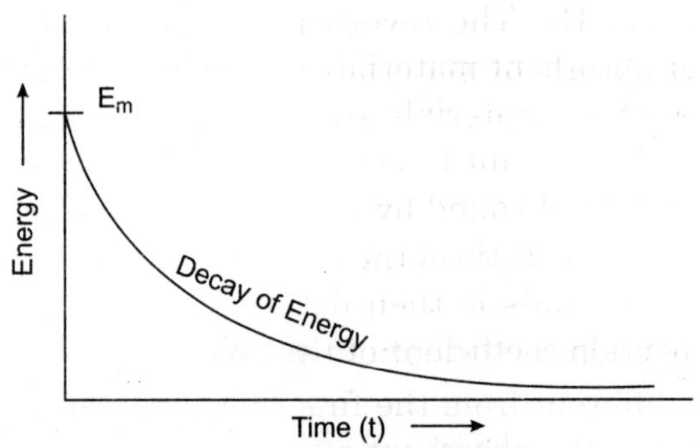

Fig. 8.5 Decay of energy

Sabine's Formula or Standard Reverberation Time

The standard time of reverberation T is defined as the time taken by a sound to reduce its intensity to one millionth $\left(\frac{1}{10^6}\right)$ of its intensity when the source is cut-off.

Hence by putting $E = \frac{E_m}{10^6}$ and $t = T$ in Equation 8.1.9, we get

$$E_m \times 10^{-6} = E_m e^{-\alpha t}$$

$$e^{\alpha t} = 10^6$$

i.e.,

$$\alpha t = 6 \times \log_e 10$$

$$\frac{CA}{4V} T = 6 \times 2.303 \times \log_{10} 10$$

$$AT = \frac{4 \times 6 \times 2.303 V}{340}$$

$$T = \frac{0.165\ V}{A} \quad \text{since } A = \Sigma as$$

Now

$$\boxed{T = \frac{0.165\ V}{\sum as}} \qquad ...(8.1.10)$$

This above equation is in good agreement with the experimental values obtained by Sabine.

Determination of Absorption Coefficient

(1) In one of the methods of determination of absorption coefficients (Reverberation Chamber method) of different materials. Sabine used as source of sound an organ pipe of frequency 512 Hz. The reverberation period of a room with cushions suspended or other absorbent materials placed in the hall is first measured. The cushions or the absorbent materials are then removed and the extent of the open window gradually adjusted until the reverberation time is the same as before. In this case the absorption of sound by the substance is equal to the absorption of sound by the window. The ratio of the area of window opened to the total area of cushions or other materials is then determined. The result is a fraction and measures the absorption coefficient of the substance.

The second method is different from the first one. Here the reverberation time in a hall is determined first without the absorbent material 'a' and then with the material. If T_1 and T_2 be these times,

$$T_1 = \frac{0.165\,V}{\sum a_1 s_1}, \text{ where } \sum a_1 s_1 \text{ is the total absorption in the}$$

hall without the material.

$$T_2 = \frac{0.165\,V}{\sum a_1 s_1 + \alpha S}$$

where α is the absorption coefficient of the substance and S its area.

Now $$\left(\frac{1}{T_2} - \frac{1}{T_1}\right) = \frac{\alpha S}{0.165\,V}$$

$$\boxed{\alpha = \frac{0.165\,V}{S}\left(\frac{1}{T_2} - \frac{1}{T_1}\right)} \quad \text{...(8.1.11)}$$

The average acoustic absorption coefficient for ordinary vocal frequencies for some types of surfaces are as follows:

Table 8.1. Absorption coefficient of some selected materials

Material	*Absorption coefficient a in O.W.U*
Marble	0.01
Brick wall (30 cm thick)	0.03
Brick wall painted	0.016
Carpets	0.15 – 0.30
Hair felt (2.5 cm thick)	0.58
Human body	4.3 – 4.7
Glass	0.02
Ordinary chair	0.17
Wooden floor	0.06
Open window	1.30

Conditions for Good Acoustical Designs of Rooms

1. *Control of Reverberation*

As reverberation is due to repeated reflections, the remedy lies in increased absorption which may be brought about by having:

(*a*) A few open windows, which by letting the entire sound energy out serve as perfect absorbers.

(*b*) Walls covered with absorbed materials such as celotex perforated card-boards, felt, asbestos, coarse cloth, maps and pictures, heavy curtain, etc. or walls engraved and roughened with decorative materials so as to diffuse the sound. Porous bodies by providing small capillaries absorb a good deal of sound energy incident on them and serve very well as sound suppressors.

(*c*) Upholtered seats which will be responsible for absorption and would avoid approximately the presence and absence of the audience.

(*d*) Presence of a good number of audience (one listener is equivalent to 4.7 sq. ft of an open window).

2. *Shape of Walls and Ceiling*

There should be no curved walls or corners bounded by two walls, firstly to avoid focussing and concentration of sound at one place and secondly to avoid 'dead' space. Concave rear walls are responsible for troublesome echos and delayed reflections in many cinema halls and auditoria. These determental reflections can be converted into useful ones by introducing a ceiling splay between the ceiling and the rear wall.

3. *Concave Surfaces and Balconies*

Concave surfaces such as domes, curved arches and barrelled ceiling should be avoided unless given careful acoustical design in order to eliminate focussing effects, non-distribution of sound or echoes. Balconies as a rule should have a shallow depth and high opening, the depth should not exceed twice the height of the opening. Under these conditions sound flows readily into the space under the balcony.

4. *Seats*

General plan of the hall with seats should be somewhat like that shown in the figure, in order to secure good distribution of sound. Here the speaker is at the focus of a paraboloid reflecting surface and a plane beam of sound is sent down the hall. Moreover, from the stand point of hearing as well as seeing the seats are gradually elevated in order to promote a force flow of a direct sound from the force to the listeners.

It follows directly from above that in the case of a room possessing bad acoustical properties, the person should speak slowly so as to produce as little volume of sound as is just sufficient to fill the room.

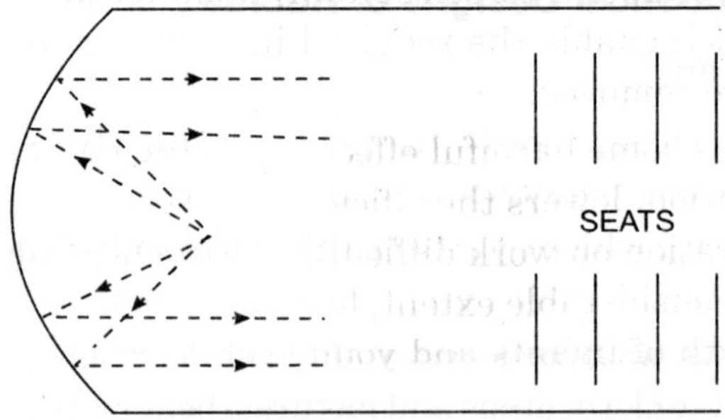

Fig. 8.6 Seating in a hall

It should however, be noted that by decreasing time of reverberation, steady value of sound also decreases, hence one would not like the sound energy to be too small to be audible. *i.e.*, to make the room 'dead'. So the time of reverberation should not be decreased beyond a certain optimum value as decided by experience.

It has been estimated that '*T*' for halls of about 10000 cubic feet capacity is 1.03 second for speech so that each syllable is heard as distinctly as possible. For purposes of music, this optimum reverberation period may be slightly increased.

5. ***Sound Amplification System***

Some points are also worth to note in public address systems so commonly in use nowadays. Such a system consists of one or more microphones, an amplifier where in turn operates one or more loud speakers. This sound amplification system is used to increase the sound level in large rooms halls and theatres to provide announcing facilities and to reproduce radio signal or recorded programmes. A good sound amplification system is characterized by uniform and adequate coverage, low noise level and negligible distortion. Here the speaker talks into the microphone and sound after being amplified led into the loudspeakers. If there are more than one loudspeaker, they should be so arranged that sound from one does not feed into another to avoid interference. The loudspeaker must be far away from the microphone otherwise forced oscillations as a consequence of mutual interactions of the microphone and the loudspeaker may cause 'howling sound' due to acoustic feed-back.

The Design of an Auditorium

Since nowadays the problem of architectural acoustics has been put on a scientific basis, the acoustic properties of an auditorium are investigated before the building is constructed. A model outline of a longitudinal section of the proposal building is placed in ripple tank and the dipper is made to operate in the position of the source of sound. When the dipper is withdrawn a short strain of waves is originated. These waves falling on the wooden boundaries of the model are reflected therefrom. To study the manner how the waves progress, an arc lamp is placed underneath the glass bottom of the tank and this casts a shadow of the waves upon a screen suitably mounted above the tank. Instantaneous snapshots are taken which enable the required information to be obtained and thus the defects in the designs are removed.

Noise: The noise has some harmful effects upon the human beings. It contributes to mental fatigue and irritation, lowers the efficiency of workers by lessening their attention and making the concentration on work difficult, impairs hearing, strains and wears down the nervous system to a considerable extent, lowers the restorative quality of sleep. It also retards the normal growth of infants and young children.

Noise Pollution

Every one with normal hearing power is continually subjected to different types of sound. In rural areas, the natural sounds of wind and animals can present a noise problem and residents of urban areas are continually disturbed (whether they are in their houses or offices by different types of sounds such as aircraft noise, automobile noises, compression noise, rail and transportation noise). There are many other systems which add to noise problem. Airconditioning, Television sets, automatic washers and even playing children contribute much to the above problem.

Fortunately man is able to reject sound within certain limits without conscious effort and carry on his normal work. But if the noise is strong it will have any one or all of the following effects (1) annoy him (2) disturb his sleep (3) damage his hearing. Residents near factories and aircraft centres are disturbed both during night and day by the noises. As regards damage caused to the hearing, these are two types (*a*) sudden damage may result from the noise of a blast or an explosion (*b*) gradual period of damage may result from continual exposure to noise over a period of years. That is why many large industrial plants give their employees periodic hearing tests to ensure against abnormal hearing losses.

Noise Control Measures

Noise control may be affected by one or more of the following methods:

(1) by redesigning the source of sound for quieter operation.

(2) by isolating the source with sound-reducing housings.

(3) by isolating noisy machines from supporting structures by means of resilent mountings and flexible pipe connections.

(4) by using special construction to reduce transmission of sound through walls and partitions.

(5) by insulating sound absorbent materials near noise sources and in areas where silence is required.

(6) by providing workers, unavoidably subjected to high intensity sounds with sound reducing ear muffs and ear plugs.

8.2 NANOTECHNOLOGY

"This piece of the review work on Nanotechnology is absolutely on the basis of data collected by the author of this book from many journals and several standard textbooks on this subject. The author does not claim any right over the subject matter presented here for the use of undergraduate students in Science and Engineering pursuing their course in Indian Universities". The field is new to him and he is away from centres doing some work on this area. Hence, the author is doubtful of the format of the consolidation of this topic and its justification.

In the line of Thirukkural "Nanotechnology is the science of understanding the structure and behaviour of materials at the atomic or molecular level". It is the world of atoms, molecules, macromolecules, quantum dots, and macromolecular assemblies.

Abstract: *Prefix denoting 10^{-9} is called nano. The symbol used is ν. Thus 1 nanosecond (ns) = 10^{-9} second. Similarly, 1 nanometre (nm) = 10^{-9} metre. Technology involving sizes of only a few nanometres is nanotechnology. A form of carbon in which the atoms are linked by a cylindrical framework with diameter of only a few nanometres is a nanotube. It is also known as a bucky tube.*

Any one of several forms of carbon made up of many atoms bonded together in a polyhedral structure of pentagons and hexagons of 60 (70, 240, 540, etc.) carbon atoms is called buckminister fullerene. The illustration shows one with 60 carbon atoms. The substance was first discovered by Richard Smalley and associates in 1985, when graphite was blasted with high-power laser. The module, C_{60}, was named buckministerfullerene in honour of Richard Buckminister Fuller, who built a geodesic dome with a similar structure at the Expo', 67 in Montreal. Informally, the molecules are known as buckyballs or fullerenes. The substance is a yellow crystalline solid called fullerite. Derivatives of fullerene have been made, the more known being buckytubes.

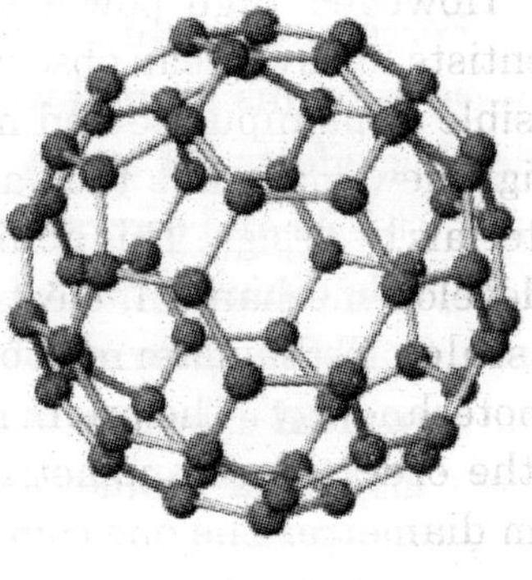

Fig. 8.7

In the IT field, Nanotechnology is expected to revolutionize the development of nanoscale components through which we can have smaller and faster computers with larger memories. This may be possible through hardware components like carbon Nanotubes, Scanning Probe Microscopes (SPMS) and Atomic Force Microscopes (AFMs). Another area which can benefit through Nanotechnology is IT Security.

Preamble: In 1959, the great physicist of our time Prof. Richard Feynman said that nothing in the laws of physics prevented us from arranging atoms the way we want: " ... it is something, in principle, that can be done; but in practice, it has not been done because we are too big. He was particularly interested in the possibility of denser computer circuitry and microscopes that could see things much smaller than is possible with "scanning tunnelling microscope", and there are other important examples. Although Feynman's talk did not explain the full concept of nanotechnology, it was K.E. Drexler who envisioned self-replicating 'nanobots', *i.e.*, self-replicating robots at the molecular scale, in engines of creation.

An Interesting Observation of Dr Pillai on Prof. Feynman

On receiving the Nobel prize Feynman visited the school where he studied. He requested "that time principal" to show him the old records of his performance during his stay in that campus. The Principal with reluctancy replied that he was poor in all the subjects at all times. Feynman joked that one of the conditions to earn Nobel prize is that one should be poor in his school career. Since it is a joke we take it easy; otherwise it is untrue in many cases.

8.2.1 INTRODUCTION

All about rearranging atoms whichever way we want is nanoscience. The technique used for manipulating atoms and molecules to fabricate materials, devices and systems is known as nanotechnology.

The common method used to understand the chemistry and physics of materials till recently was studying large and complex structures, and then investigating the size and shape of the fundamental building blocks of the structure that are usually simpler and smaller; this basic approach is sometimes called 'top-down' science.

However, high power scanning microscopes and special types of probes help the scientists to have clean observation of individual atoms and molecules. Thus, it has become possible to manipulate and move atoms and molecules to form new structures, and, thus design new materials that are built from atomic level constituents. It is also known as materials by design. This ability with utmost care in arranging atoms provides opportunities to develop mechanical, electrical, magnetic and other properties that are not otherwise possible. This is term 'bottom up' approach and the study of the materials is 'nanotechnology'. The prefix nano tells that the dimensions of these structural entities are on the order of a nanometre (10^{-9} m). Less than 100 nanometre is approximately 500 atom diametre. The one common example of a material of this type is carbon nanotube.

8.2.2 NANOSCIENCE AND NANOTECHNOLOGY

When we talk about "materials science", we normally mean ceramics or crystalline materials, glasses or non-crystalline materials, polymers or heavy chain molecular materials and metals or cohesively bonded materials with applications in diverse fields.

However, the world of materials is rapidly progressing with new and trendiest technologies, and obviously novel applications. Among modern and sophisticated technologies, nanotechnology may yield more than the expected results.

Nanoscience and nanotechnology will take you with all confidence through a world that exists at nanoscale. This technology offers us to minimise the dimensions of many devices and circuit systems to nanoscale range making them even smaller but with greater efficiency. The techniques emanating us for manufacturing products from atoms or molecules instead of from bulky materials. One common basic principle is that many of the physical properties of the products from such microscopic particles (atoms/molecules) wholly depend on how the small cluster of atoms are lined up or arranged. One may get diamonds from coal, if the atoms of this black coloured solid are arranged in a particular manner. Computer chips are got by rearranging the atoms of sand by adding a pinch of suitable impurities. In short, the one main fundamental principle of nanotechnology is high position control. No doubt at the molecular level, the proposal of holding and positioning of molecules is new and shocking. However, as long ago as 1959 Richard Feynman, the Nobel prize winning physicist, said it is possible but not with that easiness.

It is guessed that nanomechanics will be in use in the next one or two decades to manufacture consumer goods at the molecular level, piecing one atom or molecule at a time to make baseballs, telephones, etc. Nanotechnology is an umbrella term that covers many areas of research dealing with objects that are measured in nanometre. A billionth of a metre is a nanometre or a millionth of a millimetre is one nanometre.

Synthesis of nanoparticles, processing of the nanomaterials and the study of their properties and applications is given the emigration name nanotechnology. Particles of sizes between 1 and 100 nm are usually called nanoparticles or generally nanomaterials. Let us discuss one example: 10^{-9} metre is 1 nm = 10 Å. Hence, 1 μm (1 micron) = 10^{-6} metre or 1000 nm. The radius of an oxygen ion in oxide materials is 0.07 nm or of diameter 0.14 nm. Hence, seven oxygen ions in a linear direction will make alert 1 nm which is the lower side of the nano range. Therefore, about 700 oxygen ions in a spatial dimension will be the limit of the nano range materials (higher side).

In the usual and habitual language, when we talk about "materials science" and materials technology, we normally mean ceramics or crystalline materials and metals or cohesively bonded materials. In the case of the material ceramics, the trend is always to prepare finer powder for the ultimate processing and better sintering to achieve dense fine grained microstructure of the particulates with better and favourable properties for different applications. Usually, the degree of fineness goes up to atomic or molecular level *i.e.,* up to 1 nm–100 nm by special processing techniques. More is the fineness, more is the surface area which in turn increases the reactivity of the material. Thus, the densification or consolidation occurs very well at lower temperature than that of conventional ceramic systems, which is finally "cost effective" and also improves the properties of materials like abrasion resistance, corrosion resistance, mechanical properties, electrical properties, and a host of other properties for applications in diverse fields. The deviation of phase diagram of nanomaterial form that of bulk phase diagram is being exploited for forming some selected compositions of alloys. Another interesting information is that the thermal stability of interfacial regions is typically less than that of the bulk material: thus the nanophase materials are often sintered or undergo phase transformation at temperatures below those of the bulk material. This characteristic property has numerous applications to material processing. It is also found that nano phase materials have wide applications in the study of thin films, electronic devices, photonics, bioceramics, biotechnology and medical instrumentation. However, one should know that a knowledge of quantum mechanics is absolutely essential to know more about the behaviour of nanoparticles and handle them to our taste.

8.2.3 NANOSTRUCTURED MATERIALS

Nanoclusters are clusters of atoms consisting of typically hundreds to thousands on nanometre scale. Such small groups of atoms (clusters) go by different names such as nanoparticles, nanocrystals, quantum dots and quantum boxes. Extensive study of the nanomaterials over the past decade brings out remarkable variations in fundamental electrical, optical and magnetic properties that occur as one progresses from an "infinitely extended" solid to a particle of material with a countable (smaller) number of atoms or molecules. The conclusion is that nanostructured materials has opened a new chapter and several applications for short, medium and long term. Quantum wires, electron field emitters for ultra thin TV screens, high resolution tips for scanning and atomic face microscopes, nanoelectronic devices, nanoprobes, sensors, ultrahigh strength composites, gas storage, nanodevices, and parts of nanomachines are some of the devices possible using nanostructured materials.

8.2.4 GRAPHITE

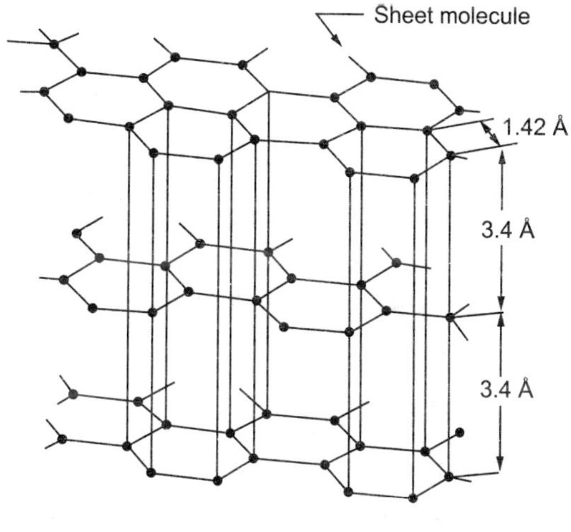

Fig. 8.8 Structure of graphite, illustrating sheet molecules connected by metalic bonds

The crystal structure of graphite is distinctly different from the structure of the glittering diamond. Also graphite is more stable than diamond at ambient temperature and pressure (see Fig. 8.8). Graphite is formed by flat hexagonal layers of carbon atoms separated by 0.34 nm. The C—C distance is about 0.142 nm. a = 0.245 nm is the lattice constant and c = 0.680 nm. The inference is that structurally its interplanar spacing (0.34 nm) is quite large compared to the plane interatomic spacing (0.142 nm). Physically, its stiffness along the plane is quite large because of strong σ bonds and in the perpendicular direction it is weak because of the van der Waal's force. The graphite structure is composed of layers of hexagonally arranged carbon atoms; within the layers, each carbon atom is bonded to three coplanar neighbour atoms by strong covalent bonds. Carbon when sp^2 hybridised contains three strong π-bonds which lie in a triangular plane with bond angle of 120° between them and one weakly bonding π-bond perpendicular to the plane. Each carbon atom in the plane is covalently bonded to three others in plane with strong σ bonds of the sp^2 hybridisation, this network forms a continuous sheet of carbon. The fourth bonding electron participates in a weak van der Waals type of bond between the layers. As a consequence of these weak interplanar bonds, interplanar cleavage is facile, which gives rise to the excellent lubricative properties of graphic. Also, the electrical conductivity is relatively high in crystallographic directions parallel to the hexagonal sheets. Graphite has metallic conductivity along the plane and semiconducting property perpendicular to the plane. The anisotropy in conductivity is about 10^3. The favourable property of graphite is, it is highly anisotropic solid.

8.2.5 SOME SELECTED PROPERTIES OF NANOMATERIALS

The sizes of the atom clusters (nanoparticles) that constitute the material decide the properties of nanomaterial. It is possible to produce new materials with unique properties, if one can control the sizes of nanoparticles. Atomic clusters, grain boundaries and interaction between the atomic clusters are the sources to have unique properties.

Mechanical Properties

Most metals used for different purposes are made up of small crystalline grains. The boundaries between the grains slow down or arrest the propagation of defects when the material is stressed, thus giving it strength. If these grains can be made very small, or even nanoscale in size, the interface area within the material greatly increases, which enhances its strength. For example, nanocrystalline nickel is as strong as hardened steel.

Chemical Properties

As the nanoparticle size decreases, the inter-particle spacing decreases in metals and the surface area increases. As a result, material properties change; for example, the melting point decreases with size. The electronic properties are also affected leading to large changes in catalytic properties. Therefore, nanoparticles can be arranged into layers on surfaces, providing a large surface area and hence enhanced activity, relevant to a range of potential applications such as catalysts.

Magnetic Properties

A normal ferromagnet contains domains each containing several thousand atomic spins; within the domains the spins are aligned in a straight direction while different domains point in different directions. However, nanostructured magnetic materials exhibit totally new class of magnetic properties.

As the particle size decreases, the formation of domain wall becomes energetically unfavourable and the particles are called single domain particles with a critical diameter. In this situation the magnetisation changes do not take place through domain wall motion but require the coherent rotation of spins resulting in large coercivities. Magnetic materials possess some type of anisotropy, which affects their magnetic behaviour. The coercivity of magnetic material is proportional to the anisotropy and so the high anisotropy of materials are preferred for high coercivity applications.

8.2.6 FULLERENES (CARBON 60)

Hary Kroto and Richard Smalley around 1980 discovered a new class of carbon material called carbon 60 (C_{60}). In the case of fullerenes, graphite acquires curvature due to the presence of pentagonal rings in the hexagonal graphite sheet. The term fullerenes is given to any closed carbon cage. Generally, C_{60} are spherical molecules about 1 nm in diameter, comprising 60 carbon atoms. In other words fullerenes exists in discrete molecular form and consists of a hallow spherical cluster of 60 carbon atoms; a single molecule is denoted by C_{60}. Each molecule is composed of groups of carbon atoms that are bonded to one another to form both hexagon (six-carbon atom) and pentagon (five-carbon atom) geometrical configurations. One such molecule (see Fig. 8.9), is found to consist of 20 hexagons and 12 pentagons which are arrayed such that no two pentagons share a common side; the molecular surface thus exhibits the symmetry of soccer ball. The material composed of C_{60} molecules that contains sixty carbon atoms in a network of sp^2 bonding which forms a spherical structure, is known as buckministerfullerene, from a number of perspectives, a molecule almost custom built for nanoscience. Discovered in 1985 by Kroto *et al.*, C_{60} is approximately 1nm in diameter. The term fullerene is used to denote the class of materials that are composed of this type of molecule. In general fullerenes are made up of hexagonal and pentagonal rings of carbon arranged exactly as of an Indian football. Today we find the presence or existence of the fullerenes C_{36}, C_{70}, C_{80}, C_{100} based on fullerenes and few more members of geometrically permitted fullerene family. These fullerenc molecules differ in the number of hexagonal rings present. When the number of carbon atoms change the shape of the molecule also changes. The C_{70} looks like a *xydy* ball and C_{120} takes a dumbbell shape. Diamond and graphite are what may be termed network solids, because all of the carbon atoms form primary bonds with neighbouring or adjacent atoms throughout the entirety of the solid. The common synthetic graphite is usually manufactured as composites.

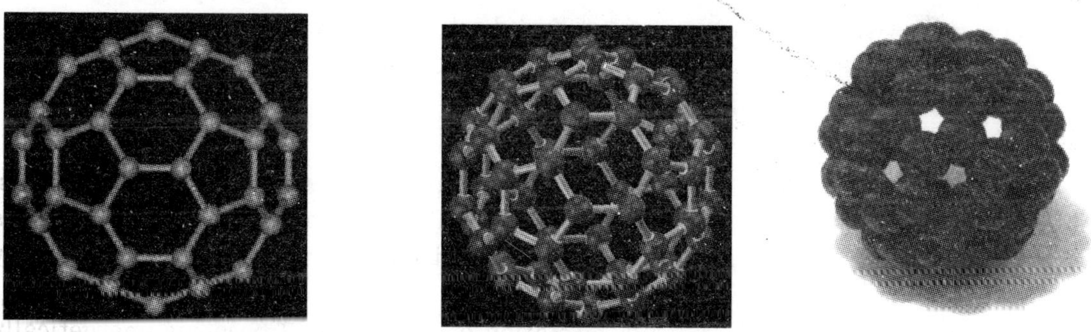

(*a*) (*b*) (*c*)

Fig. 8.9 (*a*) The structure of C_{60} molecule (*b*) Structure of fullerene C_{70} (*c*) Buckyball 60 atoms

At present, many methods are available to prepare C_{60} molecules in large quantities, so the study of this molecular form of carbon is advancing in all areas. As a pure crystalline solid, this material is electrically insulating. However, with proper impurity additions, it can be made highly conductive and semiconductive. Presently, we know that C_{60} doped with alkali metals become superconducting with $T_c \simeq 50$ K. Fullerene research has got immense scope in nanoscience and technology. They find use in catalytic chemistry, biomolecular recognition, as molecular sieves, nanoreactors, etc.

8.2.7 CARBON NANOTUBES

A form of carbon in which the atoms are linked by a cylindrical framework with diameter of only a few nanometre is known as carbon nanotube. It is also known as buckytube. In the experiment, on the arc discharge between graphite electrodes, along with the collected shoot particles there were some fine tube like structures. This on careful examination and analysis revealed that they are carbon nanotubes (CNTs). Its structure consists of a single sheet of graphite, rolled into a tube, both ends of which are capped with C_{60} fullerene hemispheres. The tubes can be either open at their ends or capped at one or both ends with half a spheroidal fullerene. The following indicates a schematic representation of a carbon nanotube:

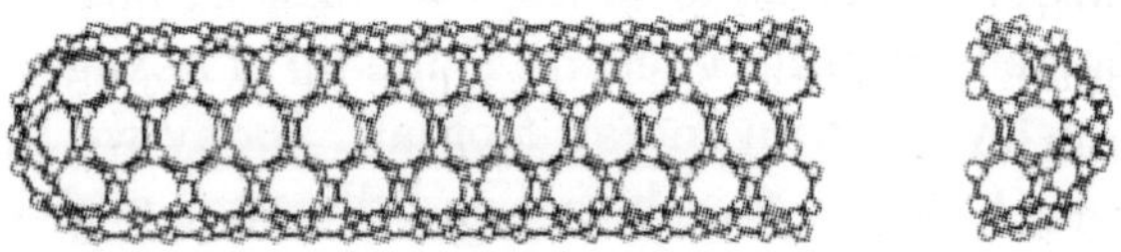

(a)

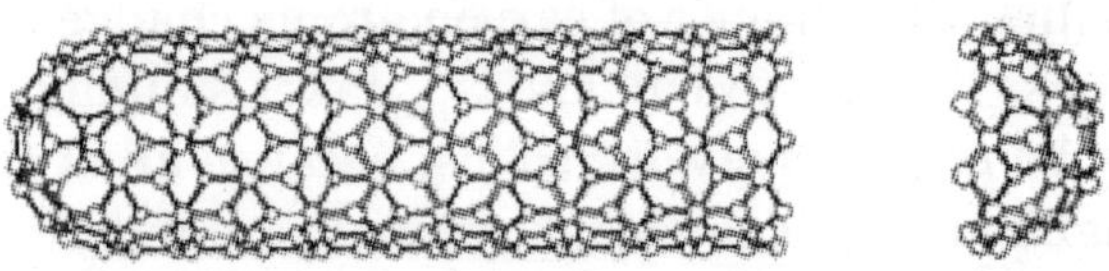

(b)

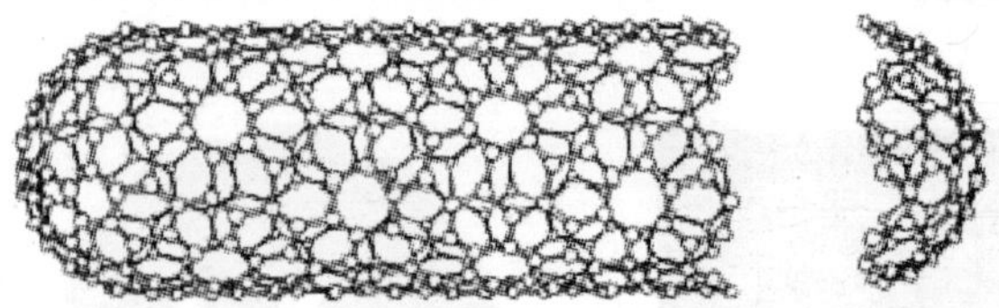

(c)

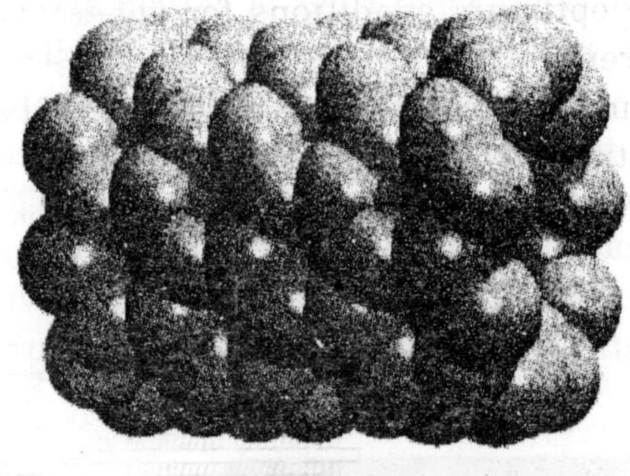

Nanotubes

Fig. 8.10 The structure of carbon nanotubes (*a*) metallic, (*b*) and (*c*) semiconducting

A nanotube is completely specified by what is referred to as a roll of vector which identifies its helical nature and diameter.

Conclusion

Carbon nanotube (CNT) is a new form of carbon configurationally equivalent to two-dimensional graphene sheet rolled into a tube. It is grown by different methods in the laboratory with a few nanometre in diameter and several microns long. CNT exhibit extraordinary mechanical and electrical properties. The Young's modulus is about 1 tetra pascal. It is very stiff like diamond. The estimated tensile strength is 200 Giga Pascal. CNT can be metallic or semiconducting and offers amazing possibilities to create future nanoelectronic devices.

Some Production Processes of CNTs

Some well-known methods of producing CNTs are discussed below:

(*a*) Arc discharge
(*b*) Pyrolysis
(*c*) Laser vaporisation
(*d*) Electrolysis.

(a) Arc discharge: The Fig. 8.11 given below supplies some details of producing CNTs by arc discharge method. Here, the pressure used is 500 torr (while it is 100 torr in fullerenes), and secondly, the nanotubes are found deposited in the cathode and not in the soot. This method produces well-ordered multi-walled CNT (MWCNT) with diameters from 2 nm up to 30 nm. The separation between graphite layers is around 0.34 nm, which is very close to that of graphite 0.335 nm. The length of these nanotubes can be up to 30 μ. The optimum conditions for generating nanotubes with this method consist of using currents of 150 amp, voltage of 25 volt and graphite electrodes between 6 mm and 8 mm in diameter with a separation between them of 1 mm under helium atmosphere. The electric arc reaction is too violent, so it is very difficult to control the formation reaction.

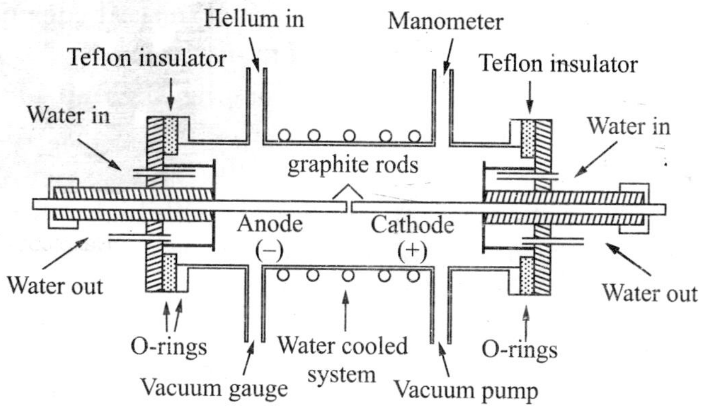

Fig. 8.11 Arc discharge set up

(b) Pyrolysis method: This method consists of heating a hydrocarbon on a precursor which contains carbon under the presence of a catalyst such as nickel, cobalt or iron. It was a kown fact that carbon fibres could be obtained by pyrolytic methods using hydrocarbons (prior to the discovery of C_{60}).

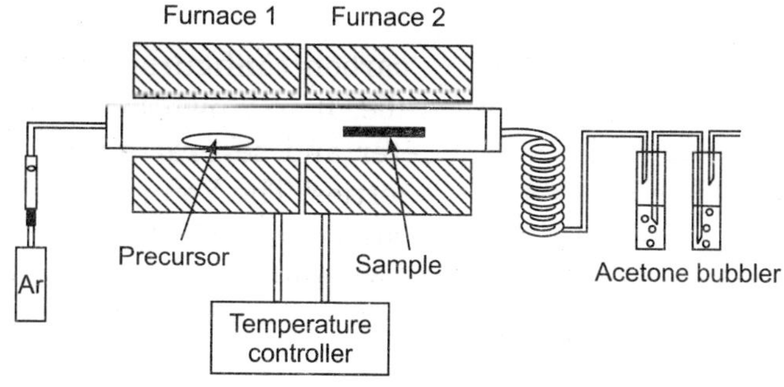

Fig. 8.12 Pyrolysis set up

It must be noted that carbon fibres are not nanotubes since the fibres are bigger in diameter and exhibit a great amount of defects and impurities. One must note that the application of carbon fibres is widespread, showing a strength comparable to steel.

It was found that (in 1994) carbon nanotubes could be aligned by embedding them in a polymer, thus forming a composite material. When the polymer was cut, it was observed an aligned arrangement of nanotubes. In this context, a better alternative is to use laser ablation of cobalt thin films and then pyrolyse them with an organic precursor such as 2-amino 4, 6-dichloro-S-tryazine. The nanotubes obtained in this way are aligned in bundles which can grow up to 40 μ and diameters between 30 nm and 50 nm.

(c) Laser vaporisation method: The laser vaporisation method involves firing a high power laser towards a graphite target inside a furnace at 1200°C. The condensation of material generated by the laser is responsible for the nanotube formation. If we add nickel or cobalt to the graphite target, we obtain Single-Walled Carbon Nanotubes (SWCNT). The SWCNT has diameter around 1.4 nm forming a two-dimensional crystal with a lattice constant of 0.17 nm.

(d) Electrolysis method: Graphite electrodes inside a molten salt such as lithium chloride (700°C) under an argon atmosphere is the basis of this method. Depending on conditions 20% to 30% consists of nanotubes. The depth of the cathode and the current (3–10 amp) play an important role in the formation of nanotubes.

Carbon: Symbol C. A non-metallic element belonging to group 14 (formerly IVA) of the periodic table; Z = 6, M_A = 12.011, m.p. = 3823 K, b.p. = 5100 K. Carbon has three main allotropic forms.

Diamond: (r.d. = 3.52) exists as very hard crystals with high refractive index and is prized as a precious stone. It is also used in many cutting tools. It occurs naturally but can be produced synthetically. The hardness is a result of the crystal covalent tetrahedron structure.

Graphite: (r.d. = 2.25) exists as a soft black slippery substance (also called black lead). It occurs naturally but can also be synthesised. It has many uses such as conductors of heat and electricity, solid lubricant, and in pencil lead (mixed with clay). The third crystalline allotrope is fullerite. There are also various amorphous forms of carbon, *e.g.*, charcoal. Carbon has two stable isotopes (nuclean number 12 and 13) and four radioactive ones (10, 11, 14, 15).

8.2.8 SEMICONDUCTOR AND SEMICONDUCTOR NANOPARTICLES

A crystalline solid (*e.g.*, silicon or germanium) with electrical conductivity between that of a conductor and insulator, *i.e.*, in the range about 10^5–10^{-4} siemen/metre. In a semiconductor, the energy bands contain a narrow forbidden band across which electrons in the valence band can cross-over to the conduction band when thermally excited. The electrons are thus mobile under the influence of an external electrical field. The crossing-over of the electron leaves a positive hole in the valence band. The hole behaves as though it were a "positively charged electron", *i.e.*, it is mobile under an electric field. In effect, electrons and holes are the charge carriers in a semiconductor.

Semiconductor nanocrystals are the most important materials amongst the nanophase materials due to their exotic band gap engineering behaviour. Silicon is the prime material in semiconductor industry and is widely used in manufacturing of integrated circuits. However, because of indirect band gap property it exhibits very weak luminescence and hence cannot be used for making optical devices. In 1990, the nanocrystalline silicon was produced successfully from bulk silicon by electrochemical anodisation. This silicon exhibits direct band gap type properties and was seen to emit light in the presence of UV excitation at room temperature. This silicon is known as porous silicon. The porous silicon emits light on excitation in the visible region corresponding to the band gap varying between 1.6 and 2.0 eV. Porous silicon finds numerous applications because of its optical properties and high surface area. Porous silicon has very interesting biological applications because it is bio-inert *i.e.*, it is biologically stable and has a low degree of biotoxicity. It has also found its use in glucose monitoring systems.

The electric current in a semiconductor is carried by both 'electrons' and 'holes' in semiconductors. These electric charge carriers move differently through semiconductors with different compositions, so we can make different sorts of semiconductor structure. Since the electrons and holes have opposite charges they can attract each other and form 'excitons'. These carriers do not move freely through the semiconductor. The crystal lattice of atoms does affect their motion by making them more likely to travel at certain speeds, but a perfect crystal will not cause them to be deflected. However, their motion is affected by imperfections in the lattice, as well as by the phonons.

8.2.9 NANOELECTRONICS

The Semiconductor Structures

If we put two different semiconductor materials side by side, it is found that the electrons and holes will have preference of one material over the other. If we visualise the electrons to be residing on a 'cage', then we will find that electrons from one cage in one material can easily fall into an adjacent partially filled cage on a lower band from another material, so that we can say that the electrons 'prefer' some materials over the others. It is a kind of 'friendship' of the electrons in a cage of a given material with the cage of a second material. The holes also can do this, but since they are voids or gaps, they simply fall upwards. By putting a piece of preferred semiconductor next to another, we can make a device where the electrons will only want to travel in the direction of the preferred material thereby creating a 'diode'. By surrounding some preferred semiconductor by a different semiconductor, the 'traps' can be created for the electrons and holes that lead us to a "quantum well", a "quantum wire", or a "quantum dot".

The Quantum Wells

A "quantum well" is a potential well that is obviously very small. This 'well' is like a cage in which the carrier particles (*i.e.,* the excitions) can be trapped, in much the same way that light can be trapped between the mirrors. By creating "layers of different semiconductors", it is possible to make particular 'layers' to act as 'traps' for excitions. These "trapped particles" can be considered to be in "quantum confinement".

Due to this particle trapping, the "quantum confinement effects" naturally restrict the motion of the carriers in same way, *i.e.,* restrictions on what the particles want to do. In a "quantum well", the excitons can move freely side ways in the plane of a "thin layer", but they might like to move in the forward and/or backward directions as well. These particles are constrained or confined in the forward- backward directions *i.e.,* a **2D** confinement. For any particular sideways motion, there could now be various ways in which the carrier is confined in the layer giving rise to quantum wells. These "quantum wells" have many useful properties, since it is possible to exactly engineer the forward-backward confinements, and since they are relatively easy to make. As a result, they are widely used to make "Semiconductor Lasers" and other useful devices. This "quantum confinement effect" can be taken further and make semiconductor structures—that brings us to the "quantum wires" and "quantum dots".

The Quantum Wires

From "quantum well", we can go ahead one step further in the semiconductor structure, *i.e.,* by making not a layer but only a thin 'wire' of the preferred semiconductors (*i.e.,* a chain like structure). These trapped excitons can now be considered to be in **ID** "quantum confinement". In this case, an exciton is only free to choose its trajectory along the wire.

Some General Properties of Carbon Nanotubes

Direct observation using high resolution transmission electron microscopy (HRTEM) reveals that CNTs are extraordinarily flexible. For example, they can be bent mechanically by an ultrasonic bath, by mechanical milling, and by using a polymeric resin. Theoretical calculations predict this flexibility. The first attempt to determine the stiffness of carbon nanotubes was when Treacy et al., used HRTEM to measure the amplitudes of the vibrating tubes, to calculate indirectly the Young's modulus. They found that MWCNTs exhibit Young's modulates of the order of 1.8 tera Pa (1 tera Pa = 1×10^{12} Pa), which is higher than conventional carbon fibres (Ca 800 Giga Pa). Direct measurement using Atomic Force Microscopy (AFM) reveal that the Young's modulus of MWCNTs is around 1.28 tera Pa. This high Young's modulus makes the nanotubes a very strong structure.

The first measurement of the electrical conductivity of CNTs were made using gold micro contacts attached to the CNTs by lithographic techniques finding that the resistance depends on the temperature and that in the range between 2 K and 300 K the tubes were semimetallic. Conductivity measurements on aligned MWCNTs show that the material behaves as a nanoconductor. Also it has been confirmed that depending on the diameter and helicity, carbon nanotubes can be conductors or semiconductors. Single-Walled Carbon Nanotubes(SWCNTs) around 1.4 nm diameter exhibit quantum conductivity. We may note that the atomic arrangement, *i.e.,* the geometry of the structure, determines the mechanical and electronic properties of CNTs. Nowadays, nanocrystals have been prepared. Ductility and strength are the central mechanical properties of any material. They are determined by the physical nature of plastic deformation, which is conventional, coarse-grained metals is mainly carried by dislocations line defects of regular crystal lattice within individual grains. Study suggest, however, that the mechanism of plastic deformation in nanocrystalline materials may be different and so may lead to novel mechanical properties. Hardness or high strength are noticed in many nanomaterials. However, in most cases nanomaterials have very low ductility, *i.e.*, they fail when their shape is changed. Some even become brittle when a deforming load (force or stress) is applied.

Luckily in 2002 Wang *et al.*, reported their finding is engineering an unusual nanocrystalline material that combines two useful properties that are often mutually exclusive strength and ductility. They created a copper nanostructure by rolling the metal at temperature below 77 K and then heating it to around 450 K. The result was a 'bimodel' structure of micrometre sized grains (at a volume fraction of 25%) embedded in a matrix of nanocrystalline grains. The material showed extraordinary high ductility, but also retained its high strength. The reason for this behaviour seems to be that while the nanocrystalline grains provide strength, the embedded larger grains stabilise the tensile deformation of the material. Processing nanomaterial to achieve both strength and ductility can enhance the material's resistance to fatigue. More than anything else, it is fatigue that at present limits the lifetime, and hence the range of applications, of many advanced materials.

Scope of Applications

Nanocrystalline materials are three-dimensional solids composed of nanometresiz crystallites or grains on account of their unique structure, which is characterised by ultrafine grains and a rather high density of crystal lattice defects, these materials have extra ordinary fundamental properties that could be exploited to make "next – generation" superstrong materials.

The central mechanical properties of any material is its mechanical properties which are determined by the physical nature of plastic deformation. The deviations from the bulk phase diagram may be exploited to form certain compositions of alloys that are otherwise unstable in the bulk form. In addition, the thermal stability of interfacial regions is typically less than that of the bulk materials: Thus the nanophase materials are often sintered or undergo phase transformation at temperatures below those of the bulk material. This is a characteristic property which has numerous application to material processing.

By improving the properties of nanomaterials, we may be able to find better applications as varied as semiconductor electronics, sensors, special polymers, magnetics, advanced ceramics, and membranes. It is necessary to improve our current understanding of particles size control and methodologies for several classes of nanophase materials and address the issues of their characterisation. We should also explore the fields such as (*i*) Electronics of thin films, electronic devices like MOSFET, JFET and in electrical ceramics (*ii*) Bionics, (*iii*) Photonics, (*iv*) Bioceramics, (*v*) Biotechnology, and (*vi*) Medical instrumentation, etc.

8.2.10 IMPORTANCE OF NANOTECHNOLOGY

The body of airplanes are made from aluminum despite the truth that the well-known diamond has a strength to weight ratio about 50 times that of aerospace aluminum. Diamond is expensive, we cannot make it in the shapes we want, and it shatters. It is untrue today. Nanotechnology will let us inexpensively make shatter proof diamond (with a structure that might resemble diamond fibres) in exactly the shapes we want. This may help us make a Boeing 747 whose unloaded weight was 50 times lighter but just as strong.

We are well aware that travel in space today is very expensive and booked for an elite few or for those who enjoy travel grants whether it is air planes or airships. But nanotechnology will amazingly reduce the costs and increase the capabilities of space ships and space flight. The strength to weight ratio and the cost of components are absolutely critical to the performance and economy of space ships. The inventions in nanoscience and their applications will improve both the above said parameters by one or two orders of magnitude. The benefits in these two parameters alone will improve the overall cost/performance ratio by over three orders of magnitude leaving other advantages of nanotechnology. Beyond inexpensively providing remarkably light and strong materials for space ships, nanotechnology will also provide extremely powerful computers with which to guide both those ships and a wide range of other activities in space. If you want the computer hardware revolution to continue at its current pace, in another five to ten years you have to move beyond lithography to some new post lithographic manufacturing technology. Making patterns on a resist and spraying chemicals around simply cannot arrange atoms with the ultimate precision that should be feasible. More precise methods will be needed. Ultimately, each logic element will be made from just a few atoms. Designs for computer gates with less than 1000 atoms have already been proposed but each atom in such a small device has to be in exactly the right place. To economically build and interconnect trillions upon trillions of such small and precise devises in a complex three-dimensional pattern one may need a manufacturing technology well beyond today's lithography: one needs nanotechnology. With it, we should be able to build mass storage devices that can store more than a hundred billion bytes in a volume equal to the volume of a sugar cube.

We know that, as on today 'smart' weapons are big and heavy *i.e.,* we have the "smart bomb" but not the "smart bullet". In the years to come, even weapons as small as a single bullet could pack more computer power than the largest supercomputer in existence today, allowing them to perform real time image analysis of their surrounding and communicate with weapons tracking systems to acquire and navigate to targets with greater precision and control. We will also be able to build weapons both in expensively and much more rapidly, at the same time taking full advantage of the remarkable materials properties of diamond. Rapid and inexpensive manufacture of great quantities of stronger more precise weapons guided by massively increased computational power will alter the way we fight wars. Coal and oil from the ground, dam across falls and river and nuclear fuels are the main sources to generate power today. But the sunshine all around us in the most part of the earth for several months in a year can provide orders of magnitude more power than we use. The principles of making efficient solar cells are of high cost. But low cost solar cells and battery is beyond our reach now. But nanotechnology will cut costs both of the solar cells and the equipment needed develop them. Solar power will then become economical.

The other interesting point is, it is not the modern medicine that does the healing (always), but the cells themselves. If we had surgical tools that were molecular both in their size and precision, we could develop a medical technology that for the first time would let us directly heal the injuries at the molecular and cellular level that are the root causes of disease and ill health. With the precision of drugs combined with the intelligent guidance of the surgeon's scalpel, we can expect a quantum leaps in our medical capabilities.

8.2.11 NANOSCIENCE, NANOTECHNOLOGY AND MEDICINE

It is the known practice of keeping clinically dead human body or brain at an extremely low temperature in the hope of later restoring it to life with the help of future medical technologies. Cryonics is the practice of preserving organisms, or at least their brains, for possible future revival by storing them at cryogenic temperatures where metabolism and decay are completely stopped. Though still controversial, it has greater potential to save lives than any other method that we can use today.

Generally, sickness (or disease) and ill health are caused largely by damage at the molecular and cellular level. Surgical tools and instruments used now are large and crude. From the view point of a cell, even a fine scalpel is a blunt instrument more suited to tear and injure than heal and cure. Modern surgery works just satisfactorily only because cells have a remarkable or wonderful ability to regroup themselves, bury their dead ones and heal the wound or injury.

Manufacturing technology of the 21st century (nothing but nanotechnology) may help us economically build a wide range of complex molecular machines. It will possible to build fleets of computer controlled molecular tools much smaller than a human cell and built with the accuracy and precision of drug molecules. Such tools will let medicine, for the first time, intervene in a sophisticated and controlled way at the cellular and molecular level. They could remove obstructions in the circulatory system, kill cancer cells, or take over the function of such cellular organelles. Just as today we have the artificial heart, so in the future we could have the artificial mitochondrion.

Bodies of beings are filled with intricate, active molecular structures. When those structures are damaged, health suffers. Modern medicine can affect the workings of the body in several ways, but from a molecular viewpoint it remains crude indeed. But molecular manufacturing can construct a range of hospital instruments, machines and devices with far greater abilities. The body is actually a huge complex world of molecules. We can learn a lot to repair it with nanotechnology. The human body can be considered as a work yard, construction site, and a war-field cum battle ground for molecular machines. It works or functions remarkably well, using systems so complex that medical science still does not understand many of them. When we chew and swallow, muscles drive our motions. Muscle fibres contain bundles of molecular fibres that shorten by sliding past one another. In the stomach and intestines, the molecular machines we call digestive enzymes break down the complex molecules in foods, forming smaller molecules for use as fuel or building blocks. Molecular devices in the lining of the digestive track carry useful molecules to the blood stream.

In the lungs, molecular storage devices called hemoglobin molecules pick up oxygen. Driven by molecular fibres, the heart pumps blood laden with fuel and oxygen to cells. In the muscles, fuel and oxygen drive contraction based on sliding molecular fibres. In the liver, they drive molecular machines that build and breakdown a whole host of molecules. The tale continues through all the work of the body. Except for tooth filling and other artificial implants, everything in the human body is constructed by molecular machines. They clear away structures that are old or out of place, sometimes using machinery like digestive enzymes to take structures apart. Failures in reconstruction increase as we become aged.

We have to remember that the immune system does not respond to all invaders or responds inadequately. Malaria, tuberculosis, herpes, and AIDS all have their strategies for evading destruction. Cancer is a special case in which the invaders are altered cells of the body itself, sometimes successfully masquerading as healthy cells and escaping detection.

The other part is that the immune system sometimes over responds attacking cells that should be left alone. Certain kinds of arthritis, as well as lupus and rheumatic fever, are caused by this mistake. Between attacking when it should, the immune system often fails, causing suffering and death even.

We are aware that diagnostic procedures vary widely, from asking questions to patient, looking at X-ray shadows, through exploratory surgery and the microscopic and chemical analysis of materials from the body. Doctors can diagnose many illness, but others remain mysteries. The two main forms of treatment are surgery and drugs. From a molecular perspective, neither is sophisticated. Surgery is direct, and manual. Surgeons torn tissues and skin to enable healing, cut out cancer, clear out clogged arteries, and even install pacemakers and replacement organs. It is direct, but not safe always; anesthetics, infections, organ rejection, and missed cancer cells can all cause failure. The body works by means of molecular machines, most working inside cells. Surgeons can see neither molecules nor cells, and can repair neither cent per cent. Antibiotics like penicillin are selective poisons. Viruses are a harder case because they are simpler and have fewer vulnerable molecular machines. Cancer is the most difficult of all and many a time they poison healthy cells of the patient while treatment. Drug molecules bind to molecules in the human body and change the behaviour. Some decrease to secrete stomach acid, others stimulate the kidneys, many affect the molecular dynamics of the brain.

Developments in nanotechnology will result in improved medical sensors. As protein chemist Bill DeGrado notes, "Probably the first use you may see would be in diagnostics: being able to take a tiny amount of blood from somebody, just a pinprick, and diagnose for a hundred different things". In the longer term, though, the story of nanotechnology in medicine will be the story of extending surgical control to the molecular level.

Conclusion

A nanometre (nm) is one-billionth of a metre; a nm is about the width of six bonded carbon atoms. About a million nanometre will reach across the head of a pin. Our fingernails grow approximately one nanometre per sec.

Some Selected Relations

- 1 inch = 25,400,000 nm
  A red blood cell is 7000 nm in diameter and 2000 nm height.
  A virus is 100 nm.
- Intel will be manufacturing devices with feature sizes about 20 nanometre across.
- In two dimensions we could stack about 250,000 components in the same space as a red blood cell.
- If the trends continue as far as 2017, which may be the end-point of "Moore's Law", we could be looking at a manufactured device the size of the red blood cell with 256,000 components.
- If we add the third dimension, that could translate into 65,536,000,000,000,000 components.
- Somewhere along the way, we are talking about the raw technical capability to produce a rather sophisticated robot small enough to wander around through your body doing whatever it has been programmed to do.
- If we make the robot $\frac{1}{10000}$ the volume of a red blood cell, we are still talking about 655, 360,000 components, which is arguably perhaps enough to embody this machine with the ability to think, move and do whatever we have programmed it.

Some Examples of Nanoparticles

1. Nanoparticles of Tin Dioxide: The details of this important work is discussed here. The emission intensity of the peak at 612 nm of the Eu^{3+} ions due to $^5D_0 \rightarrow {}^7F_2$ transition, activated by SnO_2 nanocrystals were found to be sensitive to the nanoenvironment for both doped and coated samples. The luminescent efficiencies of nanocrystals of SnO_2–'doped' by Eu_2O were compared with those of SnO_2 'coated' by Eu_2O, and the intensities were found to be significantly higher for coated materials. It was found from the measured luminescent intensity that Eu^{3+} ions occupy "low symmetry" sites in the nanocrystals of SnO_2 coated by Eu_2O with relaxation rate much higher in the coated sample than in doped sample due to the asymmetry of Eu^{3+} ions.

2. Nanoparticles of Cuprous Oxide: The Cuprous Oxide (Cu_2O) particles coated with poly-acrylamide having diameters in the range from 4.8 nm to 8.6 nm were prepared by a chemical method. The optical absorption of these particles was found to be characterized by excitons with corresponding energies varying from 2.6 to 2.25 eV in the above particles size range. A second optical absorption maximum signifying surface states within the band gap is exhibited by the specimens with relevant energies varying from 2.6 to 1.77 eV in this particle size range.

3. Nanoribbons of Zinc Sulphide: One-dimensional nanostructured materials set the pace of the recent trend in nanomaterials research due to their wide range of potential applications in many nanoscale devices. It is known that ZnS is an important semiconductor used as "phosphorus materials". Hence, some efforts have been made into controlling the size, morphology and crystallinity of the ZnS crystals in order to fine-tune their physical properties.

4. Nanoparticles of CdS: The optical and microstructural properties of nanocomposite thin films based on the systems CdS-ZnO and CdS-Al_2O_3 were studied. These nanomaterials were synthesised by solgel technique. The molar ratio of CdS with ZnO and Al_2O_3 were varied within the range of 20 : 80 to 50 : 50. The nanoparticles of CdS that are highly confined with radius 1.8 nm to 4.7 nm, which were estimated from the blue shift of the absorption edge, were obtained by using ZnO matrix, whereas for Al_2O_3 matrix the size varied from 2.8 nm to 7 nm. The microstructural characterisation by high resolution revealed well-resolved crystalline nanoparticles in both cases.

SHORT QUESTIONS

8.1.1 What is meant by time of reverberation?

8.1.2 Write down the relation for standard time of reverberation with symbols expanded.

8.1.3 How reverberation can be adjusted?

8.1.4 Mention a few conditions for good acoustical design of halls.

8.1.5 Define absorption coefficient in acoustical sound.

REVIEW QUESTIONS

8.1.1 State and explain Sabine's formula for reverberation time of a hall. Derive Sabine's formula for reverberation time.

8.1.2 Define absorption coefficient and discuss any one method of measuring it.

8.1.3 Discuss the growth and decay of energy density of sound.

8.1.4 Discuss the important factors affecting acoustics of buildings and their remedies.

8.1.5 Define absorption coefficient in acoustics. Describe with necessary theory a method of determining absorption coefficient of a material fitted in a hall.

PROBLEMS AND SOLUTIONS

8.1.1 Calculate the reverberation times of a hall 2400 m^3 having a seating capacity of 600, (*a*) when the hall is empty, and (*b*) when the hall is full with audience from the following data:

Surface description	*Area*	*Absorption coefficient (O.W.U.)*
Plasted ceiling	500 m^2	0.02
Plasted wall	600 m^2	0.03
Wooden floor	500 m^2	0.06
Wooden doors	20 m^2	0.06
Cushioned seats	400	0.1 per chair
Cane seats	200	0.01 per chair
Audience	–	0.45

Solution.

(*a*) When the hall is empty

$$T = \frac{0.16\,V}{\sum \alpha ds}$$

$$T = \frac{0.16 \times 2400}{500 \times 0.02 \times 600 \times 0.03 \times +500 \times 0.06 + 20 \times 0.06 + 400 \times 0.1 + 200 \times 0.01}$$

$$= \frac{384}{10 + 18 + 30 + 1.2 + 40 + 2}$$

$$= \frac{384}{101.2}$$

$$\boxed{T = 3.79 \text{ sec}} \quad \textbf{Ans.}$$

(*b*) When the hall is full with audience

$$T = \frac{0.16\,V}{\sum \alpha' ds'}$$

$$= \frac{384}{500 \times 0.02 + 600 \times 0.03 + 500 \times 0.06 + 20 \times 0.06 + 400 \times 0.45 + 200 \times 0.45}$$

$$= \frac{384}{10 + 18 + 30 + 1.2 + 180 + 90}$$

$$\boxed{T = 1.17 \text{ sec}} \quad \textbf{Ans.}$$

8.1.2 The reverberation time of a hall of volume 3000 m^3 is 4.6 sec. Given a material of absorption coefficient 0.69 O.W.U. How will you reduce the reverberation time of the hall to 2.7 sec?

Solution.

Let T_1 be the initial reverberation time of the hall.

Then $$T_1 = 4.6 = \frac{0.16\ V}{\sum \alpha_1\ ds_1}$$

If we increase the absorption inside the hall, the reverberation time can be brought down. Hence let an area S of a given material with absorption coefficient 0.69 O.W.U be introduced in the hall so that the final reverberation time T_2 is 2.7 sec.

$$T_2 = 2.7 = \frac{0.16V}{\sum \alpha_1 s_1 + \alpha\, s}$$

$$\left(\frac{1}{T_2} - \frac{1}{T_1}\right) = \frac{\alpha s}{0.16\ V}$$

$$s = \frac{0.16 \times 3000}{0.69}\left[\frac{1}{2.7} - \frac{1}{4.6}\right]$$

$$\boxed{s = 106.435\ \text{m}^3} \quad \textbf{Ans.}$$

i.e., By reducing 106.4 m^2 area of the given material inside the hall, the reverberation time can be reduced to 2.7 sec.

8.1.3 A class room is 4.6 × 10^3 m^3 and has a reverberation time of 1.5 sec. What is the total sound absorption of the class room?

Solution.

Data given

Volume of the hall, $V = 4 \times 6 \times 10 = 240\ m^3$

Reverberation time, T = 1.5 sec

The general expression for the reverberation time

$$T = \frac{0.167\ V}{\sum as}$$

i.e.,

Sound absorption,

$$\sum as = \frac{0.167\ V}{T} = \frac{0.167 \times 240}{1.5}$$

$$\boxed{\sum as = 26.72\ \text{Sabine}} \quad \textbf{Ans.}$$

GENERAL QUESTIONS

8.2.1 Distinguish between atomic clusters and nanoparticles.

Ans. The general practice used by scientists for investigating the materials is to break them into fundamental building blocks and grouping them (atomic clusters) and study them at each stage. From the understanding of the basic interactions of atoms, attempt is made to explain the structure and properties of bulk materials.

Nanostructured materials may be defined as those materials whose structural element, crystallites or molecules have dimensions in the 1 to 100 nm range. Clusters of atoms consisting of typically hundreds to thousands on nanometre scale are called nanoclusters (nanoparticles); others with much greater dimensions are simply called clusters.

8.2.2 What is nanotechnology?

Ans. Here manufactured products are from atoms. The properties of those products depend on how those atoms are arranged. If one rearranges the atoms in coal, he gets diamonds. If one rearranges atoms in sand (and adds a pinch of suitable impurities), he gets computer chips.

8.2.3 Discuss the special properties of nanomaterials.

Ans. Many properties of nanomaterials are drastically different from the normal bulk materials for two main reasons. First, nanomaterials have a relatively larger surface area when compared to the same mass of materials produced in bulk. This can make materials more chemically reactive and affect their strength or electrical properties. Secondly, quantum effects can begin to dominate the behaviour of matter at the nanoscale affecting the optical, electrical and magnetic behaviour of materials.

8.2.4 What are carbon nanotubes?

Ans. A carbon nanotube is a cylindrical rolled up sheet of grapheme, which is a single layer of graphite atoms arranged in hexagonal pattern. Their hexagonal structure gives them great tensile strength and elastic properties. The tubes are tough and when bent or squeezed, they spring back to their original shape. They also transfer heat very efficiently. They can be made to exhibit either metallic or semiconductor characteristics depending on the way they get folded.

8.2.5 What are nanostructured materials?

Ans. Nanostructured materials may be defined as those materials whose structural element, clusters, crystallites or molecules have dimensions in the 1 to 100 nm range. Clusters of atoms consisting of typically hundreds to thousands on nanometric scale are commonly called as nanoclusters. These very small groups of atoms, is generally go by different names such as nanoparticles, nanocrystals quantum dots or quantum boxes.

8.2.6 Give the salient features of nanotechnology.

Ans. Nanotechnology will offer greatly improved efficiency compared with electricity and computers in every facet of life. It has dual use many commercial uses and military uses for making more powerful weapons and tools of surveillance. Thus, it represents not only wonderful benefits for humanity, but is also a matter of grave risk.

The important understanding of nanotechnology is that it offers not just better products, but a vastly improved means of production. A computer can make copies of data files essentially as many copies as you want at little or not much cost.

8.2.7 Explain the role of nanorobots when injected into human body.

Ans. A typical nanomedical treatment will consist of an injection of perhaps a few cubic centimetre of micronsized nanorobots suspended in fluid (probably a water/saline suspension). A typical therapeutic dose may include up to1–10 trillion (1 trillion=10^{12}) individual nanorobots, although in some cases treatment may only require a few million or a few billion individual devices to be injected. Each nanorobot will be on the order of perhaps 0.5 micron up to perhaps 3 microns in diameter.

The adult human body has a volume perhaps 100,000 cm^3, and a blood volume of 5400 cm^3, so adding a mere 3 cm^3 dose of nanorobots is not particularly invasive. The nanorobots are going to be doing exactly what the doctor tells them to do, and nothing more. So the only physical change you will see in the patient is that he or she will very rapidly become well again. Most symptoms such as fever and itching have specific biochemical causes which can also be managed, reduced, and eliminated using the appropriate injected nanorobots. Major rashes or lesions such as those that occur when you have the measles will take a bit longer to reverse, became in this case the broken skin must also be repaired.

Explain the Following Terms:

8.2.8 Diamondoid, Assembler, Molecular manufacturing and Nanomachine.

Ans. *Diamondoid*: Structures that resemble diamond in a broad sense, strong stiff structures containing dense, three-dimensional network of covalent bonds, formed chiefly from first and second row atoms with a valence of three or more. Many of the most useful diamondoid structures will in fact be rich in tetrahedrally coordinated carbon.

Assembler: A general purpose device for molecular manufacturing capable of guiding chemical reactions by positioning molecules.

Molecular manufacturing: Manufacturing using molecular machinery, giving molecule by molecule control of products and by products via positional chemical synthesis.

Nanomachine: An artificial molecular machine of the sort made by molecular manufacturing techniques.

8.2.9 How nanorobots be retrieved from the body?

Ans. Some nanodevices will be able to exfuse themselves from the body via the usual human excretory channels; others will be designed to allow ready exfusion by medical personal using apheresis like processes or active scavenger systems. It is very design dependent.

8.2.10 Explain with crystal structures the salient features of graphite and fullerenes.

8.2.11 Discuss two methods of synthesis of nanomaterials.

8.2.12 What are the different methods for the production of CNTs? Briefly discuss them.

8.2.13 Mention some important mechanical and electronic properties of CNTs. Discuss briefly quantum wells, quantum wire and quantum dot.

Chapter 9

UNIT 9

Optics

9.1 WAVE THEORY OF LIGHT

9.1.1 INTRODUCTION

According to Prof. Huygen (1679), each point in a source of light sends out in all directions in a hypothetical medium known as ether. Though Newton's corpuscular theory could not explain the phenomenon of interference, diffraction and polarization, wave theory proposed by Huygen satisfactorily explained these phenomena.

When two waves of same frequency having constant phase difference travelling along the same direction in the same region of the medium are allowed to superimpose over each other, then there is a modification in the intensity of light in the region of superposition. This phenomenon is called the interference of light. When the resultant amplitude at certain points is sum of the amplitudes of the two waves, the interference is known as constructive interference and when the resultant amplitude at certain points is equal to the difference of two amplitudes, the interference is known as the destructive interference.

9.1.2 ANALYTICAL TREATMENT OF INTERFERENCE WITH EXPRESSION FOR BANDWIDTHS

Thomas Young was the first person to demonstrate the interference fringes on a screen. They are found to be alternate dark and bright bands or fringes equally spaced. Light rays from the source S are admitted through a sheet consisting of two vertical sources S_1 and S_2 at a distance 0.5 cm apart. S_1 and S_2 are parallel to each other. In Fig. 9. 2 A and B derived from the source S act as coherent sources and the waves originating from them produce interference pattern in the screen placed at a greater distance D of the order of

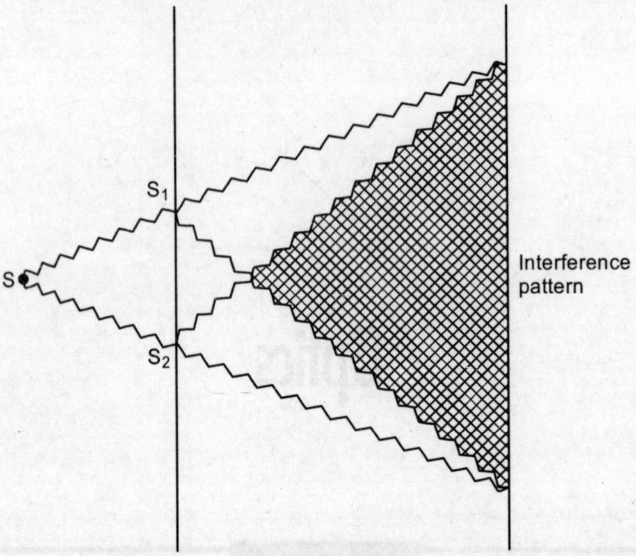

Fig. 9.1 Interference Pattern. Alternate dark and bright fringes.

1 metre or 1.5 metre. C is the middle point of the line joining A and B. Evidently A and B serve as coherent sources. Let us consider the two waves starting from the two sources A and B. Their path difference is now studied.

$$P.D = BP - AP$$

$$(BP)^2 = D^2 + \left(x + \frac{d}{2}\right)^2 = D^2 + x^2 + \frac{d^2}{4} + xd$$

$$(AP)^2 = D^2 + \left(x - \frac{d}{2}\right)^2 = D^2 + x^2 + \frac{d^2}{4} - xd$$

Thus $\quad (BP)^2 - (AP)^2 = (BP + AP)(BP - AP) = 2xd$

Thus path difference $(BP - AP) = \dfrac{2xd}{BP + AP} = \dfrac{2xd}{2D}$

assuming $\quad AP \sim BP \sim CO = D$

$$P.D = \frac{xd}{D}$$

Thus $\dfrac{xd}{D}$ must be an even multiple of $\dfrac{\lambda}{2}$ for constructive interference (brightness)

i.e., $$\frac{xd}{D} = 2n\frac{\lambda}{2}$$

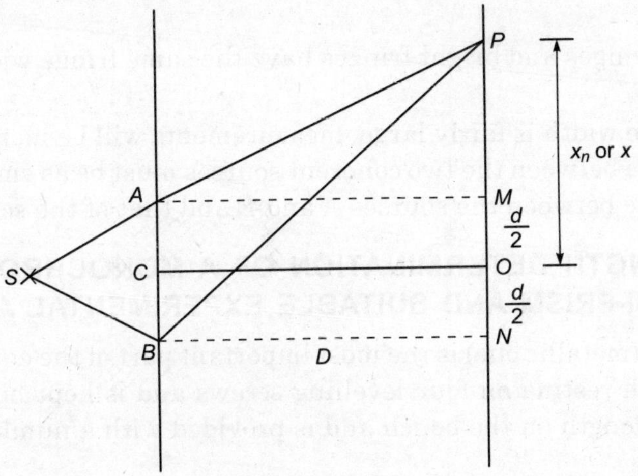

Fig. 9.2 Geometrical calculation.

or $$x = \left(\frac{D}{d}\right)n\lambda \quad \text{for } n^{\text{th}} \text{ bright fringe}$$

The general expression is

$$x_n = n\left(\frac{D\lambda}{d}\right) \quad \text{where } n \text{ is an integer}$$

Hence $$x_{n+1} = \frac{(n+1)\lambda D}{d} \quad \text{for the } (n+1)^{\text{th}} \text{ bright fringe}$$

Thus the width of the bright fringe

$$\beta = (x_{n+1} - x_n) = \frac{\lambda D}{d}(n+1-n)$$

$$\beta = \frac{D\lambda}{d} \qquad \text{...(9.1)}$$

For n^{th} dark fringe

$$x_n \frac{d}{D} = (2n-1)\frac{\lambda}{2} \text{ or } x_n = \frac{D}{d}\frac{\lambda}{2}\{(2n-1)\}$$

and the distance of $(n+1)^{\text{th}}$ dark fringe,

$$x_{n+1} = \frac{D}{d}\frac{\lambda}{2}\{2(n+1)-1\} = \frac{D\lambda}{2d}\{2n+2-1\}$$

width of a dark fringe is,

$$\beta = (x_{n+1} - x_n) = \frac{D\lambda}{2d}\{2n+1-2n+1\}$$

$$\beta = \frac{D\lambda}{d} \qquad \text{...(9.2)}$$

Conclusion

(*i*) The dark fringes and bright fringes have the same fringe widths as no n term is present.

(*ii*) If the fringe width is fairly large, measurements will be more accurate. For this the distance between the two coherent sources must be as small as possible. Also the distance between the sources A and B and that of the screen must be large.

9.1.3 WAVELENGTH DETERMINATION OF A MONOCHROMATIC SOURCE WITH A BI-PRISM AND SUITABLE EXPERIMENTAL ARRANGEMENT

A heavy optic bench (metallic one) is the most important part of the equipment. It is about 1m to 1.5m in length resting on four levelling screws and is kept horizontal. A scale is engraved along the length on the bench and is provided with a number of uprights.

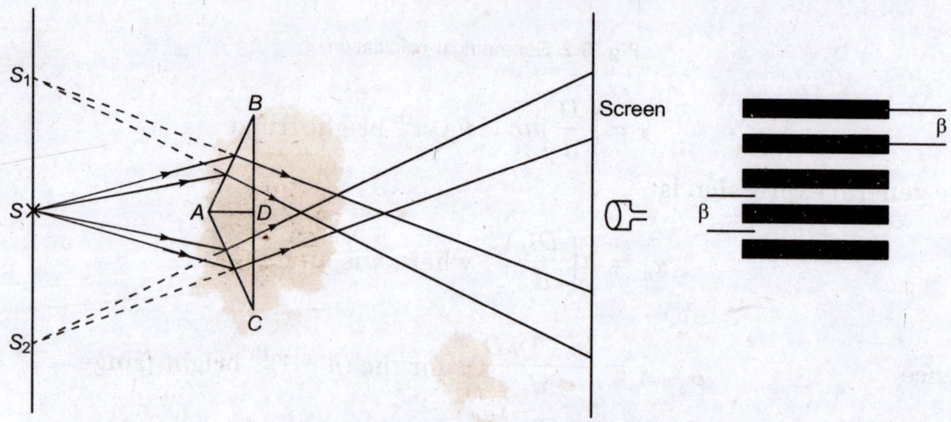

Fig. 9.3 Experimental arrangement.

These uprights can be moved from one end to the other and also can be clamped at any point. Their position can be known with the help of the index marks. The first upright is moved to the left end of the bench. It is fitted with a rectangular slit in the vertical plane and its width can be adjusted by a side screw. Also there is a provision for moving the slit in the vertical plane and its width can be adjusted.

The second upright carries a mount for the bi-prism usually it is kept about 20 cm from the slit. Here also there is a tangential screw using which the bi-prism can be rotated in the vertical plane. The third upright carries an eye-piece fitted with cross-wires. The eye-piece can be laterally moved by means of a micrometer screw attached to it. As usual there will be a pitch scale and head scale for the screw. The fourth upright is used only in the second part of the experiment and it carries amount for using lens.

The experiment consists of two parts. In the first part fringes are formed and width is determined. In the second part the distance between the coherent sources is measured by forming the real images with a convex lens and applying the principle of conjugate foci. Let us see the discussion of Part-I and Part-II.

Part I

Fringe with determination

The first upright containing the slit is moved to one end of the optic bench. The monochromatic source is held close to it. The width of the slit is first adjusted to be fairly The second upright carrying the bi-prism is moved towards the slit and kept as a distance of 10 to 15 cm. Looking from the other side of the bi-prism, the two virtual images of the illuminated slit can be seen with naked eyes.

If the images are not parallel, the tangential screw fitted to the slit is worked till they are parallel and close to each other. Now the third upright with the eye-piece is moved to a distance of 80 to 90 cm from the bi-prism. When we look through the eye-piece, we will be able to observe an yellow patch of light. At this stage if we narrow down the width of the slit, the fringes will be seen in the focal plane of the eye-piece. We get a clear pattern dark and bright interference fringes. This can be done by slightly working the tangential screw fitted to the bi-prism, if required.

Having obtained the fringes, the micrometer screw is worked till the intersection of the cross-wire coincides with a particular fringe (say bright fringe) on the left end of the pattern (see Fig. 9.3). We call this fringe as the nth bright fringe. Take the readings of the pitch scale and head scale. Let the total reading be x_n. The micrometer screw is then worked so that the cross-wire coincides with the $(n + 5)^{th}$ bright fringe. The total reading of the $(n + 5)^{th}$ fringe is obtained. Let it be x_{n+5}. Repeat this and take the readings for $(n + 10)$, $(n + 15)$, $(n + 20)$ and $(n + 25)$th frings. The readings were tabulated as given below.

Table 9.1 Fringe width calculation

Order	*Micrometer Reading (P.S.R + H.C. × L.C)*	*Width of twelve fringes*	*Mean width of fifteen fringes*	*Fringe width*
n	x_n			
$n + 5$	x_{n+5}			
$n + 10$	x_{n+10}			
$n + 15$	x_{n+15}	α_1	α	β_1
$n + 20$	x_{n+20}	α_2		
$n + 25$	x_{n+25}	α_3		

β_1 is thus the mean bright fringe width when $D = D$ and $d = d$

Now move the eye-piece away through a distance $D^1 = (D + 30)$ cm. Repeat the experiment and measure the band width β_2 for $d = d$ and D'. Let A be β_2.

Referring Eqns. (9.1) and (9.2), we can write

$$\beta_1 = \frac{D\lambda}{d} \text{ and } \beta_2 = \frac{(D+s)\lambda}{d} \text{ where } s \text{ is}$$

30cm in this case. Now

$$(\beta_2 - \beta_1) = \frac{\lambda_s}{d} \quad ...(9.3)$$

The unknown quantity λ can be determined using the above equation.

Part II

Measurement of the distance between (d) of the coherent sources

The fourth upright with a convex lens of short focal length (about 10 cm) is introduced between the bi-prism, an eye-piece without disturbing the biprism and slit. The eye-piece is moved to a distance greater than 4 *f*. The enlarged images of the coherent sources with all clarity are obtained.

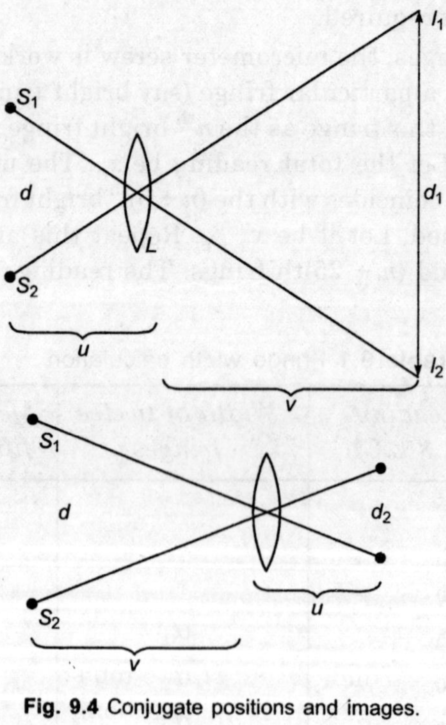

Fig. 9.4 Conjugate positions and images.

The distance between the images I_1 and I_2 is measured. Let it be d_1. Now the lens is moved to its conjugate position very near to the eye-piece when two images are seen again but diminished in the distance of separation. Measure the distance between the two images in the conjugate positions are

$$\frac{d}{u} = \frac{d_1}{v} \text{ and } \frac{d}{v} = \frac{d_2}{u}$$

or $$\frac{d^2}{uv} = \frac{d_1 d_2}{uv}$$

i.e. $$d = \sqrt{d_1 d_2} \qquad ...(9.4)$$

9.1.4 INTERFERENCE OF THIN FILMS

When a thin film of oil or thin soap bubble is viewed by reflected or transmitted light (say sunlight); it is found to exhibit different colours. This can be accounted for on the basis of the phenomenon of interference and it can be shown that bright colour at any point is due to the reinforcement of light waves of that colour. Absence of any particular colour at that point is due to interference (destructive interference)

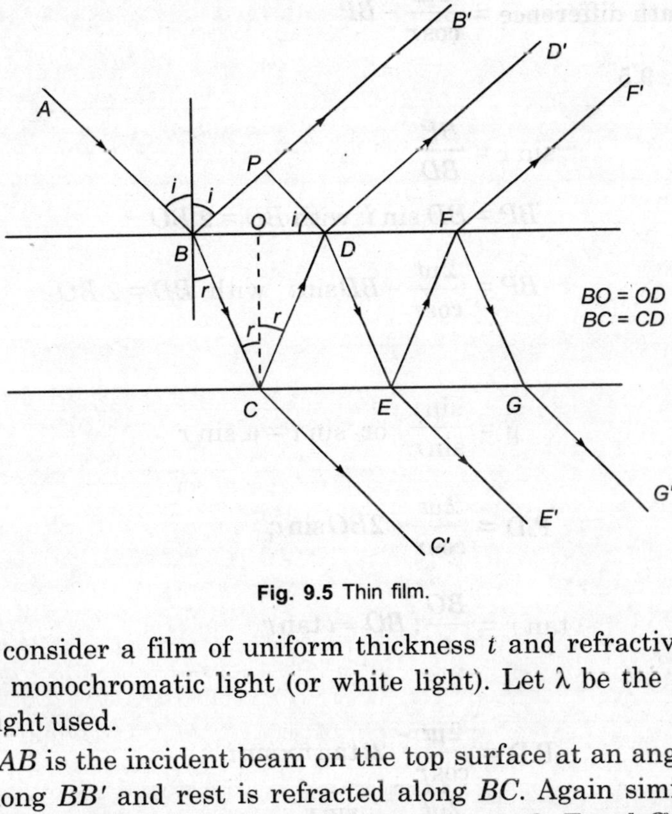

Fig. 9.5 Thin film.

Let us now consider a film of uniform thickness t and refractive index μ. This is illuminated by a monochromatic light (or white light). Let λ be the wavelength of the monochromatic light used.

In Fig. 9.5, AB is the incident beam on the top surface at an angle i. It is reflected partially at B along BB' and rest is refracted along BC. Again similar reflection and refraction take place at the lower part of the film (points C, E and G). The ray CD at D undergoes reflection and refraction at the top and bottom surfaces are indicated.

Let us now study the theory of interference under reflected light. The rays BB', DD' satisfy the conditions for producing interference fringes easily because they are derived from the same parent source and have a path difference. When these rays are collected by a lens, we get bright fringes. The path difference between the rays BB' and DD' is

$(BC + CD)$ in film – BP in air

$$= (BC + CD)\,\mu \text{ in air} - BP \text{ in air} \qquad ...(9.5)$$

Referring the Fig. 9.5.

$$\cos r = \frac{CO}{BC} = \frac{CO}{CD}$$

Substituting these values of BC and CD in Eqn. (9.5)

$$\mu\left[\left(\frac{CO}{\cos r} + \frac{CO}{\cos r}\right)\right] - BP$$

i.e.,

$$\text{Path difference} = \frac{2\mu t}{\cos r} - BP \qquad ...(9.6)$$

Referring Fig. 9.5

$$\sin i = \frac{BP}{BD}$$

$$BP = BD \sin i \text{ with } BD = 2\,BO$$

$$BP = \frac{2\mu t}{\cos r} - BD \sin i \quad \text{with } BD = 2\,BO$$

But

$$\mu = \frac{\sin i}{\sin r} \text{ or } \sin i = \mu \sin r$$

i.e.,

$$\text{P.D} = \frac{2\mu t}{\cos r} - 2BO \sin i;$$

Again

$$\tan r = \frac{BO}{t};\; BO = t \tan r$$

Now

$$\text{P.D} = \frac{2\mu t}{\cos r} - 2t \tan r \times \sin i$$

$$= \frac{2\mu t}{\cos r} - 2t \frac{\sin r}{\cos r} \mu \sin r$$

$$= \frac{2\mu t}{\cos r}(1 - \sin^2 r) = \frac{2\mu t}{\cos r} \cos^2 r$$

$$\text{P.D} = 2\,\mu t \cos r$$

This is only the geometrical path difference. According to Llyod's theory when a reflection takes palce at a rarer-denser medium boundary a phase change π corresponding to a P.D of $\frac{\lambda}{2}$ takes place while no such event at a denser-rarer medium boundary. Thus the effective P.D between the two rays is called optical path difference is given by

$$\text{O.P.D} = 2\mu t \cos r - \frac{\lambda}{2}$$

Thus

$$2\mu t \cos r - \frac{\lambda}{2} = 2n\frac{\lambda}{2} \text{ for bright fringes}$$

$$2\mu t \cos r = \frac{\lambda}{2}(2n+1) \quad ...(9.7)$$

Similarly for destructive interference (dark fringe), the condition is

$$2\mu t \cos r - \frac{\lambda}{2} = (2n-1)\frac{\lambda}{2}$$

$$2\mu t \cos r = \frac{\lambda}{2}(2n-1+1)$$

i.e.

$$2\mu t \cos r = n\lambda \quad ...(9.8)$$

9.1.5 THEORY OF NEWTON'S RINGS IN REFLECTED LIGHT

A thin plano-convex lens of large radius of curvature is placed with its convex surface on a plane glass plate. Light from a monochromatic source is allowed to incident normally on the air film enclosed between the lens and the glass plate.

The film is viewed by the reflected light by placing over eye above the system. Thus we see a set of alternate bright and dark concentric rings. The rings are called Newton's rings. The formation of these rings may be accounted for by the phenomenon of interference that takes place between the rays reflected at the top surface of the films and the rays that are reflected at the bottom surface of the film.

In Fig. 9.6 the ray AB is partly reflected at C along CBA and partly reflected at D and travels along $DCBA$. These two rays are generated from the same incident ray, they are free to interfere with each other.

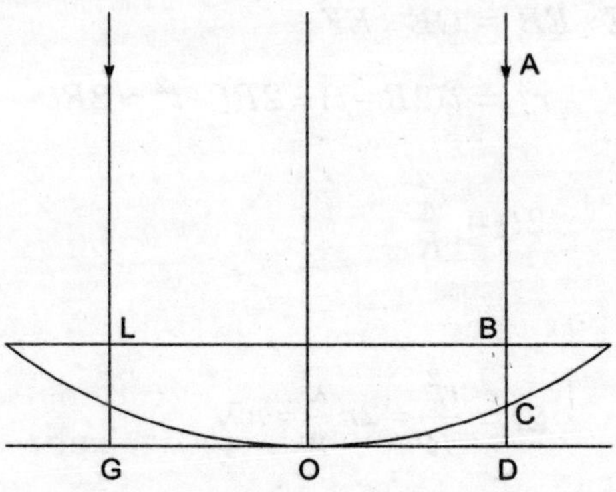

Fig. 9.6 Newton's ring arrangement.

The geometrical path difference between the two rays is $2CD$. But for the second ray reflecting at D, additional phase difference $\pi\left(=\frac{\lambda}{2}\right)$ is introduced (by following Llyod's theory). Now let us apply for destructive interference

$$2t+\frac{\lambda}{2} = (2n+1)\frac{\lambda}{2}$$

$$2t = \frac{\lambda}{2}[2n+1-\lambda] = n\lambda \quad \text{...(9.9)}$$

If this condition is satisfied, D will be always a dark point. The locus of all points on the film having the same thickness t will be a circle with OD as radius and O as centre. Thus we are able to see a dark ring through the point D.

On the other hand if the path difference at D is

$$2t+\frac{\lambda}{2} = 2n\frac{\lambda}{2}$$

i.e.,
$$2t = (2n-1)\frac{\lambda}{2} \quad \text{...(9.10)}$$

For this condition we will get a bright ring through the point D. The conclusion is between the points O and D we will be seeing a number of alternate bright and dark rings depending on the nature of the condition. The central spot corresponding to the point of contact appears dark. This accounts for the formation of Newton's rings.

9.1.6 THE GENERAL EXPRESSION THE RADII OF NEWTON'S RINGS IN THE REFLECTED SYSTEM

In Fig. 9.7 you see the formation of an air film between the plane surface and the plano-convex lens of radius of curvature. Let us assume that the nth dark ring is seen at C where the thickness of the air film is t. i.e. $BC = t$, $r_n = OC$ is the radius of the nth ring.

The law of segments is given by

$$AE \times EB = OE \times EF$$

i.e.,
$$r_n^2 = t(2R-t) \simeq 2Rt - t^2 \approx 2Rt$$

i.e.,
$$2t = \frac{r_n^2}{R}$$

For dark ring,

$$2t = \frac{r_n^2}{R} = 2n\frac{\lambda}{2} = n\lambda$$

$$\text{Hence } r_n = \sqrt{n\lambda R} \qquad \text{...(9.11)}$$

i.e., the radii of dark rings are proportional to $\sqrt{n}$, the square root of natural numbers.

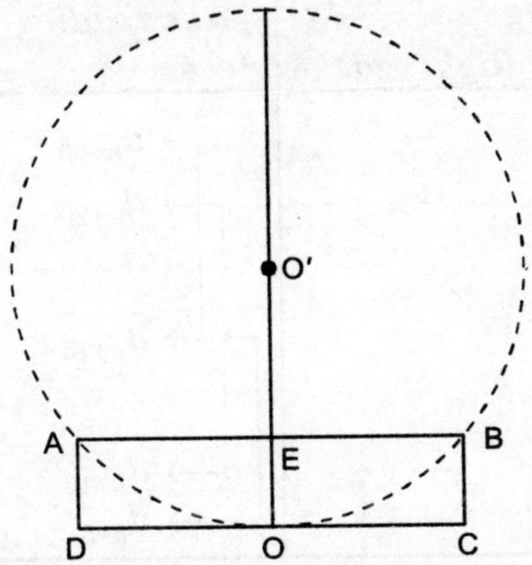

Fig. 9.7 Geometrical arrangement to study the radii of the rings.

The corresponding condition for bright rings is

$$r_n = \sqrt{R(2n-1)\frac{\lambda}{2}}$$

i.e. $r_n \propto$ square root of odd numbers

NB: If the space between the lens and the plate is filled with a transparent liquid of refractive index μ, then

$$r_n = \sqrt{\frac{Rn\lambda}{\mu}} \qquad \text{for destructive interference (dark ring)}$$

and

$$r_n = \sqrt{\frac{R}{\mu}(2n+1)\frac{\lambda}{2}} \quad \text{for constructive interference (bright ring)}$$

9.1.7 SOME NOTEWORTHY EXPERIMENTS WITH NEWTON'S RINGS

Newton's rings are formed in reflected system using a monochromatic light (sodium lamp of λ = 5893 A°) and a convex lens of large radius of curvature (R). Let us deal with the dark rings as the preferred choice. Form the ring system and take the readings using the travelling microscope corresponding to the rings $n + 25$, $n + 20$, $n + 15$, $n + 10$ in the left side of the central system keeping n as the third dark ring. Now move the microscope and take the readings corresponding to $(n + 5)$, $(n + 10)$ ..... $(n + 20)$. The readings are tabulated in the tabular column as given below:

Table 9.2: Readings of Newton's Rings Experiment

Order	*Microscope Reading (P.S.R + H.C. × L.C)* Left side	Right side	*Diameter difference of the reading*	$d_{n+s}^2 - d_n^2$ with $S = 10$	*Mean value*
$n + 25$	–	–	d_{n+25}	$d_{n+25}^2 - d_{n+15}^2$	$d_{n+5}^2 - d_n^2$
$n + 20$	– ↓	–	d_{n+20}		
$n + 15$	–	–	d_{n+15}	$d_{n+20}^2 - d_{n+10}^2$	
$n + 10$	–	–	d_{n+10}	$d_{n+15}^2 - d_{n+5}^2$	
$n + 5$	–	– ↑	d_{n+5}		

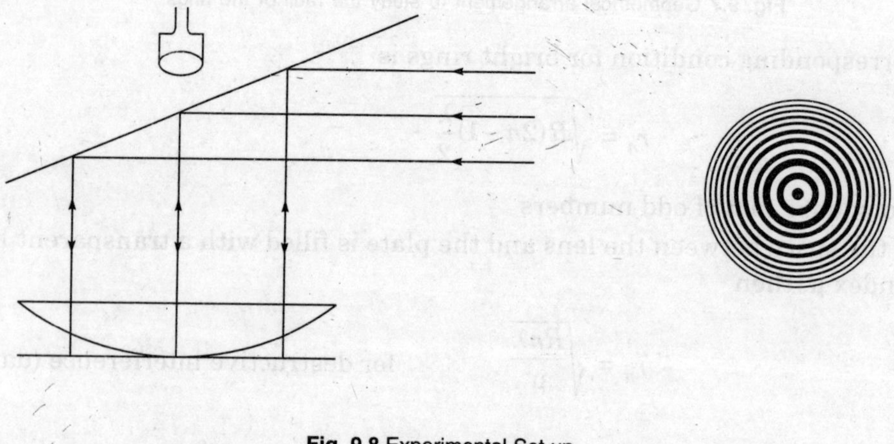

Fig. 9.8 Experimental Set up.

1. EXPERIMENTAL DETERMINATION OF THE RADIUS OF CURVATURE OF A PLANO-CONVEX LENS

As described earlier the experiment is conducted with the given plano-convex lens and the readings are tabulated. Get the mean value of

$$d_{n+s}^2 - d_n^2 \quad \text{with} \quad s = 100$$

Equation (9.11) makes us to write

$$r_n^2 = \left(\frac{d_n^2}{2}\right)^2 = \frac{d_n^2}{4} = n\lambda R$$

$$d_n^2 = 4n\lambda R$$

Similarly

$$\frac{d_{n+s}^2}{4} = (n+s)\lambda R$$

$$d_{n+s}^2 = 4R(n+s)\lambda$$

$$d_{n+s}^2 - d_n^2 = 4\lambda R(n+s) - 4n\lambda R = 4\lambda Rs$$

$$R = \frac{d_{n+s}^2 - d_n^2}{4\lambda s} \quad \text{...(9.12)}$$

Assuming the value of λ, R can be calculated.

2. COMPUTATION OF THE WAVELENGTH OF THE SOURCE USED

Here we use a lens of known radius of curvature. So the unknown quantity λ can be obtained using Eqn. (9.12).

3. CALCULATION OF REFRACTIVE INDEX OF A LIQUID

The given liquid available in small quantity is introduced between the lens and the glass plate. The rings system is obtained in the replaced system with air as the medium. Now

$$d_{n+s}^2 - d_n^2 = \frac{4\lambda RS}{1}$$

Now it is repeated, observations are made with the liquid

Thus

$$d_{n+s}'^2 - d_n'^2 = \frac{4\lambda RS}{\mu}$$

Thus

$$\frac{d_{n+s}^2 - d_n^2}{\left(d'_{n+s}\right)^2 - \left(d'_n\right)^2} = \mu \quad \text{...(9.13)}$$

Thus, the refractive index of the liquid is obtained.

9.2 DIFFRACTION OF LIGHT

This phenomenon in the case of light was first discovered by an Italian named Grimaldi. Light bends round the corners of small obstacles and spreads to some extent into the region of geometrical shadow. This bending of light round the corners of an obstacle whose size is very small and comparable to the wavelength of light or the encroachment of light within the geometrical shadow is called diffraction.

The phenomenon of diffraction supports the wave theory of light. It is one of the characteristics of all types of waves. Greater the wavelength, greater is the angle of diffraction. When light falls on an obstacle or a small aperture, there is a departure from rectilinear propagation of light. The close examination of the screen indicates the encroachment of light waves into the geometric shadow and rhythmic variation of light intensity just outside the edge of the geometric shadow.

9.2.1 FRESNEL AND FRAUNHOFER DIFFRACTION

Diffraction phenomenon can be divided into two groups:

(*i*) Fresnel diffraction (*ii*) Fraunhofer diffraction

In Fresnel diffraction, the source or the screen or both at finite distances from the obstacles causing diffraction. Diffraction at straight edge is treated as Fresnel diffraction.

In Fraunhofer diffraction, the source and the screen are at infinite distances from the obstacle. Fraunhofer diffraction can be easily observed in practice. In this case, the incident rays of light are made parallel by means of convex lens and the diffracted beam of parallel rays is focussed on the screen using another convex lens.

The diffraction at a single slit is treated as Fraunhofer diffraction.

9.2.2 DIFFRACTION DUE TO A SINGLE SLIT (FRAUNHOFER DIFFRACTION)

Experimental Set up

S is a monochromatic source of light giving spherical wave front like *XY*. It is placed

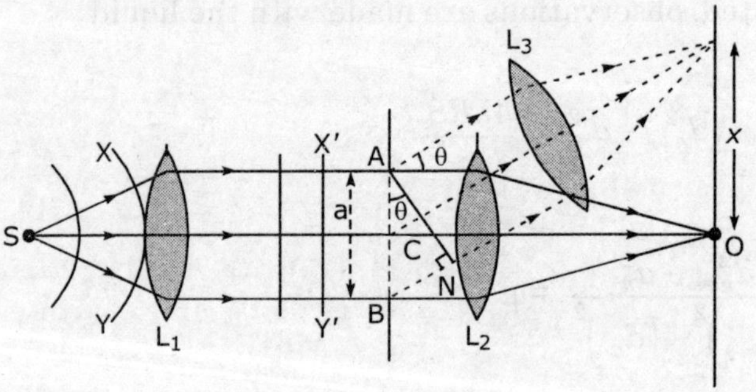

Fig. 9.9 Fraunhofer Diffraction

at a principal focus of a convex lens L_1 which makes the refracted beam parallel. $X'Y'$ is a plane incident wave front falling on the single slit *AB* of width *a* placed perpendicular to the plane of the paper. The width *a* of the slit is of the order of the wavelength λ of light. So diffraction occurs at the slit. The parallel beam of diffracted rays (*i.e.*, diffracted plane wave front) are brought to focus on the screen with the help of convex lens L_2. *O* is the foot of the perpendicular bisector *CO* of the slit *AB* on the screen.

Observations

The diffraction pattern on the screen consists of a central bright maximum (bright band) at O followed by a number of alternate minima and maxima on both sides. The secondary maxima are of much less intensity; intensity of the maxima falls off rapidly as

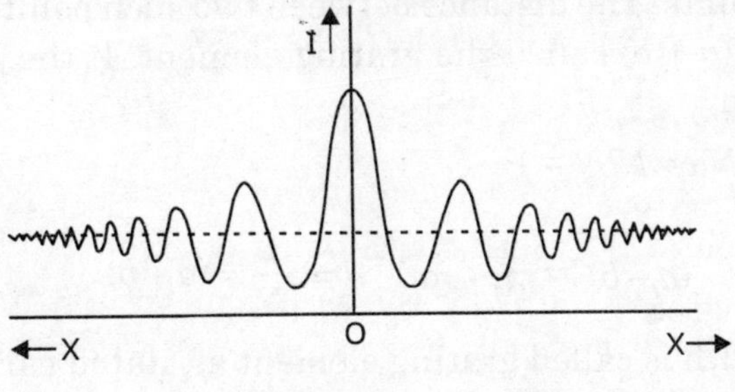

Fig. 9.10 Intensity variation

the distance from O increases on either side. The intensity at the minima is never zero; but much less than the intensity of the maxima. The width of the fringes are not same. The central maximum has maximum width; the width of the successive maxima and minima gradually decreases as distance from O increases on either side. The figure shows a graph of intensity I versus distance x from O on either side.

9.2.3 DIFFRACTION OF N SLITS.

GENERAL FORMULA FOR DETERMINATION OF THE WAVELENGTHS OF SPECTRUM

A study of the theory of diffraction grating finds many uses. The wavelengths of monochromatic source, mercury spectrum and lines of solar spectrum are possible using a spectrometer.

The plane diffraction grating is simply a plane sheet of transparent material in which opaque rulings are made with a sharp diamond pointer. Usually 15000 to 30000 opaque lines are in an inch of the material. The space between the rulings is left transparent. The width of all the rulings is the same and also the width of the transparent spaces. Thus the grating constitues a large number of parallel slits with equal widths placed at equidistant from one another. The combined width of a ruling and a slit constitutes the grating element.

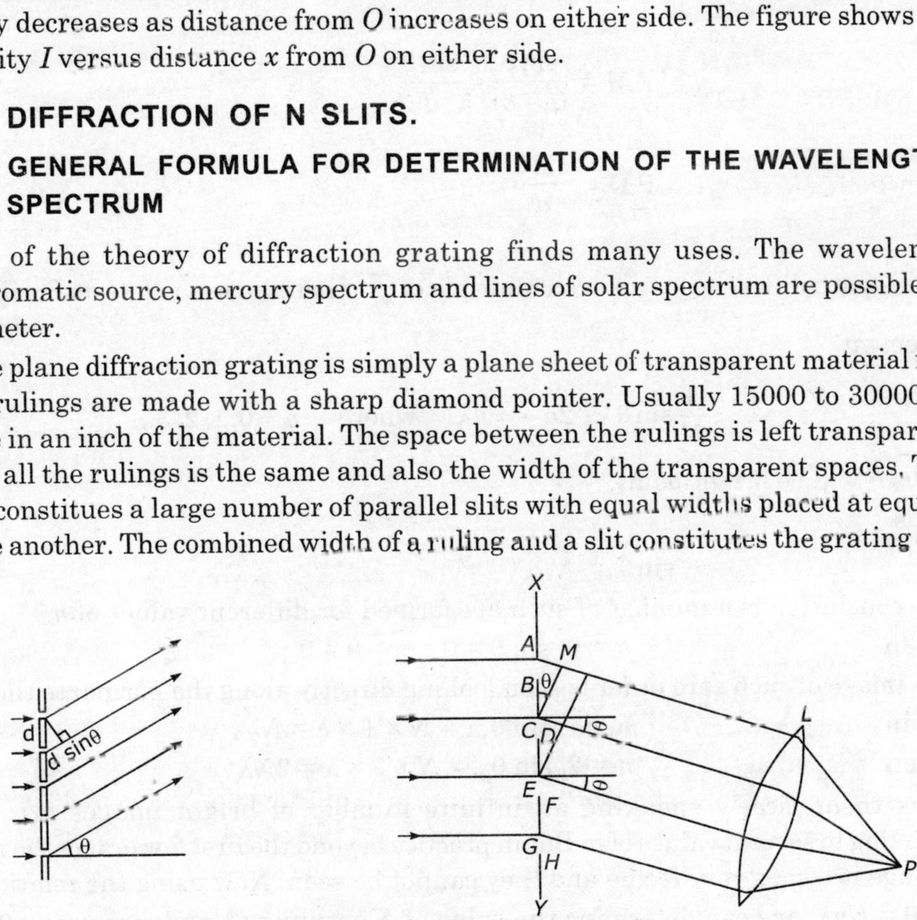

Fig. 9.10 Diffraction Grating

Let *ABCDEFG* represents the section of a grating normal to the plane of the diagram. *AB*, *CD*... represent the slit portions of *BC*, *DE*, ... represent ruled (opaque) portions. Suppose a beam of monochromatic light of wavelength λ is normally incident on the plane of the grating *XY*. The points *A* and *C* are called corresponding points. Similarly *B* and *D* are also called corresponding points. The distance between two such points (containing a slit and a opaque spacing) is say $(a + b)$ called the grating element. If the grating has *N* rulings/m and *N* interface/*m*, then

$$(Na + Nb) = 1$$

$$(a+b) = \frac{1}{N} \quad \text{or} \quad d = \frac{1}{N} = (a+b)$$

where the combined width is called grating element as stated earlier. It can be expressed as number of rulings/*m*.

Now consider light passing through several slits; they are converged by a convex lens *L* to form a bright image at *P*, the angular deviation between the diffracted and the incident beam being θ. Now the path difference between them is AM.

or

$$AM = \frac{\sin\theta}{(a+b)} = \frac{\sin\theta}{d}$$

Hence $$\text{P.D} = \frac{\sin\theta}{d}$$

If this path difference is an even multiple of $\frac{\lambda}{2}$, then the wave, lets at *P* will produce reinforcement

i.e., $$\frac{1}{N}\sin\theta = 2n\frac{\lambda}{2} = n\lambda \quad \text{where} \quad n = 0, 1, 2, 3...$$

Thus *P* will be bright point.

Thus

$$\sin\theta = Nn\lambda \qquad ...(9.14)$$

The conclusion is a number of such are formed for different values of *n*.

When $n = 0, \sin\theta = 0 \quad \text{or} \quad \theta = 0$

The image of such zero order is seen looking directly along the normal to the grating

When $n = 1, \sin\theta_1, = N \times 1 \times \lambda = N\lambda$

When $n = 2, \sin\theta_2, = N \times 2 \times \lambda = 2N\lambda$

Thus theoretically speaking an infinite number of bright images are possible corresponding to several values of *n*. But in practice beyond the first few orders the intensity of the image becomes very feeble and they cannot be seen. Now using the relation,

$\sin\theta = Nn\lambda$, one can determine the value of *N* assuming λ and measuring θ.

9.2.4 FIND THE NUMBER OF LINES DRAWN PER METER ON THE GRATING USING SODIUM SOURCE OF WAVELENGTH λ = 5893Å

(This is called standardisation of the grating)

The initial adjustments of the spectrometer are made and the grating is mounted on the circular scale. The plane of the grating is to be adjusted for normal incidence, only then $\sin\theta = Nn\lambda$ holds good.

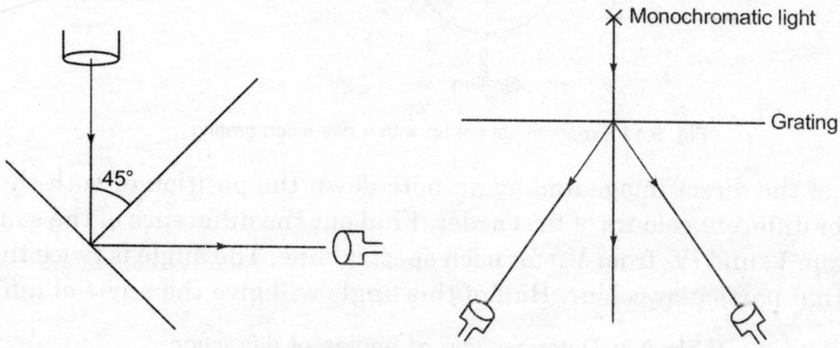

Fig. 9.11 Normal incident Adjustments

First the telescope is adjusted so that the cross-wire exactly coincides with the image of the slit illuminated by sodium light of λ = 589.3 nm. Now the reading of the circular scale is noted. Then the telescope is turned through 90°. Now the prism table with the grating is rotated and fixed so that the reflected ray is against the vertical cross-wire. *i.e.*, the angle of incidence is 45°. Now the grating with the circular scale is rotated through another 45° towards the collimator.

Now the grating is in the normal incidence position. The telescope now moved and the readings for the first order of the diffracted images of the slit is illuminated by a monochromatic light of λ are found out.

The difference is 2θ and hence θ the angle of diffraction is found out. For the first order diffraction

$$\sin\theta = Nn\lambda$$

$$N = \frac{\sin\theta}{\lambda} \text{ with } n = 1$$

using this relation one can find out the number of lines in a cm or in a metre.

9.2.5 DETERMINATION OF WAVELENGTHS OF THE VARIOUS LINES OF MERCURY SPECTRUM

The monochromatic source is replaced by a mercury source. After obtaining the mercury spectrum, rotate the telescope to the left side of the direct image and now adjust the cross-wire on violet, green, yellow and red spectral lines turn by turn for first order. For each setting, the reading of verniers should be noted down. Now rotate the telescope to

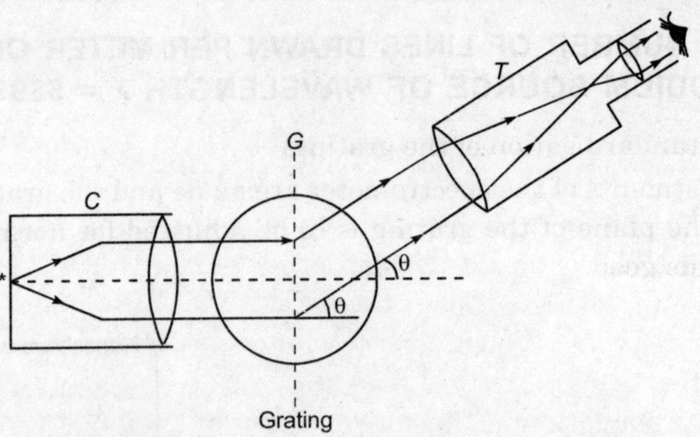

Fig. 9.12 Experimental set up with a diffraction grating

the right side of the direct image and again note down the position of both the verniers turn by turn for different colours of first order. Find out the difference of the same kind of verniers (V_1 from V) and (V_2 from V_2) for each spectral line. The angle is twice the angle of diffraction of that particular colour. Half of this angle will give the angle of diffraction.

Table 9.3: Determination of angles of diffraction

Order of spectrum	*Colour of light*	*Vernier*	*Left side*	*Right side*	*2θ*	*θ*
			M.S V.S (Total (Reading) a	*M.S V.S (Total Reading) b*	*(a – b)*	
First	Violet	V_1				
		V_2				
	Blue	V_1				
		V_2				
	Green	V_1				
		V_2				
	Yellow	V_1				
		V_2				

Calculation

$$\sin\theta = nN\lambda$$

$n = 1$

N the number of rulings/m

θ is measured

So $$\lambda = \frac{\sin\theta}{nN} = \frac{\sin\theta}{N} \qquad ...(9.15)$$

9.3 OPTICAL POLARIZATION AND NICOL PRISM

Introduction

The phenomenon of interference and diffraction are explained by the wave theory of light. However these effects give no information as to the type of wave motion *i.e.*, whether the light waves are transverse or longitudinal. The phenomenon of polarization is that light wave is transverse wave. Since light wave is a transverse wave, according to Huygen's wave theory, either particles in the medium can vibrate in infinite directions perpendicular to the direction of propagation of the wave. If this vibrations are restricted to a single plane, then it is called a plane polarized light (linearly polarised light).

Polarization by reflection, polarization by refraction and polarization by double refraction are some methods to produce plane polarised light. Some of them are

Polarisation by Reflection (Brewster's law)

In 1881 David Brewster performed a series of experiments to study the polarization of light by reflection. He observed that when a beam of ordinary light is incident on the surface of a given transparent medium, the reflected light is completely plane polarised. This particular angle of incidence is called polarising angle or Brewster's angle. It is usually denoted by i_p. If μ is the refractive index of the medium Brewster's discovered the relation,

$$\mu = \tan i_p \qquad ...(9.16)$$

i.e., the tangent of angle of polarization for a given medium is numerically equal to the refractive index of that medium. This is called Brewster's law.

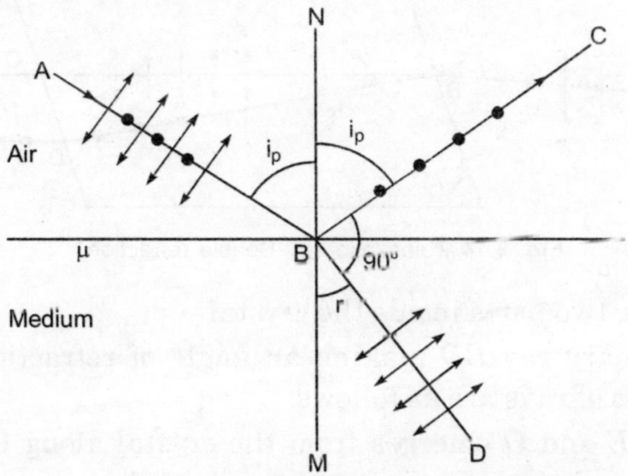

Fig. 9.13 Brewster's law

The other investigation of Brewster is to show that at the polarising angle the reflected and refracted rays are perpendicular to each other.

The beam *AB* is incident at *B* at an angle equal to the polarising angle i_p on the surface of a transparent substance. The beam *AB* is reflected along *BC* and refracted along *BD*. Let *r* be the angle of refraction.

From Brewster's law, we have

$$\mu = \tan i_p = \frac{\sin i_p}{\cos i_p} \qquad ...(9.17)$$

Snell's law points out

$$\mu = \frac{\sin i_p}{\sin r} \qquad ...(9.18)$$

Comparing the above two equations, we get.

$$\cos i_p = \sin i_r = \cos(90 - r)$$

$$i_p = (90 - r)$$

$$i_p + r = 90°$$

Hence $\qquad \angle CBD = 90°$

i.e., at the polaring angle (as the angle of incidence), the reflected and refracted rays are at right angles to each other.

9.3.1 POLARISATION AT DOUBLE REFRACTION

Let an ordinary ray AB be incident on the calcite crystal at an angle of incidence as shown in Fig. 9.14.

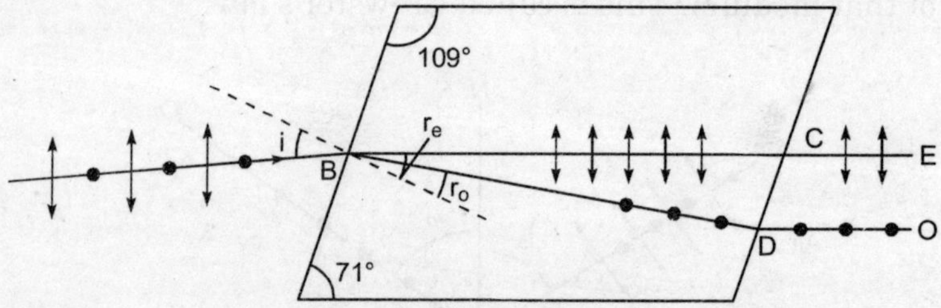

Fig. 9.14 Polarization by Double refraction

It is refracted along two parts inside the crystal

(*a*) The extraordinary ray BD making an angle of refraction r_o. The features of these two kinds of rays are as follows:

(*i*) The two rays E and O emerges from the crystal along CE and DO which are parallel

(*ii*) The refractive index of the material for the two rays are

$$\mu_o = \frac{\sin i}{\sin r_o} \text{ and } \mu_e = \frac{\sin i}{\sin r_e}$$

The refractive index for the ordinary ray remains constant while the refractive index for the extraordinary ray varies with the angle of incidence.

(iii) Both the rays are plane polarised with their vibrations perpendicular to each other. The vibration of *O*-rays are perpendicular to the principal section of the crystal with that of the E-rays and are in the plane of the principal section of the crystal.

(iv) In calcite crystal $r_e > r_o$; hence $\mu_o > \mu_e$. Therefore the velocity of light for *O*- rays inside the crystal is less than the velocity of *E*- rays. Such crystals are called negative crystals. Examples are calcite, tourmaline, etc.

(v) In the crystal for which $\mu_o < \mu_e$ or $v_o > v_e$, are called positive crystals. Examples are quartz, iron oxide, etc.

(vi) When the unpolarised light is incident along the optic axis, it will not split into *O* and *E* rays.

(vii) When the unpolarised light is incident perpendicular to the optic axis, it will not split.

9.3.2 DESCRIBE THE CONSTRUCTION AND ACTION OF A NICOL PRISM

Construction

It is an optical device made from calcite and is frequently used for producing and analysing polarized light. It is based on the phenomenon of double refraction and was invented by William Nicol. Calcite produces ordinary and extraordinary rays when natural light passes through it with some specifications. The ordinary ray is cut off by total internal reflection leaving the extraordinary ray to proceed. In a natural crystal the faces *ABCD* and *EFGH* makes angles of 71° with the longer sides *AG* and *CE* respectively.

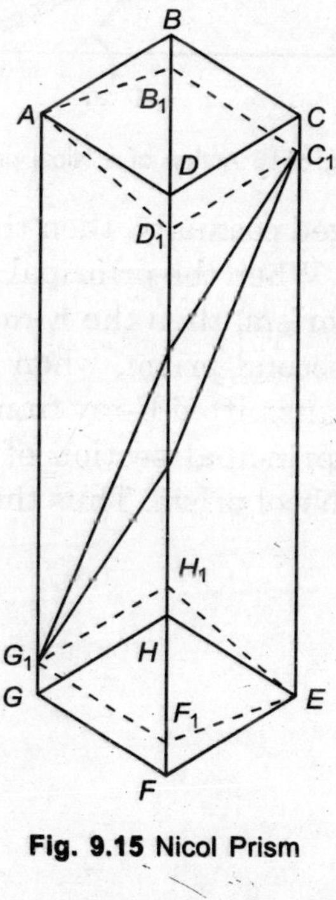

Fig. 9.15 Nicol Prism

But two new faces $A\,B_1\,C_1\,D_1$ and $E\,F_1\,G_1\,H_1$ are cut so as to make 68° with the longer sides. C_1 and G_1 are called blunt corners. Now the crystal is cut into two haeves by a plane passing through C_1G_1 this plane being perpendicular to the two new end faces as well as to the principal section of the crystal (the principal section is that which passes through the blunt corners C_1 and G_1). The cut faces are polished and connected by a flow of Canada Balsam. The refractive indices for the ordinary ray and extraordinary ray are 1.658 and 1.486 respectively, while the refractive index in Canada Balsam is 1.550 for both the rays. The end faces of the crystal are left free for the transmission of light and the remaining faces are blackened.

Action: On entering the Nicol prism, incident unpolarised light is broken into an ordinary and an extraordinary ray by calcite. When the ordinary ray reaches to Canada Balsam layer, it suffers total internal reflection because the Canada Balsam offers a rarer medium for O-rays coming from a denser medium of calcite. The extraordinary ray travels from rarer to denser medium while reaching at Canada Balsam layer and therefore E-ray is transmitted through the Nicol prism i.e. the Nicol prism is used as a polariser Fig. 9.16.

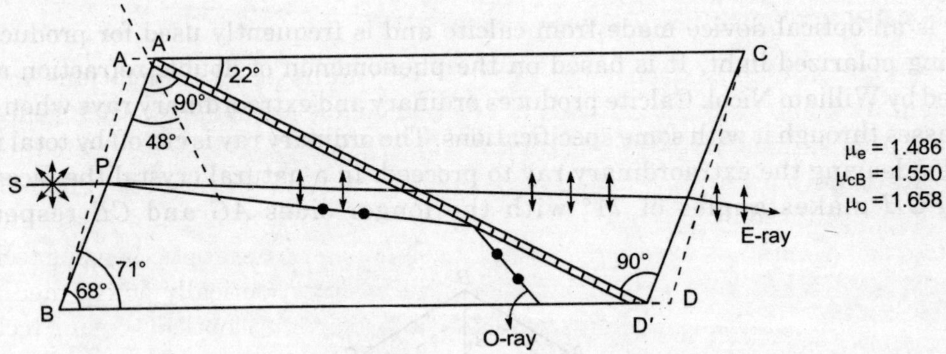

Fig. 9.16 Action of a Nicol prism

When two prisms are arranged coaxially, then the first Nicol which produces plane polarised E-ray is called analyser. When the principal section of the second Nicol prism is parallel to that of the first Nicol prism, then the E-ray transmitted by the first prism is easily transmitted through the second prism, when the second Nicol prism is rotated gradually, then the intensity of transmitted E-ray from second Nicol prism decreases and finally becomes zero when the principal section of the second Nicol prism becomes perpendicular to that of the first Nicol prism. Thus the first Nicol prism acts as polariser and the second prism as analyser.

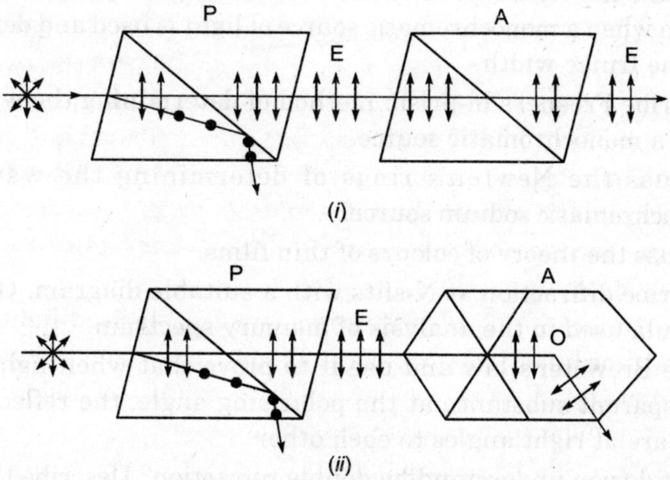

Fig. 9.17 Nicol prisms crossed

9.3.3 APPLICATIONS OF POLARIZED LIGHT

The phenomenon of polarization of light finds a number of applications in optical instrumentation and systems. Optical activity and double refraction (birefringence) find a number of applications in optical modulators, laser systems, ellipsometry etc. Polarizing devices are widely used in industries to detect strains in glasswares. Photoelasticity (birefringence induced by external stress) has become the basis of powerful techniques to study stress distribution in materials, optical accelerometers, fibre optic, sensing techniques etc. These techniques involve birefringence induced in an optically active medium and measuring it with polarization analysers. There are a variety of optical sensing techniques which are based on induced birefringence and its measurement.

Ellipsometry is a technique which applies polarimetry for determining optical properties of surfaces and interfaces. Reflectivity, extinction coefficient, refractive index for determining optical properties of surfaces and interfaces. Reflectivity, extinction coefficient, refractive index, thin film thickness etc. can be determined using this technique. 3D film projection is an example for a technique based on polarization. Concentration of sugar and hence the quality of sugar can be determined by measuring optical activity in the medium using polarizers and analysers.

REVIEW QUESTIONS

9.1.1 (*a*) What is interference of light?

(*b*) What you understand by coherent sources?

(*c*) Why it is necessary to have coherent sources for observing interference fringes. Explain why fringes formed by a bi-prism disappear when the slit is wide?

9.1.2 Explain the formation of interference fringes by means of a Fersnel's bi-prism when a monochromatic source of light is used and derive an expression for the fringe width.

9.1.3 Describe Fresnel's bi-prism method of determining the wavelength of light from a monochromatic source.

9.1.4 Discuss the Newton's rings of determining the wavelength from a monochromatic sodium source.

9.1.5 Discuss the theory of colours of thin films.

9.1.6 Describe diffraction at N-slits with a suitable diagram. Obtain the general formula used in the analysis of mercury spectrum.

9.1.7 State Brewster's law and use it to prove that when light is incident on a transparent substance at the polarising angle, the reflected and refracted rays are at right angles to each other.

9.1.8 What do you understand by double refraction? Describe the construction of a Nicol prism. Explain how it can be used as a polariser and an analyser.

Problems and Solutions

9.1.1 In a prism experiment, the eye-piece was kept at 1.2 metre from the plane containing the coherent sources. The distance between the two virtual sources was found to be 0.075 cm. Find the wavelength of the light used if the eye-piece has to be moved through a distance of 1.888 cm for 20 fringes.

Solution. Fringe width,

$$\beta = \frac{1.888}{20} = 0.0944 \text{ cm} = 0.000944 \text{ m}$$

$$\beta = \frac{D\lambda}{d} \text{ or } \lambda = \frac{Bd}{D}$$

$$\lambda = \frac{0.000944 \times 0.00075}{1.20}$$

$$\lambda = 590 \text{ nm} \quad \textbf{Ans.}$$

9.1.2 In a Young's experiment, the bandwidth produced by a monochromatic source of light measures from 0.2 mm to 0.24 mm when the screen is moved through a distance of 20 cm. The distance between the slit is 3 mm. Calculate the wavelength of light used.

Solution.

$$\beta_1 = 0.2 \times 10^{-3} \text{ m}, \; \beta_2 = 0.24 \times 10^{-3} \text{ m}$$

$$\beta_1 = \frac{\lambda D_1}{d}, \; \beta_2 = \frac{\lambda D_2}{d}$$

$$\beta_2-\beta_1 = 10^{-3}\times(0.24-0.2) = \frac{\lambda}{d}(D_2-D_1)$$

$$0.04\times10^{-3} = \frac{\lambda\times20\times10^{-2}}{3\times10^{-3}}$$

$$\lambda = \frac{0.04\times10^{-3}\times3\times10^{-3}}{20\times10^{-2}} = \frac{12\times10^{-8}}{20\times10^{-2}}$$

$$\lambda = 6\times10^{-11} \text{ metre}$$

$$\lambda = 600 \text{ nm} \quad \textbf{Ans.}$$

9.1.3 White light may be considered to have λ from 400 nm and 750 nm. If an oil film has thickness 10^{-4} cm, deduce the wavelength in the visible region for which the reflection along the normal direction will be (*i*) weak (*ii*) strong. μ for oil is 1.4.

Solution.

Given : $\mu = 1.4$, $t = 10^{-6}$ m, $n = 1$

In the case of normal incidence $i = 0$, $r = 0$. Hence $\cos r = \cos\theta = 1$.

For interference taking place in oil film, path difference is given by $2\mu t\cos r$.

i.e., $\quad 2\mu t\cos r = 2\mu t\times 1$

$2\times1.4\times10^{-6}$ m or 2.8×10^{-6} metre; when $r = 0$

i.e., $\quad 2\mu t\cos r = 2800$ nm.

(i) Weak reflection means destructive interference. The condition for weak reflection is given by

$$2\mu t\cos r = 2\mu t\cos 0 = 2\mu t = n\lambda$$

$$\lambda = \frac{2\mu t}{n} = \frac{2800}{n}\text{ nm}$$

It is given that the wavelength of light is between 400 nm to 750 nm. So the value of n must such that the wavelength lies between 400 and 700 m

When $n = 4$, $\lambda = \dfrac{2800}{4} = 700$ nm

When $n = 5$, $\lambda = \dfrac{2800}{5} = 560$ nm

When $n = 6$, $\lambda = \dfrac{2800}{6} = 466.7$ nm

When $n = 7$, $\lambda = \dfrac{2800}{7} = 400$ nm

From the calculation it is clear that the above wavelength will produce weak reflection.

(ii) Strong reflection means constructive interference. The points appear bright. For this, the condition is

$$2\mu t \cos r = (2n+1)\frac{\lambda}{2}$$

$$2\ \mu t = (2n+1)\frac{\lambda}{2}$$

$$\lambda = \frac{4\mu t}{(2n+1)} = \frac{2\times 2\mu t}{(2n+1)} = \frac{2\times 2800}{(2n+1)}$$

When $n = 4$, $\lambda = \dfrac{2\times 2800}{9} = 622.2$ nm

When $n = 5$, $\lambda = \dfrac{2\times 2800}{11} = 509.1$ nm

When $n = 6$, $\lambda = \dfrac{2\times 2800}{13} = 430.8$ nm

From the calculation we see that the above wavelengths produce strong reflection. For other values of n, the wavelength obtained will not be in the given wavelength range.

9.1.4 In a Newton's ring experiment, the wavelength of light used is 600 nm and the difference of square of diameters of successive rings are 0.125 cm^2. What will happen to this quantity if

(i) the wavelength of light is changed to 450 nm.

(ii) a liquid of refractive index 1.33 is introduced between the lens and glass plate.

(iii) the radius of curvature of the convex surface of the plano-convex lens is doubled.

Solution.

In Newton's rings, the difference of square of diameters for the nth and $(n + s)^{th}$ ring is given by

$$D_{n+s}^2 - D_n^2 = \frac{4s\lambda R}{\mu}$$

For succesive rings,

$$S = 1$$

i.e.

$$D_{n+1}^2 - D_n^2 = \frac{4\lambda R}{\mu}$$

(*i*) When light of wavelength λ is changed to λ′, then

$$D'^2_{n+1} - D'^2_n = \frac{4\lambda' R}{\mu}$$

Dividing one of the other

$$\frac{D'^2_{n+1} - D'^2_n}{D^2_{n+1} - D^2_n} = \frac{4\lambda' R \times \mu}{\mu \times 4\lambda R \times \lambda} = \frac{\lambda'}{\lambda} \text{ as } \mu = 1$$

Here,

$$\lambda = 6 \times 10^{-5} \text{ cm}, \ \lambda' = 4.5 \times 10^{-5} \text{ cm}$$

and $$D^2_{n+1} - D^2_n = 0.125 \text{ cm}^2$$

$$\frac{D'^2_{n+1} - D'^2_n}{D^2_{n+1} - D^2_n} = 0.125 \times \frac{\lambda'}{\lambda} = 0.125 \times \frac{4.5 \times 10^{-5}}{6 \times 10^{-5}}$$

$$D'^2_{n+1} - D'^2_n = 0.0037 \text{ cm}^2$$

(*ii*) When the radius curvature of the plano-convex lens is changed from *R* to *R*′, then

$$D'^2_{n+s} - D'^2_n = \frac{4\lambda R'}{\mu}$$

and $$D^2_{n+s} - D^2_n = \frac{4\lambda R}{\mu}$$

Dividing one by the other,

$$\frac{D'^2_{n+s} - D'^2_n}{D^2_{n+s} - D^2_n} = \frac{R'}{R} \text{ with } R' = 2R$$

$$D'^2_{n+s} - D'^2_n = (2R/R)(D^2_{n+s} - D^2_n)$$

$$= 2 \times 0.125$$

i.e. $$D'^2_{n+s} - D^2_n = 0.250.$$

9.1.5 In a grating spectrum, whose spectral line in the fourth order will overlap with third order line of 546 nm.

Solution.

$$(a+b)\sin\theta = n\lambda = m\lambda'$$

$$\lambda' = \frac{n\lambda}{m} = \frac{3 \times 546}{4}$$

$$\lambda' = 409.4 \text{ nm}$$

9.1.6 A ray of light is incident on the surface of a glass plate of refractive index 1.55 at the polarising angle. Calculate the angle of refraction.

Solution.

$$\mu = 1.55$$

Again

$$\mu = \tan i$$

or

$$i = \tan 1.55$$

i.e.,

$$i = 57°, 10'$$

But

$$i + r = 90$$

$$r = 90 - i = 90° - 57°, 101'$$

$$r = 32° \; 50' \textbf{ Ans.}$$

9.1.7 For a given medium, the polarisation angle is 65°. Find the critical angle of the medium.

Solution.

According to Brewster's law,

$$\mu = \tan i$$

$$\mu = \tan 65°$$

$$\mu = 2.1445$$

$$\mu = \frac{1}{\sin C}$$

or

$$\sin C = \frac{1}{\mu} = \frac{1}{2.1445}$$

$$C = \sin^{-1}(0.4663)$$

$$C = 27°, 48'$$

Appendices

- ALL ABOUT S.I. UNITS
- FEATURES OF S.I. UNITS
- CONVERSION FACTORS
- USEFUL PHYSICAL CONSTANTS
- BIBLIOGRAPHY
- SOME USEFUL CALCULUS AND INTEGRAL TABLES

As per Shakespearean usage
"Bluish writing"
means message worth reading with all curiosity.

A. 1 ALL ABOUT S.I. UNITS

Table A.1.1 S.I. Base units

Quantity	*Name*	*Symbol*
length	metre	m
mass	kilogram	kg
time	second	s
electric current	ampere	A
thermodynamic temperature	kelvin	K
luminous intensity	candela	cd
amount of substance	mole	mol

Table A.1.2 Derived units with special names

Quantity	*S.I. Units*			
	Name	*Symbol*	*Expression in terms of other Units*	*Expression in terms of S.I base Units*
force	newton	N	–	$m\ kg\ s^{-2}$
frequency	hertz	Hz	–	s^{-1}
energy, work, quantity of heat	joule	J	N m	$m^2\ kg\ s^{-2}$
pressure	pascal	Pa	N/m^2	$m^{-1}\ kg\ s^{-2}$
quantity of electricity	watt	W	J/s	$m^2\ kg\ s^{-3}$
electric charge	coulomb	C	A s	A s
electric tension, electric potential	volt	V	W/A	$m^2\ kg\ s^{-3}\ A^{-1}$
capacitance	farad	F	C/V	$m^{-2}\ kg^{-1}\ s^4\ A^2$
electrical resistance	ohm	Ω	V/A	$m^2\ kg\ s^{-3}\ A^{-2}$
electrical conductance	siemens	S	A/V	$m^{-2}\ kg^{-1}\ s^3\ A^2$
magnetic flux	weber	Wb	V s	$m^2\ kg\ s^{-2}\ A^{-1}$
magnetic flux density	tesla	T	Wb/m^2	$kg\ s^{-2}\ A^{-1}$
inductance	henry	H	Wb/A	$m^2\ kg\ s^{-2}\ A^{-2}$
luminous flux	lumen	lm	–	cd sr
illuminance	lux	lx	–	m^{-2} cd sr

Table A.1.3. Examples of S.I. derived units expressed by means of special names

Quantity	*S.I. Units*		
	Name	*Symbol*	*Expression in terms of S.I base units*
dynamic viscosity	pascal second	Pa-s	m^{-1} kg s^{-1}
moment of force	metre newton	N m	m^{-2} kg s^{-2}
surface tension	newton per metre	N/m	kg s^{-2}
heat flux density, irradiance	watt per square metre	W/m	kg s^{-3}
heat capacity, entropy	joule per kelvin	J/K	kg m^2 s^{-2} K^{-1}
specific heat capacity, specific entropy	joule per kilogram metre	J/kg K	m^2 s^{-2} K^{-1}
specific energy	joule per kilogram	J/kg	m^2 s^{-2}
thermal conductivity	watt per metre kelvin	W/m K	m kg s^{-3} K^{-1}
energy density	joule per cubic metre	J/m^3	kg m^{-1} s^{-1}
electric field strength	volt per metre	V/m	kg m s^{-3} A^{-1}
electric charge density	coulomb per cubic metre	C/m^3	m^{-3} s A
electric flux density	coulomb per square metre	C/m^2	m^{-2} s A
permittivity	farad per metre	F/m	kg^{-1} m^{-3} s^4 A^2
current density	ampere per square metre	A/m^2	–
magnetic field strength	ampere per metre	A/m	–
permeability	henry per metre	H/m	m kg s^{-2} A^{-2}
molar energy	joule per mole	J/mol	m^2 kg s^{-2} mol^{-1}
molar heat capacity	joule per mole kelvin	J/(mol K)	m^2 kg s^{-2}

Table A.1.4. Examples of S.I. derived units using supplementary units

Quantity	*S.I. Units*	
	Name	*Symbol*
angular velocity	radian per second	rad/s
angular acceleration	radian per second squared	rad/s^2
radiant intensity	watt per steradian	W/sr
radiance	watt per square metre steradian	W/m^2sr

Table A.1.5 S.I. Prefixes

Factor	*Prefix*	*Symbol*	*Factor*	*Prefix*	*Symbol*
10^{12}	tera	T	10^{-1}	deci	d
10^{9}	giga	G	10^{-2}	centi	c
10^{6}	mega	M	10^{-3}	milli	mm
10^{3}	kilo	k	10^{-6}	micro	μ
10^{2}	hecto	h	10^{-9}	nano	n
			10^{-12}	pico	p
10^{1}	deca	da	10^{-15}	femto	f
			10^{-18}	atto	a

A 2.1 Salient Features of S.I. Units

1. It is a coherent system of units, *i.e.*, product or quotient of any two base quantities results in a unit resultant quantity. For example, unit length divided by unit time gives unit velocity.
2. It is a rationalized system of units. It clearly distinguishes between the units of mass and weight (force) which are kilogram and newton respectively.
3. All the units of the system can be derived from the base and supplementary units.
4. The decimal relationship between units of same quantity makes possible to express any small or large quantity as a power of 10.
5. For any quantity there is one and only one S.I. unit. For example, joule is the unit of energy of all forms such as mechanical, heat, chemical, electrical and nuclear. However, kWh also continues as unit of electrical energy.

Advantages of S.I. Units

1. Units for many different quantities are related through a series of simple and basic relationship.
2. Being an absolute system, it avoids the use of factor 'g' *i.e.*, acceleration due to gravity in several expressions in physics and engineering which had been a nuisance in all numericals in these branches.
3. Being a rationalized system, it ensures all the advantages of rationalized MKS A system in the fields of electricity, magnetism, electrical engineering and electronics.
4. Joule is the only sole unit of energy of all forms and watt is the sole unit of power; hence lot of labour is saved in calculations.
5. It is a coherent system of units and involves only decimal coefficients. Hence it is very convenient and quick system for calculations.
6. In electricity, all the practical units like volt, ohm, ampere, henry, farad, coulomb, joule and watt accepted in industries and laboratories all over the world for well over a century have become absolute in their own right in the S.I. system, without the need for any more practical units.

Disadvantages

1. The non S.I. time units 'minute' and 'hour' will still continue to be used until the clocks and watches are all changed to kilo second and mega second.
2. S.I. units for energy, power and pressure (*i.e.*, joule, watt and pascal) are too small to be expressed in science and technology, and, therefore, in such cases the use of larger units, such MJ, kW, kPa will have to be made.

   There are difficulties with regard to developing new S.I. units for apparent and reactive energy while joule is the accepted unit for active energy in S.I. systems.

Table A.3.1 Conversion factors to convert the quantities in non S.I. units into S.I. units

To convert	*To*	*Multiply by*
angstrom	m	10^{-10}
atmosphere	kg/m^2	10332
bars	kg/m^2	1.02×10^4
Btu	joule	1054.8
Btu	kwh	2.928×10^{-4}
cubic fcet	m^3	0.02831
dyne	newton	10^{-5}
erg	joule	10^{-7}
erg	kWh	0.2778×10^{-13}
feet	m	0.3048
foot-poundal	joule	1.356
foot-pound	kg m	0.1383
gauss	tesla	10^{-4}
gram (force)	newton	9.807×10^{-3}
horse power (metric)	watt	735.5
lines/sq.inch	tesla	1.55×10^{-5}
Maxwell	weber	10^{-8}
mho	siemen	1
micron	metre	10^{-6}
mile	km	1.609
mile	cm	2.54×10^{-3}
poundals	newton	0.1383
pound	kilogram	0.494
pound (force)	newton	0.448
pound/ square feet	N/m^2	47.878
pound/ square inch	N/m^2	6894.43

Table A.4.1 Some useful physical constants

Physical quantity	*Symbol*	*Computational value*
Speed of light in a vacuum	c	3.00×10^{8} m/s
Elementary charge	e	1.60×10^{-19} C
Gravitational constant	G	6.67×10^{-11} m^3/kg s
Universal gas constant	R(or R_u)	8.31 kJ/kmol K
Avogadro's constant	N_A	6.02×10^{26} $kmol^{-1}$
Boltzmann's constant	$k(k_B)$	1.38×10^{-23} J/K
Stefan-Boltzmann's constant	σ	5.67×10^{-8} W/m^2 K^4
Molar volume of ideal gas at STP	V_m	2.27×10^{-2} m^3/mol
Permittivity constant	ϵ_o	8.85×10^{-12} F/m
Permeability constant	μ_o	$4\pi \times 10^{-7}$ ($= 1.26 \times 10^{-6}$ H/m)
Planck constant	h	6.63×10^{-34} J s
Electron mass	$m(m_e)$	9.11×10^{-31} kg (5.49×10^{-4} amu)
Proton mass	m_p	1.67×10^{-27} kg (1.0073 amu)
Ratio of proton mass to electron mass	m_p/m_e	1840
Electron charge to mass ratio	e/m_e	1.76×10^{11} C/kg
Neutron mass	m_n	1.68×10^{-27} kg (1.0087 amu)
Hydrogen atom mass	m_H	1.0078 amu
Deuterium atom mass	m_{2H}	2.0141 amu
Helium atom mass	m_{4H}	4.0026 amu
Muon mass	m_μ	1.88×10^{-28} kg
Electron magnetic moment	μ_e	9.28×10^{-24} J/T
Proton magnetic moment	μ_p	1.41×10^{-26} J/T
Bohr magneton	μ_B	9.27×10^{-24} J/T
Nuclear magneton	μ_N	5.05×10^{-27} J/T
Bohr radius	r_B	5.29×10^{-11} m
Rydberg constant	R_H	1.10×10^{7} m^{-1}
Electron Compton wavelength	λ_c	2.43×10^{-12} m

A.5.1 BIBLIOGRAPHY

1. *Modern Physics,* F.K. Richtmyer, Tata McGraw-Hill, New Delhi (1997).
2. *Electronic Materials,* L.A.A. Warnes, Macmillan Education Ltd., London (1990).
3. *Fundamentals of Physical Metallurgy,* J.D. Verhoeven, John Wiley Ltd., New York (1982).
4. *Electronic Engineering Materials and Devices,* Allison, McGraw Hill, New York (1980).
5. *Introduction to Solid State Physics,* C. Kittel, John Wiley, New York (1996).
6. *Solid State Physics,* S.O. Pillai, New Age International (P) Ltd., New Delhi (2005).
7. *Solid State Physics,* A.J. Dekker, Macmillan India Ltd., New Delhi (2002).
8. *Laser Guide Book*, J. Hecht, McGraw-Hill, New York (1989).
9. *Magnetism and Magnetic Materials,* J.C. Anderson, Chapman and Hall Ltd., London (1970).
10. *Solid State Theory in Metallurgy,* Peter Wilkes, Cambridge University Press (1983).
11. *Quantum Theory of Crystalline Solids,* A.O.E. Animalu, Prentice-Hall, New Jersey (1990).

A.6.1 CALCULUS AND INTEGRAL TABLES

Some important differentials

1. $d(x+y-z) = dx + dy - dz$
2. $d\left(\dfrac{x}{y}\right) = \dfrac{ydx - xdy}{y^2}$
3. $d(xy) = xdy + ydx$
4. $d(x^n) = nx^{n-1}dx$
5. $d(x^y) = yx^{y-1}dx + x^y(\log_e x)dy$
6. $d(e^x) = e^x dx$
7. $d(e^{ax}) = ae^{ax}dx$
8. $d(a^x) = a^x(\log_e a)dx$
9. $d \sinh x = \cosh x dx$
10. $d \cosh x = -\sinh x dx$
11. $d \tanh x = \sec h^2 x dx$

Infinite integrals

1. $\int (ax+b)^n dx = \frac{(ax+b)^{n+1}}{a(n+1)}, n \neq -1$

2. $\int \frac{dx}{ax+b} = \frac{1}{a}\log(ax+b)$

3. $\int \frac{dx}{x(ax+b)} = \frac{1}{b}\log\frac{x}{ax+b}$

4. $\int \sqrt{(ax+b)}dx = \frac{2}{3a}(ax+b)^{3/2}$

5. $\int \frac{dx}{\sqrt{(ax+b)}}dx = \frac{2\sqrt{(ax+b)}}{a}$

Exponential and logarithmic integrals

1. $\int e^{ax}dx = \frac{1}{a}e^{ax}$

2. $\int b^{ax}dx = \frac{b^{ax}}{a\log b}$

3. $\int x^n e^{ax}dx = \frac{1}{a}x^n e^{ax} - \frac{n}{a}\int x^{n-1}e^{ax}dx,$ n positive

4. $\int e^{ax}\sin bx dx = \frac{e^{ax}}{a^2+b^2}(a\sin bx - b\cos bx)$

5. $\int \log ax dx = x\log ax - x$

6. $\int x^n \log ax dx = x^{n+1}\left[\frac{\log ax}{n+1} - \frac{1}{(n+1)^2}\right], n \neq -1$

7. $\int \frac{dx}{x\log ax} = \log(\log ax)$

Definite integrals

1. $\int_0^\infty \frac{a dx}{a^2+x^2} = \frac{\pi}{2}$, if $a > 0$, 0, if $a = 0$, $-\frac{\pi}{2}$, if $a < 0$

2. $$\int_0^\infty \frac{\sin^2 x}{x^2} dx = \frac{\pi}{2}$$

3. $$\int_0^\infty \frac{\sin^2 ax}{a} dx = \frac{\pi}{2}, \text{ if } a > 0$$

4. $$\int_0^\pi \sin^2 ax dx = \int_0^\pi \cos^2 ax dx = \frac{\pi}{2}$$

5. $$\int_0^\pi \sin ax \cos bx dx = \frac{2a}{a^2 - b^2}, \text{ if } a - b \text{ is odd}$$

$$= 0 \text{ if } a - b \text{ is even}$$

6. $$\int_0^\infty \frac{\sin ax \sin bx}{x^2} dx = \frac{1}{2}\pi a, \text{ if } a < b$$

7. $$\int_0^\infty \cos(x^2) dx = \int_0^\infty \sin(x^2) dx = \frac{1}{2}\sqrt{\frac{\pi}{2}}$$

8. $$\int_0^\infty e^{-a^2x^2} dx = \sqrt{\frac{\pi}{2a}}, \text{ if } a > 0$$

9. $$\int_0^\infty x^n e^{-ax} dx = \frac{n!}{a^{n+1}} \text{ if } n \text{ is a positive integer}$$

10. $$\int x^{2n} e^{-ax^2} dx = \frac{1.3.5.....(2n-1)}{2^{n+1} a^n}\sqrt{\frac{\pi}{a}}$$

$$\int_0^\infty \sqrt{x} e^{-ax} dx = \frac{1}{2a}\sqrt{\frac{\pi}{a}}$$

11. $$\int_0^\infty \frac{e^{-ux}}{\sqrt{x}} dx = \sqrt{\frac{\pi}{a}}$$

12. $$\int_0^\infty e^{(-x^2-a^2)/x^2} dx = \frac{e^{-2a}\sqrt{\pi}}{2}, \text{ if } a > 0$$

13. $\int_0^\infty e^{-ax}\cos bx\,dx = \frac{a}{a^2+b^2}$, if $a > 0$

14. $\int_0^\infty e^{-ax}\sin bx\,dx = \frac{b}{a^2+b^2}$, if $a > 0$

15. $\int_0^1 \frac{\log x}{1-x}\,dx = -\frac{\pi^2}{6}$

16. $\int_0^1 \frac{\log x}{1+x}\,dx = -\frac{\pi^2}{12}$

17. $\int_0^1 \frac{\log x}{1-x^2}\,dx = -\frac{\pi^2}{8}$

18. $\int_0^1 \frac{\log x}{\sqrt{1-x^2}}\,dx = -\frac{\pi}{2}\log 2$

19. $\int_0^1 \log\left[\frac{1+x}{1-x}\right]\frac{dx}{x} = \frac{\pi^2}{4}$

20. $\int \log\left[\frac{e^x+1}{e^x-1}\right]dx = \frac{\pi^2}{4}$

21. $\int \frac{dx}{\log(1/x)} = \sqrt{\pi}$

22. $\int_0^\pi x\log\sin x\,dx = -\frac{\pi^2}{2}\log_e 2$

23. $\int_0^1 [\log(1/x)]^{1/2}\,dx = \frac{\sqrt{\pi}}{2}$

Model Question Paper with Answers

Engineering Physics

Answer ALL Questions

Time : Three hours Maximum Marks : 80

General Instructions

A. Answers must be precise.

B. Attempted Questions are valued and the evaluation chart is notified at the end of the answer paper.

C. Important physical constants for use.

(i) Velocity of light in air/vacuum, $c = 3 \times 10^8$ m/s

(ii) Planck's constant, $h = 6.62 \times 10^{-34}$ J-S

(iii) Avogadro number, $N_A = 6.025 \times 10^{26}$/kmol

(iv) Permittivity of free space, $\varepsilon_0 = 8.85 \times 10^{-12}$ F/m

(v) Permeability of free space, $\mu_0 = 4\pi \times 10^{-7}$ H/m

(vi) Boltzmann constant, $k = 1.38 \times 10^{-23}$ J/K

(vii) Mass of electron, $m = 9.1 \times 10^{-31}$ kg

(viii) Charge of the electron, $e = 1.6 \times 10^{-19}$ coulomb

Answer Any FIVE Questions

(All questions carry equal marks).

1. (a) Discuss the interaction between atoms with a suitable graph. Obtain the expression for potential energy of the system at equilibrium space. (4 + 7) = 11 marks

   (b) Calculate the interplanar spacing for (2 3 1) plane of an fcc structure whose atomic radius is 0.175 mm. (4 marks)

2. (a) Explain X-ray diffraction. Describe Laue method of studying the crystal structure. (4 + 7) = 11 marks

   (b) Discuss the important properties of X-rays. (4 marks)

3. (a) Enunciate the energy distribution in black body spectrum. Show that

the average energy of Planck's oscillator is given by

$$\bar{E} = \frac{h\nu}{\{(e^{h\nu/kT} - 1)\}}$$ (8 marks)

(b) State the de Broglie postulate of matter waves. (3 marks)
Describe the experiment establishing it. (4 marks)

4. (a) Give an introduction for band theory of solids. (4 marks)
(b) Discuss the motion of an electron in one-dimensional periodic potential and hence show that electron cannot possess all possible energies but its energy spectrum contains alternate allowed and forbidden energy bands. (11 marks)

5. (a) Obtain the expression for density of intrinsic semiconductors assuming the relation for the density of energy states. (11 marks)
(b) Discuss the action of a *p-n* junction as a rectifier. (4 marks)

6. (a) Deduce the equation for computing the electronic polarization. (8 marks)
(b) Discuss the general properties of magnetic materials. (4 marks)
(c) A ferromagnet becomes paramagnet and follows Curie-Weiss law. Therefore, the thermal energy just balances the magnetic potential energy. Get the value of the molecular field and express it in oersted. (3 marks)
Given: Tc = 1000 K.

7. (a) Describe the construction and working of He–Ne laser with energy level diagram. (7 marks)
(b) Derive an expression for numerical aperture for step index fibre in terms of refractive index of the core and relative refractive index difference between the core and cladding. (8 marks)

8. (a) What is meant by reverberation in a Hall or auditorium? Sabine's formula for reverberation time is

$T = \dfrac{0.165\,V}{\sum as}$. Describe a method to determine the absorption coefficient. (6 marks)

(b) A class room is 4.6×10^3 m^3 and has a reverberation time of 1.5 sec. What is the total sound absorption of the class room? (3 marks)
(c) Summarize the science of nanomaterials for the use of beginners in your style with lucid language. (6 marks)

Answers with Evaluation Report

1. (a) Let us consider a simple diatomic molecule. The two forces acting on the system are:
(i) there must be attractive forces between the atoms which keep them together.
(ii) there must also be repulsive forces between them since external energy is required

to bind them together. At equilibrium spacing (say $r = r_0$), the two forces are equal and opposite and hence the resultant force is zero.

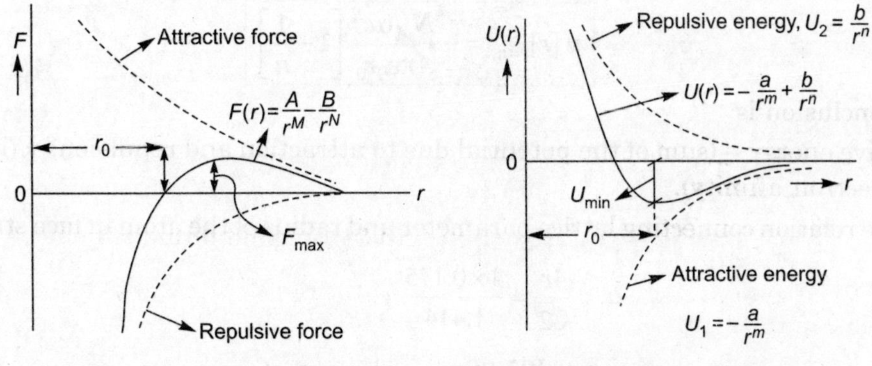

Fig. 1 Variation of interatomic forces and potential energies with interatomic spacing.

Usually the repulsive forces are of short ranges than the attractive forces. The energy at the equilibrium spacing, $u(r_0)$ is called the binding energy or the cohesive energy of the molecule. This is the energy required to separate the two atoms of the molecule such that the distance between them is infinity or they become completely free.

Calculation of Cohesive Energy

The potential energy of a crystal having both cations and anions with charge Z_1e and Z_2e is sum of the coulomb energy and the repulsive energy. It is assumed that the repulsive energy varies as an inverse power of the distance between the ions. Thus

$$u(r) = -\frac{\alpha Z_1 Z_2 e^2}{4\pi\varepsilon_0 r} + \frac{B}{r^n} \quad \text{where } B \text{ is a constant}$$

For univalent crystal, $Z_1 = Z_2 = 1$. Thus the total energy per kmol of the crystal is

$$\boxed{u(r) = N_A\left\{\frac{-\alpha e^2}{4\pi\varepsilon_0 r} + \frac{B}{r^n}\right\}}$$

where N_A is Avogadro's number. Since the P.E. is minimum at the equilibrium position,

$$\left\{\frac{du}{dr}\right\}_{r=r_0} = 0 = N_A\left[\frac{\alpha e^2}{4\pi\varepsilon_0 r_0^2} - \frac{nB}{r_0^{n+1}}\right]$$

i.e;

$$\frac{\alpha e^2}{4\pi\varepsilon_0 r_0} = \frac{nB}{r_0^n}$$

or,

$$B = \frac{\alpha e^2 r_0^n}{4\pi\varepsilon_0 r_0 n} = \frac{\alpha e^2 r_0^{n-1}}{4\pi\varepsilon_0 n}$$

Thus $$u[r.]_{r=r_0} = N_A\left[\frac{-\alpha e^2}{4\pi\varepsilon_0 r_0} + \frac{\alpha e^2 r_0^{n-1}}{4\pi\varepsilon_0 n r_0^n}\right]$$

i.e; $$\boxed{u\,[r]_{r=r_0} = -\frac{N_A \alpha e^2}{4\pi\varepsilon_0 r_0}\left[1-\frac{1}{n}\right]}$$

The conclusion is

Cohesive energy = (sum of the potential due to attraction and repulsion) + (ionisation energy – electron affinity).

1. (b) The relation connecting lattice parameter and radius of the atom in face structure is

$$a = \frac{4r}{\sqrt{2}} = \frac{4\times 0.175}{1.414}$$

Thus $$a = 0.495 \text{ mm}$$

If d is the distance between the said parallel planes, then

$$d = \frac{a}{\sqrt{h^2+k^2+l^2}} = \frac{0.495}{\sqrt{2^2+3^2+1^2}} = \frac{0.495}{\sqrt{14}} \text{ nm}$$

$$\boxed{d = 0\cdot 133 \text{ mm}}$$

2. (a) A crystal consists of a group of atoms arranged in a periodic manner. The spacing between atoms is of the order 10^{-10} m. X-rays, which are electro magnetic radiation of wavelength 10^{-12} m. See the atoms in a crystal as slits of dimensions comparable to wavelength of X-rays. Therefore when X-rays are allowed to fall on a crystal, they get refracted according to the equation

$$2\,d \sin\theta = n\lambda$$

where d is the interatomic separation between the atoms, θ is the angle of diffraction and n is the order of diffraction.

Laue Method

Here a number of polychromatic X-rays strike at 90° on the plane of the crystal under study. While passing through the crystal, the rays meet Bragg's planes having different interplanar spacings d. These different sets of planes make different angles θ with the direction of X-rays. Certain combinations of θ, and d satisfy the conditions of Bragg's law, with the resulting increase in intensity of diffracted X-rays.

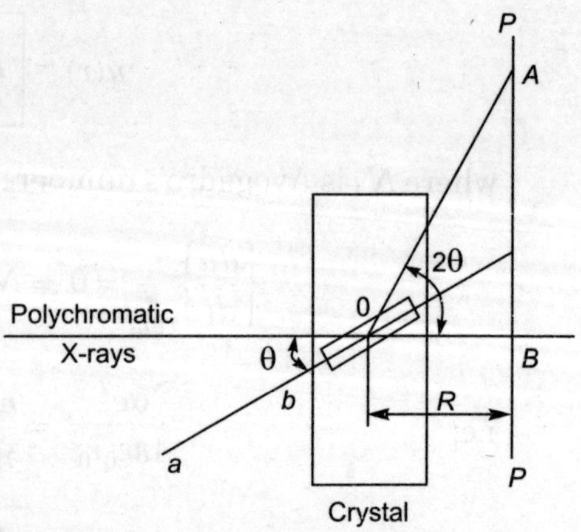

Fig. 2 X-ray diffraction (Laue method)

When primary rays pass through the crystal, they produce a black spot at point B on the photographic plate PP. As different wavelengths are included in the primary X-ray, they will produce spots, which are less pronounced, around this central spot as shown in the figure.

Let *ab* represent the position of one of the possible Bragg's planes, θ the corresponding glancing angle and *A* any spot on the photographic plate. From triangle *AOB*, we have $\tan 2\theta = \frac{AB}{R}$, or $AB = R \tan \theta$. By measuring *AB* and *R*, we can find the value of θ for corresponding plane. This method is usually used for single large size crystal.

2. (b) Main properties of X-rays may be summarised as follows:
   (i) They travel in straight lines approximately at the velocity of light (3×10^8 m/sec).
   (ii) They are not deflected by electric and magnetic fields which indicates that X-rays do not consist of charged particles.
   (iii) They are highly penetrating rays, and can pass through many solids which are opaque to ordinary light. The penetrating power depends upon the density of the solid. The penetrating power decreases when density of material increases. They can easily pass through flesh, wood, cardboard, etc. Since the penetrating power depends upon the energy of photon of X-rays. X-rays cannot penetrate through heavy metals and bones. This property has been used in detecting the fracture of bones and has also been used for decting flaws in precious castings.
   (iv) The cause fluorescence in certain materials e.g., bariumplatino cyanide, zinc sulphide, etc.
   (v) They ionise the gas through which they pass.
   (vi) They liberate electrons from certain metals on which they fall. The phenomenon is called photo electric effect.
   (vii) Like visible light, X-rays are electromagnetic radiation of very short wavelength 1 Å (or very high frequency) and show the phenomenon reflection, refraction, interference, diffraction, polarisation etc.
   (viii) They have a harmful effect on living tissues. Long exposures of human skin to X-rays causes reddening of skin and incurable sores.

3. (a) Measurement on the spectral distribution of black body radiation have been entered out by a number of workers. Figure shows the results of Lummer and Pringsheium. Their important findings are:
   (i) The energy is not uniformly distributed in the radiation spectrum of black body.
   (ii) At a given temperature, the intensity of radiation increases with increase in wavelength and become maximum at a particular wavelength. With further increase in wavelength; the intensity of radiation decreases.

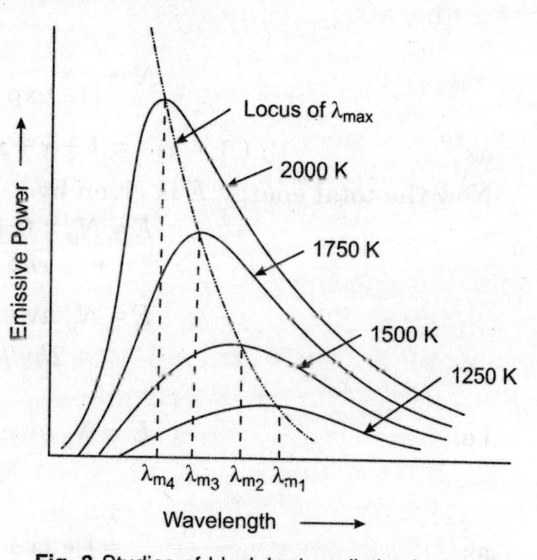

Fig. 3 Studies of black-body radiation by Lummer and Pringsheim.

(iii) The wavelength corresponding to the maximum energy represented by the peak of the curve shifts towards the shorter wavelength side.

Average Energy of the Oscillator

Planck assumed the presence of number of resonators (harmonic oscillators) inside the cavity of the black body. According to Planck's quantum hypothesis, an oscillator can have only discrete energies given by the equation

$$E_n = nh\nu \qquad \text{...(1)}$$

with $n = 0, 1, 2...$ *i.e*; an oscillator can have energies 0, $h\nu$, $2h\nu$,.... This means the oscillator cannot absorb or emit energy continuously but quanta of energy $h\nu$. If N is the total number of oscillators and E their total energy, then the energy of Planck oscillator,

$$\overline{E} = \frac{E}{N} \qquad \text{...(2)}$$

Let $N_0, N_1, ...N_r, ...$be the number of oscillators having energies 0, $h\nu$, $2h\nu$....$rh\nu$,... respectively.

According to Maxwell's law of molecular motions of $h\nu$ is a certain amount of energy, the probabilities that a system will have energies 0, $h\nu$, $2h\nu$ ... $rh\nu$ ... are in the ratio $1 : e^{-h\nu/kT} : e^{-2h\nu/kT} : ... e^{-rh\nu/kT}$ ... etc. If N_0 is the number of resonators having energy 0, then the number of resonators N_1 having energy $h\nu$ will be $N_0\, e^{-h\nu/kT}$ and, in general, the number of resonators N_r having energy $rh\nu$ will be $N_0\, e^{-rh\nu/kT}$ and so on. Now

$$N = N_0 + N_1 + ... N_r + ...$$

$$= N_0 + N_0 \exp(-h\nu/kT) + N_0 \exp(-2h\nu/kT)$$

$$+ ... N_0 \exp\left(\frac{-rh\nu}{kT}\right) + ...$$

$$N = \frac{N_0}{[1 - \exp(-h\nu/kT)]} \qquad \text{...(3)}$$

as $\quad (1 - x)^{-1} = 1 + x + x^2 + ...$

Now the total energy E is given by

$$E = N_0 \times 0 + h\nu\, N_0 \exp(-h\nu/kT) + 2h\nu\, N_0 \exp(-2h\nu/kT) + ... rh\nu\, N_0 \exp(-rh\nu/kT) + ...$$

$$E = N_0\, h\nu \exp(-h\nu/kT)\{1 + 2\exp(-h\nu/kT) + 3\exp(-2h\nu/kT) + ...\}$$

i.e, $$E = N_0\, h\nu \exp(-h\nu/kT)\frac{1}{\{1 - \exp(-h\nu/kT)\}^2}$$

as $$\frac{1}{(1-x)^2} = 1 + 2x + 3x^2 + ... r x^{r-1} + ...$$

Now the average energy of the oscillator is given by

$$\bar{E} = \frac{E}{N} = \frac{N_0\, h\nu \exp(-h\nu/kT)\,\{1-\exp(-h\nu/kT)\}^{-2}}{N_0/[1-\exp(-h\nu/kT)]}$$

$$\bar{E} = \frac{h\nu \exp(-E/kT)}{[1-\exp(-E/kT)]}$$

$$\boxed{\bar{E} = \frac{h\nu}{\{\exp(h\nu/kT)-1\}}} \quad \text{...(4)}$$

3. (b) De Broglie deduced the relation between particle and wave properties from Einstein-Planck expression for electromagnetic wave (or photons).

$$E = h\nu \text{ and } p = m\text{v}$$

$$E = m\text{v}^2 = pv = h\nu$$

i.e.

$$p = \frac{h\nu}{v} = \frac{h}{\lambda}$$

$$\lambda = \frac{h}{p} \quad \text{...(1)}$$

Also

$$\text{K.E.} = \frac{p^2}{2m}$$

i.e.

$$p^2 = 2m\,\text{K.E.}$$

Thus

$$\boxed{\lambda = \frac{h}{\sqrt{2m\text{K.E.}}}} \quad \text{...(2)}$$

Again if an electron is accelerated by a potential of V-volt, then

$$\frac{1}{2}mv^2 = \text{eV}$$

Hence

$$\lambda = \frac{h}{\sqrt{2\text{meV}}} = \frac{6.63\times10^{-34}}{\sqrt{2\times9.1\times10^{-31}\times1.6\times10^{-19}\times V}} = \frac{1.23}{\sqrt{\text{V}}}\text{ nm}$$

$$\boxed{\lambda = \frac{1.23}{\sqrt{\text{V}}}} \quad \text{...(3)}$$

Davisson-Germer Experiment

They performed an experiment to study the wave nature of electrons as suggested by de Broglie. Here electrons from a filament F are rendered into a fine beam by applying a positive potential to the cylinder *D*. A fine narrow beam of electrons is incident on a nickel crystal. The electrons are reflected and the intensity of the reflected electrons is measured at various angles with the help of a current meter *M*. Different positive potential can be applied to the cylinder *D* and the velocity of the electrons can be varied. The curve represents a polar graph indicating the intensity of the reflected electrons at various angles. It is found that the intensity is maximum at 50° at a potential difference of 54 volt. It is found that the nature of the intensity graph is similar and the spur (maximum intensity) occurs always at 50°. The intensity of the electrons scattered in any given direction is found by allowing the electrons to enter a chamber, set to an appropriate position and measuring the deflection produced by a meter connected to the chamber.

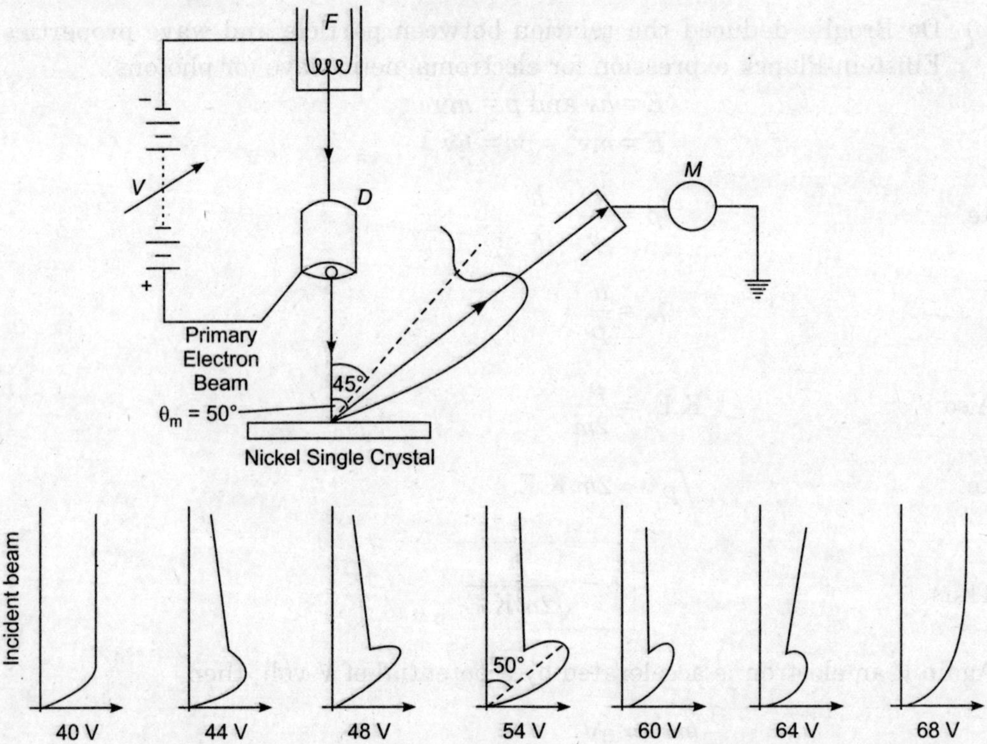

Fig. 4

Results

When the difference of potential is 40 volt between *F* and *D*, the curve is fairly smooth. At 44 volt a distinct spur appears on the curve at an angle of 60°. The spur increases as the voltage increases and reaches a maximum at 54 volt at an angle of 50°. After this the spur decreases and disappears at 68 volt. The selective reflection at 54 volt electrons at 50° is due to constructive interference. The waves from the successive layers of the crystal reinforce each other. Hence we conclude that waves must be associated with the electrons. The spur in its prominent stage is conclusive evidence that waves are associated with electrons and with matter in general. From de Broglie relation,

$$\lambda_e = \frac{1.227}{\sqrt{54}} = 0.167 \text{ nm}$$

The experimental value of interplanar distance is 0.1075 nm. By Bragg's law

$$2d \sin\theta = n\lambda_e; \lambda_e = 2 \times 0.1075 \times \sin 50°$$

$$\lambda_e = 0.166 \text{ nm}$$

The two values of λ_e are in excellent agreement.

4. (*a*) The free electron was able to explain successfully many different phenomena such as the electrical conductivity, electron emission, electron paramagnetism etc. But the model failed to explain why some solids are conductors, some are semiconductors while some others are insulators. These and many other phenomena can be satisfactorily explained by the band theory of solids. i.e. the old free electron theory must be modified.

The simplest quantum mechanical view of the potential experienced by an electron in passing through the crystal is one which is perfectly periodic as

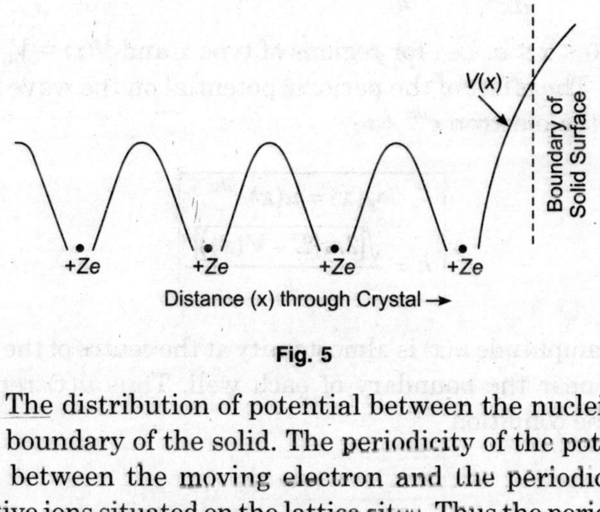

Fig. 5

shown in the figure. The distribution of potential between the nuclei is a series of humps and increases at the boundary of the solid. The periodicity of the potential is the result of coulomb interaction between the moving electron and the periodic charge distribution arising from the positive ions situated on the lattice sites. Thus the periodicity of the potential is the periodicity of the lattice.

4. (*b*) Kronig-Penny Model: Here the problem is more simplied replacing the previously said model into periodic potential with rectangular barriers as shown in the above figure and it is called Kronig-Penny model. Because of the motion of the electron in the periodic potential, the mass of the electron is effectively changed, which we call "effective mass". The effective mass of the electron may be larger or smaller than the rest mass of it, or even may be negative. Further, in solids, there may be charge carriers, having a positive charge + *e* and such charge carriers are known as "holes".

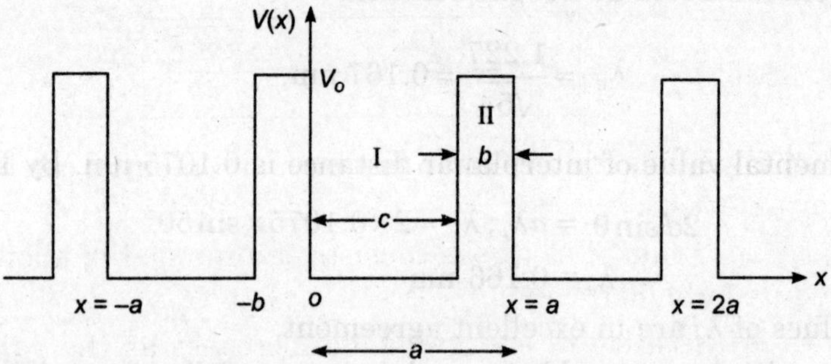

Fig. 6

The solution of Schrödinger wave equation for the single electron in this one dimensional periodic crystal provides the set of one electron states which the electron may occupy. The problem is simplified by replacing the periodic potential with potential rectangular barriers as shown in the above figure. This is so called Kronig-Penny Model. The Schrödinger wave equation to be solved is

$$\frac{d^2\psi(x)}{dx^2} + \frac{2m}{\hbar}\{(E - V(x))\}\psi(x) = 0 \qquad ...(1)$$

where $V(x) = 0$ when $0 < x < a$. i.e., for regions of type 1 and $V(x) = V_0$ when $-b < x < 0$, i.e. for regions of type 11. The effect of the periodic potential on the wave function is to change the wave function of the electron e^{ikx} to

with
$$\boxed{\begin{gathered}\psi(x) = u(x)e^{ikx} \\ k = \frac{\sqrt{[2m(E - V(x))]}}{\hbar}\end{gathered}} \qquad ...(2)$$

The modulating amplitude $u(x)$ is almost unity at the centre of the well and appreciably different from unity near the boundary of each well. Thus $u(x)$ repeats from lattice to lattice and satisfies the condition

$$\boxed{u\,(x) = u\,(x + a)} \qquad ...(3)$$

where $a = b + c$ is the lattice spacing.

Equations (2) and (3) together are known as Bloch's theorem which is a mathematical statement regarding the form of one electron wave functions for a perfectly periodic potential.

Using Bloch theorem and the periodic boundary conditions for continuity of a wave function and its derivative, we get the condition for the solutions of wave equation to exist in the form

where
$$\boxed{P\frac{\sin\alpha a}{\alpha a} + \cos\alpha a = \cos ka} \qquad ...(4)$$

$$\boxed{P = \frac{mV_0 ba}{\hbar^2} \text{ and } \alpha = \frac{\sqrt{2mE}}{\hbar^2}} \qquad \text{...(5)}$$

The allowed energy levels of an electron in a periodic potential can be obtained by the following two interpretations.

(i) The energy spectrum of electrons consists of a number of allowed energy bands separated by forbidden bands.

(ii) The width of the allowed energy bands increases as the value of α. i.e., the energy increases.

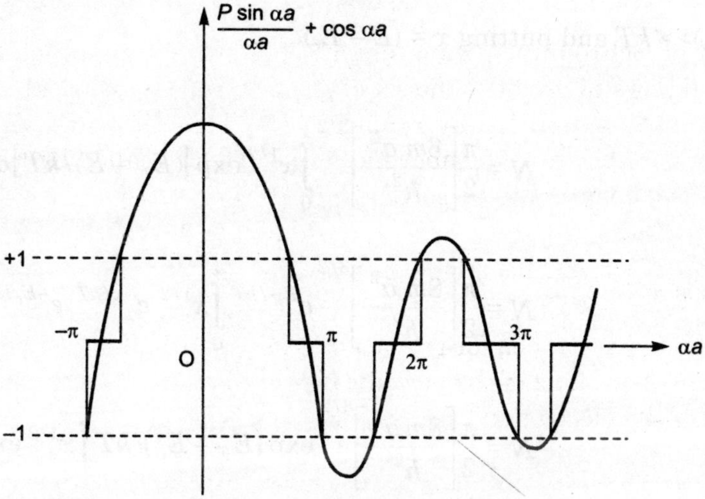

Fig. 7 Plot of $\frac{P\sin\alpha a}{\alpha a} + \cos\alpha a$

5. (*a*) The general expression for the number of electrons per unit volume in the

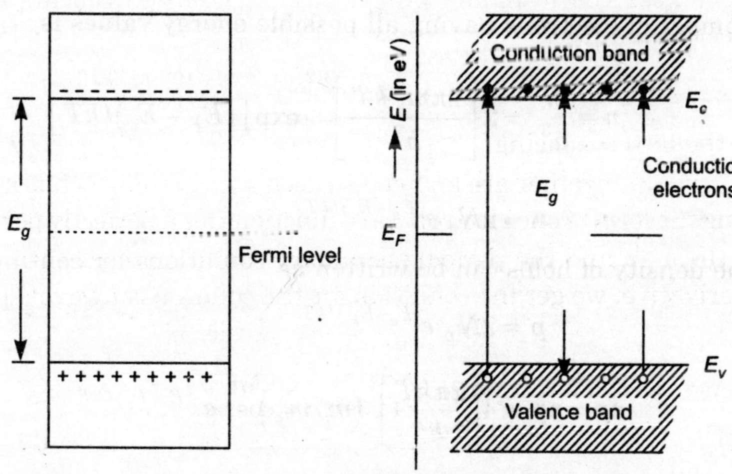

Fig. 8

conduction band having energy values lying between E and $E + dE$ is

$$N(E)\,dE = \frac{\pi}{2}\left[\frac{8ma^2}{h^2}\right]^{3/2} (E - E_c)^{1/2} dE\, F(E) \quad ...(1)$$

Integrating Eqn. (1), we get

$$N = \frac{\pi}{2}\left[\frac{8m\,a^2}{h^2}\right]^{3/2} \int_{E_c}^{\infty} (E - E_c)^{1/2} \frac{dE}{[1 + \exp(E - E_F)/kT]}$$

At 300 K, $(E - E_F) >> kT$ and putting $x = (E - E_c)$
Hence

$$N = \frac{\pi}{2}\left[\frac{8m\,a^2}{h^2}\right]^{3/2} \int_0^{x} x^{1/2} \exp[(E_F - E)/kT]\,dx$$

$$N = \frac{\pi}{2}\left[\frac{8m\,a^2}{h^2}\right]^{3/2} e^{E_F/kT} \int_0^{\infty} x^{1/2}\, e^{-x/kT}\, e^{-E_c/kT}\, dx$$

i.e.,

$$N = \frac{\pi}{2}\left[\frac{8ma^2}{h^2}\right]^{3/2} \exp(E_F - E_c)/kT \int x^{1/2} \exp(-x/kT)\,dx$$

The solution of the integral,

$$\int_0^{\infty} x^{1/2}\, e^{-\alpha x}\, dx = \frac{\sqrt{\pi}}{2}\frac{1}{\alpha^{3/2}} \text{ with } \alpha = \frac{1}{kT}$$

Thus the number of electrons having all possible energy values is

$$n = \frac{N}{a^3} = 2\left[\frac{2\pi m_e^x kT}{h^2}\right]^{3/2} \exp[(E_F - E_c)/kT]$$

or

$$n = 2N_e\, e^{(E_F - E_c)/kT}$$

Similarly the density of holes can be written as

$$p = 2N_p\, e^{(E_v - E_F)/kT}$$

Thus

$$np = n_i^2 = 4\left[\frac{2\pi kT}{h^2}\right]^3 \left(m_e^x m_p^x\right)^{3/2} e^{-(E_c - E_v)/kT}$$

If $m_e^x = m_p^x = m$, then

$$n_i^2 = 4\left[\frac{2\pi mkT}{h^2}\right]^3 e^{-\frac{E_g}{kT}}$$

$$n_i = 2\left[\frac{2\pi mkT}{h^2}\right]^{3/2} e^{-Eg/kT} \quad ...(2)$$

or

$$n_i = \left[\frac{2\pi\times 9.1\times 10^{-31}\times 1.38\times 10^{-23}\times T}{h^2}\right] e^{-E_g/2kT}$$

with $h = 6.62 \times 10^{-34}$ J-sec

$$n_i = 4.83\times 10^{21}\times T^{3/2}\, e^{-E_g/2kT}$$

$$n_i = CT^{3/2}\, e^{-E_g/2kT} \quad ...(3)$$

5. (*b*) A *p-n* junction permits easy flow of charge in one direction but restricts the flow in the opposite direction. Thus it works as a rectifier. We here study qualitatively how this rectifier action results.

Reverse Biased *p-n* Junction

When battery connections to the semiconductor are made as shown in Fig. (*a*) below, the junction is said to be reverse-biased. In this case, holes are attracted by the negative battery terminal and electrons by the positive terminal so that both holes and electrons move away from the junction and away from each other. Since there is no electron-hole combination, no current flows and the junction offers high resistance.

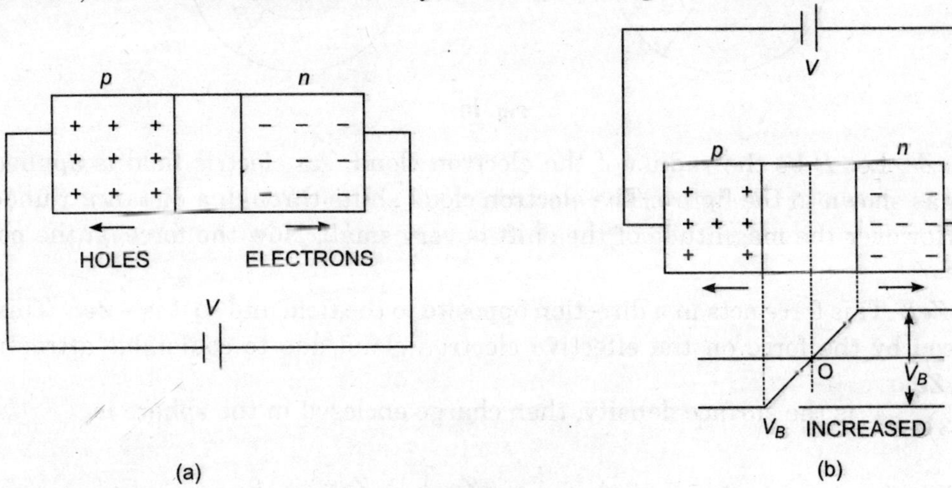

Fig. 9(a) and (b)

Another way of looking at the above process is that, in this case, the applied voltage increases the barrier there by blocking the flow of majority carriers as shown in Fig. (*b*).

Remarks

(i) In reverse bias conditions, width of the depletion layer is increased because majority charge carriers are pulled away from the junction (ii) Although under reverse-bias conditions, there is practically no current due to majority carriers, yet there is a small amount of current (a few μ A) due to the flow of minority carriers across the junction. Due to thermal energy, there are always generated some holes in the *n*-type region and some electrons in the *p*-type region of the semiconductor. The applied voltage acts as a forward bias for these minority carriers across the junction thereby producing a small current called reverse current or reverse saturation current I_s or I_0. Since, minority carriers are thermally generated, I_0 is extremely temperature dependent. I_0 is found to double for every 10ºC rise for Ge and for every 6ºC rise in the case of Si. Usually, it is of the order of μA for Ge and mA for Si.

6. (*a*) Let us consider an atom placed in the space between the plates of a capacitor. In the absence of external electric field the centres of gravity of the positive and negative charges coincide and the atom has no dipole or dipole moment. When such an atom is subjected to an electric field, the positive charge being firmly in the nucleus, remains in its original position while the electron cloud surrounding the positive charge gets shifted with respect to the positive charge.

Let Ze be the positive charge of the nucleus and a smeared out electron cloud of

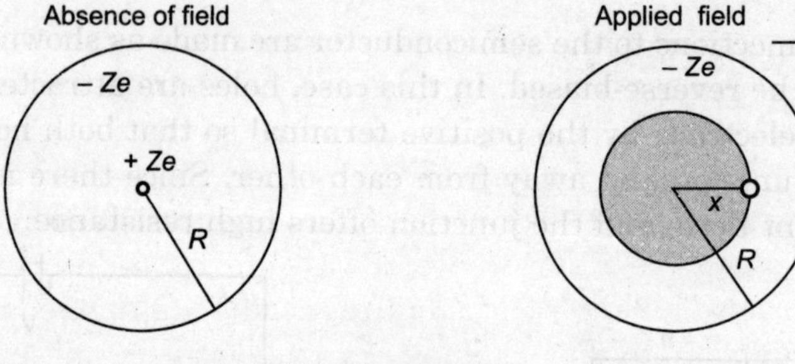

Fig. 10

charge – Ze. Let R be the radius of the electron cloud. An electric field is applied in a direction as shown in the figure. The electron cloud shifts through a distance x under the field E. However the magnitude of the shift is very small. Now the force on the positive charge is

$F = ZeE$. This force acts in a direction opposite to the field and so it is – ZeE. This force is balanced by the force on the effective electron cloud due to coulombic attraction. If $\rho = -\dfrac{Ze}{(4/3)\pi R^3}$ is the surface density, then charge enclosed in the sphere is

$$\frac{4}{3}\pi x^3 \rho = \frac{4}{3}\pi x^3 \left[-\frac{3Ze}{4\pi R^3}\right] = -\frac{Zex^3}{R^3}$$

Thus coulomb force is

$$\frac{Ze}{4\pi\varepsilon_0 x^2}\left[-\frac{Zex^3}{R^3}\right] = -\frac{Z^2e^2x}{4\pi\varepsilon_0 R^3}$$

Equating this with the restoring force, we get

$$ZeE = \frac{Z^2e^2x}{4\pi\varepsilon_0 R^3}$$

or

$$x = \frac{4\pi\varepsilon_0 R^3 E}{Ze}$$

Thus the induced electric dipole moment

$$\mu_e = Zex = 4\pi\varepsilon_0 E R^3$$

$$\mu_e = \alpha_e E\,;\, \mu_e \propto E$$

Here α_e is called electronic polarizability. Hence the electronic polarization, $P_e = N\mu_e = N\alpha_e E$ where N is number of atom/m^3.

Also

$$P_e = E\varepsilon_0(\varepsilon_r - 1) = N\alpha_e E$$

Thus

$$\boxed{(\varepsilon_r - 1) = \frac{N\alpha_e}{\varepsilon_0}} \quad ...(1)$$

6. (*b*) Materials in which mangetization can be induced are called magnetic materials. The design and practice of these materials are tailored to optimise a specific device function performed under certain operating conditions like frequency, power level, temperature range, etc. For example, in transformers and rotating machines, these materials have to serve as a medium for conversion and transformation of engines from one to another. In the field of electronics the magnetic materials have to respond to varied frequency conditions. And so selection of proper magnetic material with proper composition of alloys to suit the requirements is of prime importance. A detailed understanding of types and behaviour of the magnetic material is essential before the choice is decided for the image of material for the magnetic function, may it be electrical or electronics.

If a magnetic field is applied to a material without a magnetic moment, a magnetic field is set up (Len's law) which opposes the applied field. This phenomenon is known as diamagnetism and corresponds to a slight repulsion of a material by a magnetic field. The magnetic susceptibility; χ is small and negative; typical values range from -1×10^{-5} for copper to $-3.6. \times 10^{-5}$ for gold in SI units.

If a magnetic moment is present in the absence of a magnetic field, the magnetic moments line up with the magnetic field to decrease the total energy. The phenomenon is known as para magnetism, and corresponds to a slight attraction of a material by a magnetic field. The magnetic susceptibility χ is small and positive; typical values range from $+ 2 \times 10^{-5}$ for aluminium to $+ 98 \times 10^{-5}$ for manganese in SI units. Since thermal motion of the atoms tends to misaligm the magnetic moments, paramagnetism decreases with increasing temperature. Corresponding to ferroelectricity is ferromagnetism, the presence of a large magnetic moment that persists in the absence of a magnetic field due to spontaneous ordering of the moments by direct interaction at temperatures below some critical value called curie temperature. The magnetic susceptibility can have very large values of the order of 10^6, and is field dependent.

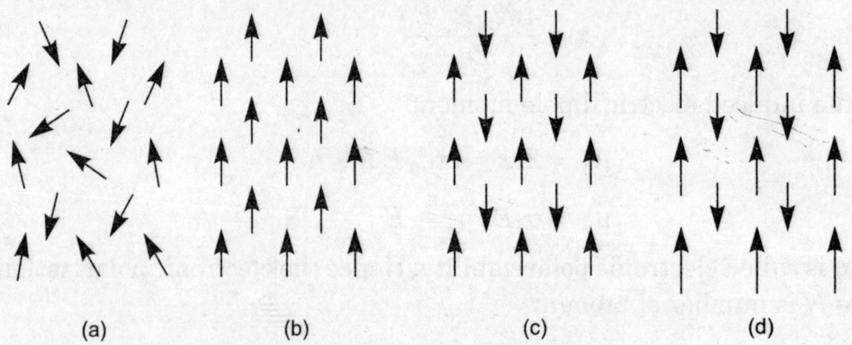

Fig. 11 Schematic illustration of the orientation of spins in (*a*) paramagnetic, (*b*) ferromagnetic, (*c*) antiferromagnetic and (*d*) ferrimagnetic materials

Two other magnetic effects can be mentioned here. They are antiferromagnetism and ferrimagnetism. Like ferromagnetism, antiferromagnetism results from internal interactions between magnetic moments, but produces in these material an ordered situation where the magnetic moments of atoms on neighbouring planes are oppositely oriented. The total moment over a finite volume is zero. The antiferromagnetic ordering is also destroyed thermally above a critical temperature, the Neel temperature.

Ferrimagnetism involves an ordered structure not unlike that of antiferromagnetism, but involves a case where the number of atoms with opposite spin are unequal, therefore yielding a net magnetic moment. The behaviour is much like that of ferromagnetism, yielding large field dependent susceptibilities below a critical temperature.

6.(*c*) $$\mu_B H_i = k\,T_c ; H_i = \frac{kT_c}{\mu_B}$$

i.e. $$H_i = \frac{1.38\times10^{-23}\times1000}{9.27\times10^{-24}} = 1.49\times10^3 \text{ A/m}$$

$$\boxed{H_i = 1.49\times10^3 \text{ A/m}}$$

7. (*a*) Many laser applications require a continuous wave. He–Ne is a gas laser which emits light continuously, rather than pulses. It employes a mixture of helium and neon gases. It differs in operation ruby laser because four energy levels are involved in it, one in helium and three in neon. The excitation of helium and neon atoms is obtained by high frequency electromagnetic field. The energy is transferred to the atoms of the gas by electron impact and collisions between atoms.

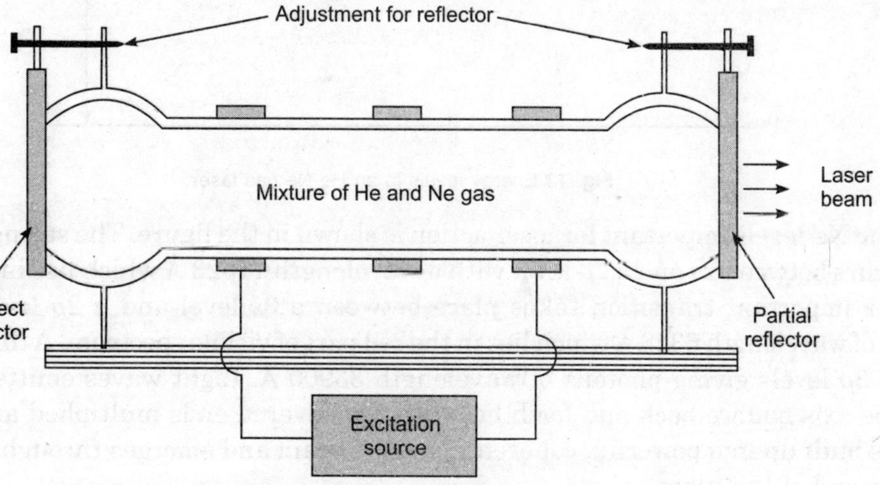

Fig. 12 He–Ne gas laser

A typical He–Ne laser is shown in the figure. It consists of:

(i) A nearly 1 metre narrow quartz tube containing He and Ne in the ratio 7:1 at a pressure of 1torr (1 mm of mercury).

(ii) An excitation source for creating a discharge in the gas. It is generally a high voltage high frequency potential difference, such as that obtained from a Tesla coil, and is applied by means of three metal bands around the outside of the tube.

(iii) End mirrors for forming the resonant cavity which can be adjusted to a high degree of parallelism. At one end of the tube, however, the mirror is fully silvered (perfect reflector) while at the other end it is partially silvered (partial reflector).

Operation

When electromagnetic energy is injected into the tube through metal bands, helium atoms are excited to metastable states by means of collision with the electron in the tube. The excited helium atoms collide with the unexcited neon atoms, thereby causing inversion in Ne atoms. Since He has metastable levels at almost the same energies as the 2*s* and 3*s* levels of Ne, it allows a resonant exchange of energy and a resultant increase in population of the corresponding N levels. The laser action takes place in neon atoms, He in the mixture serves the only purpose to enhance the excitation process.

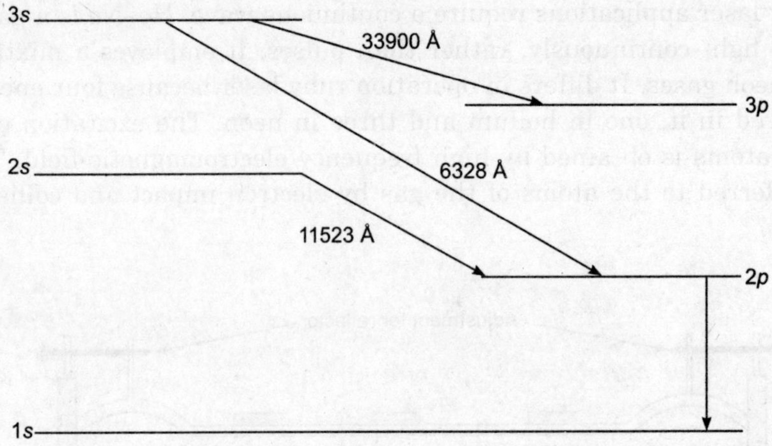

Fig. 13 Energy levels in an He-Ne gas laser

The Ne-levels important for laser action is shown in the figure. The strongest emission line occurs between 2*s* and a 2*p* level with a wavelength 11523 Å which lies in the infrared. Another important transition takes place between a 3*s* level and *a* 2*p* level giving of *f* photon of wavelength 6328 Å which lies in the red part of visible spectrum. A third transition 3*a* and 3*p* levels giving photons of wavelength 33900 Å. Light waves emitted parallel to the tube axis bounce back and forth between the silvered ends multiplied and the initial beam is built up into powerful, coherent, parallel beam and emerges through the partially silvered end of the tube.

(*b*) Light collecting capacity of a fibre is expressed in terms of numerical aperture. Sine of the acceptance angle is called the numerical aperatise (NA) of the fibre. We may thus infer that greater is the acceptance angle larger will be the size of the cone. Larger the core more number of light rays can pass through the acceptance angle. Hence more light will pass through the optical fibre.

Propagation of Light Ray through Optical Fibre

Let a highly monochromatic light ray *SA* from a laser source *S* enter an optical fibre. Let the angle of incidence at air core interface *AB* be *i*. Let *r* be the angle of refraction at the air-core interface. Let the light ray travel along *AD* and be incident at *D* at core-cladding interface. Let θ be the angle of incidence at core-cladding interface. Let n_0, n_1 and n_2 be the refractive indices of air, core and cladding materials respectively. Thus we can

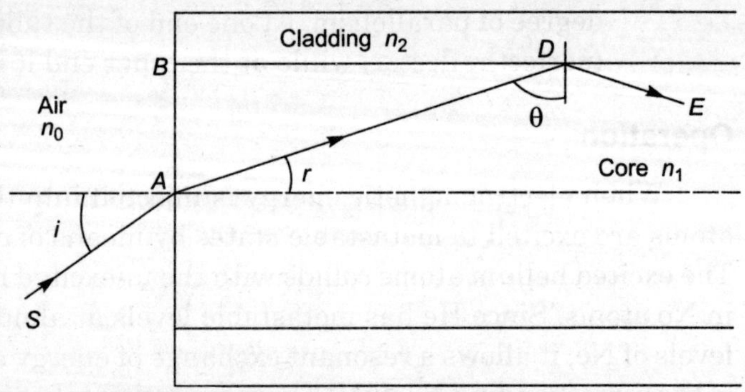

Fig. 14 Propagation of light ray through optical fibre

write,

$$\frac{\sin i}{\sin r} = \frac{n_1}{n_0}$$

$$\sin i = \left(\frac{n_1}{n_0}\right)\sin r = \left(\frac{n_1}{n_0}\right)\sin(90-\theta)$$

$$\sin i = \frac{n_1}{n_0}\cos\theta$$

We are aware that as i increases, r will also increases and hence θ will decrease. However, when the angle of incidence i at air-core interface is maximum, i.e., equal to the acceptance angle, i_{max}, the angle of incidence θ at core-cladding, interface becomes equal

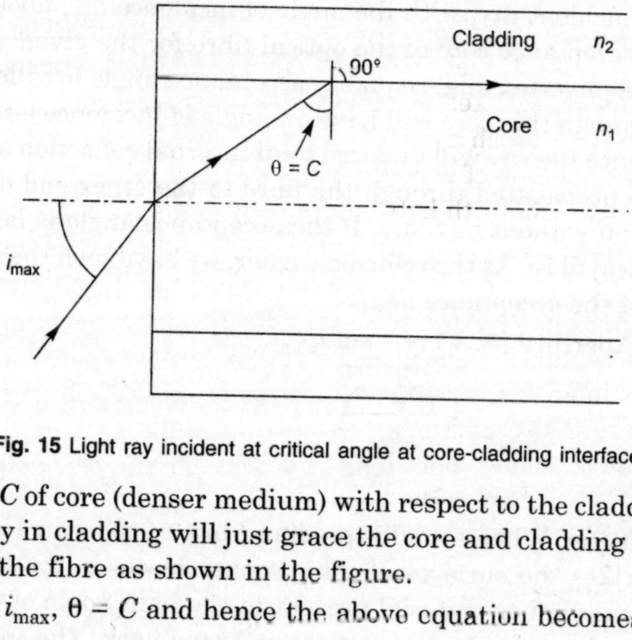

Fig. 15 Light ray incident at critical angle at core-cladding interface

to the critical angle C of core (denser medium) with respect to the cladding (rarer medium) and the refracted ray in cladding will just grace the core and cladding interface and travel along the length of the fibre as shown in the figure.

Thus when $i = i_{max}$, $\theta = C$ and hence the above equation becomes

$$\sin i_{max} = \frac{n_1}{n_0}\cos C$$

$$\sin i_{max} = \frac{n_1}{n_0}\sqrt{1-\sin^2\theta}$$

Referring the previous figure, we have

$$\frac{\sin C}{\sin 90} = \frac{n_2}{n_1}$$

Substituting this in the previous equation, we get

$$\sin i_{max} = \frac{n_1}{n_0}\sqrt{1-\frac{n_2^2}{n_1^2}} = \frac{\sqrt{n_1^2 - n_2^2}}{n_0}$$

$$i_{max} = \sin^{-1}\left[\frac{\sqrt{n_1^2 - n_2^2}}{n_0}\right]$$

For air $n_0 = 1$. In that case acceptance angle (for air-core interface) is equal to

$$\sin^{-1}\left[\frac{\sqrt{n_1^2 - n_2^2}}{1}\right]$$

By rotating the incident ray, with the angle of incidence i_{max} about the axis of fibre, we get a core called acceptance core of the optical fibre for the given wavelength of light i.e., any cone with a semivertex angle equal to acceptance angle is called acceptance cone. Any light ray sent through this core will have an angle of incidence either equal to i_{max} or less than that and hence the ray will undergo total internal reflection at the core-cladding interface and will be propagated through the fibre to the other end of the fibre by total international reflection without any loss. If the acceptance angle is large, more light can pass through the optical fibre. As the referred earlier, we have seen the numerical aperture is equal to the sine of the acceptance angle

i.e., Numerical Aperture (N.A.) is = $\sin i_{max}$

$$= \frac{\sqrt{n_1^2 - n_2^2}}{n_0}$$

Now $\qquad i_{max} = \sin^{-(N.A)}$

8. (*a*) **Reverberation:** When sound is produced in a room or an auditorium it spreads out and strikes the surfaces of the various objects in the hall such as ceiling, walls, floor and other material present in the hall. Some of the sound energy is absorbed by these objects but some is reflected back. The sound waves received by the listener are (i) direct wave (ii) reflected wave due to multiples reflection at the surfaces of the various objects. Before the direct wave received by the listener dies out, he may receive the reflected wave. This creates the impression of prolongation or persistence of audible sound even after the source has ceased to emit sound is called reverberation.

Sabine's formula for reverberation time is $T = \dfrac{0.165\,V}{\sum a\,s}$

Here *V*-is the volume of the hall *s* surface area having an absorption coefficient α.

In the simplest method the reverberation time in a hall is determined first without the absorbent material '*a*' and then with the material. If T_1 and T_2 be the reverberation times, then

$$T_1 = \frac{0.165\,V}{\sum a_1\,s_1} \text{ where } \sum a_1 s_1 \text{ is the total absorption in the}$$

hall without the material,

and
$$T_2 = \frac{0.165\,V}{\sum a_1\,s_1 + \alpha S}$$

where α is the absorption coefficient of the material and S its surface area.

Now
$$\left[\frac{1}{T_2} - \frac{1}{T_1}\right] = \frac{\alpha S}{0.165\,V}$$

i.e.,
$$\boxed{\alpha = \frac{0.165\,V}{S}\left[\frac{1}{T_2} - \frac{1}{T_1}\right]} \qquad ...(1)$$

8. (*b*)

$$T = \frac{0.165\,V}{\sum \alpha s}$$

or
$$\sum \alpha s = \frac{0.165\,V}{T} = \frac{0.167 \times 240}{1.5}$$

$$\boxed{\sum \alpha S = 26.72 \text{ Sabine}} \qquad ...(2)$$

8. (*c*) The discovery of fullerenes and carbon nanotubes has opened a new research area in physics and materials science. Nanostructured materials may be defined as those materials whose structured element clusters, crystallites or molecules-have dimensions 1 to 100 nm range. Clusters of atoms consisting of typically hundreds to thousands on nanometre (nm) scale are generally called nanoclusters. These small groups of atoms, in general, go by different names such as nano-particles, nanocrystals, quantum dots and quantum boxes.

The "nano" prefix denotes that the dimensions of the structural entities are on the order of a nanometre (10^{-9} m) as a rule, less than 100 nm (equivalent to approximately 500 atom diameters). One example of a material of this type is the carbon nanotube. Thus one nanosecond (ns) = 10^{-9} sec. Similarly 1 nano metre (nm) = 10^{-9} metre. A form of carbon in which the atoms are linked by a cylindrical frame work with diameter of only a few nanometres is a nanotube. It is also known as a bucky tube. Nanotechnology is the science of understanding the structure and behaviour of materials at and from the atomic or molecular level rather than the crystal structure. A reorientation of atoms from their original arrangement is a technique freely used here — of course with new kind of difficult processing, but the benefits out of this are unlimited.

Any one of several forms of carbon made up of many atoms bonded together in a polyhedral structure of pentagons and hexagones of 60 (70, 240, 540, etc.) carbon atoms is called buck minister fuller. The illustration shows one with 60 carbon atoms. The module C_{60} was named buck minister fulleren. Derivatives of fullerene have been made, the more known being bucky tubes.

A form of carbon in which the atoms are linked by a cylindrical frame work with diameter of only a few nanometre is known as carbon nanotubes. It is also known as bucky tube. In the experiment arc discharge between graphite electrodes, along with the collected shoot particles there were some fine tube like structures. This on careful examination and analysis revealed that they are carbon nanotubes (CNTs). CNT exhibits extra ordinary mechanical and electrical properties.

Importance of Nanotechnology

Some examples: The body of air planes are made from aluminium despite the truth that the well known diamond has a strength weight ratio 50 times that of aerospace aluminium. Diamond is expensive, we cannot make it in the shapes we want, and it shatters. It is untrue today. Nanotechnology will let us inexpensively make shatter proof diamond (using carbon) in exactly the shapes we want. This may help us make a Boeing 747 whose unloaded weight is 50 times lighter than aluminium but just as strong.

Development in nanotechnology will result in improved medical sensors. The first use you may see would be in diagnostics: being able to take a tiny amount of blood from one's body, just a pinprick, and diagnose for a hundred different things. In the longer turn, though, the story of nanotechnology will be the story of extending surgical control to the molecular level.

N B: I can write more on nanotechnology. Space is sufficient, but time is insufficient.

Evaluation chart

Q. No.		Marks
1.	$3\frac{1}{2}+6\frac{1}{2}+4$	$=14$
3.	$7\frac{1}{2}+3+3\frac{1}{2}$	$=14$
5.	$10\frac{1}{2}+4$	$=14\frac{1}{2}$
7.	$7+7\frac{1}{2}$	$=14\frac{1}{2}$

8.	$5\frac{1}{2}+3+6$	$=14\frac{1}{2}$
	Total	$71\frac{1}{2}$
	Neatness	$3\frac{1}{2}$
		75

Q. (8) Half mark is deducted for irrelevant addition of two lines.
Total marks 74.5 for 80 or 93.1%.